Fundamentals of
Matrix Computations

Fundamentals of Matrix Computations

DAVID S. WATKINS
Washington State University

JOHN WILEY & SONS

New York • Chichester • Brisbane • Toronto • Singapore

ACQUISITIONS EDITOR	Valerie Hunter
PRODUCTION MANAGER	Joe Ford
DESIGNER	Nancy Field
PRODUCTION SUPERVISORS	Nancy Prinz, Savoula Amanatidis
MANUFACTURING MANAGER	Lorraine Fumoso
COPY EDITOR	Richard Blander
ILLUSTRATION	Dean Gonzalez
COVER COMPUTER ART	Marjory Dressler

Copyright © 1991, by John Wiley & Sons, Inc.

Library of Congress Cataloging in Publication Data:

Watkins, David S.
    Fundamentals of matrix computations / David S. Watkins.
      p.   cm.
    Included bibliographical references and index.
    ISBN 0-471-61414-9 (cloth)
    1. Matrices I. Title
  QA188.W38 1991
  512.9′434–dc20   90-48895
                                                       CIP

Printed in the United States of America

10 9 8 7 6 5 4 3 2 1

to Camille

Preface

This book was written for advanced undergraduates, graduate students, and mature scientists in mathematics, computer science, engineering, and all disciplines in which numerical methods are used. At the heart of most scientific computer codes lie matrix computations, so it is important to understand how to perform such computations efficiently and accurately. This book meets that need by providing a detailed introduction to the fundamental ideas of numerical linear algebra.

The prerequisites are an elementary course in linear algebra and some familiarity with a high-level programming language, such as Fortran, Pascal, or C. From the beginning the reader will be expected to know something about systems of linear equations and matrix algebra, including inverses and determinants of matrices. Later on we will use more advanced concepts, such as eigenvalues and eigenvectors, linear transformations, and changes of bases. We will review most of these ideas as we come to them.

It is my experience that the subject comes to life for students when they write programs of their own and see them work. Therefore, I have designed the book so that algorithms are introduced early in the discussion of each major topic. To accomplish this, I have sometimes postponed the discussion of theoretical issues. I have found that this approach works well in the classroom; the idea is to get the students interested in the material by having them work some examples and write a program or two. The instructor can then discuss the theory at leisure, while the students are getting the last bugs out of their programs.

The first part of the book is easy to read. This is because a minimum of mathematical machinery is used. The philosophy is not to introduce the entire mathematical tool kit at once but rather to bring in new ideas one at a time, as they are needed. I have adhered to the same philosophy with respect to the prerequisite material: There is no preliminary chapter summarizing facts and results that the student is expected to know. Instead the material is introduced (and sometimes reviewed) as needed. This makes the book more easily accessible to readers who have no previous experience with computational linear algebra.

There are numerous exercises, most of which are placed in the body of the text. Optimally, the reader should work each exercise before going on, as this is the only way to learn the material. I realize that realistically one cannot work every single exercise, but at the very least the reader should read each one as he or she comes to it. The exercises are an integral part of the book; many of them develop ideas that will be used later. They range from routine computations to extensive programming projects and challenging proofs. Most of them are routine.

The numbering schemes used in the book are standard. Sections are numbered by chapter and section number. For example, Section 5.3 is the third section of Chapter 5. The lemmas, theorems, corollaries, examples, equations, and so on, are numbered consecutively by section and order of appearance. Thus, Equation 2.5.1 is the first item in Section 2.5. The same triple-number convention applies to exercises, which are numbered separately from the other items. Finally, tables and figures are double numbered by chapter. Table 3.2 is the second table of Chapter 3.

At the end of the book is a bibliography in two parts. The first part consists of major text and reference books that cover the area from which this book's material is drawn. The books in this list are the ones that have been most important to me over the years. They are cited in the text by initials that are mnemonics for their titles. For example, [AEP] refers to *The Algebraic Eigenvalue Problem* by J. H. Wilkinson, and [MC] refers to *Matrix Computations* by G. H. Golub and C. F. Van Loan. The second part of the bibliography is a listing of the other books and journal articles that are cited in the text. These entries are cited by author and publication date. For example, the citation Björck (1967) refers to an article by Ake Björck that was published in 1967. I have made no attempt to compile a complete bibliography, nor have I tried to document the history of the subject. I have just included references wherever they seemed appropriate.

The book contains more than enough material for a two-quarter course, but I expect that it will usually be used in a one-semester course. For this latter type of course, I recommend that at least the following material be covered: Gaussian elimination (Sections 1.1 to 1.5, 1.7, and 1.8), sensitivity and stability (Sections 2.1 to 2.7), orthogonal matrices and the least-squares problem (Sections 3.1 to 3.5), the eigenvalue problem (Sections 4.1 to 4.6), and the singular value decomposition (Sections 7.1, 7.3, and 7.4). These sections form the heart of the book. I always find that I have time to discuss more; for example, solution of banded systems (Sections 1.6 and 1.9), a close look at the *QR* algorithm for the eigenvalue problem (Sections 4.7 to 5.3), or one or more of the special methods for the symmetric eigenvalue problem (Sections 6.1 to 6.3).

When I teach the course, I require that the students write their programs in Fortran. The reason for this is that Fortran continues to be the nearly universal language of scientific computing. It seems to me worthwhile to force students to learn something about this most important scientific computing language. For those of you who think Fortran is going to go away, let me remind you that the demise of Fortran has been predicted for many years. In fact, the first computer language I learned was PL/I, when I was a college freshman in 1966. The instructor of the course assured us that this was the right choice, as PL/I would soon supplant Fortran, Cobol, and other languages. As we all know, Fortran has flourished, while PL/I never lived up to its expectations. Of course today's Fortran is much easier to use than the Fortran of 1966. The evolution of the language has been a key to its success.

While my pro-Fortran bias shows itself from time to time, this need not stop confirmed "fortranophobes" from using the book. The book is not language specific;

it contains numerous algorithms but no programs. The instructor who dislikes Fortran can have the students use some other language.

As I have already remarked, the best way to appreciate a numerical method is to program it oneself and then see it work. Unfortunately there is not enough time to program everything. For those who would like to perform experiments with matrices without having to do all the programming, I recommend MATLAB, a commercially available matrix manipulation package that makes experimentation very easy. Others prefer to use MATLAB's competitor, GAUSS. Information on how to obtain these two packages is given in Appendix B.

Even those who enjoy writing their own code eventually grow tired of reinventing the wheel. Furthermore, one might prefer to use a well-tested routine rather than one's own, which is much more likely to contain errors. Fortran versions of many of the algorithms discussed in this book are contained in LINPACK and EISPACK, two high-quality packages that are in the public domain. These are documented in [LUG] and [EG], respectively. The Fortran listings can also be obtained by electronic mail through the NETLIB facility, which also offers numerous other software packages. See Appendix B for information about NETLIB. As this book is being written, a new software project, LAPACK, which is intended to supersede both LINPACK and EISPACK, is underway. It will include updated versions of most of the LINPACK and EISPACK programs, as well as some new programs. Some of the EISPACK codes are obsolete and will be deleted. When it is finished, LAPACK will also be available through NETLIB.

No book can contain everything. The biggest omission from this book is that of iterative methods for solving systems of linear equations. I left them out because, in my opinion, iterative methods are most effectively taught within the context of the numerical solution of partial differential equations. That is where these methods arose and where they continue to find their greatest use.

The bulk of this book was written while I was on sabbatical leave at the University of Bielefeld, Federal Republic of Germany. For financial support I am indebted to Washington State University and the Fulbright Commission. I thank the University of Bielefeld for providing excellent facilities, and my host Ludwig Elsner and his colleagues Angelika Bunse-Gerstner and Volker Mehrmann for helping to make my stay both pleasant and scientifically profitable.

I am indebted to the authors of a great number of earlier books in this field; their influence is evident throughout this book. They include G. W. Stewart [IMC], G. Forsythe and C. Moler [CSLAS], A. George and J. W. Liu [SPDS], J. H. Wilkinson [AEP], G. Golub and C. Van Loan [MC], B. N. Parlett [SEP], the authors of the Handbook [HAC], the EISPACK Guide [EG], and the LINPACK Users' Guide [LUG], and many other authors of books and journal articles.

I thank the many students to whom I have taught this subject over the past ten years. From them I have learned a great deal about how this material should be presented. Two of my classes worked from preliminary versions of the book and spotted numerous misprints. Professors Dale Olesky, Kemble Yates, and Tjalling Ypma also tested portions of the book in their classes. I thank them for the numerous improvements they suggested. I also benefitted from the extensive comments of several reviewers. Finally I thank my wife, Camille Minogue, who has been a source of steady support and encouragement. I dedicate the book to her.

Contents

1

....

Gaussian Elimination and its Variants

1.1
SYSTEMS OF LINEAR EQUATIONS

You have undoubtedly already had some experience with systems of linear equations. In order to remind you of how linear systems can arise in practice, we will begin by considering a simple electrical circuit problem. In Figure 1.1 the battery between the nodes labeled E and F causes a voltage difference $F - E$. The resistances $R_1, R_2, \ldots, R_6$ are all known. The problem is to find the voltages E_1, E_2, and E_3 at the other three nodes, given the voltages E and F. It is assumed that the circuit is in a steady state; that is, all voltages and currents are constant.

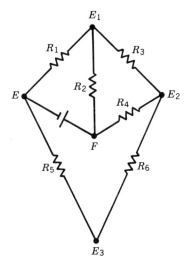

Figure 1.1

We can solve this problem easily by applying a couple of simple principles. First, we will assume that the resistors satisfy Ohm's law: the current through a resistor is directly proportional to the voltage drop. The proportionality constant is the resistance. Thus, for example, if I_1 denotes the current flowing from node E_1 to node E through resistor R_1, Ohm's law gives $E_1 - E = R_1 I_1$. Therefore $I_1 = (E_1 - E)/R_1$. Similarly the current from node E_1 to node F is $(E_1 - F)/R_2$, and the current from node E_1 to node E_2 is $(E_1 - E_2)/R_3$. The second principle is that the sum of the currents leaving each node is zero. Applying this principle to node E_1, we get the equation

$$\frac{E_1 - E}{R_1} + \frac{E_1 - F}{R_2} + \frac{E_1 - E_2}{R_3} = 0$$

Applying the same principle to nodes E_2 and E_3, we get two more equations:

$$\frac{E_2 - E_1}{R_3} + \frac{E_2 - F}{R_4} + \frac{E_2 - E_3}{R_6} = 0$$

$$\frac{E_3 - E_2}{R_6} + \frac{E_3 - E}{R_5} = 0$$

Rewriting each of these equations, keeping the unknown voltages E_1, E_2, and E_3 on the left-hand side, and placing the known voltages E and F on the right-hand side, we obtain the following system of three linear equations in the three unknowns E_1, E_2, and E_3:

$$\left(\frac{1}{R_1} + \frac{1}{R_2} + \frac{1}{R_3}\right)E_1 \qquad -\frac{1}{R_3}E_2 \qquad\qquad = \frac{E}{R_1} + \frac{F}{R_2}$$

$$-\frac{1}{R_3}E_1 + \left(\frac{1}{R_3} + \frac{1}{R_4} + \frac{1}{R_6}\right)E_2 \qquad -\frac{1}{R_6}E_3 = \frac{F}{R_4}$$

$$-\frac{1}{R_6}E_2 + \left(\frac{1}{R_6} + \frac{1}{R_5}\right)E_3 = \frac{E}{R_5}$$

This system can in turn be expressed as a single matrix equation.

$$\begin{bmatrix} \frac{1}{R_1} + \frac{1}{R_2} + \frac{1}{R_3} & -\frac{1}{R_3} & 0 \\ -\frac{1}{R_3} & \frac{1}{R_3} + \frac{1}{R_4} + \frac{1}{R_6} & -\frac{1}{R_6} \\ 0 & -\frac{1}{R_6} & \frac{1}{R_5} + \frac{1}{R_6} \end{bmatrix} \begin{bmatrix} E_1 \\ E_2 \\ E_3 \end{bmatrix} = \begin{bmatrix} \frac{E}{R_1} + \frac{F}{R_2} \\ \frac{F}{R_4} \\ \frac{E}{R_5} \end{bmatrix} \qquad (1.1.1)$$

Though we will not prove it here, the coefficient matrix is nonsingular, so the system has exactly one solution. For specified values of $R_1, R_2, \ldots, R_6$, E, and F, it would be a simple matter to solve (1.1.1) by elimination to determine the voltages E_1, E_2, and E_3.

You may be wondering why we have a linear system. Why weren't the equations nonlinear? The answer is that Ohm's law is a linear law: The voltage drop and the current are related linearly. If the resistors satisfied some nonlinear law, if, for example, the voltage drop were proportional to the square of the current, we would have $E_1 - E = R_1 I_1^2$ and $I_1 = [(E_1 - E)/R_1]^{1/2}$, which are nonlinear equations.

Now imagine that we wish to analyze a larger circuit, one involving, say, 10 unknown voltages. Proceeding as above, we can assemble a system of 10 equations in 10 unknowns that can be solved for the voltages. Although in principle solving this system is the same as solving (1.1.1), in practice much more work is involved.

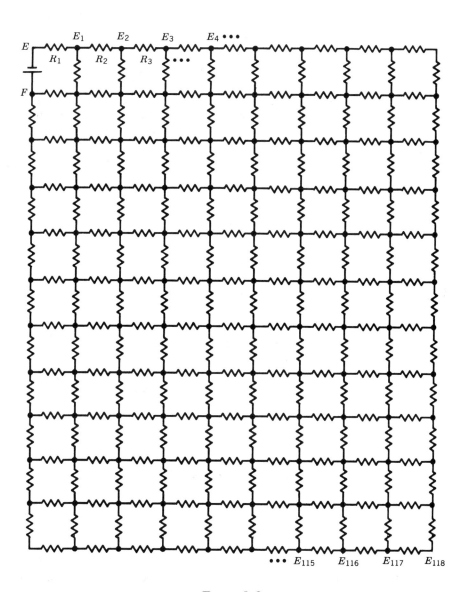

Figure 1.2

In this chapter we will see how problems of this type can be solved efficiently and accurately with the help of a computer.

An interesting feature of system (1.1.1) is that the (3, 1) and (1, 3) entries of the coefficient matrix are zero. You can easily see the reason for this: E_3 does not appear in the equation for node 1 because nodes 1 and 3 are not adjacent. Therefore the (1, 3) matrix entry is zero. For the same reason E_1 does not appear in the equation for node 3. The percentage of zeros in the coefficient matrix is typically much greater for a larger circuit such as that depicted in Figure 1.2. Once again we can set up an equation for each node, and in this case we get a system of 118 equations in 118 unknowns. Since a given node is directly connected to at most four other nodes, the equation associated with that node will involve not more than five of the unknowns (the ones associated with the node in question and the four adjacent nodes). Thus each row of the coefficient matrix will have at most five nonzeros and at least 113 zeros. Matrices such as this one, having almost all coefficients equal to zero, are called *sparse* matrices.

Although the circuit in Figure 1.2 may appear farfetched, its structure is not atypical of that of large systems in general (not just electrical systems). In a typical large system each variable is directly connected to relatively few other variables. Consequently the coefficient matrix is sparse. Since the solution of a large system is costly, special methods that take advantage of the many zeros to reduce both computer time and storage requirements have been developed. Today, thanks to them, systems of thousands of equations can be solved routinely. In Sections 1.6 and 1.9 we will examine a few of the most elementary (but effective and widely used) sparse matrix techniques.

1.2
PARTITIONING MATRICES

Before we begin our study of the solution of systems of linear equations, we pause briefly to introduce the concept of partitioned matrices, a simple but powerful idea that we will use throughout the book. If A is an m-by-n matrix and B is an n-by-p matrix, we can form the product $C = AB$, which is m-by-p. If a_{ij}, b_{ij}, and c_{ij} denote the (i, j) entry of A, B, and C, respectively, then

$$c_{ij} = \sum_{k=1}^{n} a_{ik}b_{kj} \qquad i = 1, \ldots, m, \; j = 1, \ldots, p \qquad (1.2.1)$$

Now suppose we partition A into blocks.

$$A = \begin{array}{c} \\ m_1 \\ m_2 \end{array} \overset{\begin{array}{cc} n_1 & n_2 \end{array}}{\begin{bmatrix} A_{11} & A_{12} \\ A_{21} & A_{22} \end{bmatrix}} \qquad \left\{ \begin{array}{c} m_1 + m_2 = m \\ n_1 + n_2 = n \end{array} \right. \qquad (1.2.2)$$

The labels m_1, m_2, n_1, and n_2 indicate that the block A_{ij} has m_i rows and n_j columns. We can partition B similarly.

$$B = \begin{array}{c} \\ n_1 \\ n_2 \end{array} \overset{\begin{array}{cc} p_1 & p_2 \end{array}}{\begin{bmatrix} B_{11} & B_{12} \\ B_{21} & B_{22} \end{bmatrix}} \qquad \left\{ \begin{array}{c} n_1 + n_2 = n \\ p_1 + p_2 = p \end{array} \right. \qquad (1.2.3)$$

The numbers n_1 and n_2 are the same as in (1.2.2). Thus, for example, the number of rows in the blocks B_{21} and B_{22} is the same as the number of columns in A_{12} and A_{22}. Continuing in the same spirit, we partition C as follows:

$$C = \begin{array}{c} \\ m_1 \\ m_2 \end{array} \overset{\begin{array}{cc} p_1 & p_2 \end{array}}{\begin{bmatrix} C_{11} & C_{12} \\ C_{21} & C_{22} \end{bmatrix}} \qquad \left\{ \begin{array}{c} m_1 + m_2 = m \\ p_1 + p_2 = p \end{array} \right. \qquad (1.2.4)$$

The row partition of C is the same as that of A, and the column partition of C is the same as that of B. The product $C = AB$ can now be written as

$$\begin{bmatrix} C_{11} & C_{12} \\ C_{21} & C_{22} \end{bmatrix} = \begin{bmatrix} A_{11} & A_{12} \\ A_{21} & A_{22} \end{bmatrix} \begin{bmatrix} B_{11} & B_{12} \\ B_{21} & B_{22} \end{bmatrix} \qquad (1.2.5)$$

We know that C is related to A and B by the Equations (1.2.1), but how are the blocks of C related to the blocks of A and B? We would hope to be able to multiply the blocks as if they were numbers. For example, we hope that $C_{11} = A_{11}B_{11} + A_{12}B_{21}$. Theorem 1.2.6 states that this is indeed the case.

THEOREM 1.2.6   Let A, B, and C be partitioned as in (1.2.2), (1.2.3), and (1.2.4), respectively. Then $C = AB$ if and only if

$$C_{ij} = A_{i1}B_{1j} + A_{i2}B_{2j} \qquad i, j = 1, 2$$

A careful proof of Theorem 1.2.6 would be tedious, as complicated notation would be required. However, the following exercise should be enough to convince you that it is true.

Exercise 1.2.1   Consider the matrices A, B, and C, partitioned as indicated.

$$A = \begin{bmatrix} 1 & 3 & 2 \\ 2 & 1 & 1 \\ -1 & 0 & 1 \end{bmatrix} \qquad B = \begin{bmatrix} 1 & 0 & 1 \\ 2 & 1 & 1 \\ -1 & 2 & 0 \end{bmatrix} \qquad C = \begin{bmatrix} 5 & 7 & 4 \\ 3 & 3 & 3 \\ -2 & 2 & -1 \end{bmatrix}$$

Thus, for example, $A_{12} = \begin{bmatrix} 3 & 2 \\ 1 & 1 \end{bmatrix}$ and $A_{21} = [-1]$. Show that $AB = C$ and $A_{i1}B_{1j} + A_{i2}B_{2j} = C_{ij}, i = 1, 2, j = 1, 2$. $\square$

Once you believe Theorem 1.2.6, you should have no difficulty with the following generalization. We now make a finer partition of A into r block rows and s block columns.

$$
A = \begin{array}{c} \\ m_1 \\ \vdots \\ m_r \end{array}
\begin{array}{c} n_1 \quad \cdots \quad n_s \\ \left[\begin{array}{ccc} A_{11} & \cdots & A_{1s} \\ \vdots & & \vdots \\ A_{r1} & \cdots & A_{rs} \end{array}\right] \end{array}
\qquad
\left\{ \begin{array}{l} m_1 + m_2 + \cdots + m_r = m \\[1mm] n_1 + n_2 + \cdots + n_s = n \end{array} \right.
\qquad (1.2.7)
$$

We partition B *conformably* with A; that is, we make the block row structure of B identical to the block column structure of A.

$$
B = \begin{array}{c} \\ n_1 \\ \vdots \\ n_s \end{array}
\begin{array}{c} p_1 \quad \cdots \quad p_t \\ \left[\begin{array}{ccc} B_{11} & \cdots & B_{1t} \\ \vdots & & \vdots \\ B_{s1} & \cdots & B_{st} \end{array}\right] \end{array}
\qquad
\left\{ \begin{array}{l} n_1 + n_2 + \cdots + n_s = n \\[1mm] p_1 + p_2 + \cdots + p_t = p \end{array} \right.
\qquad (1.2.8)
$$

Finally, we partition the product $C(= AB)$ conformably with both A and B.

$$
C = \begin{array}{c} \\ m_1 \\ \vdots \\ m_r \end{array}
\begin{array}{c} p_1 \quad \cdots \quad p_t \\ \left[\begin{array}{ccc} C_{11} & \cdots & C_{1t} \\ \vdots & & \vdots \\ C_{r1} & \cdots & C_{rt} \end{array}\right] \end{array}
\qquad
\left\{ \begin{array}{l} m_1 + m_2 + \cdots + m_r = m \\[1mm] p_1 + p_2 + \cdots + p_t = p \end{array} \right.
\qquad (1.2.9)
$$

THEOREM 1.2.10   Let A, B, and C be partitioned as in (1.2.7), (1.2.8), and (1.2.9), respectively. Then $C = AB$ if and only if

$$
C_{ij} = \sum_{k=1}^{s} A_{ik} B_{kj} \qquad i = 1, \ldots, r, \; j = 1, \ldots, t
$$

The special cases of Theorem 1.2.10 that we will use most frequently involve matrix-vector multiplication $Ab = c$, which is the special case of matrix multiplication obtained by taking $p = 1$. Some examples of partitions of the product $Ab = c$ are

1. $(r = 2, s = 3, t = 1)$

$$
\begin{array}{c} \\ m_1 \\ m_2 \end{array}
\begin{array}{c} n_1 \quad n_2 \quad n_3 \\ \left[\begin{array}{ccc} A_{11} & A_{12} & A_{13} \\ A_{21} & A_{22} & A_{23} \end{array}\right] \end{array}
\begin{array}{c} 1 \\ \left[\begin{array}{c} b_1 \\ b_2 \\ b_3 \end{array}\right] \begin{array}{c} n_1 \\ n_2 \\ n_3 \end{array} \end{array}
=
\begin{array}{c} \\ m_1 \\ m_2 \end{array}
\begin{array}{c} 1 \\ \left[\begin{array}{c} c_1 \\ c_2 \end{array}\right] \end{array}
\qquad
\left\{ \begin{array}{l} m_1 + m_2 = m \\[1mm] n_1 + n_2 + n_3 = n \end{array} \right.
$$

for which

$$
A_{11}b_1 + A_{12}b_2 + A_{13}b_3 = c_1
$$
$$
A_{21}b_1 + A_{22}b_2 + A_{23}b_3 = c_2
$$

2. $(r = 1, s = 2, t = 1)$

$$m \begin{array}{c} n_1 \quad n_2 \\ \begin{bmatrix} A_1 & A_2 \end{bmatrix} \end{array} \begin{bmatrix} b_1 \\ b_2 \end{bmatrix} \begin{array}{c} n_1 \\ n_2 \end{array} = c \qquad n_1 + n_2 = n$$

for which

$$A_1 b_1 + A_2 b_2 = c$$

3. $(r = 1, s = n, t = 1)$

$$m \begin{array}{c} 1 \quad 1 \qquad 1 \\ \begin{bmatrix} A_1 & A_2 & \cdots & A_n \end{bmatrix} \end{array} \begin{bmatrix} b_1 \\ b_2 \\ \vdots \\ b_n \end{bmatrix} \begin{array}{c} 1 \\ 1 \\ \vdots \\ 1 \end{array} = c$$

for which

$$A_1 b_1 + A_2 b_2 + \cdots + A_n b_n = c \qquad (1.2.11)$$

In this last example, $A_1, A_2, \ldots, A_n$ are just the columns of A. Rewriting (1.2.11) less compactly, we have

$$\begin{bmatrix} a_{11} \\ a_{21} \\ \vdots \\ a_{n1} \end{bmatrix} b_1 + \begin{bmatrix} a_{12} \\ a_{22} \\ \vdots \\ a_{n2} \end{bmatrix} b_2 + \cdots + \begin{bmatrix} a_{1n} \\ a_{2n} \\ \vdots \\ a_{nn} \end{bmatrix} b_n = \begin{bmatrix} c_1 \\ c_2 \\ \vdots \\ c_n \end{bmatrix}$$

This shows clearly that c is a linear combination of the columns of A.

1.3
THE PROBLEM

In this chapter we will study systems of n linear equations in n unknowns

$$\begin{array}{c} a_{11}x_1 + a_{12}x_2 + \cdots + a_{1n}x_n = b_1 \\ a_{21}x_1 + a_{22}x_2 + \cdots + a_{2n}x_n = b_2 \\ \vdots \qquad \vdots \qquad \vdots \qquad \vdots \\ a_{n1}x_1 + a_{n2}x_2 + \cdots + a_{nn}x_n = b_n \end{array} \qquad (1.3.1)$$

where the coefficients a_{ij} and b_i are given, and we wish to find $x_1, \ldots, x_n$ that satisfy the equations. In most applications the coefficients are real numbers, and we seek a real solution. Therefore we will confine our attention to real systems.

However, everything we will do will carry over to the complex number field. (In some situations minor modifications are required. These will be covered in the exercises.)

Since it is tedious to write out (1.3.1) again and again, we generally prefer to write it as a single matrix equation

$$Ax = b \tag{1.3.2}$$

where

$$A = \begin{bmatrix} a_{11} & a_{12} & \cdots & a_{1n} \\ a_{21} & a_{22} & \cdots & a_{2n} \\ \vdots & \vdots & & \vdots \\ a_{n1} & a_{n2} & \cdots & a_{nn} \end{bmatrix} \qquad x = \begin{bmatrix} x_1 \\ x_2 \\ \vdots \\ x_n \end{bmatrix} \qquad b = \begin{bmatrix} b_1 \\ b_2 \\ \vdots \\ b_n \end{bmatrix}$$

A and b are given, and we must solve for x. A is a square matrix; it has n rows and n columns.

Equation (1.3.2) has a unique solution if and only if A is nonsingular. Theorem 1.3.3 summarizes some of the simple characterizations of nonsingularity that we will use.

Theorem 1.3.3   Let A be a square matrix. The following five conditions are equivalent; that is, if any one holds, they all hold.

   (i) A^{-1} exists.

   (ii) $\det(A) \neq 0$.

   (iii)   There is no nonzero x such that $Ax = 0$.

   (iv)   The rows of A are linearly independent.

   (v)   The columns of A are linearly independent.

(A^{-1} denotes the inverse of A, and $\det(A)$ denotes the determinant.) If these conditions hold, A is said to be *nonsingular* or *invertible*.

For a proof of Theorem 1.3.3, see any elementary linear algebra text. If A is not nonsingular, it is called *singular* or *noninvertible*. In that case (1.3.2) has either no solution or infinitely many solutions. We will concentrate on the more important nonsingular case.

If A is nonsingular, the unique solution of (1.3.2) is given by $x = A^{-1}b$. This equation solves the problem completely in theory. It also suggests that we should calculate A^{-1} as a means of finding x. This turns out to be misleading. As we shall see, it is generally cheaper to solve $Ax = b$ directly, without calculating A^{-1}. In fact, situations in which the inverse really needs to be calculated are quite rare. This does not imply that the inverse is unimportant; it is a valuable theoretical tool.

Exercise 1.3.1   Prove that if A^{-1} exists, then there can be no nonzero x for which $Ax = 0$.   □

Exercise 1.3.2   Prove that if A^{-1} exists, then $\det(A) \neq 0$.   □

1.4
TRIANGULAR SYSTEMS

A linear system whose coefficient matrix is triangular is particularly easy to solve. Although triangular systems do not occur frequently in applications, it is a common practice to reduce more general systems to a triangular form, which can then be solved inexpensively. For this reason we will study triangular systems in detail.

A matrix $G = (g_{ij})$ is *lower triangular* if $g_{ij} = 0$ whenever $i < j$. Thus a lower-triangular matrix has the form

$$G = \begin{bmatrix} g_{11} & 0 & 0 & \cdots & 0 \\ g_{21} & g_{22} & 0 & \cdots & 0 \\ g_{31} & g_{32} & g_{33} & \ddots & \vdots \\ \vdots & \vdots & \vdots & \ddots & 0 \\ g_{n1} & g_{n2} & g_{n3} & \cdots & g_{nn} \end{bmatrix}$$

Similarly, an *upper-triangular* matrix is one for which $g_{ij} = 0$ whenever $i > j$. A *triangular* matrix is one that is either upper or lower triangular.

THEOREM 1.4.1   Let G be a triangular matrix. Then G is nonsingular if and only if $g_{ii} \neq 0$, $i = 1, \ldots, n$.

Proof   Recall from elementary linear algebra that if G is triangular, then $\det(G) = g_{11}g_{22}g_{33}\cdots g_{nn}$. Thus $\det(G) \neq 0$ if and only if $g_{ii} \neq 0$, $i = 1, \ldots, n$.   ☐

Exercise 1.4.1   Prove that if G is triangular, then $\det(G) = g_{11}g_{22}\cdots g_{nn}$.   ☐

Thus if G is triangular, the system $Gy = b$ has a unique solution if and only if the main-diagonal entries of G are all nonzero.

Lower-Triangular Systems

Consider the system

$$Gy = b \tag{1.4.2}$$

where G is a nonsingular, lower-triangular matrix. It is easy to see how to solve this system if we write it out in detail:

$$\begin{aligned} g_{11}y_1 & & & = b_1 \\ g_{21}y_1 + g_{22}y_2 & & & = b_2 \\ g_{31}y_1 + g_{32}y_2 + g_{33}y_3 & & & = b_3 \\ \vdots \qquad\qquad \vdots \qquad \ddots & & & \quad\vdots \\ g_{n1}y_1 + g_{n2}y_2 + g_{n3}y_3 + \cdots + g_{nn}y_n & & & = b_n \end{aligned}$$

The first equation involves only the unknown y_1, the second equation involves only y_1 and y_2, and so on. We can solve the first equation for y_1:

$$y_1 = b_1/g_{11}$$

The assumption that G is nonsingular guarantees that $g_{11} \neq 0$. Now that we have y_1, we can substitute its value into the second equation and solve that equation for y_2:

$$y_2 = (b_2 - g_{21}y_1)/g_{22}$$

Since G is nonsingular, $g_{22} \neq 0$. Now that we know y_2, we can use the third equation to solve for y_3, and so on. In general, once we have $y_1, y_2, \ldots, y_{i-1}$, we can solve for y_i, using the ith equation:

$$y_i = \frac{b_i - g_{i1}y_1 - g_{i2}y_2 - \cdots - g_{i,i-1}y_{i-1}}{g_{ii}}$$

which can be expressed more succinctly using sigma notation:

$$y_i = \frac{b_i - \sum_{j=1}^{i-1} g_{ij}y_j}{g_{ii}} \tag{1.4.3}$$

At each step we are guaranteed that $g_{ii} \neq 0$.

This algorithm for solving a lower-triangular system is called *forward substitution* or *forward elimination*. This is the first of two versions that we will consider. It is called the *row-oriented* version of forward substitution because it accesses G by rows; the ith row is used at the ith step. It is also called the *inner-product* form of forward substitution because the sum $\sum_{j=1}^{i-1} g_{ij}y_i$ can be regarded as an inner or dot product.

Equation 1.4.3 describes the algorithm completely; it even describes the first step ($y_1 = b_1/g_{11}$), if we agree, as we shall throughout the book, that whenever the lower limit of a sum is greater than the upper limit, the sum is zero. Thus

$$y_1 = \frac{b_1 - \sum_{j=1}^{0} g_{1j}y_j}{g_{11}} = \frac{b_1 - 0}{g_{11}} = \frac{b_1}{g_{11}}$$

Exercise 1.4.2   Use forward substitution to solve the system

$$\begin{bmatrix} 2 & 0 & 0 & 0 \\ -1 & 2 & 0 & 0 \\ 3 & 1 & -1 & 0 \\ 4 & 1 & -3 & 3 \end{bmatrix} \begin{bmatrix} y_1 \\ y_2 \\ y_3 \\ y_4 \end{bmatrix} = \begin{bmatrix} 2 \\ 3 \\ 2 \\ 9 \end{bmatrix}$$

You can easily check your answer by substituting it back into the equations. This is a simple means of checking your work that you will be able to use on many of the hand computation exercises you will be asked to perform in this chapter.   □

It would be very easy to write a computer program for forward elimination. Before we write down the algorithm, notice that b_1 is used only in calculating y_1, b_2 is only used in calculating y_2, and so on. In general, once we have calculated y_i, we no longer need b_i. It is therefore usual for a computer program to store y over b. Thus we have a single array that contains b before the program executes and y afterward. The algorithm looks like this:

$$\text{for } i = 1, \ldots, n$$
$$\begin{array}{l} \left[\begin{array}{l} \text{if } g_{ii} = 0, \text{ set error flag, exit} \\ \text{for } j = 1, \ldots, i - 1 \text{ (not executed when } i = 1) \\ \quad \left[\, b_i \leftarrow b_i - g_{ij}b_j \right. \\ b_i \leftarrow b_i/g_{ii} \end{array}\right. \end{array} \qquad (1.4.4)$$
$$\text{exit}$$

There are no references to y, since y is stored in the array named b. The check of g_{ii} is included to make the program foolproof. There is nothing to guarantee that the program will not at some time be given (accidentally or otherwise) a singular coefficient matrix. The program needs to be able to respond appropriately to this situation. A good rule of thumb is to check before each division that the divisor is not zero. In most linear algebra algorithms these checks do not contribute significantly to the cost of running the program because the division operation is executed relatively infrequently.

To get an idea of the execution time of an algorithm, it is traditional to count the number of arithmetic operations performed. The cost is usually reckoned in *flops*. A *flop* (short for *floating-point operation*) consists of one floating-point multiplication plus one floating-point addition or subtraction. The operation in the inner loop of algorithm (1.4.4) is a flop. This flop is performed $i - 1$ times on the ith time through the outer loop. The outer loop is performed n times, so the total number of flops performed in the j loop is $0 + 1 + 2 + \cdots + (n - 1) = \sum_{i=1}^{n} (i - 1)$. Calculating this sum by a well-known trick,[§] we get $n(n - 1)/2$, which is approximated well by the simpler expression $n^2/2$. Looking at the operations that are performed outside the j loop, we see that g_{ii} is compared with zero n times, and there are n divisions. Regardless of what each of these operations costs, the total cost of doing all of them is proportional to n and will therefore be small compared with the cost of $n^2/2$ flops, if n is at all large. It is usually assumed that n is large, since we are not ordinarily concerned about the cost when n is small. Making this assumption, we ignore the lesser costs and state simply that the cost of doing forward substitution is $n^2/2$ flops.

This figure gives us valuable qualitative information. Suppose we have performed some scientific simulation that required the repeated solution of 20-by-20 lower-triangular systems. We now plan to perform the simulation again with improved resolution. This time we will have to solve 40-by-40 systems, and we would like to know what it is going to cost. Knowing that the amount of arithmetic is proportional to n^2 and that we are planning to double n, we expect the finer simulation to cost about 4 times as much as the original simulation. Similarly, if we

---

[§] If you do not know the trick, see Exercise 1.4.11.

TABLE 1.1

Time (in seconds) to solve
100 $n \times n$ lower-triangular
systems

n	time (T_n)	T_n/T_{20}
20	0.27	1.0
40	1.06	3.9
60	2.38	8.8

have to solve 60-by-60 systems, we expect the cost to be about 9 times that of the 20-by-20 systems.

To test how well flop counts work, we wrote a Fortran program to do forward substitution and timed it on matrices of various sizes on a small personal computer. The results, which are shown in Table 1.1, agree quite well with our expectations. Where we expected ratios of 4 and 9, we obtained 3.9 and 8.8, respectively.

It is perhaps surprising that theory and practice agree so well, given that flop counts ignore important operations such as memory fetches, stores, and index arithmetic for the arrays. The reason for the close agreement is that the amount of time spent on these uncounted operations is approximately proportional to the number of flops.[§]

We must also realize that this experiment was performed on a conventional single-processor computer. This machine has a single central processing unit that does all the work. As you read this book, the computer market is being flooded by a bewildering variety of computers of a new generation, namely, the multiprocessor machines. These computers have numerous processing units and are therefore capable of performing many tasks simultaneously. This makes the assessment of algorithms more complicated: we must take into account not only the amount of arithmetic, but also the extent to which the operations can be performed in parallel. In addition, the problem of communication among the processors must be addressed. Section 1.10 contains an elementary discussion of these issues. For now we will simply note that the flop count gives only a crude first approximation of the performance of an algorithm on a multiprocessor machine. It is nevertheless a useful indicator, and we will count flops as a matter of policy throughout the book.

Significant savings can be achieved in forward substitution if some of the leading entries of b are zero. This observation will prove important when we study sparse matrix computations in Section 1.6. First suppose $b_1 = 0$. Then obviously $y_1 = 0$ as well, and we do not need to make the computer carry out the computation $y_1 = b_1/g_{11}$. In addition all subsequent computations involving y_1 can be skipped. Now suppose that $b_2 = 0$ also. Then $y_2 = b_2/g_{22} = 0$. It is not necessary for the computer to carry out this computation or any subsequent computation involving y_2. In general, if $b_1 = b_2 = \cdots = b_k = 0$, then $y_1 = y_2 = \cdots = y_k = 0$, and we can skip all computations involving $y_1, \ldots, y_k$. Thus (1.4.3) becomes

---

[§] Some authors include these operations in the definition of a flop.

$$y_i = \frac{b_i - \sum_{j=k+1}^{i-1} g_{ij} y_j}{g_{ii}} \qquad i = k+1, \ldots, n \qquad (1.4.5)$$

Notice that the sum begins at $j = k + 1$.

It is an enlightening exercise to look at this from another point of view, that of partitioned matrices. If $b_1 = b_2 = \cdots = b_k = 0$, we can write

$$b = \begin{bmatrix} 0 \\ \hat{b}_2 \end{bmatrix} \begin{matrix} k \\ n-k \end{matrix}$$

Partitioning G and y also, we have

$$G = \begin{matrix} k \\ n-k \end{matrix} \overset{\begin{matrix} k & n-k \end{matrix}}{\begin{bmatrix} G_{11} & 0 \\ G_{21} & G_{22} \end{bmatrix}} \qquad y = \begin{bmatrix} \hat{y}_1 \\ \hat{y}_2 \end{bmatrix} \begin{matrix} k \\ n-k \end{matrix}$$

G_{11} and G_{22} are lower triangular. The equation $Gy = b$ becomes

$$\begin{bmatrix} G_{11} & 0 \\ G_{21} & G_{22} \end{bmatrix} \begin{bmatrix} \hat{y}_1 \\ \hat{y}_2 \end{bmatrix} = \begin{bmatrix} 0 \\ \hat{b}_2 \end{bmatrix}$$

or

$$G_{11}\hat{y}_1 \qquad\qquad = 0$$
$$G_{21}\hat{y}_1 + G_{22}\hat{y}_2 = \hat{b}_2$$

The first equation says just that $\hat{y}_1 = 0$. The second equation then reduces to

$$G_{22}\hat{y}_2 = \hat{b}_2$$

Thus we have only to solve this $(n - k)$-by-$(n - k)$ lower-triangular system, which is exactly what (1.4.5) does. G_{11} and G_{21} are not used because they interact only with the unknowns $y_1, \ldots, y_k$ (i.e., $\hat{y}_1$). Since the system that is now being solved is of order $n - k$, the cost is $(n - k)^2/2$ flops.

Exercise 1.4.3   Write a modified version of algorithm (1.4.4) that checks for leading zeros in b and takes appropriate action. ☐

Now let us derive another algorithm for solving lower-triangular systems, the column-oriented version of forward substitution. We achieve this by partitioning the system $Gy = b$ as follows:

$$\begin{bmatrix} g_{11} & 0 \\ h & \hat{G} \end{bmatrix} \begin{bmatrix} y_1 \\ \hat{y} \end{bmatrix} = \begin{bmatrix} b_1 \\ \hat{b} \end{bmatrix}$$

$\hat{h}$, $\hat{y}$, and $\hat{b}$ are column vectors of length $n - 1$, and $\hat{G}$ is an $(n - 1)$-by-$(n - 1)$ lower-triangular matrix. The partitioned system can be written as

$$g_{11}y_1 = b_1$$
$$\hat{h}y_1 + \hat{G}\hat{y} = \hat{b}$$

This leads to the following algorithm:

$$
\begin{aligned}
y_1 &= b_1/g_{11}\\
\tilde{b} &= \hat{b} - \hat{h}y_1\\
\text{Solve } &\hat{G}\hat{y} = \tilde{b} \text{ for } \hat{y}
\end{aligned}
\tag{1.4.6}
$$

This algorithm reduces the problem of solving an n-by-n triangular system to that of solving the $(n - 1)$-by-$(n - 1)$ triangular system $\hat{G}\hat{y} = \tilde{b}$. This smaller problem can be reduced (by the same algorithm) to a problem of order $n - 2$, which can in turn be reduced to a problem of order $n - 3$, and so on. Eventually we get to a problem of order 1: $g_{nn}y_n = \bar{b}_n$; this can be solved by $y_n = \bar{b}_n/g_{nn}$. If you are a student of mathematics, this algorithm should remind you of proof by induction. If, on the other hand, you are a student of computer science, you might think of recursion, which is the computer science analog of mathematical induction. Recall that a *recursive* procedure is one that calls itself. If you like to program in Pascal or some other language that supports recursion, you might enjoy writing a recursive procedure that implements (1.4.6). The procedure would perform steps one and two of (1.4.6) and then call itself to solve the reduced problem. All variables named b, $\hat{b}$, $\tilde{b}$, y, and so on, can be stored in a single array. This array will contain b before execution and y afterward.

Although it is fun to write recursive programs, this algorithm can also be coded nonrecursively without difficulty. We will write down a nonrecursive formulation of the algorithm, but first it is worthwhile to work through one or two examples by hand.

Example 1.4.7 Use the column-oriented version of forward substitution to solve the lower-triangular system

$$
\begin{bmatrix} 5 & 0 & 0 \\ 2 & -4 & 0 \\ 1 & 2 & 3 \end{bmatrix}
\begin{bmatrix} y_1 \\ y_2 \\ y_3 \end{bmatrix}
=
\begin{bmatrix} 15 \\ -2 \\ 10 \end{bmatrix}
$$

First we calculate y_1.

$$y_1 = 15/5 = 3$$

Then

$$
\tilde{b} = \hat{b} - \hat{h}y_1 = \begin{bmatrix} -2 \\ 10 \end{bmatrix} - \begin{bmatrix} 2 \\ 1 \end{bmatrix} 3 = \begin{bmatrix} -8 \\ 7 \end{bmatrix}
$$

Now we have to solve $\hat{G}\hat{y} = \tilde{b}$:

$$
\begin{bmatrix} -4 & 0 \\ 2 & 3 \end{bmatrix}
\begin{bmatrix} y_2 \\ y_3 \end{bmatrix}
=
\begin{bmatrix} -8 \\ 7 \end{bmatrix}
$$

We solve this by repeating the algorithm.

$$y_2 = -8/-4 = 2$$

$$\tilde{b} = \hat{b} - \hat{h}y_2 = [7] - [2]2 = [3]$$

$$[3][y_3] = [3]$$

$$y_3 = 3/3 = 1$$

Thus

$$y = \begin{bmatrix} 3 \\ 2 \\ 1 \end{bmatrix}$$

Exercise 1.4.4 Use the column-oriented version of forward substitution to solve the system from Exercise 1.4.2. □

Exercise 1.4.5 Write a nonrecursive algorithm in the spirit of (1.4.4) that implements column-oriented forward substitution. Use a single array that contains b initially, stores intermediate results (e.g., $\hat{b}$, $\tilde{b}$) during the computation, and contains y at the end. □

Your solution to Exercise 1.4.5 should look something like this:

$$
\begin{array}{l}
\text{for } j = 1, 2, \ldots, n \\
\quad\left[\begin{array}{l}
\text{if } (g_{jj} = 0), \text{ set error flag } (G \text{ is singular}), \text{ exit} \\
b_j \leftarrow b_j/g_{jj} \text{ (this is } y_j) \\
\text{for } i = j + 1, \ldots, n \text{ (not executed when } j = n) \\
\quad\left[\; b_i \leftarrow b_i - g_{ij}b_j\right.
\end{array}\right. \\
\text{exit}
\end{array}
\qquad (1.4.8)
$$

Notice that (1.4.8) accesses G by columns, as anticipated; on the jth step, the jth column is used. Each time through the outer loop, the size of the problem is reduced by one. On the last time through, the computation $b_n \leftarrow b_n/g_{nn}$ (giving y_n) is performed, and the inner loop is skipped.

Exercise 1.4.6 Count the operations in (1.4.8) and notice that the flop count is identical to that for the row-oriented version (1.4.4). □

Exercise 1.4.7 Convince yourself that the row- and column-oriented forms of forward substitution carry out exactly the same operations but in a different order. □

Like the row-oriented version, the column-oriented version can be modified to take advantage of any leading zeros that may occur in the vector b. Notice that on the jth time through the outer loop in (1.4.8), if $b_j = 0$, then no changes are made in b. Thus the jth step can be skipped whenever $b_j = 0$. (This is true regardless of whether or not b_j is a leading zero. However, once a nonzero b_j has been encountered, all subsequent b_j's will not be the originals; they will have been altered on a previous step. Therefore they are not likely to be zero.)

We derived the column-oriented version of forward substitution by partitioning the system $Gy = b$. Different partitions should lead to different versions. For example, the partition

$$\begin{bmatrix} \hat{G} & 0 \\ h^T & g_{nn} \end{bmatrix} \begin{bmatrix} \hat{y} \\ y_n \end{bmatrix} = \begin{bmatrix} \hat{b} \\ b_n \end{bmatrix}$$

where $\hat{G}$ is $(n-1)$-by-$(n-1)$, leads to the following algorithm:

$$
\begin{aligned}
& \text{solve } \hat{G}\hat{y} = \hat{b} \text{ for } \hat{y} \\
& y_n = (b_n - h^T \hat{y})/g_{nn}
\end{aligned}
\tag{1.4.9}
$$

This is a recursive algorithm; it is again itself applied to solve the subsystem $\hat{G}\hat{y} = \hat{b}$. As the next exercise shows, it is really nothing new.

Exercise 1.4.8 Write a nonrecursive algorithm that implements (1.4.9). (*Hint:* Think about how (1.4.9) would calculate y_{i+1}, given $y_1, \ldots, y_i$). Observe that this algorithm is nothing but row-oriented forward substitution. ☐

Which of the two formulations of forward substitution is superior? The answer depends on how G is stored and accessed. This in turn depends on the programmer's choice of data structures and programming language and on the architecture of the computer. These issues will be discussed in subsequent sections of this chapter.

Upper-Triangular Systems

As you might expect, upper-triangular systems can be solved in much the same way as lower-triangular systems. Consider the system $Ux = y$, where $U = (u_{ij})$ is upper triangular. Writing the system out in detail, we get

$$
\begin{aligned}
u_{11}x_1 + u_{12}x_2 + \cdots + \quad u_{1,n-1}x_{n-1} + \quad u_{1n}x_n &= y_1 \\
u_{22}x_2 + \cdots + \quad u_{2,n-1}x_{n-1} + \quad u_{2n}x_n &= y_2 \\
\ddots \qquad\qquad \vdots \qquad\qquad \vdots \\
u_{n-1,n-1}x_{n-1} + u_{n-1,n}x_n &= y_{n-1} \\
u_{n,n}x_n &= y_n
\end{aligned}
$$

It becomes clear that we should solve the system from bottom to top. The nth equation can be solved for x_n, then the $(n-1)$th equation can be solved for x_{n-1}, and so on. The process is called *back substitution* and it has row- and column-oriented versions. The cost of back substitution is obviously $n^2/2$ flops, the same as that of forward substitution.

Exercise 1.4.9 Develop the row-oriented version of back substitution. ☐

Exercise 1.4.10 Develop the column-oriented version of back substitution. ☐

Exercise 1.4.11   This exercise is especially important for those who are weak on mathematical induction. If you belong to this group, you should definitely work this problem. When we were discussing flop counts, we noted that we can show

$$1 + 2 + 3 + \cdots + (n - 1) = \frac{n(n - 1)}{2} \qquad n = 1, 2, 3, \ldots \qquad (1.4.10)$$

by a well-known trick. If you are unfamiliar with the suggested method, you can still prove (1.4.10) by induction on n. Begin by verifying the equation in the case $n = 1$. The sum on the left-hand side is empty in this case. If you feel nervous about this, you can check the case $n = 2$ as well. Next show that if (1.4.10) holds for $n = k$, then it also holds for $n = k + 1$. Once you have done this, you will have proved by induction that (1.4.10) holds for all positive integers n. □

Exercise 1.4.12   Write Fortran subroutines to do each of the following: (a) row-oriented forward substitution, (b) column-oriented forward substitution, (c) row-oriented back substitution, (d) column-oriented back substitution. Invent some problems on which to test your programs. □

An easy way to devise a problem $Ax = b$ with a known solution is to specify the matrix A and the solution x, and then multiply A by x to get b. Give A and b to your program, and see if it can calculate x.

1.5
POSITIVE DEFINITE SYSTEMS;
CHOLESKY DECOMPOSITION

Recall that the *transpose* of a matrix $A = (a_{ij})$, denoted A^T, is the matrix whose (i, j) entry is a_{ji}. Thus the rows of A are the columns of A^T and vice versa. Since every vector is also a matrix, every vector has a transpose: A column vector x is a matrix with one column. Its transpose x^T is a row vector, a matrix with one row. The set of column n vectors with real entries will be denoted $\mathbb{R}^n$. That is, $\mathbb{R}^n$ is just the set of real n-by-1 matrices. A square matrix $A = (a_{ij})$ is *symmetric* if $A = A^T$; that is, $a_{ij} = a_{ji}$ for all i and j. If A is n-by-n, real, and symmetric and also satisfies the property

$$x^T A x > 0 \qquad (1.5.1)$$

for all nonzero $x \in \mathbb{R}^n$, then A is said to be *positive definite*.[§] The left-hand side of (1.5.1) is a matrix product. Examining the dimensions of x^T, A, and x, we find that $x^T A x$ is a 1-by-1 matrix; that is, it is a real number. Thus (1.5.1) is just an inequality between real numbers. It is also possible to define complex positive definite matrices. See Exercises 1.5.29 to 1.5.31.

Positive definite matrices arise frequently in applications. We note for example that the coefficient matrices in the electrical circuit problems in Section 1.1 are

---

[§] Some books, notably [MC], do not include symmetry as part of the definition.

positive definite, although we will not take the time to prove this. Some other areas in which positive definite matrices occur are the analysis of elastic structures, least-squares problems, and the numerical solution of partial differential equations. In the elastic structures problems, the expression $x^T A x$ has physical significance. The matrix A, called the *stiffness matrix* in that context, is derived from physical properties of the structure. The vector x represents a deformation of the structure and the number $\frac{1}{2} x^T A x$ is the *strain energy* of the structure, the potential energy stored in the structure due to the deformation x. (Think of a stretched spring or a bent beam that will return to its original configuration when released.) Inequality (1.5.1) states that a structure that has been deformed has a positive strain energy.

We begin our study of positive-definite matrices with a few basic facts.

THEOREM 1.5.2   If A is positive definite, then A is nonsingular.

Proof We will prove the contrapositive form of the theorem: If A is singular, then A is not positive definite. Assume A is singular. Then by Theorem 1.3.3 there exists a nonzero $x \in \mathbb{R}^n$ such that $Ax = 0$. But then $x^T A x = 0$, so A is not positive definite.   □

COROLLARY 1.5.3   If A is positive definite and $b \in \mathbb{R}^n$, then the linear system $Ax = b$ has exactly one solution.

THEOREM 1.5.4   Let M be any n-by-n, real, nonsingular matrix, and let $A = MM^T$. Then A is positive definite.

Proof First we must show that A is symmetric. Recalling the elementary formulas $(BC)^T = C^T B^T$ and $B^{TT} = B$, we find that $A^T = (MM^T)^T = M^{TT} M^T = MM^T = A$. Next we must show that $x^T A x > 0$ for all nonzero $x \in \mathbb{R}^n$. Let x be any nonzero vector in $\mathbb{R}^n$. Then $x^T A x = x^T MM^T x$. Let $y = M^T x$, so that $y^T = (M^T x)^T = x^T M$. Since M is nonsingular, so is M^T. Thus $x \neq 0$ implies $y \neq 0$. Letting $y_1, y_2, \ldots, y_n$ denote the components of y, we have $x^T A x = y^T y = \sum_{i=1}^{n} y_i^2$. This sum of squares is certainly nonnegative. In fact it is strictly positive because at least one of the y_i is nonzero. Thus $x^T A x > 0$, and A is positive definite.   □

Theorem 1.5.4 provides an easy means of constructing positive definite matrices: Just multiply any nonsingular matrix by its transpose.

Example 1.5.5   Let $M = \begin{bmatrix} 1 & 2 \\ 3 & 4 \end{bmatrix}$. M is nonsingular since $\det(M) = -2$. Therefore $A = MM^T = \begin{bmatrix} 5 & 11 \\ 11 & 25 \end{bmatrix}$ is positive definite.

Example 1.5.6   Let

$$M = \begin{bmatrix} 1 & 0 & 0 \\ 1 & 1 & 0 \\ 1 & 1 & 1 \end{bmatrix}$$

M is nonsingular since $\det(M) = 1$. Therefore

$$A = MM^T = \begin{bmatrix} 1 & 1 & 1 \\ 1 & 2 & 2 \\ 1 & 2 & 3 \end{bmatrix}$$

is positive definite.

The next theorem, the Cholesky decomposition theorem, is the most important result of this section. It states that every positive definite matrix is of the form MM^T for some M. Thus the recipe given by Theorem 1.5.4 generates all positive definite matrices. Furthermore M can be chosen to have a very special form.

Theorem 1.5.7 *(Cholesky Decomposition Theorem)* Let A be positive definite. Then A can be decomposed in exactly one way into a product:

$$A = GG^T \qquad (Cholesky\ decomposition)$$

such that G is lower triangular and has positive entries on the main diagonal. G is called the *Cholesky factor* of A.

The proof will be delayed until the end of the section. Right now it is more important to find out how the Cholesky decomposition can be used and how the Cholesky factor can be calculated.

Example 1.5.8 Let

$$A = \begin{bmatrix} 1 & 1 & 1 \\ 1 & 2 & 2 \\ 1 & 2 & 3 \end{bmatrix} \qquad \text{and} \qquad G = \begin{bmatrix} 1 & 0 & 0 \\ 1 & 1 & 0 \\ 1 & 1 & 1 \end{bmatrix}$$

G is lower triangular and has positive entries on the main diagonal. In Example 1.5.6 we saw that $A = GG^T$. Therefore G is the Cholesky factor of A.

The Cholesky decomposition is useful because G and G^T are triangular. Suppose we wish to solve the system of linear equations $Ax = b$, where A is positive definite. If we know the Cholesky factor G, we can write the system as $GG^T x = b$. Let $y = G^T x$. We do not know x, so we do not know y either. However, y clearly satisfies $Gy = b$. Since G is lower triangular, we can solve for y by forward substitution. Once we have y, we can solve the upper-triangular system $G^T x = y$ for x by back substitution.

It turns out that this method of solving a linear system is closely related to the familiar Gaussian elimination method. The connection between them is shown in Section 1.7, which you can read right now if you wish.

If the Cholesky decomposition is to be a useful tool, we must find a practical method for calculating the Cholesky factor. One of the easiest ways to do this is to write out the decomposition $A = GG^T$ in detail and study it:

$$
\begin{bmatrix}
a_{11} & a_{12} & a_{13} & \cdots & a_{1n} \\
a_{21} & a_{22} & a_{23} & \cdots & a_{2n} \\
a_{31} & a_{32} & a_{33} & \cdots & a_{3n} \\
\vdots & \vdots & \vdots & & \vdots \\
a_{n1} & a_{n2} & a_{n3} & \cdots & a_{nn}
\end{bmatrix}
$$

$$
=
\begin{bmatrix}
g_{11} & 0 & 0 & \cdots & 0 \\
g_{21} & g_{22} & 0 & \cdots & 0 \\
g_{31} & g_{32} & g_{33} & & 0 \\
\vdots & \vdots & \vdots & \ddots & \\
g_{n1} & g_{n2} & g_{n3} & \cdots & g_{nn}
\end{bmatrix}
\begin{bmatrix}
g_{11} & g_{21} & g_{31} & \cdots & g_{n1} \\
0 & g_{22} & g_{32} & \cdots & g_{n2} \\
0 & 0 & g_{33} & \cdots & g_{n3} \\
\vdots & \vdots & & \ddots & \vdots \\
0 & 0 & 0 & & g_{nn}
\end{bmatrix}
$$

We assume A is positive definite, so G certainly exists; we just have to find it. The element a_{ij} is the (inner) product of the ith row of G with the jth column of G^T. Noting that the first column of G^T has only one nonzero entry, we focus on this column:

$$
a_{i1} = g_{i1}g_{11} + g_{i2}0 + g_{i3}0 + \cdots = g_{i1}g_{11}
$$

In particular, when $i = 1$, we have $a_{11} = g_{11}^2$; this tells us that

$$
g_{11} = +\sqrt{a_{11}}
$$

We know that the positive square root is the right one because the main-diagonal entries of G are positive. Now that we know g_{11}, we can use the equation $a_{i1} = g_{i1}g_{11}$ to calculate the rest of the first column of G:

$$
g_{i1} = a_{i1}/g_{11} \qquad i = 2, 3, \ldots, n
$$

This is also the first row of G^T. We focus next on the second column, because the second column of G^T has only two nonzero entries:

$$
a_{i2} = g_{i1}g_{21} + g_{i2}g_{22} \tag{1.5.9}
$$

Only elements from the first two columns of G appear in this equation. In particular, when $i = 2$, we have $a_{22} = g_{21}^2 + g_{22}^2$. Since g_{21} is already known, we can use this equation to calculate g_{22}.

$$
g_{22} = +\sqrt{a_{22} - g_{21}^2}
$$

Once g_{22} is known, the only unknown left in (1.5.9) is g_{i2}. Thus we can use (1.5.9) to calculate the rest of the entries in the second column of G:

$$
g_{i2} = (a_{i2} - g_{i1}g_{21})/g_{22} \qquad i = 3, 4, \ldots, n
$$

There is no need to calculate g_{12} because $g_{12} = 0$. We now know the first two columns of G (and the first two rows of G^T). Now, as an exercise, you show how to calculate the third column of G.

If you are finished at this point, let's see how to calculate the jth column of G, assuming that we already have the first $j - 1$ columns. Since only the first j

entries in the jth column of G^T are nonzero,

$$a_{ij} = g_{i1}g_{j1} + g_{i2}g_{j2} + \cdots + g_{i,j-1}g_{j,j-1} + g_{ij}g_{jj} \qquad (1.5.10)$$

All entries of G appearing in (1.5.10) lie in the first j columns of G. Bearing in mind that the first $j - 1$ columns are known, we see that the only unknowns in (1.5.10) are g_{ij} and g_{jj}. Taking $i = j$ in (1.5.10), we have

$$a_{jj} = g_{j1}^2 + g_{j2}^2 + \cdots + g_{j,j-1}^2 + g_{jj}^2$$

which can be solved for g_{jj}:

$$g_{jj} = + \sqrt{a_{jj} - \sum_{k=1}^{j-1} g_{jk}^2} \qquad (1.5.11)$$

Again note that the positive square root is the right one. Now that we have g_{jj}, we can use (1.5.10) to solve for g_{ij}:

$$g_{ij} = \left(a_{ij} - \sum_{k=1}^{j-1} g_{ik}g_{jk} \right) \Big/ g_{jj} \qquad i = j + 1, \ldots, n \qquad (1.5.12)$$

We do not have to calculate g_{ij} for $i < j$ because those entries are all zero.

Equations 1.5.11 and 1.5.12 give a complete recipe for calculating G. They even hold for the first column of G ($j = 1$) if we recall our convention that the sums $\sum_{k=1}^{0}$ are zero. Notice also that when $j = n$, nothing is done in (1.5.12); the only nonzero entry in the nth column of G is g_{nn}.

The algorithm we have just developed is called *Cholesky's method* (or the *square-root method*). This, the first of three formulations that we will examine, is called the *inner-product* formulation because the sums in (1.5.11) and (1.5.12) can be regarded as inner products.

A number of important observations can be made. First, notice that the Cholesky decomposition theorem (which we haven't proved yet) makes two assertions: (1) G exists and (2) G is unique. In the process of developing Cholesky's method, we have proven that G is unique: the equation $A = GG^T$ and the stipulation $g_{jj} > 0$ imply (1.5.11). Thus g_{jj} is specified by (1.5.11). No other choice of g_{jj} will work; g_{jj} is uniquely determined. The equation $A = GG^T$ also implies (1.5.12). This equation specifies g_{ij} uniquely, $i = j + 1, \ldots, n$. Thus G is unique. If this argument does not convince you, think about how the algorithm calculates G one element at a time, starting with $g_{11} = + \sqrt{a_{11}}$. At each step the g_{ij} that is calculated is the only one that will work. Notice the importance of the stipulation $g_{jj} > 0$ in determining which square root to choose. Without this stipulation we would not have uniqueness.

Exercise 1.5.1 Let $A = \begin{bmatrix} 4 & 0 \\ 0 & 9 \end{bmatrix}$. (a) Prove that A is positive definite. (b) Calculate the Cholesky factor of A. (c) Find three other lower-triangular matrices G such that $A = GG^T$. (d) Let A be any positive-definite matrix of dimensions n-by-n. How many lower-triangular matrices G such that $A = GG^T$ are there? $\quad\square$

The next important observation is that Cholesky's method also serves as a test of positive definiteness. By Theorems 1.5.4 and 1.5.7, A is positive definite if and only if there exists a lower-triangular matrix G with positive main-diagonal entries, such that $A = GG^T$. Given any symmetric matrix A, we can attempt to calculate G by Cholesky's method. If A is not positive definite, the algorithm must fail, because any G that satisfies (1.5.11) and (1.5.12) must also satisfy $A = GG^T$. The algorithm fails if and only if at some step the number under the square-root sign in (1.5.11) is negative or zero. In the first case there is no real square root; in the second case $g_{jj} = 0$. Thus, if A is not positive definite, there must come a step at which the algorithm attempts to take the square root of a number that is not positive. Conversely, if A is positive definite, the algorithm cannot fail. The equation $A = GG^T$ guarantees that the number under the square-root sign in (1.5.11) is positive at every step. (After all, it equals g_{jj}^2.) Thus Cholesky's method succeeds if and only if A is positive definite. This is the best computational test of positive definiteness known.

The next thing to note is that (1.5.11) and (1.5.12) use only those a_{ij} for which $i \geq j$, the main diagonal and the entries below the main diagonal. This is not surprising since in a symmetric matrix the entries above the main diagonal are identical to those below the main diagonal. This underscores the fact that in a computer program we do not need to store all of A; there is no point in duplicating information. If space is at a premium, the programmer may choose to store A in a long one-dimensional array with $a_{11}, a_{21}, \ldots, a_{n1}$, immediately followed by $a_{22}, a_{32}, \ldots, a_{n2}$, immediately followed by $a_{33}, \ldots, a_{n3}$, and so on. This compact storage scheme makes the programming more difficult, but it is worth using if space is scarce.

Finally we note that each element a_{ij} is used only to compute g_{ij}. [A quick look at (1.5.11) and (1.5.12) confirms this.] It follows that in a computer program, g_{ij} can be stored over a_{ij}. This saves additional space by eliminating the need for separate arrays to store A and G.

Exercise 1.5.2   Write an algorithm based on (1.5.11) and (1.5.12) that checks a matrix for positive definiteness, calculates G, and stores G over A.   ☐

Your solution to Exercise 1.5.2 should look something like this:

Cholesky's algorithm (inner-product form)
for $j = 1, \ldots, n$
$\quad$ for $k = 1, \ldots, j - 1$ (not executed when $j = 1$)
$\qquad$ $a_{jj} \leftarrow a_{jj} - a_{jk}^2$
$\quad$ if $a_{jj} \leq 0$, set error flag (G is not positive definite), exit
$\quad$ $a_{jj} \leftarrow \text{sqrt}(a_{jj})$ (this is g_{jj})   (1.5.13)
$\quad$ for $i = j + 1, \ldots, n$ (not executed when $j = n$)
$\qquad$ for $k = 1, \ldots, j - 1$ (not executed when $j = 1$)
$\qquad\quad$ $a_{ij} \leftarrow a_{ij} - a_{ik}a_{jk}$
$\qquad$ $a_{ij} \leftarrow a_{ij}/a_{jj}$ (this is g_{ij})
exit

The lower part of G is stored over the lower part of A. There is no need to store the upper part of G because it consists entirely of zeros.

Example 1.5.14   Let

$$
A = \begin{bmatrix} 4 & -2 & 4 & 2 \\ -2 & 10 & -2 & -7 \\ 4 & -2 & 8 & 4 \\ 2 & -7 & 4 & 7 \end{bmatrix} \quad \text{and} \quad b = \begin{bmatrix} 8 \\ 2 \\ 16 \\ 6 \end{bmatrix}
$$

Notice that A is symmetric. We will use Cholesky's method to show that A is positive definite and calculate the Cholesky factor G. We will then use the Cholesky factor to solve the system $Ax = b$ by forward and back substitution.

$$g_{11} = \sqrt{a_{11}} = \sqrt{4} = 2$$

$$g_{21} = a_{21}/g_{11} = -2/2 = -1$$

$$g_{31} = 2$$

$$g_{41} = 1$$

$$g_{22} = \sqrt{a_{22} - g_{21}^2} = \sqrt{10 - (-1)^2} = 3$$

$$g_{32} = \frac{a_{32} - g_{31}g_{21}}{g_{22}} = \frac{-2 - (-1)(2)}{3} = 0$$

$$g_{42} = -2$$

$$g_{33} = \sqrt{a_{33} - g_{31}^2 - g_{32}^2} = \sqrt{8 - 2^2 - 0^2} = 2$$

$$g_{43} = \frac{a_{43} - g_{41}g_{31} - g_{42}g_{32}}{g_{33}} = \frac{4 - (2)(1) - (0)(-2)}{2} = 1$$

$$g_{44} = \sqrt{a_{44} - g_{41}^2 - g_{42}^2 - g_{43}^2} = \sqrt{7 - 1^2 - (-2)^2 - (1)^2} = 1$$

Thus

$$
G = \begin{bmatrix} 2 & 0 & 0 & 0 \\ -1 & 3 & 0 & 0 \\ 2 & 0 & 2 & 0 \\ 1 & -2 & 1 & 1 \end{bmatrix}
$$

Since we were able to calculate G, A is positive definite.

To solve $Ax = b$, we first solve $Gy = b$ by forward substitution and obtain $y = [4 \ 2 \ 4 \ 2]^T$. We then solve $G^T x = y$ by back substitution and obtain $x = [1 \ 2 \ 1 \ 2]^T$.

An interesting and worthwhile exercise is to calculate G by the *erasure method*. Start with an array that has A penciled in initially (main diagonal and lower triangle only). As you calculate each entry g_{ij}, erase a_{ij} and replace it by g_{ij}. Do all operations in your head, using only the single array for reference. When you are done, you will have G where you once had A. This procedure is surprisingly easy, once you get the hang of it.

Example 1.5.15 Let

$$A = \begin{bmatrix} 1 & 2 & 3 \\ 2 & 5 & 10 \\ 3 & 10 & 16 \end{bmatrix}$$

We will use Cholesky's method to determine whether or not A is positive definite. Proceeding as in Example 1.5.14, we find that $g_{11} = 1$, $g_{21} = 2$, $g_{31} = 3$, $g_{22} = 1$, $g_{32} = 4$, and finally $g_{33} = \sqrt{a_{33} - g_{31}^2 - g_{32}^2} = \sqrt{-9}$. In attempting to calculate g_{33}, we encounter a negative number under the square-root sign. Thus A is not positive definite.

Exercise 1.5.3 Let

$$A = \begin{bmatrix} 16 & 4 & 8 & 4 \\ 4 & 10 & 8 & 4 \\ 8 & 8 & 12 & 10 \\ 4 & 4 & 10 & 12 \end{bmatrix} \quad \text{and} \quad b = \begin{bmatrix} 32 \\ 26 \\ 38 \\ 30 \end{bmatrix}$$

Notice that A is symmetric. (a) Use Cholesky's method to show that A is positive definite and find its Cholesky factor. (b) Use forward and back substitution to solve $Ax = b$ for x. ☐

Exercise 1.5.4 Determine whether or not each of the following matrices is positive definite.

$$A = \begin{bmatrix} 9 & 3 & 3 \\ 3 & 10 & 5 \\ 3 & 7 & 9 \end{bmatrix} \quad B = \begin{bmatrix} 4 & 2 & 6 \\ 2 & 2 & 5 \\ 6 & 5 & 29 \end{bmatrix}$$

$$C = \begin{bmatrix} 4 & 4 & 8 \\ 4 & -4 & 1 \\ 8 & 1 & 6 \end{bmatrix} \quad D = \begin{bmatrix} 1 & 1 & 1 \\ 1 & 2 & 2 \\ 1 & 2 & 1 \end{bmatrix}$$ ☐

Although Cholesky's method generally works very well, a word of caution is appropriate here. Most matrix computations are performed by computer, in which case each arithmetic operation is subject to a roundoff error. In Chapter 2 we will find that the performance of Cholesky's method in the face of roundoff errors is as good as we could possibly expect. However, there exist linear systems, called ill-conditioned systems, that simply cannot be solved accurately in the presence of errors. Naturally we cannot expect Cholesky's method (performed with roundoff errors) to solve ill-conditioned systems accurately. See Chapter 2 for details.

Our next task will be to count the operations in Cholesky's method. In order to do this we need to know that

$$\sum_{j=1}^{n} j^2 = \frac{n(n + \frac{1}{2})(n + 1)}{3} \tag{1.5.16}$$

Exercise 1.5.5 Prove (1.5.16) by induction on n. ☐

From (1.5.16) we get the approximation

$$\sum_{j=1}^{n} j^2 \approx \frac{n^3}{3} \tag{1.5.17}$$

valid for large n, which is usually used instead of (1.5.16). This estimate can be derived directly, bypassing (1.5.16), by approximating the sum in (1.5.17) by an integral:

$$\sum_{j=1}^{n} j^2 \approx \int_0^n j^2 \, dj = \left.\frac{j^3}{3}\right|_0^n = \frac{n^3}{3}$$

Exercise 1.5.6   Draw a picture that shows that for large n, $\int_0^n j^2 \, dj$ is a good estimate of $\sum_{j=1}^{n} j^2$.   □

   Let us now count the operations in Cholesky's algorithm (1.5.13). In each of the two k loops in (1.5.13), one flop is executed each time we pass through the loop. The first of these k loops is executed $j - 1$ times on the jth time through the outer loop. Thus, the total number of flops attributable to this loop is $\sum_{j=1}^{n}(j - 1) = n(n - 1)/2 \approx n^2/2$. Another way to express this number is

$$\sum_{j=1}^{n} \sum_{k=1}^{j-1} 1$$

which was found by examining the limits on the loop indices. We sum the number 1 because one flop is done each time through the inner loop. Applying this idea to the second of the k loops, we find that the total number of flops attributable to that loop is

$$\sum_{j=1}^{n} \sum_{i=j+1}^{n} \sum_{k=1}^{j-1} 1$$

We have a triple sum this time, because the loops are nested three deep.

$$\sum_{j=1}^{n} \sum_{i=j+1}^{n} \sum_{k=1}^{j-1} 1 = \sum_{j=1}^{n} \sum_{i=j+1}^{n} (j - 1)$$

$$= \sum_{j=1}^{n} (n - j)(j - 1)$$

$$= n \sum_{j=1}^{n} (j - 1) - \sum_{j=1}^{n} j^2 + \sum_{j=1}^{n} j$$

$$= \frac{nn(n - 1)}{2} - \frac{n(n + \frac{1}{2})(n + 1)}{3} + \frac{n(n + 1)}{2}$$

$$\approx \frac{n^3}{2} - \frac{n^3}{3} = \frac{n^3}{6}$$

Thus approximately $n^3/6$ flops are performed in the second of the k loops. Like the flop count of the previous section, this estimate assumes that n is large. Hence the terms of order n^2 or less are negligible in comparison with $n^3/6$ and can be ignored. Since only about $n^2/2$ flops are done in the first k loop, this number is negligible compared to $n^3/6$ and can be ignored as well. In addition to the flops, some divisions are done in the i loop. The exact number is

$$\sum_{j=1}^{n} \sum_{i=j+1}^{n} 1 = \sum_{j=1}^{n} (n - j) = \frac{n(n-1)}{2}$$

which is also negligible. Finally $\sum_{i=1}^{n} 1 = n$ error checks and square roots are done. We conclude that the cost of Cholesky's algorithm is about $n^3/6$ flops.

Recall that the combined cost of doing forward and back substitution is about n^2 flops. Therefore, when we use Cholesky's method to solve the positive definite system $Ax = b$, the bulk of the time is spent calculating the Cholesky factors. The cost of the forward and back substitution is negligible by comparison. Ignoring terms of order n^2 and less, the cost of solving a positive-definite system by Cholesky's method is $n^3/6$ flops. This means that each time the size of the system is doubled, the cost of solving it by Cholesky's method is multiplied by 8.

We will consider two other forms of the Cholesky algorithm, both of which can be derived by partitioning the matrices A and G in the equation $A = GG^T$. The *outer-product* formulation arises from the partitioned equation

$$\begin{bmatrix} a_{11} & b^T \\ b & \hat{A} \end{bmatrix} = \begin{bmatrix} g_{11} & 0 \\ h & \hat{G} \end{bmatrix} \begin{bmatrix} g_{11} & h^T \\ 0 & \hat{G}^T \end{bmatrix}$$

which implies

$$a_{11} = g_{11}^2 \qquad b = h g_{11} \qquad \hat{A} = h h^T + \hat{G} \hat{G}^T \qquad (1.5.18)$$

A fourth equation is $b^T = g_{11} h^T$, but this is the same as $b = h g_{11}$. Equations (1.5.18) suggest the following procedure for calculating g_{11}, h, and $\hat{G}$ (and hence G):

$$
\begin{aligned}
g_{11} &= \sqrt{a_{11}} \\
h &= \frac{1}{g_{11}} b \\
\tilde{A} &= \hat{A} - h h^T \\
&\text{solve } \tilde{A} = \hat{G} \hat{G}^T \text{ for } \hat{G}
\end{aligned}
\qquad (1.5.19)
$$

This procedure reduces the n-by-n problem to that of finding the Cholesky factor of an $(n-1)$-by-$(n-1)$ matrix $\tilde{A}$. This problem can be reduced to an $(n-2)$-by-$(n-2)$ problem by the same algorithm, and so on. Eventually the problem is reduced to the trivial 1-by-1 case. This is called the *outer-product form*, because at each step of the reduction, an outer product $h h^T$ is subtracted from the remaining submatrix. It can be implemented recursively or nonrecursively with no difficulty.

Exercise 1.5.7   Use the outer-product formulation of Cholesky's method to calculate the Cholesky factor of the matrix A of Example 1.5.14. ☐

Exercise 1.5.8   Use the outer-product formulation of Cholesky's method to work Example 1.5.15. ☐

Exercise 1.5.9   Use the outer-product form to work part (a) of Exercise 1.5.3. ☐

Exercise 1.5.10   Use the outer-product form to work Exercise 1.5.4. ☐

Exercise 1.5.11   Write a nonrecursive algorithm that implements the outer-product formulation (1.5.19). Your algorithm should exploit the symmetry of A by referencing only the main diagonal and lower part of A, and it should store G over A. Be sure to put in the necessary check before taking the square root. ☐

Exercise 1.5.12   (a) Do a flop count for the outer-product formulation of Cholesky's method. You will find that approximately $n^3/6$ flops are performed, the same number as for the inner-product formulation. If you do an exact flop count for the two algorithms, you will find that the counts are exactly equal. (b) Convince yourself that the two algorithms perform exactly the same operations, but not in the same order. ☐

The third and final form of Cholesky's method that we will consider is the *bordered form*. First we will introduce some new notation. For $i = 1, \ldots, n$, let A_i be the i-by-i submatrix of A consisting of the intersection of the first i rows and columns of A. A_i is called the ith *leading principal submatrix* of A. We will find in Exercise 1.5.22 that if A is positive definite, then all of its leading principal submatrices are positive definite. Suppose A is positive definite, and let G be the Cholesky factor of A. Then G has leading principal submatrices G_i, $i = 1, \ldots, n$, which are lower triangular and have positive entries on the main diagonal.

Exercise 1.5.13   By partitioning the equation $A = GG^T$ appropriately, show that G_i is the Cholesky factor of A_i. ☐

It is easy to construct $G_1 = [g_{11}]$, since $a_{11} = g_{11}^2$. Thinking inductively, if we can figure out how to construct G_i given G_{i-1}, then we will be able to construct $G_n = G$ in n steps. Suppose therefore that we have calculated G_{i-1} and wish to find G_i. Partitioning the equation $A_i = G_i G_i^T$ as follows:

$$\begin{bmatrix} A_{i-1} & c \\ c^T & a_{ii} \end{bmatrix} = \begin{bmatrix} G_{i-1} & 0 \\ h^T & g_{ii} \end{bmatrix} \begin{bmatrix} G_{i-1}^T & h \\ 0 & g_{ii} \end{bmatrix} \tag{1.5.20}$$

we get the equations

$$A_{i-1} = G_{i-1} G_{i-1}^T \qquad c = G_{i-1} h \qquad a_{ii} = h^T h + g_{ii}^2 \tag{1.5.21}$$

Since we already have G_{i-1}, we have only to calculate h (or equivalently h^T) and g_{ii}. Equations (1.5.21) show that this can be done as follows:

$$\begin{aligned} &\text{solve } G_{i-1} h = c \text{ for } h \\ &g_{ii} = \sqrt{a_{ii} - h^T h} \end{aligned} \tag{1.5.22}$$

The first of these steps can be done by forward substitution. You may already have written a subroutine to perform this task (Exercise 1.4.12). The algorithm that can be built using these ideas is called the *bordered form* of Cholesky's method. This form will be used in the next section, in which we discuss the solution of banded positive definite systems.

Exercise 1.5.14   Use the bordered form of Cholesky's method to calculate the Cholesky factor of the matrix A of Example 1.5.14.   □

Exercise 1.5.15   Use the bordered form of Cholesky's method to work Example 1.5.15.   □

Exercise 1.5.16   Use the bordered form to work part (a) of Exercise 1.5.3.   □

Exercise 1.5.17   Use the bordered form to work Exercise 1.5.4.   □

Exercise 1.5.18   (a) Do a flop count for the bordered form of Cholesky's method. Again you will find that approximately $n^3/6$ flops are done. If you do an exact flop count, you will find that exactly the same number of flops are performed by this algorithm as are done by the other two formulations of Cholesky's method. (b) Convince yourself that this algorithm carries out exactly the same operations as the other two forms of Cholesky's method.   □

Exercise 1.5.19   Write Fortran subroutines to implement Cholesky's method in (a) inner-product form, (b) outer-product form, and (c) bordered form. If you have already written a forward-substitution routine, you can use it in part (c). Your subroutines should operate only on the main diagonal and lower-triangular part of A, and they should overwrite A with G. They should either return G or set a warning flag indicating that G is not positive definite. Try out your routines on the following examples.

$$\text{(i)} \quad A = \begin{bmatrix} 36 & 30 & 24 \\ 30 & 34 & 26 \\ 24 & 26 & 21 \end{bmatrix}$$

$$\text{(ii)} \quad A = \begin{bmatrix} 1 & -1 & 1 & -1 & 1 \\ -1 & 2 & -2 & 2 & -2 \\ 1 & -2 & 3 & -3 & 3 \\ -1 & 2 & -3 & 4 & -4 \\ 1 & -2 & 3 & -4 & 5 \end{bmatrix}$$

$$\text{(iii)} \quad A = \begin{bmatrix} 1 & 2 & 3 \\ 2 & 5 & 10 \\ 3 & 10 & 16 \end{bmatrix} \quad \text{(cf., Example 1.5.15)}$$

You might like to devise some additional examples. The easy way to do this is to write down G first and then multiply G by G^T to get A.   □

Exercise 1.5.20   Write a program that solves positive definite systems $Ax = b$ by calling subroutines to (a) calculate the Cholesky factor, (b) perform forward substitution, and (c) perform back substitution. Try out your program on the following examples.

(i) $\begin{bmatrix} 36 & -30 & 24 \\ -30 & 34 & -26 \\ 24 & -26 & 21 \end{bmatrix} \begin{bmatrix} x_1 \\ x_2 \\ x_3 \end{bmatrix} = \begin{bmatrix} 0 \\ 12 \\ -7 \end{bmatrix}$

(ii) $\begin{bmatrix} 1 & 1 & 1 & 1 & 1 \\ 1 & 2 & 2 & 2 & 2 \\ 1 & 2 & 3 & 3 & 3 \\ 1 & 2 & 3 & 4 & 4 \\ 1 & 2 & 3 & 4 & 5 \end{bmatrix} \begin{bmatrix} x_1 \\ x_2 \\ x_3 \\ x_4 \\ x_5 \end{bmatrix} = \begin{bmatrix} 5 \\ 9 \\ 12 \\ 14 \\ 15 \end{bmatrix}$ □

Proof of the Cholesky Decomposition Theorem

Having seen how the Cholesky decomposition can be used, and having developed three ways to compute the Cholesky factor, we will now take the time to prove the Cholesky decomposition theorem. We begin with a few simple exercises.

Exercise 1.5.21   Show that if A is positive definite, then $a_{ii} > 0$, $i = 1, \ldots, n$. Do not use the Cholesky decomposition, since we have not yet proven that it exists. (*Hint:* Find a nonzero $x \in \mathbb{R}^n$ such that $x^T A x = a_{ii}$.)   □

Armed with this new knowledge you can now work part of Exercise 1.5.4 more easily.

Exercise 1.5.22   Let $A = \begin{array}{c} \\ j \\ k \end{array} \overset{\begin{array}{cc} j & k \end{array}}{\begin{bmatrix} A_{11} & A_{12} \\ A_{21} & A_{22} \end{bmatrix}}$ be positive definite. Show that A_{11} and A_{22} are positive definite. Do not use the Cholesky decomposition; use the fact that $x^T A x > 0$ for all nonzero $x \in \mathbb{R}^n$.

Solution   We will show that A_{11} is positive definite; a similar argument works for A_{22}. Clearly A_{11} is symmetric. We must show that $y^T A_{11} y > 0$ for all nonzero $y \in \mathbb{R}^j$. Given such a y, define $x \in \mathbb{R}^n$ by $x = \begin{bmatrix} y \\ 0 \end{bmatrix} \begin{array}{c} j \\ k \end{array}$. Then since $x \neq 0$ and A is positive definite, $x^T A x > 0$. But

$$x^T A x = \begin{bmatrix} y^T & 0^T \end{bmatrix} \begin{bmatrix} A_{11} & A_{12} \\ A_{21} & A_{22} \end{bmatrix} \begin{bmatrix} y \\ 0 \end{bmatrix} = y^T A_{11} y$$

Thus $y^T A_{11} y > 0$.   □

The result of Exercise 1.5.22 is crucial to the proof of the Cholesky decomposition theorem. The following exercise will allow us to introduce some notation that will be used in the proof.

Exercise 1.5.23   Let B be any nonsingular matrix. Prove that $(B^T)^{-1} = (B^{-1})^T$.   □

The symbol B^{-T} will denote $(B^{-1})^T$. By Exercise 1.5.23 this is the same as $(B^T)^{-1}$.

Recall that the Cholesky decomposition theorem (Theorem 1.5.7) states that given any positive definite matrix A, there exists a unique lower-triangular matrix G with positive elements on the main diagonal such that $A = GG^T$. We have already proved uniqueness. A proof of the existence of G can be based on the outer-product formulation or the bordered formulation of the Cholesky algorithm. Just for fun, we will do something more general. In preparation for our existence proof, let us see how the decomposition $A = GG^T$ looks when the matrices have been partitioned. Let

$$A = \begin{array}{c} j \\ k \end{array} \overset{\displaystyle j \quad k}{\left[\begin{array}{cc} A_{11} & A_{21}^T \\ A_{21} & A_{22} \end{array} \right]} \qquad G = \begin{array}{c} j \\ k \end{array} \overset{\displaystyle j \quad k}{\left[\begin{array}{cc} G_{11} & 0 \\ G_{21} & G_{22} \end{array} \right]}$$

where j and k are any two positive integers such that $j + k = n$. A_{11} and A_{22} are positive definite (Exercise 1.5.22), and G_{11} and G_{22} are lower triangular with positive main-diagonal entries. The equation $A = GG^T$ is equivalent to

$$A_{11} = G_{11}G_{11}^T$$

$$A_{21} = G_{21}G_{11}^T \qquad (\text{and } A_{21}^T = G_{11}G_{21}^T)$$

$$A_{22} = G_{21}G_{21}^T + G_{22}G_{22}^T$$

These equations suggest the following sequence of operations:

calculate G_{11}, the Cholesky factor of A_{11} (1.5.23)

calculate $G_{21}\ (= A_{21}G_{11}^{-T})$ (1.5.24)

let $\hat{A}_{22} = A_{22} - G_{21}G_{21}^T$ (1.5.25)

calculate G_{22}, the Cholesky factor of $\hat{A}_{22}$ (1.5.26)

An inductive proof of the existence of G can be based on this construction. (Why?) The only difficult point is to prove that $\hat{A}_{22}$ is positive definite. Notice that when $j = 1$ the construction (1.5.23) to (1.5.26) is just the outer-product form of Cholesky's method, and when $k = 1$, it is the bordered form.

Proof of Theorem 1.5.7   Let $A \in \mathbb{R}^{n \times n}$ be positive definite. We must show that there exists a lower-triangular G with positive main-diagonal entries such that $A = GG^T$. We will use induction on n. The case $n = 1$ is trivial: If $A = [a_{11}]$ is positive definite, then $a_{11} > 0$. Clearly, if we define $G = [g_{11}]$ by $g_{11} = +\sqrt{a_{11}}$, then $A = GG^T$ and $g_{11} > 0$. It is also clear that this is the only choice of G that works. Now let $n > 1$ and make the inductive hypothesis that every positive definite matrix of dimension less than n has a Cholesky factor. Partition A as above; that is,

$$A = \begin{array}{c} j \\ k \end{array} \overset{\displaystyle j \quad k}{\left[\begin{array}{cc} A_{11} & A_{21}^T \\ A_{21} & A_{22} \end{array} \right]}$$

where j and k can be any positive integers whose sum is n. By Exercise 1.5.22, A_{11} is positive definite. Since A_{11} is of dimension less than n, the induction hypothesis implies that there exists a lower-triangular matrix G_{11} with positive main-diagonal entries such that $A_{11} = G_{11}G_{11}^T$. The matrix G_{11} is nonsingular, so G_{11}^{-T} exists. Using (1.5.24) as a guide, define a k-by-j matrix G_{21} by $G_{21} = A_{21}G_{11}^{-T}$. Following (1.5.25), define a k-by-k matrix $\hat{A}_{22}$ by $\hat{A}_{22} = A_{22} - G_{21}G_{21}^T$. Clearly $\hat{A}_{22}$ is symmetric. To prove that it is positive definite, we must show that for all nonzero $z \in \mathbb{R}^k$, $z^T\hat{A}_{22}z > 0$; that is,

$$z^TA_{22}z - z^TG_{21}G_{21}^Tz > 0 \tag{1.5.27}$$

We are given that A is positive definite. Thus $x^TAx > 0$ for all nonzero $x \in \mathbb{R}^n$. Partitioning x conformably with A, we have $x^T = [y^T, z^T]$, where $y \in \mathbb{R}^j$ and $z \in \mathbb{R}^k$.

$$x^TAx = [y^T \ z^T] \begin{bmatrix} A_{11} & A_{21}^T \\ A_{21} & A_{22} \end{bmatrix} \begin{bmatrix} y \\ z \end{bmatrix}$$

$$= y^TA_{11}y + z^TA_{21}y + y^TA_{21}^T z + z^TA_{22}z$$

Thus

$$y^TA_{11}y + z^TA_{21}y + y^TA_{21}^Tz + z^TA_{22}z > 0 \tag{1.5.28}$$

as long as y and z are not both zero. Our approach will be, given a nonzero $z \in \mathbb{R}^k$, to try to find a $y \in \mathbb{R}^j$ for which the left-hand side of (1.5.28) reduces to the left-hand side of (1.5.27). If we can do this, then (1.5.27) will follow from (1.5.28). In order to determine the conditions that such a y must satisfy, we equate the left-hand side of (1.5.28) with that of (1.5.27) to obtain

$$y^TA_{11}y + z^TA_{21}y + y^TA_{21}^Tz + z^TG_{21}G_{21}^Tz = 0$$

Recalling that $A_{11} = G_{11}G_{11}^T$ and $A_{21} = G_{21}G_{11}^T$, we get

$$y^TG_{11}G_{11}^Ty + z^TG_{21}G_{11}^Ty + y^TG_{11}G_{21}^Tz + z^TG_{21}G_{21}^Tz = 0 \tag{1.5.29}$$

From here on the notation is simplified considerably if we make the substitutions $v = G_{11}^Ty$ and $w = G_{21}^Tz$. Equation 1.5.29 then reduces to

$$v^Tv + w^Tv + v^Tw + w^Tw = 0$$

which further compresses to

$$0 = (v + w)^T(v + w) = \sum_{i=1}^{j} (v_i + w_i)^2$$

This last equation implies that $v = -w$, from which $G_{11}^Ty = -G_{21}^Tz$ or, solving for y,

$$y = -G_{11}^{-T}G_{21}^Tz \tag{1.5.30}$$

We have just shown that if (1.5.28) is to reduce to (1.5.27), then y must be given by (1.5.30). Since every step we have made can be reversed, it follows that if we define y by (1.5.30), then (1.5.28) must reduce to (1.5.27). You can check directly that it does. We conclude that $\hat{A}_{22}$ is positive definite. Since the dimension of $\hat{A}_{22}$ is less than n, the induction hypothesis implies that there exists a lower-triangular matrix G_{22} with positive main-diagonal entries, such that $\hat{A}_{22} = G_{22}G_{22}^T$. [Compare with (1.5.26).] Finally, define G by

$$G = \begin{bmatrix} G_{11} & 0 \\ G_{21} & G_{22} \end{bmatrix}$$

G is n by n, lower triangular, and has positive entries on the main diagonal. You can verify that $GG^T = A$. Thus A has a Cholesky factor. $\square$

Exercise 1.5.24 Verify that if the value of y given by (1.5.30) is substituted into (1.5.28), then (1.5.28) reduces to (1.5.27). $\square$

Exercise 1.5.25 Verify that the matrix G defined in the proof satisfies $GG^T = A$. $\square$

Exercise 1.5.26 Using (1.5.23) to (1.5.26) as a guide, prove by induction on n that the Cholesky factor is unique: Suppose $A = GG^T = HH^T$, where G and H are both lower triangular with positive main-diagonal entries. Partition G and H and then show that $G_{11} = H_{11}$, $G_{21} = H_{21}$, and $G_{22} = H_{22}$. $\square$

Exercise 1.5.27 Prove that if A is positive definite, then $\det(A) > 0$. $\square$

Exercise 1.5.28 This exercise generalizes Exercise 1.5.22. Let A be an n-by-n positive definite matrix. Let $j_i, j_2, \ldots, j_k$ be integers such that $1 \le j_1 < j_2 < \cdots < j_k \le n$, and let $\hat{A}$ be the k-by-k matrix obtained by intersecting rows $j_1, \ldots, j_k$ with columns $j_1, \ldots, j_k$. Prove that $\hat{A}$ is positive definite. $\square$

Complex Positive Definite Matrices

The next three exercises show how the results of this section can be extended to complex matrices. The set of complex column n vectors will be denoted $\mathbb{C}^n$. The *conjugate transpose* A^* of a complex m-by-n matrix $A = (a_{ij})$ is the n-by-m matrix whose (i, j) entry is $\overline{a}_{ji}$. The bar denotes the complex conjugate. A is *Hermitian* if $A = A^*$; that is, $a_{ij} = \overline{a}_{ji}$ for all i and j.

Exercise 1.5.29 (a) Prove that if A and B are m-by-n and n-by-p complex matrices, respectively, then $(AB)^* = B^*A^*$. (b) Prove that if A is Hermitian, then x^*Ax is real for every $x \in \mathbb{C}^n$. [*Hint:* x^*Ax can be viewed as a 1-by-1 matrix or as a real number. Viewing x^*Ax as a 1-by-1 matrix, examine $(x^*Ax)^*$, which is the same as the number $\overline{(x^*Ax)}$.] $\square$

If A is Hermitian and satisfies $x^*Ax > 0$ for all nonzero $x \in \mathbb{C}^n$, then A is said to be *positive definite*.

Exercise 1.5.30   Prove that if M is any n-by-n complex, nonsingular matrix, then MM^* is positive definite.   □

Exercise 1.5.31   Prove that if A is positive definite, then there exists a unique matrix G such that G is lower triangular and has positive (real) main-diagonal entries, and $A = GG^*$.   □

All the algorithms that we have developed in this section can easily be adapted to the complex case.

1.6
BANDED POSITIVE DEFINITE SYSTEMS

We noted in Section 1.1 that large systems occur frequently in applications, and large systems are usually sparse. In this section we will study a simple yet very effective scheme for solving large, positive-definite systems that are banded or have an envelope structure. This scheme is in widespread use and, as we shall see, it can yield enormous savings in computer time and storage space. However, it is not necessarily the most efficient scheme, especially for extremely large problems or problems with very irregular structures. For such problems a more sophisticated approach may be worthwhile. See, for example, [SPDS] or Duff et al. (1986).

A *banded* matrix is one whose nonzero entries are confined to a narrow band of diagonals near the main diagonal. The structure of a banded matrix is shown schematically in Figure 1.3. Each entry within the band can be either zero or nonzero; the important point is that all entries outside the band are zero. Since the matrices that we are considering in this section are positive definite, hence symmetric, we need only consider half the band. A symmetric matrix A has *semiband*

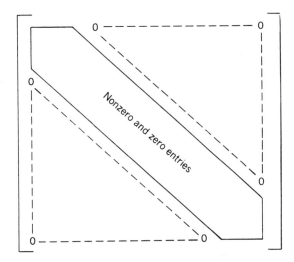

Figure 1.3   Structure of a banded matrix.

width s if the nonzero entries below the main diagonal are restricted to the first s subdiagonals. This is the same as saying that $a_{ij} = 0$ whenever $i - j > s$. (By symmetry we also have that $a_{ij} = 0$ whenever $j - i > s$.) A lower-triangular matrix G has *semiband width* s if $g_{ij} = 0$ whenever $i - j > s$. Similarly, an upper-triangular matrix U has *semiband width* s if $u_{ij} = 0$ whenever $j - i > s$.

Every large, regular network gives rise to a banded system. Consider for example the positive definite matrix associated with the electrical circuit of Figure 1.2, and recall that $a_{ij} \neq 0$ if and only if nodes i and j are adjacent. If the nodes are numbered from left to right and then top to bottom, as indicated in the figure, any two nodes i and j for which $i - j > 10$ cannot be adjacent. Conversely, there are many pairs of adjacent nodes for which $i - j = 10$. Thus the semiband width of the matrix is 10.

Exercise 1.6.1 Make a rough sketch of the matrix associated with Figure 1.2, noting where the zeros and nonzeros lie. Notice that even within the band most of the entries are zero. This is typical of the banded problems that occur in applications.

□

Notice that the bandedness depends on how the nodes are numbered. If, for example, nodes 2 and 118 in Figure 1.2 are interchanged, the resulting matrix is not banded, since its $(118, 1)$ entry is nonzero. However it is still sparse; the number of nonzero entries in the matrix does not depend on the ordering of the nodes. If a network is regular, it is easy to see how to number the nodes to obtain a narrow band. Irregular networks can also lead to banded systems; however, it will usually be more difficult to decide how the nodes should be numbered.

Banded positive definite systems can be solved economically because it is possible to ignore the entries that lie outside of the band. For this it is crucial that the Cholesky factor of a banded matrix A is also banded and has the same band width as A. (We will not prove this now because we are going to prove a more general result below.) Thus we can save computer storage by employing a data structure that stores only the (semi) band of A; G can be stored over A. More important, computer time is saved because all operations involving entries outside the band can be skipped. As we shall soon see, these savings are substantial.

Instead of analyzing banded systems, we will introduce a more general idea, that of the envelope of a matrix. This will both increase the generality of the discussion and simplify the analysis. The *envelope* of a symmetric or lower-triangular matrix A is a set of ordered pairs (i, j), $i > j$, representing element locations in the lower triangle of A, defined as follows: (i, j) is in the envelope of A if and only if $a_{ik} \neq 0$ for some $k \leq j$. Thus if the first nonzero entry of the ith row is a_{im} and $i > m$, then $(i, m), (i, m + 1), \ldots, (i, i - 1)$ are the members of the envelope of A from the ith row. Figure 1.4 shows a sparse matrix and its envelope.

The crucial theorem about envelopes (Theorem 1.6.1) states that if G is the Cholesky factor of A, then G has the same envelope as A. Thus A can be stored in a data structure that stores only its main diagonal and the entries in its envelope, and G can be stored over A. All operations involving off-diagonal entries lying outside the envelope can be skipped. If the envelope is small, substantial savings in computer time and storage space are realized. Banded matrices have small

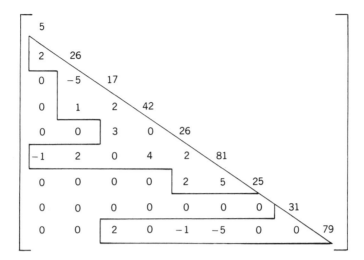

Figure 1.4   The envelope of a matrix.

envelopes. An example of an unbanded matrix with a small envelope is shown in Figure 1.5.

Like the band width, the envelope of a matrix depends on the order in which the equations and unknowns are numbered. Often it is easy to see how to number the nodes to obtain a reasonably small envelope. For those cases in which it is hard to tell how the nodes should be ordered, there exist algorithms that attempt to minimize the envelope in some sense. For example, see the discussion of the reverse Cuthill–McKee algorithm in [SPDS].

THEOREM 1.6.1   Let A be positive definite, and let G be the Cholesky factor of A. Then the envelope of G is the same as the envelope of A.

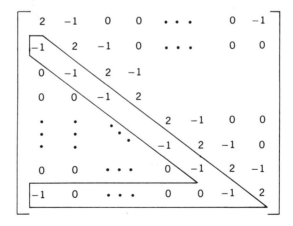

Figure 1.5   Matrix obtained from discretization of an ordinary differential equation with a periodic boundary condition.

Proof Consider the bordered form of the Cholesky decomposition algorithm. At the ith step we solve a system

$$G_{i-1}h = c$$

where $c^T \in \mathbb{R}^{i-1}$ is the portion of the ith row of A lying to the left of the main diagonal, and h^T is the corresponding portion of G. (See Eqs. 1.5.20 to 1.5.22.) Let $\hat{c} \in \mathbb{R}^{s_i}$ be the portion of c that lies in the envelope of A. Then

$$c = \begin{bmatrix} 0 \\ \hat{c} \end{bmatrix}$$

In Section 1.4 we observed that if c has leading zeros, then so does h:

$$h = \begin{bmatrix} 0 \\ \hat{h} \end{bmatrix}$$

where $\hat{h} \in \mathbb{R}^{s_i}$. (See Eq. 1.4.5 and the accompanying discussion.) It follows immediately that the envelope of G is contained in that of A. Furthermore it is not hard to show that the first entry of $\hat{h}$ is nonzero. Suppose the first entry of $\hat{h}$ is g_{ij}. Then the first entry of $\hat{c}$ is a_{ij} ($\neq 0$). As you can easily verify, $g_{ij} = a_{ij}/g_{jj} \neq 0$. (For example, you can use (1.5.12) and the fact that all entries of the ith row of G before g_{ij} are zero.) Thus the envelope of G is exactly the same as the envelope of A. $\square$

COROLLARY 1.6.2 Let A be a banded, positive definite matrix. Then its Cholesky factor G is also banded. If A has semiband width s, then so does G.

Referring to the notation in the proof of Theorem 1.6.1, if $\hat{c} \in \mathbb{R}^{s_i}$, then the cost of the arithmetic in the ith step of Cholesky's method is essentially equal to that of solving an s_i-by-s_i lower-triangular system, that is $s_i^2/2$. (See the discussion following Eq. 1.4.5.) If the envelope is not exploited, the cost of the ith step is about $i^2/2$. To get an idea of the magnitude of the savings that can be realized by exploiting the envelope structure of a matrix, consider the special case of a banded matrix. If A has semiband width s, then the portion of the envelope that lies in the ith row has at most s entries, so the flop count for the ith step is not more than about $s^2/2$. Since there are n steps in the decomposition, the total flop count is about $ns^2/2$.

Exercise 1.6.2 Let G be an n-by-n lower-triangular matrix with semiband width s. Show that the cost of solving the system $Gy = b$ by forward substitution is about ns flops. An analogous result holds for upper-triangular banded matrices. $\square$

Example 1.6.3 The matrix associated with the electrical circuit in Figure 1.2 has $n = 118$ and $s = 10$. If we perform a Cholesky decomposition, using a program that does not exploit the band structure of the matrix, the cost of the arithmetic is about $n^3/6 \approx 2.7 \times 10^5$ flops. By contrast, if we do exploit the band structure, the cost is about $ns^2/2 \approx 5.9 \times 10^3$ flops, which is about 2.2 percent of the previous figure. In the forward and back substitution steps substantial but less spectacular savings can be achieved. The combined arithmetic cost of forward and back substitution without exploiting the band structure is about $n^2 \approx 1.4 \times 10^4$ flops. If the band structure is exploited, the flop count is about $2ns \approx 2.4 \times 10^3$, which is approximately 17 percent of the preceeding figure. We can also examine the savings in storage space. If symmetry is exploited but not the band structure, $n(n + 1)/2 \approx 7020$ storage locations (words) are needed to store the coefficient matrix. If only the half band is stored, $n(s + 1) \approx 1300$ words are needed.

Example 1.6.4   The results are even more impressive for larger systems. A typical system arising from the numerical solution of a partial differential equation might have $n = 1200$ and $s = 31$. (This is a medium-small problem by today's standards.) For the Cholesky decomposition the flop counts are $n^3/6 \approx 2.9 \times 10^8$ and $ns^2/2 \approx 5.8 \times 10^5$. The ratio of the two counts is about $0.002 = 0.2$ percent. For the forward and back substitution the flop counts are $n^2 \approx 1.4 \times 10^6$ and $2ns \approx 7.4 \times 10^4$. The ratio is about 5.2 percent. The respective storage requirements for the coefficient matrix are $n(n + 1)/2 = 720{,}600$ words and $n(s + 1) = 38{,}400$ words. The ratio is about 5.3 percent.

A fairly simple data structure can be used to store the coefficient matrix in an envelope storage scheme. We will describe the one recommended by George and Liu [SPDS]. A one-dimensional real array DIAG of length n is used to store the main diagonal of the matrix. A second one-dimensional real array ENV is used to store the envelope by rows, one row after the other. A third array IENV, an integer array of length $n + 1$, is used to store pointers to ENV. Usually IENV(I) names the position in ENV of the first (nonzero) entry of row I of the matrix. However, if row I contains no nonzero entries to the left of the main diagonal, then IENV(I) points to row $I + 1$ instead. Thus the absence of nonzero entries in row I left of the main diagonal is signaled by IENV(I) = IENV($I + 1$). IENV($N + 1$) points to the first storage location after the envelope. These rules can be expressed more succinctly (*and* more accurately) as follows: IENV(1) = 1 and IENV($I + 1$) − IENV(I) equals the number of elements from row I of the matrix that lie in the envelope.

Example 1.6.5   The matrix

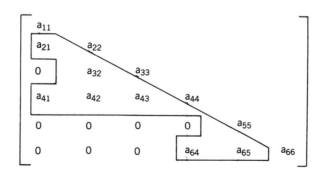

is stored as follows using the envelope scheme:

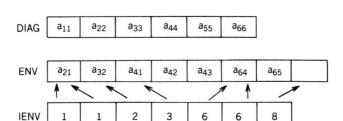

When the envelope storage scheme is used, certain formulations of the Cholesky decomposition, forward-substitution, and back-substitution algorithms are much more appropriate than others. For example, we would not want to use the outer-product formulation of the Cholesky decomposition because that algorithm operates on (the lower triangle of) A by columns. In the envelope storage scheme A is stored by rows; columns are hard to access. The inner-product formulation is inappropriate for the same reason.[§] From the proof of Theorem 1.6.1 it is clear that the bordered form of the Cholesky decomposition algorithm is appropriate. At each step virtually all the work goes into solving a lower-triangular system $G_{i-1}h = c$ [cf., (1.5.22)], where $c = [0\ \hat{c}^T]^T$ and $h = [0\ \hat{h}^T]^T$. Partitioning G_{i-1} conformably,

$$G_{i-1} = \begin{bmatrix} H_{11} & 0 \\ H_{21} & H_{22} \end{bmatrix}$$

the equation $G_{i-1}h = c$ reduces to $H_{22}\hat{h} = \hat{c}$. H_{22} is a lower-triangular matrix consisting of rows and columns $i - s_i$ through $i - 1$ of G. A subroutine can be used to solve $H_{22}\hat{h} = \hat{c}$. What is needed is a forward-elimination routine that solves systems of the form $\hat{G}\hat{h} = \hat{c}$, where $\hat{G}$ is a submatrix of G consisting of rows and columns j through k, where j and k can be any integers satisfying $1 \le j \le k \le n$. Since G, and hence $\hat{G}$, is stored by rows, the appropriate formulation of forward substitution is the row-oriented version. This subroutine can also be used with $j = 1$ and $k = n$ to perform the forward-elimination step ($Gy = b$) after the Cholesky decomposition has been completed. Finally a back-substitution subroutine is needed to solve $G^T x = y$. The rows of G are the columns of G^T, so we use the column-oriented version.

Exercise 1.6.3   Write a set of three Fortran subroutines to solve positive definite systems, using the envelope storage scheme:

(a)   Row-oriented forward-substitution routine, capable of solving systems $\hat{G}\hat{h} = \hat{c}$, where $\hat{G}$ is a submatrix of G consisting of rows and columns j through k, $1 \le j \le k \le n$.

[§] The inner-product formulation accesses A by both rows and columns.

(b) Cholesky decomposition routine, bordered form. Calls the forward-substitution routine to do the bulk of the work.

(c) Column-oriented back-substitution routine.

Write a main program that allocates storage, handles input and output, and calls the subroutines to solve positive definite systems $Ax = b$. Test your programs using the test problems given below.

For storage you will need the arrays DIAG, ENV, and IENV discussed above and one additional real array of length n that holds b initially, gets changed to y during the forward-substitution step, and finally gets changed to x during the back-substitution step. The arrays DIAG and ENV contain A initially and get changed to G during the Cholesky decomposition step. Except for a small amount of additional space for counters such as DO loop indices and the like, this is all the storage space that is needed. Test problems:

(a) $\quad A = \begin{bmatrix} 4 & 0 & 6 & 0 & 2 \\ 0 & 1 & 3 & 0 & 2 \\ 6 & 3 & 19 & 2 & 6 \\ 0 & 0 & 2 & 5 & -5 \\ 2 & 2 & 6 & -5 & 16 \end{bmatrix}$ $\quad (i) \quad b = \begin{bmatrix} 12 \\ 6 \\ 36 \\ 2 \\ 21 \end{bmatrix}$ $\quad (ii) \quad b = \begin{bmatrix} 12 \\ 4 \\ 26 \\ -8 \\ 27 \end{bmatrix}$

The Cholesky decomposition only has to be done once. Solution:

$$G = \begin{bmatrix} 2 & & & & \\ 0 & 1 & & & \\ 3 & 3 & 1 & & \\ 0 & 0 & 2 & 1 & \\ 1 & 2 & -3 & 1 & 1 \end{bmatrix}$$

$(i) \quad y = \begin{bmatrix} 6 \\ 6 \\ 0 \\ 2 \\ 1 \end{bmatrix} \quad x = \begin{bmatrix} 1 \\ 1 \\ 1 \\ 1 \\ 1 \end{bmatrix} \quad (ii) \quad y = \begin{bmatrix} 6 \\ 4 \\ -4 \\ 0 \\ 1 \end{bmatrix} \quad x = \begin{bmatrix} 1 \\ -1 \\ 1 \\ -1 \\ 1 \end{bmatrix}$

(b) $\quad A = \begin{bmatrix} 16 & 4 & 4 & 0 & 0 & 0 & 0 & 0 & 0 & 0 \\ 4 & 5 & -3 & 6 & 0 & 0 & 0 & 10 & 0 & 0 \\ 4 & -3 & 6 & -8 & 0 & 2 & 0 & -9 & 0 & 0 \\ 0 & 6 & -8 & 17 & 8 & -2 & 0 & 13 & 0 & 0 \\ 0 & 0 & 0 & 8 & 17 & 3 & 2 & -1 & 0 & 0 \\ 0 & 0 & 2 & -2 & 3 & 15 & 1 & -3 & 3 & 0 \\ 0 & 0 & 0 & 0 & 2 & 1 & 21 & -4 & -7 & 0 \\ 0 & 10 & -9 & 13 & -1 & -3 & -4 & 32 & -1 & 4 \\ 0 & 0 & 0 & 0 & 0 & 3 & -7 & -1 & 10 & 0 \\ 0 & 0 & 0 & 0 & 0 & 0 & 0 & 4 & 0 & 24 \end{bmatrix} \quad b = \begin{bmatrix} 36 \\ 109 \\ -76 \\ 188 \\ 141 \\ 113 \\ 68 \\ 281 \\ 51 \\ 272 \end{bmatrix}$

(c) Use your program to show that

$$\begin{bmatrix} 1 & 2 & 4 \\ 2 & 13 & 16 \\ 4 & 16 & 10 \end{bmatrix}$$

is not positive definite.

(d)   Think about how your subroutines could be used to calculate the inverse of a positive definite matrix. Calculate A^{-1}, where

$$A = \begin{bmatrix} 1 & 1/2 & 1/3 \\ 1/2 & 1/3 & 1/4 \\ 1/3 & 1/4 & 1/5 \end{bmatrix}$$

It turns out that the entries of A^{-1} are all integers. Notice that your computed inverse suffers from significant roundoff errors. This is because A is mildly ill conditioned. A is the 3-by-3 member of a famous family of ill-conditioned matrices called Hilbert matrices; the condition gets worse as the size of the matrix increases. We will study ill-conditioned matrices in Chapter 2.

Fortran Programming Tips   Write clear, structured code by relying on DO loops, the IF . . . THEN . . . ELSE construction, and a coherent indentation policy. Avoid GO TO statements and the accompanying statement labels.

Include plenty of comments to achieve the following ends: (1) The subroutines should be self-documenting, so that an intelligent person who is not familiar with the algorithm can figure out how to use them. In particular, the function of each parameter in each subroutine's parameter list must be explained clearly. For good examples of self-documentation, see the subroutines in Appendix C of [LUG]. (2) You should endeavor to document each block of code clearly, so that an intelligent person who is familiar with the algorithm can modify the code if necessary. Avoid trivial comments that convey no useful information.

Subroutines have a number of useful features. The most obvious is that they can be called from different points in a program. For example, the forward substitution routine is used as a work horse by Cholesky decomposition routine and it is also called by the main program to solve $Gy = b$. Also, subroutines can be used over and over again as modules in larger programs to help solve a variety of problems. In addition, subroutines can be compiled and tested separately, so you can write and debug each of your subroutines before you start on the next one. Once a subroutine has been debugged, it can be compiled once and for all and used again and again without having to be recompiled. Therefore, you should write subroutines that can be used for various-sized problems without having to be altered (and hence recompiled) for each problem.

This means that COMMON blocks should not be used; all information should be passed to and from the subroutine via the parameters in the subroutine's call sequence. Subroutines should not do any READing or PRINTing; those are tasks for the main program. For example, if a subroutine discovers that a matrix is (unexpectedly) not positive definite, it should not print a message, rather it should set a flag (parameter) to inform the calling program of what has occurred. A subroutine should never STOP; it should always RETURN control to the calling program.   □

1.7
GAUSSIAN ELIMINATION
AND THE *LU* DECOMPOSITION

In this and the next section, we will consider the problem of solving a system of n linear equations in n unknowns $Ax = b$ by Gaussian elimination. The algorithms developed here produce the unique solution whenever A is nonsingular. A is not assumed to have any special properties such as symmetry or positive definiteness. Initially we will not even assume that A is nonsingular.

Our strategy will be to transform the system $Ax = b$ to an equivalent system $Ux = y$ whose coefficient matrix U is upper triangular. The system $Ux = y$ can then be solved easily by back substitution if U is nonsingular. To say that two systems are *equivalent* is to say that they have the same solutions.

We will transform the system by means of *elementary operations* on the equations, of which there are three types:

1. Add a multiple of one equation to another equation.

2. Interchange two equations.

3. Multiply an equation by a nonzero constant.

Exercise 1.7.1   Show that each of these operations transforms a system to an equivalent system. Discussion: Suppose that the system $Ax = b$ is transformed to $\hat{A}x = \hat{b}$ by an operation of type 1. You must show that (a) every solution of $Ax = b$ is a solution of $\hat{A}x = \hat{b}$ and (b) every solution of $\hat{A}x = \hat{b}$ is a solution to $Ax = b$. Part (a) should be easy. Part (b) becomes easy when you realize that the system $Ax = b$ can be recovered from $\hat{A}x = \hat{b}$ by an operation of type 1: If $\hat{A}x = \hat{b}$ was obtained from $Ax = b$ by adding m times the jth row to the ith row, then $Ax = b$ can be recovered from $\hat{A}x = \hat{b}$ by adding $(-m)$ times the jth row to the ith row. Analogous remarks apply to operations of types 2 and 3.   □

It is convenient to represent the system $Ax = b$ by an *augmented matrix* $[A\,|\,b]$. Each equation in the system $Ax = b$ corresponds to a row of the matrix $[A\,|\,b]$. The elementary operations on the equations amount to the following *elementary row operations* on $[A\,|\,b]$:

1. Add a multiple of one row to another row.

2. Interchange two rows.

3. Multiply a row by a nonzero constant.

Although each type of operation is of theoretical importance, not all are of equal importance in practice. In fact we will not use type 3 operations at all.[§] Mostly we will use type 1 operations.

It is both interesting and important to note that each type of row operation is equivalent to premultiplication (multiplication on the left) by a matrix of simple

---

[§] An exception to this statement is the discussion of scaling in Chapter 2.

form. Suppose for example that we wish to add m times the jth row to the ith row, transforming $[A\,|\,b]$ to $[\hat{A}\,|\,\hat{b}]$. You can easily verify that

$$[\hat{A}\,|\,\hat{b}] = M[A\,|\,b]$$

where

$$
M = \quad\begin{matrix} & & & j \\ & & & \downarrow \end{matrix}
\begin{bmatrix}
1 & & & & & & & \\
 & \ddots & & & & & & \\
 & & 1 & & & & & \\
 & & & \ddots & & & & \\
i \rightarrow & & m & & 1 & & & \\
 & & & & & \ddots & & \\
 & & & & & & 1 &
\end{bmatrix}
\qquad (1.7.1)
$$

is the matrix that differs from the identity matrix only in that it has an m in the (i, j) position. M is called an *elementary matrix* of type 1.

Exercise 1.7.2   Verify the assertion of the previous paragraph. $\square$

Exercise 1.7.3   Show that if M is given by (1.7.1), then M^{-1} is the matrix that differs from M only in that it has a $-m$ instead of an m in the (i, j) position. Notice that M^{-1} is the elementary matrix associated with the elementary operation that subtracts m times the jth row from the ith row. $\square$

Since $\hat{A} = MA$, we have $\det(\hat{A}) = \det(MA) = \det(M)\det(A)$. Since obviously $\det(M) = 1$, we have $\det(\hat{A}) = \det(A)$. Thus operations of type 1 do not alter the determinant of the coefficient matrix.

Now suppose we perform an operation of type 2 to transform $[A\,|\,b]$ to $[\hat{A}\,|\,\hat{b}]$. If the operation consists of an interchange of rows i and j, then $[\hat{A}\,|\,\hat{b}] = M[A\,|\,b]$, where

$$
M = \quad\begin{matrix} & & j & & i \\ & & \downarrow & & \downarrow \end{matrix}
\begin{bmatrix}
1 & & & & & & & \\
 & \ddots & & & & & & \\
j \rightarrow & & 0 & & 1 & & & \\
 & & & \ddots & & & & \\
i \rightarrow & & 1 & & 0 & & & \\
 & & & & & \ddots & & \\
 & & & & & & 1 &
\end{bmatrix}
\qquad (1.7.2)
$$

is the matrix obtained by interchanging the ith and jth rows of the identity matrix. This is called an *elementary matrix* of type 2.

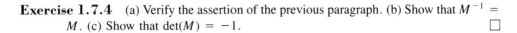

Exercise 1.7.4   (a) Verify the assertion of the previous paragraph. (b) Show that $M^{-1} = M$. (c) Show that $\det(M) = -1$.   □

Part (c) of Exercise 1.7.4 implies that each row operation of type 2 reverses the sign of the determinant of the coefficient matrix, a fact that should be familiar to you from elementary linear algebra.

Exercise 1.7.5   Determine the form of an elementary matrix of type 3. Specifically find the matrix M such that $[\hat{A}\,|\,\hat{b}] = M[A\,|\,b]$, where $[\hat{A}\,|\,\hat{b}]$ is obtained from $[A\,|\,b]$ by multiplying the ith row of $[A\,|\,b]$ by the nonzero constant c. Find M^{-1} and notice that M^{-1} is the matrix of the elementary row operation that undoes the original row operation. Calculate $\det(M)$ and determine the effect of a row operation of type 3 on the determinant of the coefficient matrix.   □

Exercise 1.7.6   We noted that an elementary matrix of type 2 is obtained by interchanging two rows of the identity matrix. More precisely, the matrix corresponding to a given elementary row operation of type 2 is obtained by performing that row operation on the identity matrix. Verify that this recipe works for row operations of types 1 and 3 as well.   □

Exercise 1.7.7   Let M be any nonsingular matrix, and suppose that $[\hat{A}\ |\ \hat{b}] = M[A\ |\ b]$. Show that the system $Ax = b$ is equivalent to the system $\hat{A}x = \hat{b}$. Some elementary linear algebra texts prove the converse as well: If the systems $Ax = b$ and $\hat{A}x = \hat{b}$ are equivalent, then there exists a nonsingular matrix M such that $[\hat{A}\,|\,\hat{b}] = M[A\,|\,b]$. We will not take the time to prove this.   □

Gaussian Elimination without Row Interchanges

We will approach the problem of solving $Ax = b$ in two stages. In the first stage, which will occupy us for the rest of this section, we will assume that A satisfies a certain property that makes it possible to transform A to upper-triangular form using row operations of type 1 only. Recall that for $k = 1, 2, \ldots, n$, the kth *leading principal submatrix* of A is the matrix A_k obtained by intersecting the first k rows and columns of A. For the rest of this section the following assumption will hold:

$$A_k \text{ is nonsingular} \qquad k = 1, 2, \ldots, n \qquad (1.7.3)$$

In particular, $A_n = A$, so we are assuming that A is nonsingular. Hence the system $Ax = b$ has a unique solution.

The reduction to triangular form is carried out in $n - 1$ steps. In the first step appropriate multiples of the first row are subtracted from each of the other rows to create zeros in positions $(2, 1), (3, 1), \ldots, (n, 1)$. It is clear that in order to do this we must have $a_{11} \neq 0$. This is guaranteed by the assumption that $A_1 = [a_{11}]$ is nonsingular. The appropriate multiplier for the ith row is

$$m_{i1} = a_{i1}/a_{11} \qquad i = 2, \ldots, n \qquad (1.7.4)$$

Thus we carry out the operations

$$a_{ij}^{(1)} = a_{ij} - m_{i1}a_{1j} \qquad j = 2, \ldots, n, i = 2, \ldots, n$$
$$b_i^{(1)} = b_i - m_{i1}b_1 \qquad\qquad i = 2, \ldots, n$$

(1.7.5)

to reduce $[A \,|\, b]$ to the form

$$\begin{bmatrix} a_{11} & a_{12} & \cdots & a_{1n} & b_1 \\ 0 & a_{22}^{(1)} & \cdots & a_{2n}^{(1)} & b_2^{(1)} \\ \vdots & \vdots & & \vdots & \vdots \\ 0 & a_{n2}^{(1)} & \cdots & a_{nn}^{(1)} & b_n^{(1)} \end{bmatrix}$$

It is not necessary to calculate the entries $a_{i1}^{(1)}$, $i = 2, \ldots, n$, explicitly, because we know in advance that they are all zero.

In a computer implementation of this step it is not necessary to store the zeros in the first column. Those storage spaces can be used for something else. As we will soon see, it turns out to be a good idea to store the multipliers $m_{21}, m_{31}, \ldots, m_{n1}$ there. Thus the array that initially contained A and b would look as follows after the first step:

$$\begin{bmatrix} a_{11} & a_{12} & \cdots & a_{1n} & b_1 \\ m_{21} & a_{22}^{(1)} & \cdots & a_{2n}^{(1)} & b_2^{(1)} \\ \vdots & \vdots & & \vdots & \vdots \\ m_{n1} & a_{n2}^{(1)} & \cdots & a_{nn}^{(1)} & b_n^{(1)} \end{bmatrix}$$

The cost of the arithmetic in the first step is $n - 1$ divisions (1.7.4) and $(n - 1)n$ flops (1.7.5). Ignoring terms of order n or less, the cost is about n^2 flops.

The second step operates on rows $2, 3, \ldots, n$. Any operations that are performed on these rows will leave the zeros in column 1 undisturbed, because subtraction of a multiple of zero from zero gives zero. Thus the second step ignores both the first row and the first column. Appropriate multiples of the second row are subtracted from rows $3, 4, \ldots, n$ to create zeros in positions $(3, 2), (4, 2), \ldots, (n, 2)$. Thus the second step is identical to the first except that it operates on the submatrix

$$\begin{bmatrix} a_{22}^{(1)} & \cdots & a_{2n}^{(1)} & b_2^{(1)} \\ \vdots & & \vdots & \vdots \\ a_{n1}^{(1)} & \cdots & a_{nn}^{(1)} & b_n^{(1)} \end{bmatrix}$$

The operations are

$$m_{i2} = a_{i2}^{(1)} / a_{22}^{(1)} \qquad i = 3, \ldots, n$$

and

$$a_{ij}^{(2)} = a_{ij}^{(1)} - m_{i2}a_{2j}^{(1)} \qquad j = 3, \ldots, n, i = 3, \ldots, n$$
$$b_i^{(2)} = b_i^{(1)} - m_{i2}b_2^{(1)} \qquad\qquad i = 3, \ldots, n$$

As in the first step there is no need to calculate $a_{i2}^{(2)}$ explicitly, $i = 3, \ldots, n$, because the multipliers m_{i2} were chosen so that $a_{i2}^{(2)} = 0$.

In order to carry out this step, we need $a_{22}^{(1)} \neq 0$. This follows from the assumption that A_2 is nonsingular [$\det(A_2) \neq 0$], and

$$\det(A_2) = \begin{vmatrix} a_{11} & a_{12} \\ a_{21} & a_{22} \end{vmatrix} = \begin{vmatrix} a_{11} & a_{12} \\ 0 & a_{22}^{(1)} \end{vmatrix} = a_{11}a_{22}^{(1)} \tag{1.7.6}$$

Thus $a_{22}^{(1)} \neq 0$. The second equation in (1.7.6) holds because the matrix $\begin{bmatrix} a_{11} & a_{12} \\ 0 & a_{22}^{(1)} \end{bmatrix}$ can be obtained from $\begin{bmatrix} a_{11} & a_{12} \\ a_{21} & a_{22} \end{bmatrix}$ by a row operation of type 1, under which the determinant is unchanged. After the second step the augmented matrix $[A \,|\, b]$ will have been transformed to

$$\left[\begin{array}{ccccc|c} a_{11} & a_{12} & a_{13} & \cdots & a_{1n} & b_1 \\ 0 & a_{22}^{(1)} & a_{23}^{(1)} & \cdots & a_{2n}^{(1)} & b_2^{(1)} \\ 0 & 0 & a_{33}^{(2)} & \cdots & a_{3n}^{(2)} & b_3^{(2)} \\ \vdots & \vdots & \vdots & & \vdots & \vdots \\ 0 & 0 & a_{n3}^{(2)} & \cdots & a_{nn}^{(2)} & b_n^{(2)} \end{array} \right]$$

In a computer implementation the zeros will be replaced by the multipliers $m_{21}, \ldots, m_{n1}$ and $m_{32}, \ldots, m_{n2}$. Since the second step is the same as the first, but on a matrix with one less row and one less column, the flop count for the second step is about $(n - 1)^2$ flops.

The third step is identical to the previous two, except that it operates on the smaller matrix

$$\left[\begin{array}{ccc|c} a_{33}^{(2)} & \cdots & a_{3n}^{(2)} & b_3^{(2)} \\ \vdots & & \vdots & \vdots \\ a_{n3}^{(2)} & \cdots & a_{nn}^{(2)} & b_n^{(2)} \end{array} \right]$$

In order to carry out the step, we need to have $a_{33}^{(2)} \neq 0$. This is guaranteed by the assumption that A_3 is nonsingular, since then

$$0 \neq \det(A_3) = \begin{vmatrix} a_{11} & a_{12} & a_{13} \\ a_{21} & a_{22} & a_{23} \\ a_{31} & a_{32} & a_{33} \end{vmatrix} = \begin{vmatrix} a_{11} & a_{12} & a_{13} \\ 0 & a_{22}^{(1)} & a_{23}^{(1)} \\ 0 & 0 & a_{33}^{(2)} \end{vmatrix} = a_{11}a_{22}^{(1)}a_{33}^{(2)}$$

The arithmetic in the third step costs about $(n - 2)^2$ flops.

After $n - 1$ steps the system will be reduced to $[U \,|\, y]$, where U is upper triangular. For each k, the possibility of carrying out step k is guaranteed by the assumption that A_k is nonsingular. In the end the fact that $0 \neq \det(A) = \det(U)$ guarantees that the system $Ux = y$ can be solved by back substitution to yield x, the unique solution of the system $Ax = b$.

The total flop count for the reduction to triangular form is about

$$n^2 + (n-1)^2 + (n-2)^2 + \cdots + 2^2 \approx \sum_{k=1}^{n} k^2$$

As in Section 1.5 we can approximate this sum by an integral:

$$\sum_{k=1}^{n} k^2 \approx \int_0^n k^2 \, dk = \frac{n^3}{3}$$

Thus the total cost of the arithmetic in the reduction is about $n^3/3$ flops. The additional cost of the back substitution is $n^2/2$ flops, which is relatively insignificant for large n. Thus the total cost of solving $Ax = b$ by this method is about $n^3/3$ flops. Notice that this is about twice the cost of solving a positive definite system by Cholesky's method, which saves a factor of 2 by exploiting the symmetry of the coefficient matrix.

Example 1.7.7 Let

$$A = \begin{bmatrix} 2 & 1 & 1 \\ 2 & 2 & -1 \\ 4 & -1 & 6 \end{bmatrix} \quad \text{and} \quad b = \begin{bmatrix} 9 \\ 9 \\ 16 \end{bmatrix}$$

Notice that $\det(A_1) = 2$, $\det(A_2) = 2$, and $\det(A_3) = -4$; this guarantees that it will be possible to reduce A to upper-triangular form by row operations of type 1 only. It also guarantees that the system $Ax = b$ has a unique solution. In order to solve this system, we form the augmented matrix

$$[A \,|\, b] = \begin{bmatrix} 2 & 1 & 1 & 9 \\ 2 & 2 & -1 & 9 \\ 4 & -1 & 6 & 16 \end{bmatrix}$$

The multipliers for the first step are $m_{21} = a_{21}/a_{11} = 1$ and $m_{31} = a_{31}/a_{11} = 2$. Thus we subtract 1 times the first row from the second row and 2 times the first row from the third row to create zeros in the $(2, 1)$ and $(3, 1)$ positions. Performing these row operations, we obtain

$$\begin{bmatrix} 2 & 1 & 1 & 9 \\ 0 & 1 & -2 & 0 \\ 0 & -3 & 4 & -2 \end{bmatrix}$$

The multiplier for the second step is $m_{32} = a_{32}^{(1)}/a_{22}^{(1)} = -3$. Thus we subtract -3 times the second row from the third row to obtain

$$\begin{bmatrix} 2 & 1 & 1 & 9 \\ 0 & 1 & -2 & 0 \\ 0 & 0 & -2 & -2 \end{bmatrix}$$

After two steps the reduction is complete. If we save the multipliers in place of the zeros, the array looks like this:

$$\left[\begin{array}{ccc|c} 2 & 1 & 1 & 9 \\ \underline{1} & 1 & -2 & 0 \\ 2 & -3 & -2 & -2 \end{array}\right] \qquad (1.7.8)$$

We can now solve the system by solving

$$\begin{bmatrix} 2 & 1 & 1 \\ 0 & 1 & -2 \\ 0 & 0 & -2 \end{bmatrix} \begin{bmatrix} x_1 \\ x_2 \\ x_3 \end{bmatrix} = \begin{bmatrix} 9 \\ 0 \\ -2 \end{bmatrix}$$

by back substitution. Doing so, we find that $x_3 = 1$, $x_2 = 2$, and $x_1 = 3$.

Exercise 1.7.8   Let

$$A = \begin{bmatrix} 2 & 1 & -1 & 3 \\ -2 & 0 & 0 & 0 \\ 4 & 1 & -2 & 6 \\ -6 & -1 & 2 & -3 \end{bmatrix} \qquad \text{and} \qquad b = \begin{bmatrix} 13 \\ -2 \\ 24 \\ -14 \end{bmatrix}$$

(a)   Calculate the appropriate (four) determinants to show that A can be transformed to upper-triangular form by operations of type 1 only. (By the way, this is strictly an academic exercise. In practice one never calculates these determinants in advance.)

(b)   Carry out the row operations of type 1 to transform the system $Ax = b$ to an equivalent system $Ux = y$, where U is upper triangular. Save the multipliers for use in Exercise 1.7.10.

(c)   If you had not calculated the determinants in advance and one of them was zero, how would you have been able to detect this in the process of doing the row operations?

(d)   Carry out back substitution on the system $Ux = y$ to solve the system $Ax = b$. Don't forget to check your answer.   □

Exercise 1.7.9   Let A be a symmetric matrix with $a_{11} \neq 0$. Suppose A has been reduced to the form

$$\left[\begin{array}{c|c} a_{11} & a_{12} \cdots a_{1n} \\ \hline 0 & \\ \vdots & A^{(1)} \\ 0 & \end{array}\right]$$

by row operations of type 1 only. Prove that $A^{(1)}$ is symmetric. (It follows that by exploiting symmetry we can cut the arithmetic almost in half provided that A_k is nonsingular, $k = 1, \ldots, n$. This is the case if A is positive definite, for example.)   □

In a real-life problem it will not be practical to verify that the conditions (1.7.3) are satisfied, so it is not possible to know in advance whether the method will work. More important, the numbers involved will usually not be integers, and if the computations are done on a computer, there will be roundoff errors. Consequently, even if the method does work in principle, it might produce an inaccurate answer. We will see how to deal with these problems in the next section.

Interpretation of the Multipliers

It was stated previously that it is a good idea to save the multipliers m_{ij}. Now let us see why this is so. Suppose we have solved the system $Ax = b$, and we now wish to solve another system $Ax = \hat{b}$, which has the same coefficient matrix but a different right-hand side. We could form the augmented matrix $[A \mid \hat{b}]$ and solve the system from scratch, but this would be inefficient. Since the coefficient matrix is the same as before, all the multipliers and row operations will also be the same as before. If the multipliers have been saved, we can perform the row operations on the $\hat{b}$ column only and save a good many computations. Let's see how this works. The operations on b were as follows:

$$
\begin{aligned}
b_i^{(1)} &= b_i - m_{i1}b_1 & i &= 2, 3, \ldots, n \\
b_i^{(2)} &= b_i^{(1)} - m_{i2}b_2^{(1)} & i &= 3, 4, \ldots, n \\
&\vdots \qquad \vdots \\
b_i^{(j)} &= b_i^{(j-1)} - m_{ij}b_j^{(j-1)} & i &= j+1, \ldots, n \\
&\vdots \\
b_i^{(n-1)} &= b_i^{(n-2)} - m_{i,n-1}b_{n-1}^{(n-2)} & i &= n
\end{aligned}
\tag{1.7.9}
$$

At the end of these operations b had been transformed to

$$
\begin{bmatrix} b_1 \\ b_2^{(1)} \\ b_3^{(2)} \\ \vdots \\ b_n^{(n-1)} \end{bmatrix} = \begin{bmatrix} y_1 \\ y_2 \\ y_3 \\ \vdots \\ y_n \end{bmatrix} = y
\tag{1.7.10}
$$

The same operations applied to $\hat{b}$ would yield a vector $\hat{y}$. The system $Ax = \hat{b}$ could then be solved by solving the equivalent upper-triangular system $Ux = \hat{y}$ by back substitution, where U is the upper-triangular matrix that was obtained in the original reduction of A.

Example 1.7.11   Suppose we wish to solve the system $Ax = \hat{b}$, where

$$
A = \begin{bmatrix} 2 & 1 & 1 \\ 2 & 2 & -1 \\ 4 & -1 & 6 \end{bmatrix} \quad \text{and} \quad \hat{b} = \begin{bmatrix} 7 \\ 3 \\ 20 \end{bmatrix}
$$

The coefficient matrix is the same as in Example 1.7.7, so the multipliers are as shown in (1.7.8), that is, $m_{21} = 1$, $m_{31} = 2$, and $m_{32} = -3$. This means that A can be transformed to upper-triangular form by subtracting 1 times the first row from the second row, 2 times the first row from the third row, and -3 times the (new) second row from the third row. Rather than perform these operations on the augmented matrix $[A \,|\, \hat{b}]$, we apply them to the column $\hat{b}$ only and get

$$\hat{b}_2^{(1)} = \hat{b}_2 - m_{21}\hat{b}_1 = 3 - 1 \cdot 7 = -4$$

$$\hat{b}_3^{(1)} = \hat{b}_3 - m_{31}\hat{b}_1 = 20 - 2 \cdot 7 = 6$$

$$\hat{b}_3^{(2)} = \hat{b}_3^{(1)} - m_{32}\hat{b}_2^{(1)} = 6 - (-3)(-4) = -6$$

After these transformations, the new right-hand side is

$$\hat{y} = \begin{bmatrix} \hat{b}_1 \\ \hat{b}_2^{(1)} \\ \hat{b}_3^{(2)} \end{bmatrix} = \begin{bmatrix} 7 \\ -4 \\ -6 \end{bmatrix}$$

Now we can get the solution by solving $Ux = \hat{y}$ by back substitution, where

$$U = \begin{bmatrix} 2 & 1 & 1 \\ 0 & 1 & -2 \\ 0 & 0 & -2 \end{bmatrix}$$

as in Example 1.7.7. Doing so, we find that $x_3 = 3$, $x_2 = 2$, and $x_1 = 1$.

A quick count shows that (1.7.9) requires about $n^2/2$ flops. Adding this to the $n^2/2$ flops for the back substitution, we find that the total arithmetic cost of solving $Ax = b$ is about n^2 flops. When we compare this with the $n^3/3$ flops that were needed to reduce A to upper-triangular form initially, we conclude that once we have solved $Ax = b$, we can solve additional systems with the same coefficient matrix at little extra cost.

A closer look at the transformation of b to y yields an important interpretation of Gaussian elimination. By (1.7.10) we can rewrite (1.7.9) in terms of the components of y as follows:

$$
\begin{aligned}
b_i^{(1)} &= b_i - m_{i1}y_1 & i &= 2, 3, \ldots, n \\
b_i^{(2)} &= b_i^{(1)} - m_{i2}y_2 & i &= 3, \ldots, n \\
&\;\;\vdots & &\;\;\vdots \\
b_i^{(n-1)} &= b_i^{(n-2)} - m_{i,n-1}y_{n-1} & i &= n
\end{aligned}
\tag{1.7.12}
$$

Together with (1.7.10), these equations can be used to derive expressions for $y_1, y_2, \ldots, y_n$. First of all $y_1 = b_1$. By the first equation of (1.7.12), $y_2 = b_2^{(1)} = b_2 - m_{21}y_1$. By the first two equations of (1.7.12), $y_3 = b_3^{(2)} = b_3^{(1)} - m_{32}y_2 = b_3 - m_{31}y_1 - m_{32}y_2$. Similarly $y_4 = b_4 - m_{41}y_1 - m_{42}y_2 - m_{43}y_3$, and in general

$$y_i = b_i - \sum_{j=1}^{i-1} m_{ij}y_j \qquad i = 1, 2, \ldots, n \tag{1.7.13}$$

One can use (1.7.13) to calculate $y_1, y_2, \ldots, y_n$. Clearly the operations are the same as those of (1.7.9), but in a different order. (1.7.13) can be interpreted as a matrix operation if we rewrite it as

$$\sum_{j=1}^{i-1} m_{ij} y_j + y_i = b_i \qquad i = 1, 2, \ldots, n$$

This can be expressed as the matrix equation

$$
\begin{bmatrix}
1 & 0 & 0 & \cdots & 0 \\
m_{21} & 1 & 0 & \ddots & \vdots \\
m_{31} & m_{32} & 1 & \ddots & \\
\vdots & & & \ddots & 0 \\
m_{n1} & m_{n2} & m_{n3} & \cdots & 1
\end{bmatrix}
\begin{bmatrix}
y_1 \\ y_2 \\ y_3 \\ \vdots \\ y_n
\end{bmatrix}
=
\begin{bmatrix}
b_1 \\ b_2 \\ b_3 \\ \vdots \\ b_n
\end{bmatrix}
\qquad (1.7.14)
$$

Thus we see that y is just the solution of a linear system $Ly = b$, where L is lower triangular. L is in fact *unit lower triangular*, which means that it has ones on the main diagonal. We learned in Section 1.4 that any lower-triangular system can be solved by forward substitution. In fact (1.7.13) is just row-oriented forward substitution. The divisions that are generally required [cf., (1.4.3)] are absent from (1.7.13) because the main-diagonal entries in (1.7.14) are ones. You can easily check that (1.7.9) is nothing but column-oriented forward substitution.

A brief summary of what we have done so far will lead to an interesting and important conclusion: L and U can be interpreted as factors of A. In order to solve the system

$$Ax = b$$

we reduced it to the form

$$Ux = y$$

where U is upper triangular and y is the solution of a unit-lower-triangular system

$$Ly = b$$

Combining these last two equations, we find that $LUx = b$. Thus $LUx = b = Ax$. These equations hold for any choice of b, and hence for any choice of x. (For a given x, the appropriate b is obtained by the calculation $b = Ax$.) Since the equation $LUx = Ax$ holds for all $x \in \mathbb{R}^n$, it must be the case that

$$A = LU$$

We conclude that the process of transforming A to upper-triangular form (saving the multipliers) can be viewed as a process of decomposing A into a product, $A = LU$, where L is unit lower triangular and U is upper triangular. In fact it is usual not to form an augmented matrix $[A \mid b]$ but to do row operations on A alone. (We took the augmented matrix approach because that is what most students are accustomed to.) A is reduced (saving multipliers) to the form

$$\begin{bmatrix} u_{11} & u_{12} & u_{13} & \cdots & u_{1n} \\ m_{21} & u_{22} & u_{23} & \cdots & u_{2n} \\ m_{31} & m_{32} & u_{33} & \cdots & u_{3n} \\ \vdots & \vdots & \vdots & & \vdots \\ m_{n1} & m_{n2} & m_{n3} & & u_{nn} \end{bmatrix}$$

which contains all information about L and U. The system $LUx = b$ can then be solved by first solving $Ly = b$ for y by forward substitution and then solving $Ux = y$ for x by back substitution.

Example 1.7.15   Solve the system $Ax = b$, where

$$A = \begin{bmatrix} 2 & 1 & 1 \\ 2 & 2 & -1 \\ 4 & -1 & 6 \end{bmatrix} \quad \text{and} \quad b = \begin{bmatrix} 3 \\ 0 \\ 11 \end{bmatrix}$$

The coefficient matrix is the same as in Example 1.7.7. From (1.7.8) we know that $A = LU$, where

$$L = \begin{bmatrix} 1 & 0 & 0 \\ 1 & 1 & 0 \\ 2 & -3 & 1 \end{bmatrix} \quad \text{and} \quad U = \begin{bmatrix} 2 & 1 & 1 \\ 0 & 1 & -2 \\ 0 & 0 & -2 \end{bmatrix}$$

Solving $Ly = b$ by forward substitution, we get $y = [3, -3, -4]^T$. Solving $Ux = y$ by back substitution, we get $x = [0, 1, 2]^T$.

Exercise 1.7.10   Solve the linear system $Ax = \hat{b}$, where A is as in Exercise 1.7.8 and $\hat{b} = [12, -8, 21, -26]^T$. Use the L and U that you calculated in Exercise 1.7.8.   □

We have already proved most of the following theorem.

Theorem 1.7.16 *(LU Decomposition Theorem)*   Let A be an n-by-n matrix whose leading principal submatrices are all nonsingular. Then A can be decomposed in exactly one way into a product

$$A = LU$$

where L is unit lower triangle and U is upper triangular.

Proof   We have already shown that L and U exist. It remains only to show that they are unique. Our uniqueness proof will yield a second algorithm for calculating the LU decomposition. Look at the equation $A = LU$ in detail:

$$\begin{bmatrix} a_{11} & a_{12} & a_{13} & \cdots & a_{1n} \\ a_{21} & a_{22} & a_{23} & \cdots & a_{2n} \\ a_{31} & a_{32} & a_{33} & \cdots & a_{3n} \\ \vdots & \vdots & \vdots & & \vdots \\ a_{n1} & a_{n2} & a_{n3} & \cdots & a_{nn} \end{bmatrix}$$

$$= \begin{bmatrix} 1 & 0 & 0 & \cdots & 0 \\ l_{21} & 1 & 0 & \cdots & 0 \\ l_{31} & l_{32} & 1 & \cdots & 0 \\ \vdots & \vdots & \vdots & & \vdots \\ l_{n1} & l_{n2} & l_{n3} & \cdots & 1 \end{bmatrix} \begin{bmatrix} u_{11} & u_{12} & u_{13} & \cdots & u_{1n} \\ 0 & u_{22} & u_{23} & \cdots & u_{2n} \\ 0 & 0 & u_{33} & \cdots & u_{3n} \\ \vdots & \vdots & \vdots & & \vdots \\ 0 & 0 & 0 & \cdots & u_{nn} \end{bmatrix}$$

The first row of L is known completely, and it has only one nonzero entry. Multiplying the first row of L by the jth column of U, we find that

$$a_{1j} = 1u_{1j} + 0u_{2j} + 0u_{3j} + \cdots + 0u_{nj}$$

That is, $u_{1j} = a_{1j}$. Thus the first row of U is uniquely determined. Now that we know the first row of U, we see that the first column of U is also known, since its only nonzero entry is u_{11}. Multiplying the ith row of L by the first column of U we find that

$$a_{i1} = l_{i1}u_{11} \qquad i = 2, 3, \ldots, n \qquad (1.7.17)$$

The assumption that A is nonsingular implies that U also is nonsingular. (Why?) Hence, $u_{kk} \neq 0$, $k = 1, \ldots, n$. In particular $u_{11} \neq 0$. Therefore, (1.7.17) determines l_{i1} uniquely:

$$l_{i1} = \frac{a_{i1}}{u_{11}} \qquad i = 2, \ldots, n$$

Thus the first column of L is uniquely determined. Now that the first row of U and first column of L have been determined, it is not hard to show that the second row of U is also uniquely determined. As an exercise, determine a formula for u_{2j} ($j \geq 2$) in terms of a_{2j} and entries of the first row of U and column of L. Once u_{22} is known, it is possible to determine the second column of L. Do this also as an exercise.

Now suppose the first $k - 1$ rows of U and columns of L have been shown to be uniquely determined. We will show that the kth row of U and column of L are uniquely determined; this will prove uniqueness by induction. The kth row of L is $[l_{k1} \ l_{k2} \ \cdots \ l_{k,k-1} \ 1 \ 0 \ \cdots \ 0]$. Since $l_{k1}, l_{k2}, \ldots, l_{k,k-1}$ are in the first $k - 1$ columns of L, they are uniquely determined. Multiplying the kth row of L by the jth column of U ($j \geq k$), we find that

$$a_{kj} = \sum_{m=1}^{k-1} l_{km}u_{mj} + u_{kj} \qquad (1.7.18)$$

All the u_{mj} (aside from u_{kj}) lie in the first $k - 1$ rows of U and are therefore known (that is, uniquely determined). Therefore u_{kj} is uniquely determined by (1.7.18):

$$u_{kj} = a_{kj} - \sum_{m=1}^{k-1} l_{km} u_{mj} \qquad j = k, k + 1, \ldots, n \qquad (1.7.19)$$

This proves that the kth row of U is uniquely determined and provides a way of calculating it. Now that u_{kk} is known, the entire kth column of U is determined. Multiplying the ith row of L ($i > k$) by the kth column of U, we find that

$$a_{ik} = \sum_{m=1}^{k-1} l_{im} u_{mk} + l_{ik} u_{kk} \qquad (1.7.20)$$

All the l_{im} (aside from l_{ik}) lie in the first $k - 1$ columns of L and are therefore uniquely determined. Furthermore $u_{kk} \neq 0$. Thus Eq. (1.7.20) determines l_{ik} uniquely:

$$l_{ik} = \frac{a_{ik} - \sum_{m=1}^{k-1} l_{im} u_{mk}}{u_{kk}} \qquad i = k + 1, k + 2, \ldots, n \qquad (1.7.21)$$

This proves that the kth column of L is uniquely determined and provides a way of calculating it. The proof that L and U are unique is now complete. $\square$

Equations 1.7.19 and 1.7.21, applied in the correct order, provide a means of calculating L and U. Because both (1.7.19) and (1.7.21) require inner-product computations, we will call this algorithm the *inner-product formulation* of the *LU* decomposition.[§] (You have undoubtedly already noticed the similarity of this algorithm to the inner-product form of the Cholesky decomposition.) Historically it has been known as the *Doolittle reduction*.[‖] In the calculation of L and U by row operations, the entries of A are gradually replaced by entries of L and U. The same can be done in the inner-product form of the *LU* decomposition: a_{kj} ($k \leq j$) is used only to compute u_{kj}, and a_{ik} ($i > k$) is used only to compute l_{ik}. Therefore each entry of L or U can be stored in place of the corresponding entry of A as soon as it is computed. You should convince yourself that the two methods perform exactly the same operations but not in the same order. In Gaussian elimination by row operations a typical entry is modified numerous times before the final result is obtained. Each modification involves one flop. In the inner-product formulation the entire modification of each entry is done at once. The Doolittle and Crout reductions were popular in the era of hand computation because they require the storage of fewer intermediate results.

[§] By contrast, the calculation of L and U by row operations is an *outer-product* formulation. See Exercise 1.7.12.

[‖] A well-known variant is the *Crout reduction*, which is very similar but produces L and U for which U, instead of L, has ones on the main diagonal.

Example 1.7.22 Let

$$
A = \begin{bmatrix} 2 & 4 & 2 & 3 \\ -2 & -5 & -3 & -2 \\ 4 & 7 & 6 & 8 \\ 6 & 10 & 1 & 12 \end{bmatrix} \quad \text{and} \quad b = \begin{bmatrix} -3 \\ 3 \\ -1 \\ -16 \end{bmatrix}
$$

We will calculate L and U such that $A = LU$ by two different methods. We will then solve the system $Ax = b$. First let's do Gaussian elimination by row operations.

Step 1:
$$
\begin{bmatrix} 2 & 4 & 2 & 3 \\ -1 & -1 & -1 & 1 \\ 2 & -1 & 2 & 2 \\ 3 & -2 & -5 & 3 \end{bmatrix}
$$

Step 2:
$$
\begin{bmatrix} 2 & 4 & 2 & 3 \\ -1 & -1 & -1 & 1 \\ 2 & 1 & 3 & 1 \\ 3 & 2 & -3 & 1 \end{bmatrix}
$$

Step 3:
$$
\begin{bmatrix} 2 & 4 & 2 & 3 \\ -1 & -1 & -1 & 1 \\ 2 & 1 & 3 & 1 \\ 3 & 2 & -1 & 2 \end{bmatrix}
$$

Now let's try the inner-product formulation.

Step 1:
$$
\begin{bmatrix} 2 & 4 & 2 & 3 \\ -1 & -5 & -3 & -2 \\ 2 & 7 & 6 & 8 \\ 3 & 10 & 1 & 12 \end{bmatrix}
$$

The first row of U and column of L have been calculated. The rest of the matrix remains untouched.

Step 2:
$$
\begin{bmatrix} 2 & 4 & 2 & 3 \\ -1 & -1 & -1 & 1 \\ 2 & 1 & 6 & 8 \\ 3 & 2 & 1 & 12 \end{bmatrix}
$$

Step 3:

$$
\begin{bmatrix}
2 & 4 & 2 & 3 \\
-1 & -1 & -1 & 1 \\
2 & 1 & 3 & 1 \\
3 & 2 & -1 & 12
\end{bmatrix}
$$

Now only u_{44} remains to be calculated.

Step 4:

$$
\begin{bmatrix}
2 & 4 & 2 & 3 \\
-1 & -1 & -1 & 1 \\
2 & 1 & 3 & 1 \\
3 & 2 & -1 & 2
\end{bmatrix}
$$

Note that both reductions yield the same result. You might find it instructive to try the inner-product reduction by the erasure method. Begin with the entries of A entered in pencil. As you calculate each U or L entry, erase the corresponding entry of A and replace it with the new result. Do the arithmetic in your head.

Now that we have the *LU* decomposition of A, we perform forward substitution on

$$
\begin{bmatrix}
1 & 0 & 0 & 0 \\
-1 & 1 & 0 & 0 \\
2 & 1 & 1 & 0 \\
3 & 2 & -1 & 1
\end{bmatrix}
\begin{bmatrix}
y_1 \\
y_2 \\
y_3 \\
y_4
\end{bmatrix}
=
\begin{bmatrix}
-3 \\
3 \\
-1 \\
-16
\end{bmatrix}
$$

to get $y = [-3, 0, 5, -2]^T$. We then perform back substitution on

$$
\begin{bmatrix}
2 & 4 & 2 & 3 \\
0 & -1 & -1 & 1 \\
0 & 0 & 3 & 1 \\
0 & 0 & 0 & 2
\end{bmatrix}
\begin{bmatrix}
x_1 \\
x_2 \\
x_3 \\
x_4
\end{bmatrix}
=
\begin{bmatrix}
-3 \\
0 \\
5 \\
-2
\end{bmatrix}
$$

to get $x = [4, -3, 2, -1]^T$.

Exercise 1.7.11 Use the inner-product formulation to calculate the *LU* decomposition of the matrix A of Exercise 1.7.8. ☐

Exercise 1.7.12 Develop an outer-product formulation of the *LU* decomposition algorithm in the spirit of the outer-product formulation of the Cholesky decomposition algorithm. Show that this algorithm is identical to Gaussian elimination by row operations of type 1. ☐

Exercise 1.7.13 (a) Develop a bordered form of the *LU* decomposition algorithm analogous to the bordered form of the Cholesky decomposition algorithm. (b) Suppose A is sparse, its lower part is stored in a row-oriented envelope and its upper part is stored in a column-oriented envelope. Prove that the envelope of L lies within the envelope of the lower part of A, and the envelope of U (by columns) lies within the envelope of the upper part of A. $\qquad\square$

Exercise 1.7.14   We have seen that if A_k is nonsingular, $k = 1, 2, \ldots, n$, then A has an LU decomposition. Prove the following converse result. If A is nonsingular and has an LU decomposition, then A_k is nonsingular, $k = 1, 2, \ldots, n$. (*Hint:* Partition the matrices in the decomposition $A = LU$.) $\qquad\square$

The next three exercises establish some basic properties of triangular matrices that will be used in the proof of Theorem 1.7.23, as well as in Exercise 1.7.18.

Exercise 1.7.15   (a) Prove that if V is nonsingular and upper triangular, then V^{-1} is upper triangular. (b) Prove that if V is unit upper triangular, then V^{-1} is unit upper triangular.

Solution   There are many ways to work this problem. Here is a solution based on blocks. First suppose V is block upper triangular:

$$
V = \begin{array}{c} \\ k \\ n-k \end{array} \begin{array}{c} \overset{\displaystyle k \qquad n-k}{\left[\begin{array}{cc} V_{11} & V_{12} \\ 0 & V_{22} \end{array}\right]} \end{array}
$$

and let $W = V^{-1}$. Partition W in conformity with V: $W = \begin{bmatrix} W_{11} & W_{12} \\ W_{21} & W_{22} \end{bmatrix}$. We will show that $W_{21} = 0$; that is, W is also block upper triangular. The equation $VW = I$ implies that $V_{22}W_{22} = I$; therefore V_{22} is nonsingular and $V_{22}^{-1} = W_{22}$. A second relationship that follows from $VW = I$ is $V_{22}W_{21} = 0$. Thus $W_{21} = V_{22}^{-1}0 = 0$. Part (a) now follows from the fact that a matrix (V or W) is upper triangular if and only if it is block upper triangular, with a k-by-k upper left-hand block, for $k = 1, \ldots, n-1$. Part (b) is related to the observation that $V_{22}^{-1} = W_{22}$. Since we now know that $W_{21} = 0$, you can easily show that $V_{11}^{-1} = W_{11}$ as well. Applying this fact inductively you can show that if V is nonsingular and upper triangular, then the main diagonal entries of V^{-1} are $w_{ii} = v_{ii}^{-1}$, $i = 1, \ldots, n$. Part (b) is a corollary of this fact. $\qquad\square$

Exercise 1.7.16   (a) Prove that if V and W are upper triangular, then VW is upper triangular. (b) Prove that if V and W are unit upper triangular, then VW is unit upper triangular. $\qquad\square$

Exercise 1.7.17   Prove lower-triangular analogues of the results in Exercises 1.7.15 and 1.7.16. (*Hint:* The easy way to do this is to take transposes and invoke the upper-triangular results.) $\qquad\square$

Exercise 1.7.18   Our proof of the existence of the LU decomposition (Theorem 1.7.16) used the equation $Ax = LUx$ (for all $x \in \mathbb{R}^n$) to infer that $A = LU$. Another proof, which does not use the auxiliary vector x, can be based on the fact that performing an elementary row operation of type 1 is equivalent to premultiplication by a unit-lower-triangular matrix (1.7.1). Using the results of Exercise 1.7.17, construct such a proof.   ☐

Variants of the *LU* Decomposition

An important variant of the *LU* decomposition is the *LDV* decomposition, which has a diagonal matrix sandwiched between two unit-triangular matrices. A *diagonal* matrix is one all of whose entries off the main diagonal are zero.

THEOREM 1.7.23 *(LDV Decomposition Theorem)*   Let A be an n-by-n matrix whose leading principal submatrices are nonsingular. Then A can be decomposed in exactly one way into a product

$$A = LDV$$

where L is unit lower triangular, D is diagonal, and V is unit upper triangular.

Proof   By Theorem 1.7.16 there exist unit-lower-triangular L and upper-triangular U such that $A = LU$. Since U is nonsingular, $u_{kk} \neq 0$, $k = 1, \ldots, n$. Let D be the diagonal matrix whose main-diagonal entries are $u_{11}, u_{22}, \ldots, u_{nn}$. Then D is nonsingular; D^{-1} is the diagonal matrix whose main-diagonal entries are $u_{11}^{-1}, u_{22}^{-1}, \ldots, u_{nn}^{-1}$. Let $V = D^{-1}U$. You can easily check that V is unit upper triangular. Clearly $A = LDV$.

To complete the proof, we must show that the decomposition is unique. Suppose $A = L_1 D_1 V_1 = L_2 D_2 V_2$. Let $U_1 = D_1 V_1$ and $U_2 = D_2 V_2$. Then obviously U_1 and U_2 are upper triangular and $A = L_1 U_1 = L_2 U_2$. By the uniqueness of LU decompositions, $L_1 = L_2$ and $U_1 = U_2$. The latter equation implies that $D_1 V_1 = D_2 V_2$; therefore

$$D_2^{-1} D_1 = V_2 V_1^{-1} \tag{1.7.24}$$

Since V_1 is unit upper triangular, so is V_1^{-1}. Since V_2 and V_1^{-1} are unit upper triangular, so is $V_2 V_1^{-1}$. On the other hand $D_2^{-1} D_1$ is obviously diagonal. Thus by (1.7.24) $V_2 V_1^{-1}$ is both unit upper triangular and diagonal; that is. $V_2 V_1^{-1} = I$. Therefore $V_2 = V_1$ and $D_2 = D_1$.   ☐

Because of the symmetric roles played by L and V, the *LDV* decomposition is of special interest when A is symmetric.

THEOREM 1.7.25   Let A be a symmetric matrix whose leading principal submatrices are nonsingular. Then A can be expressed in exactly one way as a product $A = LDL^T$, where L is unit lower triangular and D is diagonal.

Proof   A has an *LDV* decomposition: $A = LDV$. We need only show that $V = L^T$. Now $A = A^T = (LDV)^T = V^T D^T L^T$. V^T is unit lower triangular, $D^T (= D)$ is diagonal, and L^T is unit upper triangular. Thus $V^T D^T L^T$ is an *LDV* decomposition of A. By the uniqueness of the *LDV* decomposition, $V = L^T$.   ☐

In Section 1.5 we proved that if A is positive definite, then A_k is also positive definite and hence nonsingular, $k = 1, \ldots, n$. Therefore every positive definite matrix satisfies the hypotheses of Theorem 1.7.25.

Exercise 1.7.19 (a) Let A be positive definite, and suppose that $A = LDL^T$, where L is unit lower triangular and D is diagonal. Prove that the main-diagonal entries of D are all positive. (*Hint:* Use the equation $D = L^{-1}AL^{-T}$ to prove that D is positive definite.)
(b) Conversely, suppose $A = LDL^T$, where L is unit lower triangular and D is diagonal. Prove that if the main-diagonal entries of D are positive, then A is positive definite. $\square$

The results of Exercise 1.7.19, part (a), leads us to a second proof of the existence part of the Cholesky decomposition theorem: Let $E = D^{1/2}$. That is, E is the diagonal matrix whose (i, i) entry is $+\sqrt{d_{ii}}$. Then $A = LDL^T = LE^2L^T = LEE^TL^T = (LE)(LE)^T$. Let $G = LE$. Then G is lower triangular with positive main-diagonal entries, and $A = GG^T$. This proves the existence of a Cholesky factor.

The LDL^T decomposition is sometimes used instead of the Cholesky decomposition to solve positive definite systems. Algorithms analogous to the various formulations of the Cholesky decomposition algorithm exist. The LDL^T decomposition is preferred by some because it does not require the extraction of square roots.

Exercise 1.7.20 Develop algorithms to calculate the LDL^T decomposition of a positive definite matrix: (a) inner-product formulation; (b) outer-product formulation, (c) bordered formulation. (d) Count the operations for each algorithm. You may find that $n^3/3$ multiplications are required, twice as many as for Cholesky's method. In this case, show how half the multiplications can be moved out of the inner loop to cut the flop count to $n^3/6$. Notice that a little extra storage space is needed to store the intermediate results. (e) Which of the three formulations is the same as the one suggested by Exercise 1.7.9? $\square$

A third decomposition for positive definite matrices has the form $A = MD^{-1}M^T$, where M is lower triangular, D is diagonal with positive main-diagonal entries, and M and D have the same main-diagonal entries. The relationship with the LDL^T decomposition is that $M = LD$.

Exercise 1.7.21 Develop algorithms to calculate the $MD^{-1}M^T$ decomposition of a positive definite matrix: (a) inner-product formulation, (b) outer-product formulation, and (c) bordered formulation. Again no square roots are required, and the flop count is approximately $n^3/6$ if the algorithms are written carefully. $\square$

The LDL^T and $MD^{-1}M^T$ algorithms can be applied to symmetric matrices that are not positive definite provided A_k is nonsingular for $k = 1, \ldots, n$. However, it is difficult to check in advance whether these conditions are satisfied, and even if they are satisfied, the computation may be spoiled by roundoff error. Stable, efficient algorithms for symmetric, nonpositive definite linear systems do exist, but they will not be discussed in this text. See the discussion in [MC, Section 4.4].

1.8
GAUSSIAN ELIMINATION WITH PIVOTING

We now begin the second phase of our study of the equation $Ax = b$, in which we drop the assumption that the leading principal submatrices are nonsingular. No assumptions about A are made, except that it is an n-by-n matrix. We will develop an algorithm that uses elementary row operations of types 1 and 2 to either solve the system $Ax = b$ or determine that A is singular. The algorithm is identical to the Gaussian elimination algorithm that we have already developed, except that at each step a row interchange can be made. Let us consider the kth step of the algorithm. After $k - 1$ steps, the array that originally contained A has been transformed to the following form:

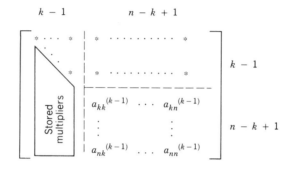

In order to calculate the multipliers for the kth step, we must divide by $a_{kk}^{(k-1)}$. If $a_{kk}^{(k-1)} = 0$, we will have to use a type 2 row operation (row interchange) to get a nonzero entry into the (k, k) position. In fact, even if $a_{kk}^{(k-1)} \neq 0$, we may still choose to do a row interchange.

Consider the following possibility. If $|a_{kk}^{(k-1)}|$ is very small, it may be that $a_{kk}^{(k-1)}$ would have been exactly zero except for roundoff errors in previous steps. If we now calculate multipliers by dividing by this number, we will certainly get an erroneous result. For this and other reasons, we will always carry out row interchanges in such a way as to avoid having a small entry in the (k, k) position. The effects of roundoff error will be studied in detail in Chapter 2.

Returning to the description of our algorithm, we examine the entries $a_{kk}^{(k-1)}, a_{k+1,k}^{(k-1)}, \ldots, a_{nk}^{(k-1)}$. If all are zero, then A is singular (Exercise 1.8.1). Set a flag to warn that this is the case. At this point we can either stop or go to step $k + 1$. If not all of $a_{kk}^{(k-1)}, \ldots, a_{nk}^{(k-1)}$ are zero, let $a_{mk}^{(k-1)}$ be the one whose absolute value is largest. Interchange rows m and k, including the previously stored multipliers. Keep a record of the row interchange. This is easily done in an integer array of length n. Store the integer m in the kth position of the array to indicate that at step k, rows k and m were interchanged. Now subtract the appropriate multiples of the new kth row from rows $k + 1, \ldots, n$ to produce zeros in positions $(k + 1, k), \ldots, (n, k)$. Of course we actually store the multipliers $m_{k+1,k}, \ldots, m_{nk}$ in those positions instead of zeros. This concludes the description of step k.

The eventual (k, k) entry, by which we divided to form the multipliers for step k, is called the *pivot* for step k. The kth row, multiples of which were subtracted from each of the remaining rows at step k, is called the *pivotal row* for step k. Our strategy was to make the pivot at step k (that is, at each step) as large as possible in order to protect against disasters due to roundoff errors. This strategy is called *partial pivoting*. (Later in this section we will briefly discuss *complete pivoting*, in which both rows and columns are interchanged.) Notice that the pivots end up on the main diagonal of the matrix U of the LU decomposition. Also, the choice of pivots implies that all multipliers will satisfy $|m_{ij}| \leq 1$. Thus in the matrix L of the LU decomposition, all entries will have absolute values less than or equal to 1.

After $n - 1$ steps the decomposition is complete. One final check must be made: If $a_{nn}^{(n-1)} = 0$, then A is singular; set a flag. This is the last pivot. It is not used to create zeros in other rows, but being the (n, n) entry of U, it is used as a divisor in the back-substitution process.

It is clear that the effect of the row interchanges is the same as if the rows of A had been interchanged initially. That is, if we were to make the appropriate row interchanges in A to form a new matrix $\hat{A}$ and then carry out Gaussian elimination on $\hat{A}$ without row interchanges, we would get exactly the same result as we got by making the row interchanges during the elimination process. Thus the result of our new algorithm is an LU decomposition not of A, but of $\hat{A}$.

Solving the system $Ax = b$ is the same as solving a system $\hat{A}x = \hat{b}$ obtained by interchanging the equations. Since we have saved a record of the row interchanges, it is easy to permute the entries of b to obtain $\hat{b}$. We then solve $Ly = \hat{b}$ by forward substitution and $Ux = y$ by back substitution to get the solution vector x.

Example 1.8.1   Solve the system

$$
\begin{bmatrix} 0 & 4 & 1 \\ 1 & 1 & 3 \\ 2 & -2 & 1 \end{bmatrix} \begin{bmatrix} x_1 \\ x_2 \\ x_3 \end{bmatrix} = \begin{bmatrix} 9 \\ 6 \\ -1 \end{bmatrix}
$$

by Gaussian elimination with partial pivoting.

In step 1 the pivotal position is the $(1, 1)$ position. Since there is a zero there, a row interchange is absolutely necessary. Since the largest potential pivot is the 2 in the $(3, 1)$ position, we interchange rows 1 and 3 to get

$$
\begin{bmatrix} 2 & -2 & 1 \\ 1 & 1 & 3 \\ 0 & 4 & 1 \end{bmatrix}
$$

The multipliers for the first step are 1/2 and 0. Subtracting 1/2 the first row from the second row and storing the multipliers, we have

$$
\begin{bmatrix} 2 & -2 & 1 \\ \hline 1/2 & 2 & 5/2 \\ 0 & 4 & 1 \end{bmatrix}
$$

The pivotal position for the second step is the $(2, 2)$ position. Since the 4 in the

$(3, 2)$ position is larger than the 2 in the $(2, 2)$ position, we interchange rows 2 and 3 (including the multipliers) to get the 4 into the pivotal position:

$$\begin{bmatrix} 2 & -2 & 1 \\ 0 & 4 & 1 \\ 1/2 & 2 & 5/2 \end{bmatrix}$$

The multiplier for the third step is $1/2$. Subtracting $1/2$ the second row from the third row and storing the multiplier, we have

$$\begin{bmatrix} 2 & -2 & 1 \\ 0 & 4 & 1 \\ 1/2 & 1/2 & 2 \end{bmatrix}$$

This completes the *LU* decomposition. Noting that the pivot in the $(3, 3)$ position is nonzero, we conclude that A is nonsingular, and the system has a unique solution. The *LU* decomposition is that of

$$\hat{A} = \begin{bmatrix} 2 & -2 & 1 \\ 0 & 4 & 1 \\ 1 & 1 & 3 \end{bmatrix}$$

You can check that $\hat{A} = LU$, where

$$L = \begin{bmatrix} 1 & 0 & 0 \\ 0 & 1 & 0 \\ 1/2 & 1/2 & 1 \end{bmatrix} \qquad U = \begin{bmatrix} 2 & -2 & 1 \\ 0 & 4 & 1 \\ 0 & 0 & 2 \end{bmatrix}$$

To solve $Ax = b$, we first transform b to $\hat{b}$; $b = [9, 6, -1]^T$. Since we interchanged rows 1 and 3 of A at step 1, we must interchange components 1 and 3 of b to get $[-1, 6, 9]^T$. At step 2 we interchanged rows 2 and 3, so we must interchange components 2 and 3 to get $\hat{b} = [-1, 9, 6]^T$. We now solve

$$\begin{bmatrix} 1 & 0 & 0 \\ 0 & 1 & 0 \\ 1/2 & 1/2 & 1 \end{bmatrix} \begin{bmatrix} y_1 \\ y_2 \\ y_3 \end{bmatrix} = \begin{bmatrix} -1 \\ 9 \\ 6 \end{bmatrix}$$

to get $y = [-1, 9, 2]^T$. Finally we solve

$$\begin{bmatrix} 2 & -2 & 1 \\ 0 & 4 & 1 \\ 0 & 0 & 2 \end{bmatrix} \begin{bmatrix} x_1 \\ x_2 \\ x_3 \end{bmatrix} = \begin{bmatrix} -1 \\ 9 \\ 2 \end{bmatrix}$$

to get $x = [1, 2, 1]^T$. You can easily check that this answer is correct. □

Gaussian elimination with partial pivoting works very well in practice. However, it is important to realize that an accurate answer is not absolutely guaranteed. There exist ill-conditioned systems (see Chapter 2) that simply cannot be solved accurately in the presence of roundoff errors. The most extreme case of an ill-conditioned system is one whose coefficient matrix is singular; there is no unique solution to such a system. Our algorithm is supposed to detect this case, but unfortunately it usually will not. If the coefficient matrix is singular, there will come a step at which all potential pivots $(a_{kk}^{(k-1)}, \ldots, a_{nk}^{(k-1)})$ are zero. However, due to roundoff errors on previous steps, the actual computed quantities will not be exact zeros. Consequently, the singularity of the matrix will not be detected, and the algorithm will march ahead and produce a nonsensical "solution." An obvious precautionary measure would be to issue a warning whenever the algorithm is forced to use a very small pivot. A better way to detect inaccuracy will be discussed in Chapter 2.

The additional costs associated with row interchanges in the partial-pivoting strategy are not great. There are two costs to consider, that of finding the largest pivot at each step and that of physically interchanging the rows. At the kth step, $n - k + 1$ numbers must be compared to find the one that is largest in magnitude. This does not involve any arithmetic, but it requires the comparison of $n - k$ pairs of numbers. The total number of comparisons made in the $n - 1$ steps is therefore

$$\sum_{k=1}^{n-1} (n - k) = \frac{n(n - 1)}{2} \approx \frac{n^2}{2}$$

Regardless of the cost of comparisons, this cost, being of order n^2, is small compared with the $O(n^3)$ arithmetic cost of the algorithm. The interchange of rows does not require any arithmetic either, but time is required to fetch and store numbers. At each step, at most one row interchange is carried out. This involves fetches and stores of $2n$ numbers. Since there are $n - 1$ steps, the total number of fetches and stores is of order n^2. Thus the cost of interchanging rows is also negligible compared with the cost of the arithmetic in the algorithm.

Exercise 1.8.1   After $k - 1$ steps of the Gaussian elimination process the coefficient matrix has been transformed to the form

$$B = \begin{bmatrix} B_{11} & B_{12} \\ 0 & B_{22} \end{bmatrix}$$

where B_{11} is $(k - 1)$ by $(k - 1)$ and upper triangular. Prove that B is singular if the first column of B_{22} is zero. (*Remark:* The fact that B_{11} is upper triangular is of no consequence.)   ☐

Exercise 1.8.2   Let

$$A = \begin{bmatrix} 2 & 2 & -4 \\ 1 & 1 & 5 \\ 1 & 3 & 6 \end{bmatrix} \qquad \text{and} \qquad b = \begin{bmatrix} 10 \\ -2 \\ -5 \end{bmatrix}$$

Use Gaussian elimination with partial pivoting to find matrices L and U such that U is upper triangular, L is unit lower triangular with $|l_{ij}| \leq 1$ for all $i > j$, and $LU = \hat{A}$, where $\hat{A}$ can be gotten from A by interchanging rows. Use your LU decomposition to solve the system $Ax = b$.   ☐

Exercise 1.8.3 Write an algorithm for Gaussian elimination with partial pivoting. ☐

Your solution to Exercise 1.8.3 should look something like this:

$\text{Gauss}(A, intch, flag)$
clear *flag*
for $k = 1, 2, \ldots, n - 1$
$\quad$ $amax \leftarrow 0$
$\quad$ $imax \leftarrow 0$
$\quad$ for $i = k, \ldots, n$ $\qquad$ (find maximal pivot)
$\quad\quad$ if $(|a_{ik}| > amax)$ then
$\quad\quad\quad$ $amax \leftarrow |a_{ik}|$
$\quad\quad\quad$ $imax \leftarrow i$
$\quad$ if $(imax = 0)$ then
$\quad\quad$ set flag $\qquad$ (*A* is singular)
$\quad\quad$ $intch(k) \leftarrow 0$
$\quad$ else
$\quad\quad$ if $(imax \neq k)$ then
$\quad\quad\quad$ for $j = 1, \ldots, n$ $\qquad$ (interchange rows *imax* and *k*,
$\quad\quad\quad\quad$ $temp \leftarrow a_{kj}$ $\qquad$ including multipliers)
$\quad\quad\quad\quad$ $a_{kj} \leftarrow a_{imax,j}$
$\quad\quad\quad\quad$ $a_{imax,j} \leftarrow temp$
$\quad\quad$ $intch(k) \leftarrow imax$ $\qquad$ (record row interchange)
$\quad\quad$ for $i = k + 1, \ldots, n$
$\quad\quad\quad$ $a_{ik} \leftarrow -a_{ik}/a_{kk}$ $\qquad$ (calculate multipliers)
$\quad\quad$ for $i = k + 1, \ldots, n$
$\quad\quad\quad$ for $j = k + 1, \ldots, n$ $\qquad$ (row operations)
$\quad\quad\quad\quad$ $a_{ij} \leftarrow a_{ij} + a_{ik}a_{kj}$
if $(a_{nn} = 0)$ then
$\quad$ set flag $\qquad$ (*A* is singular)
$\quad$ $intch(n) \leftarrow 0$
else
$\quad$ $intch(n) \leftarrow n$
exit

If A is found to be singular, the algorithm sets a flag but does not stop. It finishes the LU decomposition, but U is singular.

Notice that the multipliers that are calculated are $-m_{ik}$ rather than m_{ik}. Thus in effect $-L$ is stored instead of L. In the subsequent row operations the appropriate multiple of the kth row is added to, rather than subtracted from, the ith row. Most Gaussian elimination programs are organized this way.

The bulk of the work is done in the segment labeled "row operations":

$$\text{for } i = k + 1, \ldots, n$$
$$\quad \text{for } j = k + 1, \ldots, n \qquad (1.8.2)$$
$$\quad\quad a_{ij} \leftarrow a_{ij} + a_{ik}a_{kj}$$

Each time through the outer loop, a complete row operation is performed. However there is no logical reason why the j loop should be on the inside. Clearly the code segment

$$\text{for } j = k + 1, \ldots, n$$
$$\left[\begin{array}{l} \text{for } i = k + 1, \ldots, n \\ \left[a_{ij} \leftarrow a_{ij} + a_{ik} a_{kj} \right. \end{array} \right. \tag{1.8.3}$$

performs exactly the same operations, and there is no a priori reason to prefer (1.8.2) over (1.8.3). If (1.8.2) is used, the algorithm is said to be *row oriented*, because in the inner loop the row index i is held fixed. As the column index j is incremented, a_{ij} and a_{kj} traverse the ith and kth rows. If (1.8.2) is used, the algorithm is said to be *column oriented*. In this case the elements a_{ij} and a_{ik} traverse the jth and kth columns as the row index i is incremented.

The appropriate choice of orientation depends on the language in which the algorithm is implemented. If Fortran is used, then the column-oriented code is better, because Fortran stores arrays by columns. As each column is traversed, elements are taken from consecutive locations in the computer memory. By contrast in Pascal and PL/I, arrays are stored by rows, so a row-oriented code is better. This is important because modern operating systems employ *virtual memory* or *paging*. In this scheme programs are divided into *pages*. If the program (including storage space for arrays) is at all long, it will occupy many pages. Only those pages that are currently being used or have been used recently are kept in the computer's high-speed main memory. The rest are kept in a large, slower-speed peripheral storage area. If a page that is not in the high-speed memory is suddenly needed, it is moved into the high-speed area after a page that is not being used has been moved out to make room for it. Of course, these swapping operations are time-consuming, so it is important to try to minimize them. A large matrix will be spread over several pages of memory. In a Fortran program consecutive elements in a column lie in consecutive storage locations, so a typical column will lie on a single page. By contrast, consecutive elements of a row do not lie in consecutive storage locations, and a single row can be spread over a number of pages. Therefore a column-oriented Fortran Gaussian elimination routine will require much less page swapping than a row-oriented one will. In Pascal and PL/I programs the situation is reversed.

Exercise 1.8.4 How would you expect the inner-product formulation of the *LU* decomposition algorithm to perform in a virtual-memory environment? □

In order to solve systems of linear equations, we need not only an *LU* decomposition algorithm, but also an algorithm to scramble the b vector and perform forward and back substitution. Such an algorithm would look something like this:

Solve $(A, b, intch, flag)$
for $k = 1, \ldots, n - 1$
$$\left[\begin{array}{l} i \leftarrow intch(k) \\ temp \leftarrow b_k \\ \quad b_k \leftarrow b_i \\ \quad b_i \leftarrow temp \end{array} \right\} \quad \text{(interchange entries of } b\text{)}$$

$$
\begin{aligned}
&\text{for } j = 1, \ldots, n - 1 \\
&\quad \left[\begin{aligned}
&\text{for } i = j + 1, \ldots, n \\
&\quad [b_i \leftarrow b_i + a_{ij}b_j]
\end{aligned}\right. \qquad \text{(forward substitution, column oriented)} \\[1em]
&\text{clear flag} \\
&\text{for } j = n, n - 1, \ldots, 1 \\
&\quad \left[\begin{aligned}
&\text{if } (a_{jj} = 0) \text{ then} \\
&\quad \left[\begin{aligned} &\text{set flag} \\ &\text{exit} \end{aligned}\right\} \ (U \text{ is singular}) \\
&b_j \leftarrow b_j/a_{jj} \\
&\text{for } i = 1, \ldots, j - 1 \\
&\quad \left[b_i \leftarrow b_i - a_{ij}b_j \right.
\end{aligned}\right. \qquad \text{(back substitution, column oriented)} \\[1em]
&\text{exit}
\end{aligned}
$$

This algorithm takes as inputs the outputs A and $intch$ from the algorithm Gauss. The array A now contains an LU decomposition, and $intch$ contains a record of the row interchanges. The array b contains the vector b initially and the solution vector x afterward. The forward substitution segment is particularly simple because L is unit lower triangular. The operation $b_i \leftarrow b_i + a_{ij}b_j$ has a plus sign instead of a minus sign because $-L$ is stored instead of L. The column-oriented formulation is suitable for languages, such as Fortran, that store arrays by columns.

A main program to drive the two subroutines might be organized as follows:

```
read A
call Gauss (A, intch, flag)
if (flag is set) then
    [ print 'coefficient matrix is singular'
    [ stop
read num (number of right-hand side vectors)
for count = 1, ..., num
    [ read b
    [ call solve (A, b, intch, flag)
    [ if (flag is set) then
    [     [ print 'singular matrix detected
    [     [        during back substitution'
    [     [ stop
    [ print 'solution number' count 'is' b (prints solution)
stop
```

(this never happens unless program Gauss has bugs)

This program allows for the possibility that the user would like to solve several problems $Ax = b^{(1)}, Ax = b^{(2)}, \ldots, Ax = b^{(k)}$ with the same coefficient matrix. The relatively expensive LU decomposition only has to be done once.

Exercise 1.8.5 Write a Fortran program and subroutines to solve linear systems $Ax = b$ by Gaussian elimination with partial pivoting. Your subroutines should be column oriented. Try out your program on the test problems $Ax = b$ and $Ax = c$, where

$$
A = \begin{bmatrix} 2 & 10 & 8 & 8 & 6 \\ 1 & 4 & -2 & 4 & -1 \\ 0 & 2 & 3 & 2 & 1 \\ 3 & 8 & 3 & 10 & 9 \\ 1 & 4 & 1 & 2 & 1 \end{bmatrix} \qquad b = \begin{bmatrix} 52 \\ 14 \\ 12 \\ 51 \\ 15 \end{bmatrix} \qquad c = \begin{bmatrix} 50 \\ 4 \\ 12 \\ 48 \\ 12 \end{bmatrix}
$$

After the decomposition your transformed matrix should be

$$
\begin{bmatrix}
3 & 8 & 3 & 10 & 9 \\
-2/3 & 14/3 & 6 & 4/3 & 0 \\
-1/3 & -2/7 & -33/7 & 2/7 & -4 \\
-1/3 & -2/7 & -4/11 & -20/11 & -6/11 \\
0 & -3/7 & 1/11 & 4/5 & 1/5
\end{bmatrix}
$$

assuming that you are storing $-L$. The row interchanges are given by *intch* = $[4, 4, 4, 5, 5]$. The solution vectors are $[1, 2, 1, 2, 1]^T$ and $[2, 1, 2, 1, 2]^T$. Once you have your program working this problem correctly, try it out on some other problems of your devising. (See the Fortran programming tips at the end of Section 1.6.) □

Exercise 1.8.6   Write a program in your favorite computer language to solve systems of linear equations by Gaussian elimination with partial pivoting. Determine whether your language stores arrays by rows or by columns, and write your subroutines accordingly. Test your program on the problem from Exercise 1.8.5 and other problems of your own. □

Exercise 1.8.7   Same as Exercise 1.8.6, but use your instructor's favorite computer language. □

Calculating A^{-1}

Your program to solve $Ax = b$ can be used to calculate the inverse of a matrix. Letting $X = A^{-1}$, we have $AX = I$. Rewriting this equation in partitioned form as

$$
A[x_1 \; x_2 \; \cdots \; x_n] = [e_1 \; e_2 \; \cdots \; e_n]
$$

where $x_1, \ldots, x_n$ are the columns of A^{-1} and $e_1, \ldots, e_n$ are the columns of I, we find that the equation $AX = I$ is equivalent to the n equations

$$
Ax_i = e_i \qquad i = 1, \ldots, n \tag{1.8.4}
$$

Solving these n systems by Gaussian elimination with partial pivoting, we obtain A^{-1}.

Exercise 1.8.8   Find A^{-1}, where A is the matrix given in Exercise 1.8.5. □

How much does it cost to calculate A^{-1}? The LU decomposition has to be done once, at a cost of $n^3/3$ flops. Each of the n systems in (1.8.4) can then be solved by forward and back substitution at a cost of n^2 flops. Thus the total flop cost is $4n^3/3$. Actually a more careful count shows that the job can be done for a bit less.

Exercise 1.8.9 The forward substitution phase requires the solution of $Ly_i = e_i$, $i = 1, \ldots, n$. Some operations can be saved by exploiting the leading zeros in e_i (see Section 1.4). Do a flop count that takes these savings into account, and conclude that A^{-1} can be found in n^3 flops. ☐

Exercise 1.8.10 Modify your Gaussian elimination program so that the forward substitution segment exploits leading zeros in the right-hand side. You now have a program that can calculate an inverse in n^3 flops. ☐

You may be wondering how the procedure recommended here compares with the algorithm for calculating A^{-1} that is usually presented in elementary linear algebra classes; that is, the algorithm in which one begins with an augmented matrix $[A \mid I]$ and performs row operations that transform the augmented matrix to the form $[I \mid X]$. That algorithm also takes n^3 flops provided it is organized efficiently. In fact it is basically the same algorithm as has been proposed here. You may also recall the cofactor method, which is summarized by the equation

$$A^{-1} = \frac{1}{\det(A)} \, \text{adj}(A)$$

That method requires the computation of many determinants. If the determinants are calculated in the classical way (row or column expansions), the cost is more than $n!$ flops. Since $n!$ grows much more rapidly than n^3, the cofactor method is not competitive unless n is very small.

There are of course ways to calculate determinants in fewer than $n!$ flops. One very good way is to calculate an LU decomposition of A. Then $\det(A) = \pm \det(U) = \pm u_{11} u_{22} \cdots u_{nn}$, where the sign is plus or minus, depending on whether an even or odd number of row interchanges was made. This method obviously costs about $n^3/3$ flops.

Exercise 1.8.11 Verify that this procedure does indeed yield the determinant of A. ☐

Now that we know how much it costs to calculate A^{-1}, we can easily see that it is inefficient to solve $Ax = b$ by first calculating A^{-1} and then calculating $A^{-1}b$. This is true even if we wish to solve many systems with the same coefficient matrix.

Exercise 1.8.12 (a) How many flops are required to calculate $A^{-1}b$, given A^{-1}?
(b) How many flops are required to solve $Ax_i = b_i$, $i = 1, \ldots, k$ (k large) by

   (i) calculating $A^{-1}, A^{-1}b_1, \ldots, A^{-1}b_k$?
   (ii) doing an LU decomposition and performing forward and back substitution k times? ☐

The moral of this exercise is that having the LU decomposition is just as good as having the inverse. Since it costs only one-third as much to calculate the LU decomposition, it is usually better not to compute the inverse.

Complete Pivoting

A more conservative pivoting strategy known as *complete pivoting* deserves at least brief mention. This strategy allows both row and column interchanges. At the first step the entire matrix is searched. The element of largest magnitude is found and moved to the $(1, 1)$ position by a row interchange and a column interchange. (Note that the effect of an interchange of columns i and j is to interchange unknowns x_i and x_j.) The maximal element is then used as a pivot to create zeros below it. The second step is the same as the first, but it operates on the $(n - 1)$-by-$(n - 1)$ submatrix obtained by ignoring the first row and column, and so on. The complete pivoting strategy gives extra protection against the effects of roundoff errors and is quite satisfactory both in theory and in practice. Its disadvantage is that it is somewhat expensive. In the first step n^2 pairs of numbers have to be compared in order to find the largest entry. In the second step $(n - 1)^2$ comparisons are required, in the third step $(n - 2)^2$, and so on. Thus the total number of comparisons made during the pivot searches is $\sum_{k=1}^{n-1}(n - k)^2 \approx n^3/3$. Since the cost of making a comparison is not insignificant, this means that the cost of the pivot searches is roughly comparable to the cost of the arithmetic. By contrast the total cost of the pivot searches in the partial-pivoting strategy is of order n^2 and is therefore insignificant. The extra cost of complete pivoting would be worth paying if it gave significantly better results. However, it has been found that partial pivoting works very satisfactorily in practice, so partial pivoting is much more widely used.

On Competing Methods

A variant of Gaussian elimination that is taught in many elementary linear algebra texts is *Gauss–Jordan elimination*. In this variant the augmented matrix $[A \,|\, b]$ is converted to the form $[I \,|\, x]$ by elementary row operations. At the kth step the pivot is used to create zeros in column k both above and below the main diagonal. The disadvantage of this method is that it is more costly than the variants that were discussed in this chapter.

Exercise 1.8.13   (a) Do a flop count for Gauss–Jordan elimination and show that about $n^3/2$ flops are required. (b) Why is Gauss–Jordan elimination more expensive than transformation to the form $[U \,|\, y]$, followed by back substitution? How could the operations in Gauss–Jordan elimination by reorganized so that only $n^3/3$ flops are needed to get to the form $[I \,|\, x]$? $\square$

Another method for solving $Ax = b$ that is covered in most elementary linear algebra texts is Cramer's rule, which is closely related to the cofactor method for calculating A^{-1}. Cramer's rule states that the solution to $Ax = b$ is given by

$$
x_i = \frac{\det \begin{bmatrix} a_{11} & \cdots & \overset{\displaystyle \overset{i}{\downarrow}}{b_1} & \cdots & a_{1n} \\ a_{21} & \cdots & b_2 & \cdots & a_{2n} \\ \vdots & & \vdots & & \vdots \\ a_{n1} & \cdots & b_n & \cdots & a_{nn} \end{bmatrix}}{\det(A)} \qquad i = 1, \ldots, n
$$

where the matrix in the numerator is obtained by replacing the ith column of A by b. Because Cramer's rule requires the computation of determinants, it is too expensive to be a practical computational tool, except when $n = 2$ or 3. However, Cramer's rule will continue to be taught in linear algebra courses because of its theoretical interest.

All the methods that we have considered are *direct methods*. This means that after a prespecified, finite sequence of operations, the solution is obtained. If there are no roundoff errors, the solution is exact. Another class is that of *iterative methods*. An iterative method produces a sequence $x^{(1)}, x^{(2)}, x^{(3)}, \ldots, x^{(m)}, \ldots,$ that converges to the solution x as $m \to \infty$. We have chosen to ignore this class completely, at least in connection with the solution of linear systems $Ax = b$. To include a reasonably up-to-date treatment of these methods would have meant a great increase in the length of the book. Beside that, iterative methods for solving linear systems are perhaps best studied in conjunction with the numerical solution of partial differential equations, as this is the context in which they usually arise. For information on this large and important class of methods see, for example, Hageman and Young (1981) and McCormick (1987).

1.9
BANDED SYSTEMS

In Section 1.6 we discussed banded matrices and an envelope storage scheme in the positive-definite case. It is natural to ask whether these ideas can be applied to systems that are not positive definite. We shall see that they can, in part, but row interchanges make the situation more complicated.

Let A be any square matrix. The *lower envelope* of A is the set of all ordered pairs (i, j) such that $i > j$ and $a_{ik} \neq 0$ for some $k \leq j$. Thus the lower envelope of A is the same as the envelope of a symmetric matrix, as defined in Section 1.6. The *upper envelope* of A is the set of ordered pairs (i, j) such that $i < j$ and $a_{kj} \neq 0$ for some $k \leq j$. Thus the upper envelope is the set of all element locations above the main diagonal, excluding those containing the leading zeros in each column. Figure 1.6 illustrates the lower and upper envelopes of a matrix. In

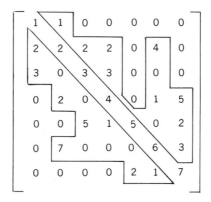

Figure 1.6 Lower and upper envelopes of a matrix.

Section 1.6 we defined the envelope of a lower-triangular matrix. It is the same as the lower envelope, as defined here. The *envelope* of an upper-triangular matrix is defined to be its upper envelope.

THEOREM 1.9.1 Suppose A has an LU decomposition, $A = LU$. Then the envelope of L equals the lower envelope of A, and the envelope of U equals the upper envelope of A.

Proof (This is the solution to Exercise 1.7.13.) We begin by developing a bordered form of the LU decomposition algorithm. Suppose we have an LU decomposition of A_{i-1}: $A_{i-1} = L_{i-1}U_{i-1}$. Then we can obtain an LU decomposition of A_i as follows:

$$A_i = \begin{bmatrix} A_{i-1} & c \\ b^T & a_{ii} \end{bmatrix} = \begin{bmatrix} L_{i-1} & 0 \\ l^T & 1 \end{bmatrix} \begin{bmatrix} U_{i-1} & v \\ 0 & u_{ii} \end{bmatrix}$$

where $A_{i-1} = L_{i-1}U_{i-1}$, $b^T = l^T U_{i-1}$, $c = L_{i-1}v$, and $a_{ii} = l^T v + u_{ii}$. By assumption we already have L_{i-1} and U_{i-1}. We can get l and v by solving the lower-triangular systems

$$U_{i-1}^T l = b \qquad \text{and} \qquad L_{i-1} v = c \tag{1.9.2}$$

by forward substitution. Then $u_{ii} = a_{ii} - l^T v$.

With this bordered form it is very easy to prove the theorem. From Section 1.4 we know that if $b = [0 \ \hat{b}^T]^T$, where $\hat{b} \in \mathbb{R}^{s_i}$ and $c = [0 \ \hat{c}^T]^T$, where $\hat{c} \in \mathbb{R}^{t_i}$, then $l = [0 \ \hat{l}^T]^T$ and $u = [0 \ \hat{u}^T]^T$, where $\hat{l} \in \mathbb{R}^{s_i}$ and $\hat{u} \in \mathbb{R}^{t_i}$. Theorem 1.9.1 follows immediately. $\square$

Exercise 1.9.1 Show that if $b \in \mathbb{R}^{s_i}$ and $c \in \mathbb{R}^{t_i}$, then the ith step of Gaussian elimination requires at most about $\left(s_i^2 + t_i^2\right)/2$ flops. $\square$

The importance of Theorem 1.9.1 is diminished by the fact that row interchanges are usually needed. It is clear that row interchanges alter the envelopes in a complicated way and thus make an envelope scheme impractical. However, there do exist classes of problems for which it can be guaranteed that Gaussian elimination without row interchanges will deliver an accurate solution. For such systems an envelope scheme can be used to advantage.

Now let us turn to banded schemes. The matrix A is *lower s-banded* if $a_{ij} = 0$ whenever $i - j > s$. This means that the nonzero entries below the main diagonal of A lie in the first s subdiagonals. A is *upper t-banded* if $a_{ij} = 0$ whenever $j - i > t$. The following theorem is an immediate corollary of Theorem 1.9.2.

THEOREM 1.9.3 Suppose A is lower s-banded and upper t-banded and has an LU decomposition, $A = LU$. Then L is lower s-banded and U is upper t-banded.

Exercise 1.9.2 Examine the Gaussian elimination process (row operations of type 1) for a banded matrix A. Show that no nonzeros are created outside the band. This yields a second proof of Theorem 1.9.3. $\square$

Exercise 1.9.3 Suppose A is lower s-banded and upper t-banded, and suppose A has an LU decomposition. (a) Use the result of Exercise 1.9.1 to show that an LU decomposition can be computed in not more than about $n(s^2 + t^2)/2$ flops. (b) Examine the Gaussian elimination process (row operations) directly to get the flop count nst. (c) Prove that $st \le (s^2 + t^2)/2$, with equality if and only if $s = t$. [*Hint:* Consider the expression $(s - t)^2$.] Show that if $s \ll t$ or $t \ll s$, then $st \ll (s^2 + t^2)/2$. Thus the flop count from part (b) is never greater than the count from part (a). When $s = t$, the two counts coincide, and when $s \ll t$ or $t \ll s$, the count from part (a) is much greater than the count from part (b). (d) Explain why the count from part (a) gives such a severe overestimate when $s \ll t$ or $t \ll s$. ☐

The next theorem shows that row interchanges do not seriously damage the band structure of a matrix if the lower band is not too wide.

THEOREM 1.9.4 Suppose A has lower band width s and upper band width t. Let L and U be the LU factors obtained from A by Gaussian elimination with row interchanges. Then L is lower s-banded, and U is upper r-banded, where $r \le s + t$.

Proof The LU factors satisfy $\hat{A} = LU$, where $\hat{A}$ is obtained from A by interchanging rows. You can easily check that, at step i, the only rows with which row i can be interchanged are rows $i + 1, \ldots, i + s$ and that such an interchange does not increase the lower band width. Thus $\hat{A}$ is lower s-banded. These interchanges do affect the upper band width, but they can increase it by at most s. Thus $\hat{A}$ is upper r-banded, where $r \le s + t$. Therefore, by Theorem 1.9.3, L is lower s-banded and U is upper r-banded. ☐

Exercise 1.9.4 A matrix that is lower 1-banded is called an *upper Hessenberg* matrix. Theorem 1.9.4 shows that the upper Hessenberg form is not disturbed by Gaussian elimination with row interchanges. Show that the LU decomposition of an upper Hessenberg matrix costs about $n^2/2$ flops. ☐

Applying the result of Exercise 1.9.3, part (b), we see that not more than about $ns(t + s)$ flops are needed for the LU decomposition (with row interchanges) of a matrix that is lower s-banded and upper t-banded.

Example 1.9.5 Let A be a 1200-by-1200 matrix that is both lower and upper 31-banded. The LU decomposition can be calculated in about $(1200)(31)(62) \approx 2.3 \times 10^6$ flops. If the band structure is not exploited, the flop count is $(1200)^3/3 \approx 5.7 \times 10^8$. Thus by exploiting the band structure, we cut the flop count to less than half of 1 percent of what it would otherwise be.

A banded matrix can be stored conveniently by diagonals in an array of dimension $(2s + t + 1)$ by n, as shown in Figure 1.7. The (i, j) entry of A is stored in position $((i - j) + s + t + 1, j)$ of the array. The array contains some unused space, indicated by asterisks. If $s \ll n$ and $t \ll n$, this space is negligible. Extra diagonals of zeros are included at the top of the array in order to accommodate band-width expansion that can occur because of row interchanges during the

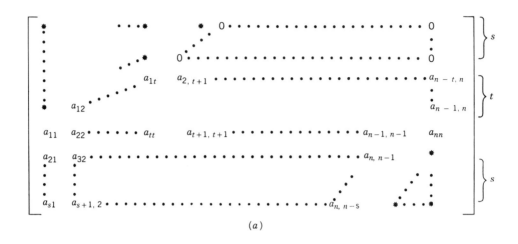

(a)

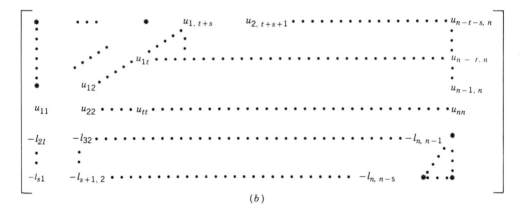

(b)

Figure 1.7 (a) Storage scheme for a banded matrix. (b) Storage of L and U after Gaussion elimination.

elimination procedure. After Gaussian elimination the LU factors are stored in the array as shown in Figure 1.7.

   The use of the storage scheme shown in Figure 1.7 makes the task of writing a Gaussian elimination routine more difficult, but the obstacles are hardly insurmountable. One must keep in mind the transformation $(i, j) \to ((i - j) + s + t + 1, j)$ and remember to limit the loop indices so that they do not run outside the bounds of the array. The storage scheme suggested here is suitable for Fortran because columns of A are stored in columns of the array. (By contrast, rows of A are stored in diagonals of the array.) A suitable formulation of Gaussian elimination is the column-oriented outer-product formulation.

Exercise 1.9.5   Devise a storage scheme for banded matrices that is suitable for programming languages that store arrays by rows.   ☐

Exercise 1.9.6   Write an LU decomposition routine and forward- and back-substitution routines that use a banded storage scheme.   ☐

1.10
THE INFLUENCE OF COMPUTER ARCHITECTURE
ON ALGORITHM SELECTION

Since the beginning of the computer era, the evolution of computing machines has been rapid and steady, and it continues today. In the never-ending quest for more speed, computer designers have turned increasingly to parallel architectures, which achieve improvements in overall speed by increasing the number of processing units. This development has had a significant impact both on our choice of algorithms and on how we implement them. In order to obtain performance approaching the theoretical maximum on a computer of this type, we must keep most of the many processing units occupied simultaneously. This means that we prefer *parallelizable* algorithms; that is, algorithms that can be broken into independent pieces of roughly equal size, that can be performed simultaneously by different processors. Of course no algorithm is completely parallelizable, but some are distinctly better than others from this standpoint. This means that our assessment of the efficiency of an algorithm can no longer be based on the flop count alone; we must also consider the extent to which the algorithm can be implemented in parallel. In some cases it has happened that algorithms that have been rejected as too slow in the past have come back into favor because of their high degree of parallelism. An example is the Jacobi method for calculating the eigenvalues of a symmetric matrix, which is discussed in Chapter 6.

This section provides a brief and very general overview of parallel computers and discusses the effects they have had on the way we implement Gaussian elimination. For simplicity we will restrict our attention to the Cholesky decomposition of positive definite matrices, but the ideas apply to the general problem as well.

Although a number of parallel algorithms have been proposed as alternatives, Gaussian elimination has held its ground rather well so far. Some of the alternatives are discussed in the review articles by Heller (1978) and Ortega and Voigt (1985), each of which has an extensive bibliography. See also the book by Ortega (1988).

In the remainder of this book we will consider many algorithms. We will see that some of them are highly parallelizable, whereas others are not. Algorithms that are highly parallel will be referred to as *parallel* algorithms; those that are not will be called *serial*. While parallelism is very desirable, it would be wrong to conclude that all serial algorithms are now obsolete. For one thing, some algorithms have such a low flop count that they are worth keeping around, even though they must be executed serially. Furthermore, some serial algorithms will find use as components of larger algorithms. Many problems can be at least partially broken up into smaller problems that can be solved simultaneously, each by a serial algorithm on a single processor. Finally, we should not forget that most large computers are shared by many users at the same time. A user with a serial algorithm can keep only one processor busy, but the other processors need not sit idly by. They can be working on problems for other users.

Vector Pipeline Machines

The first type of parallelism that we will consider is called pipelining. This is the simplest form of parallelism, and it has been with us the longest. In a sense it is not

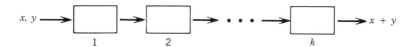

Figure 1.8   Schematic diagram of a pipeline.

true parallelism, for a pipeline computer need not have more than one processor.
The parallelism is achieved by having several operations performed concurrently by
a single processor. Most operations are performed in several stages. For example,
if we wish to add two floating-point numbers, we must first make an adjustment so
that they have the same exponent. Only then can they be added together. The actual
addition operation can itself be broken into stages. Once the addition is complete,
the exponent of the sum must be readjusted so that the result is a normalized
floating-point number. Floating-point multiplication can be broken up in a similar
manner.

A *pipeline* is nothing but an assembly line for carrying out an operation in
stages. Suppose we can break an operation, say floating-point addition, into k
stages, each of which takes one time unit to perform. If we just want to add two
numbers, the whole operation requires k time units, during which the data pass
from one end of the pipeline to the other, as shown schematically in Figure 1.8.
Now suppose we have more than one pair of numbers to add. There is no need
to wait for the first pair to pass all the way through the pipe before initiating the
second addition. It can be begun as soon as the first pair has passed into the second
stage. Next suppose we have many pairs of numbers to add, as we do whenever
we wish to add a pair of long vectors. Then the pairs can be fed into the pipeline
one after the other, one at each time unit. After an initial pipe-filling period of k
time units, one result per time unit will emerge from the pipe. See Figure 1.9.
Thus if the vectors are long compared with the length of the pipeline, we can
ignore the initial startup time and say that the rate at which sums are produced is
k times what it would have been without pipelining. The operation of multiplying
a vector by a scalar can be performed in the same efficient way, using a pipeline
for multiplication.

In Gaussian elimination the most common operation is the elementary row
operation of type 1, in which a multiple of one vector is added to another. The
vectors are partial rows of the matrix. If the code is oriented by columns, as was
recommended in Section 1.8 for Fortran programs, the same type of operation is
performed, but the vectors are partial columns instead of rows. In either case the
operation has the form $x \leftarrow x + cy$, where c is a scalar and x and y are vectors.
Both the multiplication and the addition can be performed efficiently by pipelining. Furthermore, the operation can be made even faster by overlapping the two
steps, a practice known as *chaining*. As soon as each product cy_i emerges from the

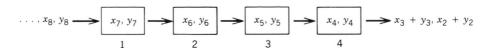

Figure 1.9   A full pipeline of length four.

product pipeline, it can be fed directly into the sum pipeline for addition to x_i. This procedure effectively merges the two pipelines into a single long pipe.

Many modern computers can perform arithmetic so rapidly that the time it takes to carry out an arithmetic operation is comparable to the time needed to move data to and from memory. If each datum is loaded from memory, participates in one operation, and then is immediately stored in memory, the time spent on memory access will exceed the time spent on arithmetic. It is thus clear that if an algorithm can be arranged so that each datum undergoes numerous arithmetic operations before being stored back into memory, the performance will be improved considerably.

Most (but not all) vector pipeline computers have vector registers. Entire vectors of data are loaded from memory into the vector registers, and vector operations are performed on the vectors in the registers. Consider an algorithm (such as Gaussian elimination) in which a vector x is updated repeatedly by operations of the form $x \leftarrow x + cy$. If these updates are mixed in with other operations, it will be necessary to store x after each update, and then load it again later. If, on the other hand, the algorithm can be arranged so that all the updates can be performed at once, then x can be loaded once, kept in a vector register until all updates have been done, and then stored once. The whole operation has the form $x \leftarrow x + c_1 y^{(1)} + \cdots + c_k y^{(k)}$. Of course not only x has to be loaded; the vectors $y^{(1)}, \ldots, y^{(k)}$ must be loaded as well. But these loading operations can be chained to the arithmetic operations. That is, as soon as the components of a vector begin to arrive from memory, they can be fed into the arithmetic pipelines.

It is significant that the only store operation that has to be done is that of x after the operations have been completed. This is especially important because some pipeline computers do not chain the store operation; that is, they do not initiate the store until the arithmetic operation is complete. If this is not done, if the load, arithmetic, and store operations are all chained together, it can occasionally happen that a datum that is to be loaded, used in an arithmetic operation, and then altered (by a store operation) will accidentally be altered before it is loaded, causing an erroneous result. A pipeline computer must either have mechanisms to ensure that this cannot occur or not initiate the store operation until the arithmetic operation is complete.

Now let us see how these considerations affect the organization of Gaussian elimination routines. We will consider the special case of the Cholesky decomposition, since this is sufficient to illustrate the main ideas. For a discussion of the general problem see the excellent article by Dongarra et al. (1984) or the book by Ortega (1988). Let A be an n-by-n positive definite matrix. Then A has a unique Cholesky decomposition $A = GG^T$, where G is lower triangular and has positive entries on the main diagonal. In Section 1.5 we studied a number of very similar algorithms for calculating G. All are variants of Gaussian elimination and have a flop count of about $n^3/6$. In fact they all perform exactly the same operations, but not in the same order. We would like to find out which of these variants perform best in a vector pipeline environment. It may seem as though we have thought of all possible variants in Section 1.5, but it will turn out that the best variants are ones that we have not yet considered.

In what follows we will assume that the matrix is stored in the conventional manner by columns, as in Fortran. In Section 1.8 we saw that if the matrix is stored by columns, it is best to access it by columns, particularly in a virtual memory environment. This advice also holds for vector pipeline machines, regardless of whether or not they have virtual memory, because memories are designed so that vectors whose components are stored in contiguous locations can be accessed rapidly. It can be much more difficult to retrieve a vector whose components are scattered. Thus we will immediately eliminate from consideration the bordered form of the Cholesky algorithm, as it accesses the matrix by rows. Now consider the inner-product form. In this variant the bulk of the work is concentrated in the formation of sums of products, that is, inner products, of the form $\sum_{k=1}^{j-1} a_{ik} a_{jk}$. This eliminates the inner-product form as a serious contender, for inner products cannot be pipelined efficiently. The problem lies with the accumulation of the sum. If we wish to add a long string of numbers $p_1, p_2, p_3, \ldots, p_j$ that is emerging in a stream from some pipeline, we must perform the operation $p_1 + p_2$ in its entirety before we begin the operation $(p_1 + p_2) + p_3$, and so on. Thus we cannot make good use of an addition pipeline. Of course we can dream up other ways of organizing an inner-product calculation; perhaps you can think of some yourself. But, none of them has proven completely satisfactory. Now consider the outer-product form.

Exercise 1.10.1 Turn to Section 1.5 and review the outer-product form of the Cholesky decomposition algorithm. Rework Exercise 1.5.11.   □

Your solution to Exercise 1.10.1 (= Exercise 1.5.11) should look something like this:

$$
\begin{aligned}
&\text{for } k = 1, \ldots, n \\
&\quad \left[\begin{array}{l}
\text{if } (a_{kk} \le 0) \text{ give up} \\
a_{kk} \leftarrow \sqrt{a_{kk}} \\
\text{for } i = k+1, \ldots, n \\
\quad \left[a_{ik} \leftarrow a_{ik}/a_{kk} \right. \\
\text{for } j = k+1, \ldots, n \\
\quad \left[\begin{array}{l}
\text{for } i = j, \ldots, n \\
\quad \left[a_{ij} \leftarrow a_{ij} - a_{ik} a_{jk} \right.
\end{array} \right.
\end{array} \right.
\end{aligned}
\qquad (1.10.1)
$$

You may have written the code with the innermost i and j loops reversed:

$$
\begin{aligned}
&\text{for } i = k+1, \ldots, n \\
&\quad \left[\begin{array}{l}
\text{for } j = k+1, \ldots, i \\
\quad \left[a_{ij} \leftarrow a_{ij} - a_{ik} a_{jk} \right.
\end{array} \right.
\end{aligned}
$$

The two variants are technically the same. We chose the one with the i loop innermost here because it is column oriented: as i runs through its range, a_{ij} and a_{ik} each run down a column of A. The operation performed by the innermost loop in (1.10.1) is a vector operation of the form $x \leftarrow x + cy$. Indeed, let $a^{(j)}$ denote the vector consisting of that portion of the jth column from the main diagonal down (cf., Figure 1.10). For $k < j$, let $\hat{a}^{(k)}$ denote the subvector obtained by discarding just the right number of components from the top of $a^{(k)}$ to make $\hat{a}^{(k)}$ have the same number of components as $a^{(j)}$. (This notation is somewhat imprecise, but it should not cause any confusion.) Then the operations performed in the i loop can be described by the vector operation $a^{(j)} \leftarrow a^{(j)} - a_{jk}\hat{a}^{(k)}$. This is the important part of the computation, since the bulk of the flops are performed in this innermost

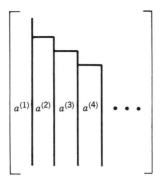

Figure 1.10

loop. Note, however, that the other i loop, in which the kth column is divided by a_{kk}, can also be expressed as a vector operation: $a^{(k)} \leftarrow a_{kk}^{-1} a^{(k)}$. Thus (1.10.1) can be rewritten in terms of vector operations as

$$
\begin{array}{l}
\text{for } k = 1, \ldots, n \\
\left[
\begin{array}{l}
\text{if } (a_{kk} \leq 0) \text{ give up} \\
a_{kk} \leftarrow \sqrt{a_{kk}} \\
a^{(k)} \leftarrow a_{kk}^{-1} a^{(k)} \\
\text{for } j = k+1, \ldots, n \\
\left[a^{(j)} \leftarrow a^{(j)} - a_{jk} \hat{a}^{(k)} \right.
\end{array}
\right.
\end{array}
\qquad (1.10.2)
$$

In this algorithm the arithmetic can be pipelined quite nicely. But notice that the memory traffic is somewhat high. On the kth time through the outer loop, each of the columns $a^{(j)}$, $j = k+1, \ldots, n$, is loaded, modified once, and then stored. On the next time through the loop, these same columns, with the exception of $a^{(k+1)}$, are loaded again, modified once more, and stored, and so on. On a machine with vector registers the memory traffic can be decreased substantially if the modifications of a given column are postponed until the point at which all modifications can be carried out at once. Then each column can be loaded once, modified numerous times, and stored only once.

Exercise 1.10.2   Reformulate algorithm (1.10.2) so that loops j and k are interchanged. Then on the jth time through the outer loop, column j is changed from its initial value to its final value all at once. Work this exercise carefully; it requires more than the simple interchange of two statements.   □

Your solution to Exercise 1.10.2 should look like this:

$$
\begin{array}{l}
\text{for } j = 1, \ldots, n \\
\left[
\begin{array}{l}
\text{for } k = 1, \ldots, j-1 \\
\left[a^{(j)} \leftarrow a^{(j)} - a_{jk} \hat{a}^{(k)} \right. \\
\text{if } (a_{jj} \leq 0) \text{ give up} \\
a_{jj} \leftarrow \sqrt{a_{jj}} \\
a^{(j)} \leftarrow a_{jj}^{-1} a^{(j)}
\end{array}
\right.
\end{array}
\qquad (1.10.3)
$$

Algorithm (1.10.3) is very good, but let us consider just one more complication. So far we have assumed that the vector registers are long enough to hold any vector. In fact they are not; they have finite length. If the matrix is so large that a column cannot fit into a single vector register, the operations on that column must be performed in segments. To see how this can be done efficiently, we will consider yet another point of view.

The operations in the k loop of algorithm (1.10.3) are

$$a^{(j)} \leftarrow a^{(j)} - \sum_{k=1}^{j-1} a_{jk} \hat{a}^{(k)}$$

which can be expressed more succinctly as the subtraction of a matrix-vector product from $a^{(j)}$; that is, $a^{(j)} \leftarrow a^{(j)} - My$, where $M = [\hat{a}^{(1)} \ \hat{a}^{(2)} \ \cdots \ \hat{a}^{(j-1)}]$ is a matrix of order $(n - j + 1)$ by $(j - 1)$, and $y = [a_{j1}, a_{j2}, \ldots, a_{j,j-1}]^T$. The vectors that are to be manipulated in vector registers are $a^{(j)}$ and the columns of M, which are vectors of length $n - j + 1$. Let l denote the length of the computer's vector registers. If $n - j + 1 > l$, the operation $a^{(j)} \leftarrow a^{(j)} - My$ will have to be done in segments. It is easy to see how this can be done. Just partition M and $a^{(j)}$ into a number of blocks:

$$M = \begin{bmatrix} M_1 \\ M_2 \\ \vdots \\ M_b \end{bmatrix} \qquad a^{(j)} = \begin{bmatrix} x_1 \\ x_2 \\ \vdots \\ x_b \end{bmatrix}$$

where no block has more than l rows. The operation

$$\text{for } i = 1, \ldots, b$$
$$\left[x_i \leftarrow x_i - M_i y \right.$$

is equivalent to $a^{(j)} \leftarrow a^{(j)} - Mx$ and acts on vectors that can fit into the registers.

Exercise 1.10.3   Rewrite algorithm (1.10.3) so that the vector operations are broken into segments that can fit into vector registers of length l.   ☐

The modifications that we have considered allow for the very efficient execution of Gaussian elimination on a vector pipeline machine with vector registers. However, it still needs to be pointed out that we have not by any means achieved perfection. Notice that as j gets larger in (1.10.3), the vector lengths get shorter (cf., Figure 1.10). As j approaches n, the vectors become very short indeed. This means that the pipeline filling time will become relatively more and more significant, and thus the efficiency will fall off. There seems to be no way around this; it is a shortcoming of Gaussian elimination.

Generic Gaussian Elimination   Dongarra, Gustavson, and Karp (1984) introduced the very interesting and enlightening concept of a *generic* Gaussian elimination algorithm. Here we will discuss this idea in the context of the Cholesky decomposition algorithm.

Near the end of Section 1.7 we introduced the LDL^T decomposition, a square-root-free variant of the Cholesky decomposition. Here L is unit lower triangular and D is diagonal with positive entries on the main diagonal. Algorithm (1.10.1) can be turned into an algorithm for the LDL^T decomposition by leaving out the square-root step. Then the kth column is divided by d_k (the kth main-diagonal entry of D) instead of its square root. Of course the instruction $a_{ij} \leftarrow a_{ij} - a_{ik}a_{jk}$ has to be modified as well because a_{ik} and a_{jk} are now scaled differently than they were before. We will skip over this detail for now because we are going to carry the modifications somewhat further. Suppose we leave out not only the square root, but also the division of the kth column by d_k. Then instead of getting L, we get the matrix $M = LD$ of the $MD^{-1}M^T$ decomposition, which was also introduced in Section 1.7. To see how the instruction $a_{ij} \leftarrow a_{ij} - a_{ik}a_{jk}$ must be modified, consider that in (1.10.1) each of a_{ik} and a_{jk} would have been divided by $\sqrt{a_{kk}}$ by now. To compensate for the fact that they have not, we must replace $a_{ij} \leftarrow a_{ij} - a_{ik}a_{jk}$ by $a_{ij} \leftarrow a_{ij} - a_{ik}a_{jk}/a_{kk}$. With these modifications (1.10.1) becomes

$$
\begin{array}{l}
\text{for } k = 1, \ldots, n \\
\left\lceil \begin{array}{l}
\text{for } j = k+1, \ldots, n \\
\left\lceil \begin{array}{l}
\text{for } i = j, \ldots, n \\
\left\lceil a_{ij} \leftarrow a_{ij} - a_{ik}a_{jk}/a_{kk} \right.
\end{array} \right.
\end{array} \right.
\end{array}
\qquad (1.10.4)
$$

We have left out the square-root check in order to remove all distractions; a practical implementation would have to include it. Beside that, (1.10.4) should not be implemented as it stands because of the division operation in the inner loop, which would cause an extra $n^3/6$ divisions in all. The reason for writing down (1.10.4) is not to get a better algorithm but to get some insight into Gaussian elimination. You should think of (1.10.4) as an uncluttered version of (1.10.1). Looking at (1.10.4), we see that the basic structure of the algorithm is very simple; it consists of three nested loops. We can apply the same treatment to any form of the algorithm.

Exercise 1.10.4 Modify (1.10.3) in the same spirit as our modification of (1.10.1). Write the vector operation as a loop with index i. $\square$

When you worked Exercise 1.10.4, you undoubtedly noticed that the solution looks just like (1.10.4). Only the order of the loop indices is changed. It is thus natural to ask whether one can obtain a working Cholesky decomposition algorithm with the loop indices in any desired order. The answer turns out to be yes. It is just a matter of choosing the ranges of the loop indices carefully. Thus we can write down a generic Cholesky decomposition algorithm:

$$
\begin{array}{l}
\text{for } \underline{\qquad} \\
\left\lceil \begin{array}{l}
\text{for } \underline{\qquad} \\
\left\lceil \begin{array}{l}
\text{for } \underline{\qquad} \\
\left\lceil a_{ij} \leftarrow a_{ij} - a_{ik}a_{jk}/a_{kk} \right.
\end{array} \right.
\end{array} \right.
\end{array}
\qquad (1.10.5)
$$

The blanks can be filled in with i, j, and k in any order. We can refer to (1.10.4) [and also its parent (1.10.1)] as the kji algorithm, because kji is the order of the loops. If we modify (1.10.1) by interchanging the i and j loops, as one does if one wants row-oriented code, we get the kij algorithm. Similarly, the algorithm of Exercise 1.10.4 [that is (1.10.3)] is the jki algorithm.

Exercise 1.10.5   Show that the inner-product form of the Cholesky decomposition algorithm is the jik algorithm. Write an uncluttered version of this algorithm and compare it with your result from Exercise 1.10.4. □

Exercise 1.10.6   Write down the ijk algorithm. First write it in the generic form (1.10.5), being careful to get the ranges of the loop indices right. Then rewrite the algorithm so that square roots of the main-diagonal entries are taken and each entry is scaled by division by the appropriate main-diagonal entry, thereby eliminating the need to perform a division in the innermost loop. Since the k loop is innermost, this is another inner-product form. □

Exercise 1.10.7   Review the bordered form of Cholesky's method in Section 1.5. Recall that in this variant the bulk of the work for each step is performed by a subroutine that performs forward substitution on a submatrix of the Cholesky factor that has already been computed.

> **(a)** Show that if the column-oriented version of the forward-substitution algorithm is used, then the bordered form is the ikj algorithm.
>
> **(b)** Show that if row-oriented forward substitution is used, then the bordered form is the ijk algorithm. Compare this algorithm with the algorithm you developed in Exercise 1.10.6. Resolve any differences between them. They should be essentially the same. □

Exercise 1.10.8   We have now examined all six variants of the generic Cholesky algorithm. From the discussion of vector pipeline machines we know that if the matrix is stored by columns, the best variant for use on a pipeline machine with vector registers is the jki algorithm (1.10.3).

> **(a)** What is the best variant for use on a pipeline machine with vector registers if the matrix is stored by rows?
>
> **(b)** Write a version of this algorithm that takes square roots of the main-diagonal entries and divides each other entry by the appropriate main-diagonal entry, so that the division in the innermost loop is not necessary. Express the innermost loops as vector operations on the rows, in analogy with (1.10.3). □

Parallel Machines

We now turn to the consideration of truly parallel machines, computers that have more than one processing unit. There are a great many parallel architectures. First of all, the individual processors can be slow or fast, simple or complex. It is not uncommon for each processor to be a vector pipeline machine, in which case the considerations of the previous subsection remain important. The number of processors can be anywhere from two to thousands, and there are many ways to interconnect them. The interconnections are important, since processors that are working on pieces of the same problem need to be able to communicate. In most parallel applications communication is the primary bottleneck, so rapid and efficient communications are essential.

Parallel machines can be grouped into two broad categories: shared memory and distributed memory. Shared-memory machines have a single large memory, which is accessed by all processors. The processors can communicate through the memory, since each processor can read what other processors have written there. The biggest problem with shared-memory machines is memory contention: If two processors want to access the same memory bank at the same time, one must wait. Even if there is no memory contention, the time it takes to access the memory is significant. In a distributed-memory machine each processor has its own small memory. The processors communicate by passing messages. Here there is no memory contention, but processors are sometimes obliged to wait to receive information from other processors. Not every parallel computer falls cleanly into one of the two categories. For example, a computer could have a large, shared memory in addition to small, local memories belonging to the processors.

This brings us to the topic of memory hierarchies. Some computers have a small, very fast cache memory in addition to the main memory. Each processor can have its own local cache, or there can be a shared cache for all processors. If a cache memory is available, the most effective way to use it is to transfer a body of data to the cache and perform as many operations as possible on that data before returning it to main memory.

The idea of memory hierarchies is not new. Early computers had a "fast" memory consisting of an array of ferrite cores and a "slow" memory on magnetic tape. The core memory gave way to solid-state devices, and the tape was replaced by magnetic drums and disks and eventually by solid-state devices, but the general concept has survived. The cache memory just adds one more level to the hierarchy. Notice that the vector registers in a pipeline processor can be viewed as a sort of cache. Also, the way the cache is used is not unlike the way the main memory is used by an operating system with virtual memory.

Because of the bewildering variety of architectures and the relatively immature state of the discipline, it would be pointless to attempt a detailed discussion of various implementations of Gaussian elimination on parallel machines. Instead we will discuss the one general strategy that seems to have been most effective so far in parallel matrix computations, the use of block algorithms. First let us consider the relatively uncomplicated problem of matrix multiplication. Suppose we wish to compute a product $A = BC$, where B and C are n by n. (The only reason for assuming that the matrices are square is to keep the discussion simple.) We must compute

$$a_{ij} = \sum_{k=1}^{n} b_{ik} c_{kj}$$

for $i, \; j = 1, \ldots, n$; that is,

$$
\begin{array}{l}
A \leftarrow 0 \\
\text{for } i = 1, \ldots, n \\
\quad \left[\begin{array}{l}
\text{for } j = 1, \ldots, n \\
\quad \left[\begin{array}{l}
\text{for } k = 1, \ldots, n \\
\quad \left[\; a_{ij} \leftarrow a_{ij} + b_{ik} c_{kj} \right.
\end{array} \right.
\end{array} \right.
\end{array}
\qquad (1.10.6)
$$

First of all, notice that this algorithm is highly parallelizable. If n^2 processors are available, each processor can be assigned to compute a single a_{ij}. Each processor performs n flops, after which the whole operation is complete. Unfortunately this approach has a severe data access problem. The processor that calculates a_{ij} has to access the entire ith row of B and jth column of C, that is, $2n$ data. Each datum participates in exactly one flop. This is quite an unfavorable ratio of flops to data compared with the overall ratio for the algorithm: In all there are $2n^2$ input data and n^2 output data, and the total flop count is n^3. Each input datum participates in n flops. Thus it should be possible to reorganize the algorithm so that a processor uses each datum several times instead of once. Assuming the processors have some local memory in which to store the data temporarily, such a rearrangement would make the algorithm more efficient.

There are lots of ways to reorganize (1.10.6). It is clear that the loops can be rearranged in any order, just as for the Cholesky algorithm. The various rearrangements suggest different ways of organizing the algorithm. We will not follow this line of thought, as it does not solve the data access problem. Instead, we will consider the effect of calculating A by blocks. Assume (for simplicity) that $n = pq$; partition A into blocks:

$$
A = \begin{bmatrix}
A_{11} & \cdots & A_{1q} \\
\vdots & & \vdots \\
A_{q1} & \cdots & A_{qq}
\end{bmatrix}
$$

where each block A_{ij} is p by p. Partition B and C the same way. The block version of (1.10.6) is

$$
\begin{array}{l}
A \leftarrow 0 \\
\text{for } i = 1, \ldots, q \\
\quad \left[\begin{array}{l}
\text{for } j = 1, \ldots, q \\
\quad \left[\begin{array}{l}
\text{for } k = 1, \ldots, q \\
\quad \left[\; A_{ij} \leftarrow A_{ij} + B_{ik} C_{kj} \right.
\end{array} \right.
\end{array} \right.
\end{array}
\qquad (1.10.7)
$$

If we assign a processor to compute each block A_{ij}, we can keep q^2 processors busy. The operation $A_{ij} \leftarrow A_{ij} + B_{ik} C_{kj}$ is basically a p-by-p matrix multiplication. It uses $2p^2$ input data and requires p^3 flops. It is thus clear that the ratio of flops to data is improved by increasing the block size p. This suggests that we should make the blocks as large as possible. Of course there are practical limits to how big p can be. For one thing, blocks have to be small enough to fit into a processor's local memory. Furthermore the larger p is, the fewer processors we can use. The most extreme cases are $p = n$, in which a single processor treats the full matrices as blocks, and $p = 1$, in which n^2 processors treat the individual entries as blocks. The optimum value of p will usually lie somewhere in between.

The block approach to matrix multiplication turns out to be useful for a wide range of parallel architectures. In the previous paragraph we have assumed that each A_{ij} is assembled by a single processor. In fact it might sometimes be advantageous to split the task among several processors. Indeed we do not rule out the possibility that each block matrix multiplication is spread over several processors. This would be useful in an architecture in which the processors are grouped into clusters, each of which has a local memory shared by the members of the cluster. The possibilities are endless, and the block viewpoint is a valuable tool for exploring them.

The Cholesky decomposition algorithm can also be performed by blocks. Consider a partition of an n-by-n positive definite matrix A into blocks:

$$
A = \begin{bmatrix}
A_{11} & & & \\
A_{21} & A_{22} & & \\
\vdots & \vdots & \ddots & \\
A_{q1} & A_{q2} & \cdots & A_{qq}
\end{bmatrix}
$$

We have shown only the lower half because of symmetry. It is simplest to think of all the blocks as square and of the same size, but in fact that is not necessary. All that is required is that the blocks along the main diagonal be square. We can develop a block Cholesky algorithm along the same lines as the ordinary Cholesky method. (This was already suggested by the development of the outer product and bordered forms of the Cholesky algorithm and the proof of the Cholesky decomposition theorem at the end of Section 1.5.)

Let G denote the lower-triangular Cholesky factor of A, and partition G identically with A. Treating the blocks more or less as if they were numbers, we can develop a block outer-product form (cf., Exercises 1.5.11 and 1.10.1) analogous to (1.10.1):

$$
\begin{aligned}
&\text{for } k = 1, \ldots, q \\
&\quad\left\lfloor\begin{aligned}
&A_{kk} \leftarrow \mathrm{chol}(A_{kk}) \\
&\text{for } i = k+1, \ldots, q \\
&\quad\left\lfloor\; A_{ik} \leftarrow A_{ik} A_{kk}^{-1} \right. \\
&\text{for } j = k+1, \ldots, q \\
&\quad\left\lfloor\begin{aligned}
&\text{for } i = j, \ldots, q \\
&\quad\left\lfloor\; A_{ij} \leftarrow A_{ij} - A_{ik} A_{jk}^T \right.
\end{aligned}\right.
\end{aligned}\right.
\end{aligned}
\qquad (1.10.8)
$$

The square root is replaced by the *chol* operator, which takes the Cholesky decomposition of the block A_{kk}. The bulk of the work is concentrated in the instruction $A_{ij} \leftarrow A_{ij} - A_{ik} A_{jk}^T$ in the innermost loop. This is just a matrix product. Assuming for simplicity that all blocks are p by p, we see that this operation performs p^3 flops on $2p^2$ data. Thus the data access situation is improved by increasing the block size. The operation $A_{ij} \leftarrow A_{ij} - A_{ik} A_{jk}^T$ can be performed independently for $i, j = k+1, \ldots, q$, subject to $i \geq j$, so about $(q-k)^2/2$ processors can be kept busy on the kth step, assuming that each block operation is performed by a single processor. Thus the large-grain parallelism is greater if the block size is smaller.

At the beginning of the kth step the Cholesky decomposition of A_{kk} is computed. This is a potential bottleneck in the algorithm. If the block size is small, it is an insignificant bottleneck; a single processor can do the job in negligible time.

However, if p is large, it might be worth splitting the job over several processors. One might even consider splitting each A_{kk} into smaller blocks and applying the block Cholesky algorithm recursively. The instruction $A_{ik} \leftarrow A_{ik}A_{kk}^{-1}$ can be performed simultaneously for $i = k + 1, \ldots, q$. The A_{kk} referred to here is actually G_{kk}, so it is lower triangular. Thus the operation can be performed by a back-substitution process. Indeed, it has the form $Z = XY^{-1}$, where X and Y are given, and Y is lower triangular. Rewriting this equation as $Y^T Z^T = X^T$, we see that we can solve for the columns of Z^T by back substituting the columns of X^T into the upper-triangular matrix Z^T. Notice that these back substitutions can be done simultaneously or pipelined. There are numerous ways to organize the computation. Which organization is best depends on the architecture. Algorithm (1.10.8) is the kji version of the block Cholesky algorithm; it is just one of six. The other versions suggest other ways to organize the computations to exploit parallelism.

Exercise 1.10.9 Suppose a block Cholesky decomposition is performed on a matrix of order $n = qp$ and block size p. What fraction of the computational effort, measured in flops, is spent calculating the Cholesky decompositions of the main-diagonal blocks? In particular consider the cases $q = 10$ and $q = 20$. $\square$

Basic Linear Algebra Subprograms (BLAS)

Certain operations occur over and over again in matrix computations. For example, the operation of adding a multiple of one vector to another is the basic operation of Gaussian elimination. Other common operations are the swapping of the contents of two vectors (row interchanges), determination of the component of a vector that has largest magnitude (pivot selection), and calculation of the dot (inner) product of two vectors. Lawson et al. (1979) had the idea of providing a set of subroutines to perform these basic operations. They called them *basic linear algebra subprograms* or BLAS. The use of BLAS has at least two advantages. First, it simplifies the writing of linear algebra codes. The inner loops are replaced by standard subroutine calls. More important, the BLAS can be modified for optimum performance on each computer system. All the programs in the widely used Linpack package are written with BLAS. The *Linpack Users' Guide* [LUG] contains complete listings of all the Linpack programs, so you can look there to see exactly how BLAS are used. Also contained in [LUG] is a complete description of the original BLAS, as well as Fortran listings. The BLAS given in [LUG] are commonly known as *vanilla* BLAS because they are not optimized for any particular machine. Your computer may have optimized BLAS, possibly written in assembly language, designed to run as fast as possible on that particular machine. The use of optimized BLAS can substantially improve the performance of the Linpack subroutines and all other programs written with BLAS.

The introduction of vector pipeline computers changed the way we look at matrix computations. Since these machines perform best when they are performing vector operations, it is natural to expect that the BLAS concept will be useful in this context. This is indeed the case. However, as we noted above, the best performance is obtained not just by performing vector operations, but by performing whole sequences of vector operations at once, for example,

$$a^{(j)} \leftarrow a^{(j)} - \sum_{k=1}^{j-1} a_{jk} \hat{a}^{(k)}$$

As we have seen, this operation is essentially a matrix-vector multiplication: $a^{(j)} \leftarrow a^{(j)} - My$. Therefore, a good way to obtain high efficiency on a vector pipeline machine is to formulate algorithms in terms of operations of this type. In recognition of this fact Dongarra et al. (1988) proposed an extended set of BLAS that includes matrix-vector operations. These new BLAS became known as level-2 BLAS or BLAS 2, and the original BLAS became known as BLAS 1. With optimized level-2 BLAS it is much easier to write efficient code for vector pipeline machines, for the programming details that must be considered for the efficient execution of, for example, $x \leftarrow x - My$ are already incorporated into the subroutine.

As researchers began to accumulate experience with parallel computers, it became clear that the level-2 extensions of the BLAS were not enough. For parallel computing it is often convenient to think in terms of matrix-matrix operations, as we have seen. Thus Dongarra et al. (1990) proposed a set of level-3 BLAS that includes matrix-matrix operations such as $A \leftarrow A + BC$, where A, B, and C are all matrices. Notice that this is the main operation in the algorithm (1.10.8). Another operation in the proposed BLAS 3 is $A \leftarrow AB^{-1}$, where B is lower triangular. This is another important operation in (1.10.8). The incorporation of level-3 BLAS promises to simplify programming and improve the performance of matrix computation algorithms on a variety of parallel architectures.

2

Sensitivity of Linear Systems; Effects of Roundoff Errors

2.1
INTRODUCTION

When we solve a system of linear equations, we seldom solve exactly the system we intended to solve; rather we solve one that approximates it. In a system $Ax = b$, the coefficients in A and b will typically be known from some experiment or measurement and will therefore be subject to some measurement error. For example, in the electrical circuit problems with which this book began, the entries of the coefficient matrix depend on the values of the resistances, numbers that are known only approximately in practice. Thus the A and b with which we work are slightly different from the true A and b. Assuming that the problem is to be solved on a computer, additional approximations must be made when the numbers are entered into the machine; the real or complex entries of A and b must be approximated by numbers from the computer's finite set of floating-point numbers. (However, this error is usually much smaller than the measurement error.) It is important therefore to ask what effect small changes or perturbations in the coefficients have on the solution of a system. This is the question of *sensitivity* of linear systems.

A second question we will consider is one that was already mentioned in Chapter 1. If we solve a system by Gaussian elimination on a computer, the result will be contaminated by the roundoff errors made during the computation. How do these errors affect the accuracy of the computed solution? We will see that this question is closely related to the sensitivity issue.

Although this chapter does contain some theorems, much of the material is heuristic in nature and not necessarily true in all cases. That is the nature of the subject.

2.2
VECTOR AND MATRIX NORMS

In order to study the effects of perturbations in vectors (such as b) and matrices (such as A), we need to be able to measure them. For this purpose we introduce vector and matrix norms. The vectors used in this book are generally n-tuples of real numbers. The set of all such n-tuples will be denoted R^n. It is useful to visualize the members of R^2 as points in a plane or as geometric vectors (arrows) in a plane with their tails at the origin. Likewise the elements of R^3 can be viewed as points or vectors in space. Any two elements of R^n can be added in the obvious manner to yield an element of R^n, and any element of R^n can be multiplied by any real number (scalar) to yield an element of R^n. The vector whose components are all zero will be denoted 0. Thus the symbol 0 can stand for either a number or a vector. The careful reader will not be confused by this.

The set of all n-tuples of complex numbers is denoted C^n. In this chapter, as in Chapter 1, we will restrict our attention to real numbers, but everything that we will do will carry over to complex numbers.

A *norm* (or *vector norm*) on R^n is a function that assigns to each $x \in R^n$ a real number $\| x \|$, called the norm of x, such that the following three properties are satisfied for all $x, y \in R^n$ and all $\alpha \in R$:

$$\| x \| > 0 \text{ if } x \neq 0 \qquad \text{and} \qquad \| 0 \| = 0 \qquad \text{(positive definite property)} \qquad (2.2.1)$$

$$\| \alpha x \| = |\alpha| \, \| x \| \qquad \text{(absolute homogeneity)} \qquad (2.2.2)$$

$$\| x + y \| \leq \| x \| + \| y \| \qquad \text{(triangle inequality)} \qquad (2.2.3)$$

Exercise 2.2.1   (a) In the equation $\| 0 \| = 0$, what is the nature of each zero (number or vector)?

(b) Show that the equation $\| 0 \| = 0$ actually follows from (2.2.2). Thus it need not have been stated explicitly in (2.2.1).

Any norm can be used to measure the *length* or *magnitude* (in a generalized sense) of vectors in R^n. In other words we think of $\| x \|$ as the (generalized) length of x. The (generalized) *distance* between two vectors x and y is defined to be $\| x - y \|$.   □

Example 2.2.4   The *Euclidean norm* is defined by[§]

$$\| x \|_2 = \left(\sum_{i=1}^{n} |x_i|^2 \right)^{1/2}$$

You can easily verify that this function satisfies (2.2.1) and (2.2.2). The triangle inequality (2.2.3) is not so easy. It follows from the Cauchy–Schwarz inequality,

---

[§] Notice that the absolute-value signs in the formula for $\| x \|_2$ are redundant, as $|x_i|^2 = x_i^2$ for any real number x_i. However, in the complex case it is not generally true that $|x_i|^2 = x_i^2$, and the absolute-value signs would be needed. Thus the inclusion of the absolute-value signs gives a formula for $\| x \|_2$ that is correct for both real and complex vectors.

which we will prove shortly (Theorem 2.2.5). The distance between two vectors x and y is given by

$$\| x - y \|_2 = \sqrt{\sum_{i=1}^{n} |x_i - y_i|^2}$$

In the cases $n = 1$, 2, and 3, this measure coincides with our usual notion of distance between points in a line, in a plane, and in space, respectively.

THEOREM 2.2.5 *(Cauchy–Schwarz Inequality)*   For all x, $y \in \mathbb{R}^n$

$$\left| \sum_{i=1}^{n} x_i y_i \right| \leq \left(\sum_{i=1}^{n} x_i^2 \right)^{1/2} \left(\sum_{i=1}^{n} y_i^2 \right)^{1/2}$$

Proof   For every real number t we have

$$0 \leq \sum_{i=1}^{n} (x_i + t y_i)^2 = \sum_{i=1}^{n} x_i^2 + 2t \sum_{i=1}^{n} x_i y_i + t^2 \sum_{i=1}^{n} y_i^2$$

$$= c + bt + at^2$$

where $a = \sum_{i=1}^{n} y_i^2$, $b = 2\sum_{i=1}^{n} x_i y_i$, and $c = \sum_{i=1}^{n} x_i^2$. Since $at^2 + bt + c \geq 0$ for all real t, the quadratic polynomial $at^2 + bt + c$ cannot have two distinct real zeros. Therefore the discriminant satisfies $b^2 - 4ac \leq 0$. Rewriting this inequality as

$$(\tfrac{1}{2}b)^2 \leq ac$$

and taking square roots, we obtain the desired result.   □

THEOREM 2.2.6   For all x, $y \in \mathbb{R}^n$, $\| x + y \|_2 \leq \| x \|_2 + \| y \|_2$.

Proof   We will show instead that $\| x + y \|_2^2 \leq (\| x \|_2 + \| y \|_2)^2$.

$$\| x + y \|_2^2 = \sum_{i=1}^{n} (x_i + y_i)^2 = \sum_{i=1}^{n} x_i^2 + 2 \sum_{i=1}^{n} x_i y_i + \sum_{i=1}^{n} y_i^2$$

Applying the Cauchy–Schwarz inequality to the middle term, we find that

$$\| x + y \|_2^2 \leq \sum_{i=1}^{n} x_i^2 + 2 \left(\sum_{i=1}^{n} x_i^2 \right)^{1/2} \left(\sum_{i=1}^{n} y_i^2 \right)^{1/2} + \sum_{i=1}^{n} y_i^2$$

$$= \left[\left(\sum_{i=1}^{n} x_i^2 \right)^{1/2} + \left(\sum_{i=1}^{n} y_i^2 \right)^{1/2} \right]^2$$

$$= (\| x \|_2 + \| y \|_2)^2$$ □

Example 2.2.7 Generalizing Example 2.2.4, we introduce the *p*-norms. For any real number $p \geq 1$, we define

$$\| x \|_p = \left(\sum_{i=1}^{n} |x_i|^p \right)^{1/p}$$

Again it is easy to verify (2.2.1) and (2.2.2), but not (2.2.3). This is the *Minkowski inequality*, which we will not prove because we are not going to use it.

Example 2.2.8 An important special case of the previous example is the 1-norm

$$\| x \|_1 = \sum_{i=1}^{n} |x_i|$$

In this case it is not hard to prove (2.2.3); it follows directly from the triangle inequality for real numbers.

Exercise 2.2.2 Prove that the 1-norm is a norm. ☐

Exercise 2.2.3 Let x, $y \in \mathbb{R}^2$. With respect to the 1-norm the "distance" between x and y is $\| x - y \|_1 = |x_1 - y_1| + |x_2 - y_2|$. Explain why the 1-norm is sometimes called the *taxicab norm* (or *Manhattan metric*). ☐

Example 2.2.9 The ∞-norm is defined by

$$\| x \|_\infty = \max_{1 \leq i \leq n} |x_i|$$

Exercise 2.2.4 Prove that the ∞-norm is a norm. ☐

Exercise 2.2.5 Given any norm on $\mathbb{R}^2$, the *unit circle* with respect to that norm is the set $\{ x \in \mathbb{R}^2 \mid \| x \| = 1 \}$. Thinking of the members of $\mathbb{R}^2$ as points in the plane, the unit circle is just the set of points whose distance from the origin is 1. On a single set of coordinate axes, sketch the unit circle with respect to the *p*-norm for $p = 1, 3/2, 2, 3, 10$, and ∞. ☐

The analytically inclined reader might like to prove that for all $x \in \mathbb{R}^n$, $\| x \|_\infty = \lim_{p \to \infty} \| x \|_p$.

Example 2.2.10 Let A be any *n*-by-*n* positive definite matrix. Define a norm on $\mathbb{R}^n$ by

$$\| x \|_A = \left(x^T A x \right)^{1/2}$$

You can easily verify that in the special case $A = I$, this norm is just the Euclidean or 2-norm.

Exercise 2.2.6   (a) Let A be a positive definite matrix, and let G be its Cholesky factor. Verify that for all $x \in R^n$, $\| x \|_A = \| G^T x \|_2$.

(b) Using the fact that the 2-norm is indeed a norm on R^n, prove that the A-norm is a norm on R^n. $\square$

Let us now turn our attention to matrix norms. The set of m-by-n real matrices will be denoted $R^{m \times n}$. Like vectors, the matrices in $R^{m \times n}$ can be added and multiplied by scalars in the obvious manner. In fact the matrices in $R^{m \times n}$ can be viewed simply as vectors in R^{mn} with the components arranged differently. In the case $m = n$, the theory becomes somewhat richer. Unlike vectors, two matrices in $R^{n \times n}$ can be multiplied together (using the usual matrix multiplication) to yield a product in $R^{n \times n}$. A *matrix norm* is a function that assigns to each $A \in R^{n \times n}$ a real number $\| A \|$, called the norm of A, such that the three norm properties hold, as well as one additional property (consistency), which relates the norm function to the operation of matrix multiplication. Specifically, for all $A, B \in R^{n \times n}$ and $\alpha \in R$,

$$\| A \| > 0 \text{ if } A \neq 0 \tag{2.2.11}$$

$$\| \alpha A \| = |\alpha| \| A \| \tag{2.2.12}$$

$$\| A + B \| \leq \| A \| + \| B \| \tag{2.2.13}$$

$$\| AB \| \leq \| A \| \| B \| \quad \text{(consistency)} \tag{2.2.14}$$

In (2.2.11) we have introduced yet another use for the symbol 0: We have used it to denote the zero matrix.

Example 2.2.15   The *Frobenius norm* is defined by

$$\| A \|_F = \left(\sum_{i=1}^{n} \sum_{j=1}^{n} |a_{ij}|^2 \right)^{1/2}$$

Because it is the same as the Euclidean norm on vectors, we already know that it satisfies properties (2.2.11), (2.2.12), and (2.2.13). The consistency property (2.2.14) can be deduced from the Cauchy–Schwarz inequality as follows. Let $C = AB$. Then $c_{ij} = \sum_{k=1}^{n} a_{ik} b_{kj}$. Thus

$$\| AB \|_F^2 = \| C \|_F^2 = \sum_{i=1}^{n} \sum_{j=1}^{n} |c_{ij}|^2 = \sum_{i=1}^{n} \sum_{j=1}^{n} \left| \sum_{k=1}^{n} a_{ik} b_{kj} \right|^2$$

Applying the Cauchy–Schwarz inequality to the expression $\sum_{k=1}^{n} a_{ik} b_{kj}$, we find that

$$\| AB \|_F^2 \leq \sum_{i=1}^{n} \sum_{j=1}^{n} \left(\sum_{k=1}^{n} |a_{ik}|^2 \sum_{k=1}^{n} |b_{kj}|^2 \right) = \left(\sum_{i=1}^{n} \sum_{k=1}^{n} |a_{ik}|^2 \right) \left(\sum_{j=1}^{n} \sum_{k=1}^{n} |b_{kj}|^2 \right)$$

$$= \| A \|_F^2 \| B \|_F^2$$

Thus the Frobenius norm is a matrix norm.

Exercise 2.2.7 Define $\|A\|_{\max} = \max_{1 \le i, j \le n} |a_{ij}|$. Clearly this function satisfies (2.2.11), (2.2.12), and (2.2.13). Show by example that it violates (2.2.14) and is therefore not a matrix norm. □

Every vector norm on $\mathbb{R}^n$ can be used to define a matrix norm on $\mathbb{R}^{n \times n}$ in a natural way. Given the vector norm $\| \cdot \|_v$, the matrix norm *induced* by $\| \cdot \|_v$ is defined by

$$\|A\|_M = \max_{x \neq 0} \frac{\|Ax\|_v}{\|x\|_v}$$

In Theorem 2.2.18 we will see that the induced norm is indeed a matrix norm on $\mathbb{R}^n$. Another name for the induced norm is the *operator norm*.

The induced norm has geometric significance that can be understood by viewing A as a linear transformation (operator) mapping $\mathbb{R}^n$ into $\mathbb{R}^n$: Each $x \in \mathbb{R}^n$ is mapped by A to the vector $Ax \in \mathbb{R}^n$. (See Figure 2.1.) The ratio $\|Ax\|_v / \|x\|_v$ is the magnification undergone by x when acted upon by A. The number $\|A\|_M$ is then the maximum magnification caused by A.

It is a common practice not to use distinguishing suffixes v and M, but to use the same symbol for both vector and matrix norms and simply to write

$$\|A\| = \max_{x \neq 0} \frac{\|Ax\|}{\|x\|}$$

We will adopt this practice. This need not lead to confusion, because the meaning of the norm can always be deduced from the context.

Before proving that the induced norm is indeed a matrix norm, it is useful to make note of the following simple but important fact.

THEOREM 2.2.16 A vector norm and its induced matrix norm satisfy the inequality

$$\|Ax\| \le \|A\| \|x\| \tag{2.2.17}$$

for all $A \in \mathbb{R}^{n \times n}$ and $x \in \mathbb{R}^n$. This inequality is sharp; that is, for all $A \in \mathbb{R}^{n \times n}$ there exists a nonzero $x \in \mathbb{R}^n$ for which equality holds.

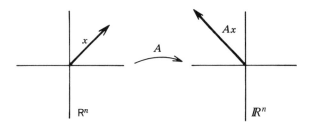

Figure 2.1

Proof If $x = 0$, equality holds trivially. Otherwise

$$\frac{\|Ax\|}{\|x\|} \leq \max_{\hat{x} \neq 0} \frac{\|A\hat{x}\|}{\|\hat{x}\|} = \|A\|$$

Thus $\|Ax\| \leq \|A\|\|x\|$. Equality holds if and only if x is a vector for which the maximum magnification is attained. (That such a vector exists is actually not obvious. It follows from a compactness argument that works because $\mathbb{R}^n$ is a finite-dimensional space. We omit the details.) $\qquad\square$

The fact that equality is attained in (2.2.17) is actually less important than the fact that there exist vectors for which equality is approached. The latter fact is obvious.

THEOREM 2.2.18 The induced norm is a matrix norm.

Proof This proof is not particularly difficult. You are encouraged to provide your own proof before reading further.

Each of the first three norm properties follows from the corresponding property of the vector norm. To prove (2.2.11), we must show that $\|A\| > 0$ if $A \neq 0$. If $A \neq 0$, then there exits $\hat{x} \in \mathbb{R}^n$ such that $A\hat{x} \neq 0$. Since the vector norm satisfies (2.2.11), this implies $\|A\hat{x}\| > 0$. Of course it must also be the case that $\hat{x} \neq 0$; therefore $\|\hat{x}\| > 0$. Thus $\|A\hat{x}\|/\|\hat{x}\| > 0$. We conclude that

$$\|A\| = \max_{x \neq 0} \frac{\|Ax\|}{\|x\|} \geq \frac{\|A\hat{x}\|}{\|\hat{x}\|} > 0$$

To prove (2.2.12), we note first that for every $x \in \mathbb{R}^n$ and $\alpha \in \mathbb{R}$, $\|\alpha(Ax)\| = |\alpha|\|Ax\|$. This is because the vector norm satisfies (2.2.12). (*Remember*: Ax is a vector, not a matrix.) Thus

$$\|\alpha A\| = \max_{x \neq 0} \frac{\|(\alpha A)x\|}{\|x\|} = \max_{x \neq 0} \frac{\|\alpha(Ax)\|}{\|x\|} = \max_{x \neq 0} \frac{|\alpha|\|Ax\|}{\|x\|}$$

$$= |\alpha| \max_{x \neq 0} \frac{\|Ax\|}{\|x\|} = |\alpha|\|A\|$$

Applying similar ideas, we prove (2.2.13).

$$\|A + B\| = \max_{x \neq 0} \frac{\|(A + B)x\|}{\|x\|} = \max_{x \neq 0} \frac{\|Ax + Bx\|}{\|x\|}$$

$$\leq \max_{x \neq 0} \frac{\|Ax\| + \|Bx\|}{\|x\|} \leq \max_{x \neq 0} \frac{\|Ax\|}{\|x\|} + \max_{x \neq 0} \frac{\|Bx\|}{\|x\|}$$

$$= \|A\| + \|B\|$$

Finally we prove (2.2.14). Replacing x by Bx in (2.2.17), we have $\|ABx\| \leq \|A\|\|Bx\|$ for any x. Applying (2.2.17), again we have $\|Bx\| \leq \|B\|\|x\|$. Thus

$$\|ABx\| \le \|A\|\|B\|\|x\|$$

For nonzero x we can divide both sides by $\|x\|$ and conclude that

$$\|AB\| = \max_{x \neq 0} \frac{\|ABx\|}{\|x\|} \le \|A\|\|B\| \qquad \square$$

Exercise 2.2.8   (a) Show that for any nonzero vector x and scalar c, $\|A(cx)\|/\|cx\| = \|Ax\|/\|x\|$. Thus rescaling a vector does not change the amount by which it is magnified under multiplication by A. In geometric terms, the magnification undergone by x depends only on its direction, not on its length.
(b) Prove that the induced matrix norm satisfies

$$\|A\| = \max_{\|x\|=1} \|Ax\|$$

This alternative characterization is often useful. $\square$

Some of the most important matrix norms are induced by p-norms. For $1 \le p \le \infty$, the norm induced by the p-norm is called the *matrix p-norm*:

$$\|A\|_p = \max_{x \neq 0} \frac{\|Ax\|_p}{\|x\|_p}$$

The matrix 2-norm is also known as the *spectral* norm. As we shall see in Chapter 3, this norm has great theoretical importance. Its drawback is that it is hard to compute; it is *not* the Frobenius norm. We will develop the tools to compute the spectral norm in Chapter 7.

Exercise 2.2.9   (a) Calculate $\|I\|_F$ and $\|I\|_2$, where I is the n-by-n identity matrix, and notice that they are different. (b) Use the Cauchy–Schwarz inequality (Theorem 2.2.5) to show that for all $A \in \mathbb{R}^{n \times n}$, $\|A\|_2 \le \|A\|_F$. $\square$

The other important cases are $p = 1$ and $p = \infty$. These norms can be computed easily.

THEOREM 2.2.19

$$\|A\|_1 = \max_{1 \le j \le n} \sum_{i=1}^{n} |a_{ij}|$$

THEOREM 2.2.20

$$\|A\|_\infty = \max_{1 \le i \le n} \sum_{j=1}^{n} |a_{ij}|$$

Thus $\|A\|_1$ is found by summing the absolute values of the entries in each column of A and then taking the largest of these column sums. Therefore the matrix 1-norm is sometimes called the *column sum norm*. Similarly, the matrix ∞-norm is sometimes called the *row sum norm*.

Proof of Theorem 2.2.19 We show first that $\|A\|_1 \le \max_j \left(\sum_{i=1}^{n} |a_{ij}| \right)$. For all $x \in \mathbb{R}^n$,

$$\|A\|_1 = \sum_{i=1}^{n} |(Ax)_i| = \sum_{i=1}^{n} \left| \sum_{j=1}^{n} a_{ij} x_j \right| \le \sum_{i=1}^{n} \sum_{j=1}^{n} |a_{ij}| |x_j|$$

$$= \sum_{j=1}^{n} \sum_{i=1}^{n} |a_{ij}| |x_j| \le \sum_{j=1}^{n} \left(\max_k \sum_{i=1}^{n} |a_{ik}| \right) |x_j|$$

$$= \left(\max_k \sum_{i=1}^{n} |a_{ik}| \right) \left(\sum_{j=1}^{n} |x_j| \right) = \left(\max_k \sum_{i=1}^{n} |a_{ik}| \right) \|x\|_1$$

Therefore $\|Ax\|_1 / \|x\|_1 \le \max_k \sum_{i=1}^{n} |a_{ik}|$ for all $x \ne 0$; from this $\|A\|_1 \le \max_k \left(\sum_{i=1}^{n} |a_{ik}| \right)$. To prove equality, we must simply find an $x \in \mathbb{R}^n$ for which

$$\frac{\|Ax\|_1}{\|x\|_1} = \max_k \left(\sum_{i=1}^{n} |a_{ik}| \right)$$

Suppose that the maximum is attained in the mth column of A. Let $\hat{x}$ be the vector with a 1 in position m and a 0 in all other positions. Then $\|\hat{x}\|_1 = 1$, $A\hat{x} = [a_{1m} \ a_{2m} \ \cdots \ a_{nm}]^T$, and $\|A\hat{x}\|_1 = \sum_{i=1}^{n} |a_{im}|$. Thus

$$\frac{\|A\hat{x}\|_1}{\|\hat{x}\|_1} = \sum_{i=1}^{n} |a_{im}| = \max_j \sum_{i=1}^{n} |a_{ij}| \qquad \square$$

Exercise 2.2.10 Prove Theorem 2.2.20. *Hint:* The argument is generally similar to that of the proof of Theorem 2.2.19, but your special vector $\hat{x}$ should be chosen to have either 1 or -1 in each component.

2.3
SENSITIVITY OF LINEAR SYSTEMS; CONDITION NUMBERS

In this section we introduce and discuss the *condition number* of a matrix. The condition number of A is a simple but useful measure of the sensitivity of the linear system $Ax = b$.

Consider a linear system $Ax = b$, where A is nonsingular, and b is nonzero. The system has a unique solution x, which is nonzero. Now suppose we add a small vector δb to b and consider the perturbed system $A\hat{x} = b + \delta b$. This system also has a unique solution $\hat{x}$, which is hoped to be not too far from x. Let δx denote the difference between $\hat{x}$ and x, so that $\hat{x} = x + \delta x$. We would like to say that if δb is small, then δx is also small. A more precise statement would involve relative terms: when we say that δb is small, we really mean that it is small in comparison with b; when we say that δx is small, we mean small compared with x. In order to quantify the size of vectors, we introduce a vector norm $\|\cdot\|$. The size of δb relative to b is then given by $\|\delta b\|/\|b\|$, and the size of δx relative to x is given by $\|\delta x\|/\|x\|$. We would like to say that if $\|\delta b\|/\|b\|$ is small, then $\|\delta x\|/\|x\|$ is also small.

The equations $Ax = b$ and $A(x + \delta x) = b + \delta b$ imply that $A\delta x = \delta b$; that is, $\delta x = A^{-1}\delta b$. Whatever vector norm we have chosen, we will use the induced matrix norm to measure matrices. Theorem 2.2.16 and the equation $\delta x = A^{-1}\delta b$ imply that

$$\|\delta x\| \le \|A^{-1}\|\|\delta b\| \tag{2.3.1}$$

Similarly the equation $b = Ax$ implies that $\|b\| \le \|A\|\|x\|$, or equivalently

$$\frac{1}{\|x\|} \le \|A\|\frac{1}{\|b\|} \tag{2.3.2}$$

Multiplying inequality (2.3.1) by (2.3.2), we arrive at the important inequality

$$\frac{\|\delta x\|}{\|x\|} \le \|A\|\|A^{-1}\|\frac{\|\delta b\|}{\|b\|} \tag{2.3.3}$$

which provides a bound for $\|\delta x\|/\|x\|$ in terms of $\|\delta b\|/\|b\|$. The factor $\|A\|\|A^{-1}\|$ is called the *condition number* of A and denoted $\kappa(A)$. With this new notation (2.3.3) becomes

$$\frac{\|\delta x\|}{\|x\|} \le \kappa(A)\frac{\|\delta b\|}{\|b\|} \tag{2.3.3$'$}$$

Since inequalities (2.3.1) and (2.3.2.) are sharp, (2.3.3$'$) is also sharp; that is, there exist b and δb (and associated x and δx) for which equality holds in (2.3.3$'$).

Exercise 2.3.1   (a) Show that $\kappa(A) = \kappa(A^{-1})$.
(b) Show that for any nonzero scalar c, $\kappa(cA) = \kappa(A)$.   ☐

From (2.3.3$'$) we see that if $\kappa(A)$ is not too large, then small values of $\|\delta b\|/\|b\|$ imply small values of $\|\delta x\|/\|x\|$. That is, the system is not overly sensitive to perturbations in b. Thus if $\kappa(A)$ is not too large, we say that A is *well conditioned*. By contrast, if $\kappa(A)$ is large, a small value of $\|\delta b\|/\|b\|$ does not guarantee that $\|\delta x\|/\|x\|$ will be small. Since (2.3.3$'$) is sharp, we know that there definitely are choices of b and δb for which the resulting $\|\delta x\|/\|x\|$ is much larger than $\|\delta b\|/\|b\|$. In other words the system is potentially very sensitive to perturbations in b. Thus if $\kappa(A)$ is large, we say that A is *ill conditioned*.

PROPOSITION 2.3.4   For any induced matrix norm (a) $\|I\| = 1$ and (b) $\kappa(A) \ge 1$.

Proof Part (a) follows immediately from the definition of the induced matrix norm. To prove part (b), we note that $I = AA^{-1}$, so $1 = \|I\| = \|AA^{-1}\| \le \|A\|\|A^{-1}\| = \kappa(A)$. $\qquad\square$

Thus the best possible condition number is 1. Of course the condition number depends on the choice of norm. While it is possible to concoct bizarre norms such that a matrix has a large condition number with respect to one norm and a small condition number with respect to the other, we will use mainly the 1-, 2-, and ∞-norms, which tend to give comparable values for the condition numbers of matrices. We will use the notation $\kappa_p(A) = \|A\|_p\|A^{-1}\|_p$ for $1 \le p \le \infty$.

It is useful to develop a geometric picture of the condition number. We begin by introducing some new terms. The *maximum* and *minimum magnification* by A are defined by

$$\text{maxmag}(A) = \max_{x \ne 0} \frac{\|Ax\|}{\|x\|}$$

$$\text{minmag}(A) = \min_{x \ne 0} \frac{\|Ax\|}{\|x\|}$$

Of course maxmag(A) is nothing but the induced matrix norm $\|A\|$.

Exercise 2.3.2 Prove that if A is a nonsingular matrix, then

$$\text{maxmag}(A) = \frac{1}{\text{minmag}(A^{-1})} \qquad \text{and} \qquad \text{maxmag}(A^{-1}) = \frac{1}{\text{minmag}(A)} \qquad \square$$

From this exercise it follows easily that $\kappa(A)$ is just the ratio of the maximum magnification to the minimum magnification.

PROPOSITION 2.3.5 $\kappa(A) = \dfrac{\text{maxmag}(A)}{\text{minmag}(A)}$ for all nonsingular A.

Exercise 2.3.3 Prove Proposition 2.3.5. $\qquad\square$

An ill-conditioned matrix is one for which the maximum magnification is much larger than the minimum magnification.

If the matrix A is singular, then there exists $x \ne 0$ such that $Ax = 0$. Thus minmag(A) = 0, so it is reasonable to say that $\kappa(A) = \infty$. That is, we view singularity as the extreme case of ill conditioning. Reversing the point of view, we can say that an ill-conditioned matrix is one that is "nearly" singular.

Since a matrix A is singular if and only if $\det(A) = 0$, it is natural to expect that the condition number of a matrix will have something to do with its determinant. This turns out not to be the case. As the following exercise demonstrates, there is no useful relationship between the determinant and the condition number.

Exercise 2.3.4 Let ϵ be a small positive number, and define

$$A_\epsilon = \begin{bmatrix} \epsilon & 0 \\ 0 & \epsilon \end{bmatrix}$$

Show that for any induced matrix norm, we have $\|A_\epsilon\| = \epsilon$, $\|A_\epsilon^{-1}\| = 1/\epsilon$, and $\kappa(A_\epsilon) = 1$. Thus A_ϵ is well conditioned. On the other hand, $\det(A_\epsilon) = \epsilon^2$, so we can make the determinant as close to zero as we please by taking ϵ sufficiently small.   □

Actually the phenomenon demonstrated in Exercise 2.3.4 can be produced using any nonsingular matrix. Recall from Exercise 2.3.1 that $\kappa(cA) = \kappa(A)$ for any nonsingular A and nonzero constant c. In other words, rescaling a matrix does not change its condition number. On the other hand, the determinant behaves much differently under rescaling: for any $A \in \mathbb{R}^{n \times n}$, $\det(cA) = c^n \det(A)$. Thus we can make the determinant as large or as small as we please by rescaling.

So far we have said that a matrix that has a large condition number is ill conditioned, but we have not said anything about where the cutoff line between well-conditioned and ill-conditioned matrices lies. Of course there is no point in looking for a precise boundary; there is none. Furthermore the (fuzzy) boundary depends on the accuracy of the data being used, the computer on which the problem is being solved, and the amount of error we are willing to tolerate in our computed solution. Suppose for example that the components of b are correct to about four decimal places. We do not know the exact value of b; in the computation we actually use $b + \delta b$, where $\|\delta b\|/\|b\| \approx 10^{-4}$. (Of course this estimate is not valid for any norm, but it is roughly true for the p-norms, for example.) If we solve the problem accurately, we get not x but $x + \delta x$, where an upper bound on $\|\delta x\|/\|x\|$ is given by (2.3.3′).

Now suppose $\kappa(A) \leq 10^2$. Then by (2.3.3′) the worst that can happen is $\|\delta x\|/\|x\| \approx 10^{-2}$. That is, the error in x is not bigger than about one-hundredth the size of x. In many problems this much error in the solution is acceptable. By contrast, if $\kappa(A) \approx 10^4$, then (2.3.3′) tells us that it can happen that $\|\delta x\|/\|x\| \approx 1$; that is, the error could be as big as the solution itself. In this case we would have to say that the condition number is unacceptably high. Thus it appears that in this problem the boundary between well-conditioned and ill-conditioned matrices lies somewhere in the range 10^2 to 10^4.

Occasionally the accuracy of the computer can be the deciding factor. It may be that we know b with extreme accuracy, but if the computer can only store seven decimal places, then we will be forced to work with $b + \delta b$, where $\|\delta b\|/\|b\| \approx 10^{-7}$. If we then have $\kappa(A) \approx 10^7$, we cannot expect to get a reasonable answer even if we solve the system very accurately. On the other hand a condition number of 10^3, 10^4, or even 10^5 may be small enough, depending on how accurate we require the solution to be.

We are long overdue for an example of an ill-conditioned matrix.

Example 2.3.6 Let $A = \begin{bmatrix} 1000 & 999 \\ 999 & 998 \end{bmatrix}$. You can easily verify that $A^{-1} =$

$\begin{bmatrix} -998 & 999 \\ 999 & -1000 \end{bmatrix}$. Thus $\|A\|_\infty = \|A\|_1 = 1999 = \|A^{-1}\|_\infty = \|A^{-1}\|_1$, and

$$\kappa_\infty(A) = \|A\|_\infty \|A^{-1}\|_\infty = (1999)^2 \approx 3.996 \times 10^6$$

$$\kappa_1(A) = \|A\|_1 \|A^{-1}\|_1 = (1999)^2 \approx 3.996 \times 10^6$$

(Incidentally, $\kappa_2(A) \approx 3.992 \times 10^6$.) This matrix would be considered ill conditioned by most standards.

Notice that

$$A\begin{bmatrix} 1 \\ 1 \end{bmatrix} = \begin{bmatrix} 1999 \\ 1997 \end{bmatrix} \tag{2.3.7}$$

If we use the ∞-norm to measure lengths, the magnification factor $\|Ax\|_\infty / \|x\|_\infty$ is 1999, which equals $\|A\|_\infty$. Thus $\begin{bmatrix} 1 \\ 1 \end{bmatrix}$ is a vector that is magnified maximally by A. Since the amount by which a vector is magnified depends only on its direction and not its length, we say that $\begin{bmatrix} 1 \\ 1 \end{bmatrix}$ is in a *direction of maximum magnification* by A. Equivalently we can say that $\begin{bmatrix} 1999 \\ 1997 \end{bmatrix}$ lies in a *direction of minimum magnification* by A^{-1}. Looking now at A^{-1}, we note that

$$A^{-1}\begin{bmatrix} -1 \\ 1 \end{bmatrix} = \begin{bmatrix} 1997 \\ -1999 \end{bmatrix}$$

The magnification factor $\|A^{-1}x\|_\infty / \|x\|_\infty$ is 1999, which equals $\|A^{-1}\|_\infty$, so $\begin{bmatrix} -1 \\ 1 \end{bmatrix}$ is in a direction of maximum magnification by A^{-1}. Equivalently

$$A\begin{bmatrix} 1997 \\ -1999 \end{bmatrix} = \begin{bmatrix} -1 \\ 1 \end{bmatrix} \tag{2.3.8}$$

and $\begin{bmatrix} 1997 \\ -1999 \end{bmatrix}$ is in a direction of minimum magnification by A. We will use the vectors in (2.3.7) and (2.3.8) to construct a spectacular example.

Suppose we wish to solve the system

$$\begin{bmatrix} 1000 & 999 \\ 999 & 998 \end{bmatrix} \begin{bmatrix} x_1 \\ x_2 \end{bmatrix} = \begin{bmatrix} 1999 \\ 1997 \end{bmatrix} \tag{2.3.9}$$

that is, $Ax = b$, where $b = \begin{bmatrix} 1999 \\ 1997 \end{bmatrix}$. Then by (2.3.7) the unique solution is $x = \begin{bmatrix} 1 \\ 1 \end{bmatrix}$.

Now suppose that we solve instead the slightly perturbed problem

$$\begin{bmatrix} 1000 & 999 \\ 999 & 998 \end{bmatrix} \begin{bmatrix} \hat{x}_1 \\ \hat{x}_2 \end{bmatrix} = \begin{bmatrix} 1998.99 \\ 1997.01 \end{bmatrix} \tag{2.3.10}$$

This is $A\hat{x} = b + \delta b$, where $\delta b = \begin{bmatrix} -0.01 \\ 0.01 \end{bmatrix} = 0.01\begin{bmatrix} -1 \\ 1 \end{bmatrix}$, which is in a direction of maximum magnification by A^{-1}. By (2.3.8) $A\,\delta x \,=\, \delta b$, where $\delta x \,=\,$ $\begin{bmatrix} 19.97 \\ -19.99 \end{bmatrix}$. Therefore $\hat{x} \,=\, x + \delta x \,=\, \begin{bmatrix} 20.97 \\ -18.99 \end{bmatrix}$. Thus the practically identical problems (2.3.9) and (2.3.10) have very different solutions.

It is important to recognize that this example was concocted in a very special way. The vector b was chosen to be in a direction of minimum magnification by A^{-1}, so that the resulting x is in a direction of maximum magnification by A, and equality is attained in (2.3.2). The vector δb was chosen in a direction of maximum magnification by A^{-1}, so that equality holds in (2.3.1). As a consequence equality also holds in (2.3.3'). Had we not chosen b and δb so carefully, the result would very likely have been much less spectacular. In fact it can happen with an ill-conditioned system that $\|\delta x\|/\|x\|$ is much smaller than $\|\delta b\|/\|b\|$. Inequality (2.3.3') has a companion inequality

$$\frac{\|\delta b\|}{\|b\|} \leq \kappa(A)\,\frac{\|\delta x\|}{\|x\|} \tag{2.3.11}$$

which can be obtained by interchanging the roles of x and δx with b and δb, respectively, and which is also sharp.

Exercise 2.3.5   Prove inequality (2.3.11). Under what conditions does equality hold in (2.3.11)? $\square$

Example 2.3.12   Let $A \,=\, \begin{bmatrix} 1000 & 999 \\ 999 & 998 \end{bmatrix}$, as in Example 2.3.6. Let $x \,=\,$ $\begin{bmatrix} 1997 \\ -1999 \end{bmatrix}$. Then by (2.3.8) $Ax \,=\, \begin{bmatrix} -1 \\ 1 \end{bmatrix}$. Now let $\delta x \,=\, \begin{bmatrix} 0.01 \\ 0.01 \end{bmatrix} \,=\, 0.01\begin{bmatrix} 1 \\ 1 \end{bmatrix}$. Then by (2.3.7) $A\,\delta x \,=\, \begin{bmatrix} 19.99 \\ 19.97 \end{bmatrix}$. Thus $A\begin{bmatrix} 1997.01 \\ -1998.99 \end{bmatrix} \,=\, A(x+\delta x) \,=\, \begin{bmatrix} 18.99 \\ 20.97 \end{bmatrix}$, which is very different from Ax.

This example points to a second problem encountered with ill-conditioned systems. Usually one wants a (computer) solution that is not only close to the true solution, but that also comes close to satisfying the equation $Ax \,=\, b$. Suppose $\hat{x}$ is our computed solution. The *residual* $r(\hat{x}) \,=\, A\hat{x} - b$ gives a measure of how close the equations are to being satisfied. The fit is good if and only if $r(\hat{x})$ [more precisely $\|r(\hat{x})\|/\|b\|$] is small: $r(\hat{x}) \,=\, 0$ if and only if $\hat{x}$ is the true solution. In the situation of Example 2.3.12, suppose we are interested in solving $Ax \,=\, b$, where $b \,=\, \begin{bmatrix} -1 \\ 1 \end{bmatrix}$. Then the true solution is $x \,=\, \begin{bmatrix} 1997 \\ -1999 \end{bmatrix}$. If it happens that $\hat{x} \,=\, x + \delta x \,=\, \begin{bmatrix} 1997.01 \\ -1998.99 \end{bmatrix}$, we have $r(\hat{x}) \,=\, A\,\delta x \,=\, \delta b \,=\, \begin{bmatrix} 19.99 \\ 19.97 \end{bmatrix}$, so that $\|r(\hat{x})\|_\infty/\|b\|_\infty \,=\, 19.99$. Thus a vector that is only slightly perturbed from the true solution can very badly fail to satisfy the equations. Since (2.3.11) is sharp, examples of this nature can be constructed using any ill-conditioned matrix. With well-conditioned matrices this phenomenon cannot occur, as (2.3.11) shows.

Exercise 2.3.6   Interpret (2.3.11) as a statement about residuals and verify the remarks made in the previous two sentences.   □

Exercise 2.3.7   The vector $\hat{x} = \begin{bmatrix} 20.97 \\ -18.99 \end{bmatrix}$ is a very poor approximation to the solution of the system $Ax = b$ of Example 2.3.6. Show that the residual $r(\hat{x})$ is quite small. Thus we see that in an ill-conditioned system there is almost no connection between the size of the residual and the accuracy of the solution.   □

Because of their great sensitivity, it is generally futile, even meaningless, to try to solve ill-conditioned systems in the presence of uncertainty in the data.

Exercise 2.3.8   Let $A = \begin{bmatrix} 375 & 374 \\ 752 & 750 \end{bmatrix}$.

(a) Calculate A^{-1} and $\kappa_\infty(A)$.

(b) Find b, δb, x, and δx such that $Ax = b$, $A(x + \delta x) = b + \delta b$, $\|\delta b\|_\infty / \|b\|_\infty$ is small, and $\|\delta x\|_\infty / \|x\|_\infty$ is large.

(c) Find b, δb, x, and δx such that $Ax = b$, $A(x + \delta x) = b + \delta b$, $\|\delta x\|_\infty / \|x\|_\infty$ is small, and $\|\delta b\|_\infty / \|b\|_\infty$ is large.   □

Ill Conditioning Caused by Poor Scaling

Some linear systems are ill conditioned simply because they are out of scale. Consider the following example.

Example 2.3.13   The system $\begin{bmatrix} 1 & 0 \\ 0 & \epsilon \end{bmatrix} \begin{bmatrix} x_1 \\ x_2 \end{bmatrix} = \begin{bmatrix} 1 \\ \epsilon \end{bmatrix}$ has the unique solution $x = [1\ 1]^T$. You can easily check that if $\epsilon \ll 1$, then the coefficient matrix A is ill conditioned with respect to the usual norms. In fact $\kappa_1(A) = \kappa_2(A) = \kappa_\infty(A) = 1/\epsilon \gg 1$. This system is subject to everything we have said so far about ill-conditioned systems. For example, one can find a small perturbation in b that causes a large perturbation in x: Just take $b + \delta b = \begin{bmatrix} 1 \\ 2\epsilon \end{bmatrix}$, for which $\|\delta b\|_\infty / \|b\|_\infty = \epsilon$, to get $x + \delta x = \begin{bmatrix} 1 \\ 2 \end{bmatrix}$, which is far from $x = \begin{bmatrix} 1 \\ 1 \end{bmatrix}$. Of course this small perturbation of b requires a perturbation of the second component of b that is large relative to that component. If we multiply the second equation of the system by $1/\epsilon$, we get a new system

$$\begin{bmatrix} 1 & 0 \\ 0 & 1 \end{bmatrix} \begin{bmatrix} x_1 \\ x_2 \end{bmatrix} = \begin{bmatrix} 1 \\ 1 \end{bmatrix}$$

which is clearly well conditioned. Thus the ill conditioning was just a consequence of poor scaling.

THEOREM 2.3.14 Let A be any nonsingular matrix, and let $a_1, a_2, \ldots, a_n$ denote its columns. Then for all i and j,

$$\kappa_p(A) \geq \frac{\|a_i\|_p}{\|a_j\|_p} \qquad 1 \leq p \leq \infty$$

Proof Clearly $a_i = Ae_i$, where e_i is the vector with a 1 in the ith position and zeros elsewhere. Thus

$$\text{maxmag}(A) = \max_{x \neq 0} \frac{\|Ax\|_p}{\|x\|_p} \geq \frac{\|Ae_i\|_p}{\|e_i\|_p} = \|a_i\|_p$$

$$\text{minmag}(A) = \min_{x \neq 0} \frac{\|Ax\|_p}{\|x\|_p} \leq \frac{\|Ae_j\|_p}{\|e_j\|_p} = \|a_j\|_p$$

and

$$\kappa_p(A) = \frac{\text{maxmag}(A)}{\text{minmag}(A)} \geq \frac{\|a_i\|_p}{\|a_j\|_p} \qquad \square$$

Thus any matrix that has columns whose norms differ by several orders of magnitude is ill conditioned. The same can be said of the rows, since A is ill conditioned if and only if A^T is. [You can easily verify that $\kappa_\infty(A) = \kappa_1(A^T)$. In Chapter 7 we will show that $\kappa_2(A) = \kappa_2(A^T)$.] Thus a necessary condition for a matrix to be well conditioned is that the norms of all of its rows and columns be of roughly the same magnitude. This condition is not sufficient, as the matrices in Example 2.3.6 and Exercise 2.3.8 show.

Exercise 2.3.9 Prove that $\kappa_\infty(A) = \kappa_1(A^T)$. $\square$

If a system is ill conditioned because its rows or columns are badly out of scale, one must refer back to the underlying physical problem in order to determine whether the ill conditioning is inherent in the problem or simply a consequence of poor choices of measurement units. The system in Example 2.3.13 was easy to handle only because it really consists of two independent problems

$$[1][x_1] = [1] \qquad \text{and} \qquad [\epsilon][x_2] = [\epsilon]$$

each of which is well conditioned. In general a more careful analysis is required. Although the rows and columns of any matrix can easily be rescaled so that all rows and columns have about the same norm, there is no unique way of doing it, nor is it guaranteed that the resulting matrix is well conditioned. Issues associated with scaling will be discussed in Section 2.9, but no definite advice will be given. Any decision about whether to rescale or not, and how to rescale, should be guided by the underlying physical problem.

Geometric Picture of Ill Conditioning

For those matrices whose rows and columns are not badly out of scale, a useful geometric picture of ill conditioning can be developed. Recall (Theorem 1.3.3) that a matrix is singular if and only if its columns are linearly dependent. We will show that the columns of an ill-conditioned matrix are "nearly" linearly dependent. This is consistent with the idea that ill-conditioned matrices are "nearly" singular. Let A be an ill-conditioned matrix whose rows and columns are not severely out of scale, and suppose A has been normalized so that $\|A\| = 1$. That is, if $\|A\| \neq 1$, we multiply A by the scalar $1/\|A\|$ to obtain a new matrix whose norm is 1. We have already seen (Exercise 2.3.1) that multiplication of an entire matrix by a scalar does not affect the condition number. This normalization procedure is not essential to our argument, but it makes it simpler and clearer. Since $\|A\| = \text{maxmag}(A)$, we have

$$1 \ll \kappa(A) = \frac{\text{maxmag}(A)}{\text{minmag}(A)} = \frac{1}{\text{minmag}(A)}$$

Thus $\text{minmag}(A) \ll 1$. This means that there exists a $c \in \mathbb{R}^m$ such that $\|Ac\|/\|c\| \ll 1$. Since the ratio $\|Ac\|/\|c\|$ depends only on the direction of c and not on the length, we can choose c so that $\|c\| = 1$, and consequently $\|Ac\| \ll 1$. Letting $a_1, a_2, \ldots, a_n$ denote the columns of A,

$$Ac = \sum_{i=1}^{n} a_i c_i$$

(cf., Eq. 1.2.11). Thus we see that there is a linear combination of the columns of A that adds up to something small (Ac). If there were a linear combination that added up exactly to zero, the columns would be linearly dependent. Since A is nonsingular, this cannot occur. But since there is a linear combination that adds up to something that is "almost" zero, we say that the columns of A are "nearly" linearly dependent.

A singular matrix has not only linearly dependent columns but also linearly dependent rows. This suggests that an ill-conditioned matrix should have rows that are "nearly" linearly dependent. That this is in fact the case follows directly from the fact that A is ill conditioned if and only if A^T is.

Example 2.3.15   It is evident that the matrices from Example 2.3.6 and Exercise 2.3.8 have rows and columns that are nearly linearly dependent.

The geometric interpretation of ill conditioning is based on the idea that the rows of an ill-conditioned matrix are nearly linearly dependent. Consider the case $n = 2$, in which there are two equations in two unknowns

$$a_{11}x_1 + a_{12}x_2 = b_1$$
$$a_{21}x_1 + a_{22}x_2 = b_2$$

The solution set of each of these equations is a line in the (x_1, x_2) plane. The solution of the system is the point at which the two lines intersect. The first line is perpendicular to the (row) vector $[a_{11} \ a_{12}]$, and the second line is perpendicular to $[a_{21} \ a_{22}]$.

If A is ill conditioned, then these two vectors are nearly linearly dependent; that is, they point in nearly the same (or opposite) direction(s). Therefore the lines determined by them are nearly parallel. See Figure 2.2. The point labeled p is the solution of the system. A small perturbation in b_1 (for example) causes a small parallel shift in the first line. The perturbed line is represented by the dashed line in Figure 2.2. Since lines 1 and 2 are nearly parallel, a small shift in one of the lines causes a large shift in the solution from point p to point q. By contrast, in the well-conditioned case the rows are not nearly linearly independent, and the lines determined by the two equations are not nearly parallel (see Figure 2.3). A small perturbation in one (or both) of the lines gives rise to a small perturbation in the solution.

Example 2.3.16   Consider the system

$$1000x_1 + 999x_2 = b_1$$
$$999x_1 + 998x_2 = b_2$$

for which the coefficient matrix is the same as in Example 2.3.6. The slopes of the two lines are

$$m_1 = -\frac{1000}{999} \cong -1.001001 \qquad m_2 = -\frac{999}{998} \cong -1.001002$$

Thus the lines are nearly parallel. In the vicinity of their intersection point the lines are virtually indistinguishable. Therefore the exact solution is hard to find.

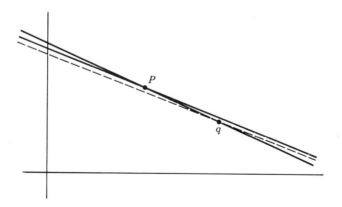

Figure 2.2   Ill-conditioned system.

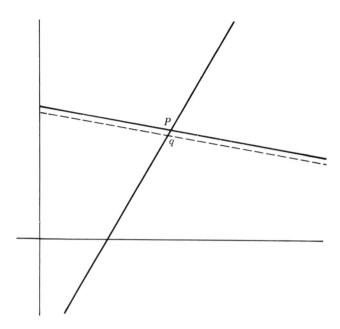

Figure 2.3   Well-conditioned system.

The system depicted in Figure 2.2 is actually not very ill conditioned. It is not possible to draw a good picture of a severely ill-conditioned system; the lines in such a system are so nearly parallel that they would be indistinguishable.

It is also possible and useful to visualize the case of three equations in three unknowns. In such a system the set of solutions of each equation is a plane in three-dimensional space. The plane determined by the ith equation is perpendicular to the (row) vector $[a_{i1}\ a_{i2}\ a_{i3}]$, $i = 1, 2, 3$. Each pair of planes intersects in a line, and the three planes together have a common intersection point, which is the solution of the system.

In the ill-conditioned case, the rows of the matrix are nearly linearly dependent, so one of the rows is nearly a linear combination of the other two rows. For the sake of argument let us say that the third row of A is nearly a linear combination of the other rows. This means that the vector $[a_{31}\ a_{32}\ a_{33}]$ nearly lies in the plane spanned by vectors $[a_{11}\ a_{12}\ a_{13}]$ and $[a_{21}\ a_{22}\ a_{23}]$. Therefore the plane of solutions of the third equation is nearly parallel to the line of intersection of the first and second planes. In the vicinity of the solution this line appears nearly to lie in the third plane. Thus the exact point of the solution is hard to distinguish, and a small perturbation of any of the planes will cause a large perturbation in the solution.

A better description would treat all equations equally rather than distinguishing the third equation. Such a description is harder to write, but the situation is not hard to visualize. Think first of a singular system in which the solution is not unique, rather there is a whole line (or plane) of solutions. This situation is represented by three planes that intersect in a line. Now perturb the picture slightly so that there is only one intersection point, but the three lines determined by intersecting the planes in pairs remain nearly parallel. This is the ill-conditioned case.

The Effect of Perturbing A

So far we have considered only the effect of perturbing b. We must also consider perturbations of A, as it also is represented only approximately. Thus, let us compare the two systems $Ax = b$ and $(A + \delta A)\hat{x} = b$, where $\|\delta A\|/\|A\|$ is small. Our first task is to establish a condition that guarantees that the system $(A + \delta A)\hat{x} = b$ has a unique solution, given that the system $Ax = b$ does. This is given by the following theorem, which, along with the two subsequent theorems, is valid for any vector norm and its induced matrix norm and condition number.

THEOREM 2.3.17 If A is nonsingular, and

$$\frac{\|\delta A\|}{\|A\|} < \frac{1}{\kappa(A)}$$

then $A + \delta A$ is also nonsingular.

Proof The hypothesis $\|\delta A\|/\|A\| < 1/\kappa(A)$ can be rewritten in various ways, for example, $\|\delta A\| < 1/\|A^{-1}\|$ and $\|A^{-1}\|\|\delta A\| < 1$. We'll use this last form of the inequality, and we'll prove the contrapositive form of the theorem: If $A + \delta A$ is singular, then $\|A^{-1}\|\|\delta A\| \geq 1$. Suppose $A + \delta A$ is singular. Then, by Theorem 1.3.3, there is a nonzero vector z such that $(A + \delta A)z = 0$. Reorganizing this equation, we obtain $z = -A^{-1}\,\delta A\,z$, which implies $\|z\| = \|A^{-1}\delta A\,z\| \leq \|A^{-1}\|\|\delta A\|\|z\|$. Since $\|z\| > 0$, we can divide both sides of the inequality by $\|z\|$ to obtain $1 \leq \|A^{-1}\|\|\delta A\|$, which is the desired result. $\square$

Theorem 2.3.17 demonstrates another important function of the condition number; it gives us an idea of the distance from A to the nearest singular matrix. If $A + \delta A$ is singular, then $\|\delta A\|/\|A\|$ must be at least $1/\kappa(A)$. It turns out that for the spectral norm, this result is exact: if $A + \delta A$ is the singular matrix nearest to A (in the sense that $\|\delta A\|_2$ is as small as possible), then $\|\delta A\|_2/\|A\|_2$ is exactly $1/\kappa_2(A)$. We will establish this fact in Corollary 7.3.10.

Now let us consider the relationship between the solutions of $Ax = b$ and $(A + \delta A)\hat{x} = b$. Let $\delta x = \hat{x} - x$, so that $\hat{x} = x + \delta x$. Under what conditions can we conclude that $\|\delta x\|/\|x\|$ is small? The following theorem shows that again the condition number plays the decisive role.

THEOREM 2.3.18 If A is nonsingular, $\|\delta A\|/\|A\| < 1/\kappa(A)$, $Ax = b$, and $(A + \delta A)(x + \delta x) = b$, then

$$\frac{\|\delta x\|}{\|x\|} \leq \frac{\kappa(A)\dfrac{\|\delta A\|}{\|A\|}}{1 - \kappa(A)\dfrac{\|\delta A\|}{\|A\|}} \tag{2.3.19}$$

Before we prove this result, let us make a few observations. If A is well conditioned and $\|\delta A\|/\|A\|$ is sufficiently small, then $\|\delta A\|/\|A\| \ll 1/\kappa(A)$; in this case the denominator on the right-hand side of (2.3.19) is approximately 1.

Thus (2.3.19) states *roughly* that

$$\frac{\|\delta x\|}{\|x\|} \leq \kappa(A)\frac{\|\delta A\|}{\|A\|}$$

Thus $\|\delta x\|/\|x\|$ is small.

If A is ill conditioned, the inequality $\|\delta A\|/\|A\| < 1/\kappa(A)$ will be satisfied only for very small δA. Unless $\|\delta A\|/\|A\|$ is very much smaller than $1/\kappa(A)$, no useful result is obtained. Suppose for example that $\|\delta A\|/\|A\| = 2 \times 10^{-7}$ and $\kappa(A) = 10^6$, so that $\|\delta A\|/\|A\| = \frac{1}{5}[1/\kappa(A)]$. Then by (2.3.19) $\|\delta x\|/\|x\| \leq \frac{1}{5}/\frac{4}{5} = 1/4$. This is not a very helpful result, since it shows only that the magnitude of the perturbation in the solution is not greater than one-fourth the magnitude of the solution itself. It is important to realize that this result does not guarantee that the perturbation is large; it merely warns that it might be large. Since inequality (2.3.19) is not sharp, it is not even certain that there exists a δA for which $\|\delta x\|/\|x\| \approx 1/4$.

Proof of Theorem 2.3.18   The equation $(A + \delta A)(x + \delta x) = b$ can be rewritten as $Ax + A\,\delta x + \delta A(x + \delta x) = b$. Using the fact that $Ax = b$ and rearranging the terms, we find that $\delta x = -A^{-1}\delta A(x + \delta x)$. Using the various properties of the vector norm and its induced matrix norm, we deduce that

$$\|\delta x\| \leq \|A^{-1}\|\|\delta A\|(\|x\| + \|\delta x\|)$$

$$= \kappa(A)\frac{\|\delta A\|}{\|A\|}(\|x\| + \|\delta x\|)$$

We now rewrite this inequality so that all terms involving $\|\delta x\|$ are on the left-hand side.

$$\left(1 - \kappa(A)\frac{\|\delta A\|}{\|A\|}\right)\|\delta x\| \leq \kappa(A)\frac{\|\delta A\|}{\|A\|}\|x\|$$

The assumption $\|\delta A\|/\|A\| < 1/\kappa(A)$ guarantees that the factor that multiplies $\|\delta x\|$ is positive, so we can divide by it without reversing the inequality. If we also divide through by $\|x\|$, we obtain the desired result.   □

So far we have considered the effects of perturbations of b and A separately. We have done this not out of necessity but from a desire to keep the analysis simple. The combined effects of perturbations in A and b can be expressed in a single inequality.

THEOREM 2.3.20   If A is nonsingular, $\|\delta A\|/\|A\| < 1/\kappa(A)$, $Ax = b$, and $(A + \delta A)(x + \delta x) = b + \delta b$, then

$$\frac{\|\delta x\|}{\|x\|} \leq \frac{\kappa(A)\left(\dfrac{\|\delta A\|}{\|A\|} + \dfrac{\|\delta b\|}{\|b\|}\right)}{1 - \kappa(A)\dfrac{\|\delta A\|}{\|A\|}}$$

You should have no difficulty combining elements of the proofs of inequality (2.3.3) and Theorem 2.3.18 to construct a proof of Theorem 2.3.20.

Exercise 2.3.10   Prove Theorem 2.3.20.   □

Geometric pictures of two- and three-dimensional ill-conditioned systems such as those we developed in order to visualize the effects of perturbations in b are also useful for visualizing perturbations in A. Whereas perturbations in b cause parallel shifts of the lines or planes, perturbations in A cause nonparallel perturbations of the lines or planes.

A Posteriori Error Analysis Using the Residual

Any time we solve a system of linear equations, we are interested in knowing whether or not our computed solution is accurate. This requires an error analysis, which attempts to determine the effects of various errors, such as roundoff errors, on the accuracy of the computation. Error analyses can be subdivided into two broad categories—a priori and a posteriori. An *a priori* error analysis attempts to determine, before solving the problem, whether a given method is going to produce an accurate solution. By contrast, an *a posteriori* error analysis is carried out after the problem has been solved, so it can make use of the computed solution and any other information that was produced during the computation. This makes a posteriori analysis much easier than a priori analysis.

For systems of linear equations there is a very simple a posteriori error analysis based on the residual and the condition number. Suppose $\hat{x} = x + \delta x$ is a computed solution of $Ax = b$, obtained by any method whatsoever. Earlier in the section we introduced the residual $r = A\hat{x} - b$, which measures how well $\hat{x}$ fits the equations. We also observed (Exercise 2.3.7) that the size of r is not necessarily a good indicator of how close $\hat{x}$ is to the true solution x. However, if the system is well conditioned, then a small residual does guarantee an accurate solution: $\hat{x}$ is the exact solution of the perturbed problem $A\hat{x} = b + \delta b$, where $\delta b = r$. If $\| r \|/\| b \|$ is small [and $\kappa(A)$ is not large], inequality (2.3.3′) guarantees that $\hat{x}$ is accurate.

The following exercise draws essentially the same conclusion via a different approach, in which the residual is associated with a perturbation in A rather than b.

Exercise 2.3.11   Define $\delta A \in \mathbb{R}^{n \times n}$ by $\delta A = -\alpha r \hat{x}^T$, where $r = A\hat{x} - b$ and $\alpha = \| \hat{x} \|_2^{-2}$.

   (a)   Show that $\hat{x}$ is the exact solution of $(A + \delta A)\hat{x} = b$.

   (b)   Show that $\| \delta A \|_F = \| r \|_2 / \| \hat{x} \|_2$ and

$$\frac{\| \delta A \|_2}{\| A \|_2} \leq \frac{\| r \|_2}{\| A \|_2 \| \hat{x} \|_2}$$

                                                                                □

Combining the result of Exercise 2.3.11 with Theorem 2.3.18, we find that if $\| r \|_2/(\| A \|_2 \| \hat{x} \|_2)$ is small and A is well conditioned, then $\hat{x}$ is a good approximation to x.

2.4
ESTIMATING THE CONDITION NUMBER

Section 2.3 shows clearly the importance of being able to estimate the condition number of a matrix. In principle the condition number is not hard to calculate: one simply finds A^{-1} and then calculates $\|A\|\|A^{-1}\|$. The problem with this is that it costs a lot to find A^{-1}. What is needed is an inexpensive estimate of $\kappa(A)$. An order of magnitude estimate is good enough.

Let us suppose that we wish to calculate $\kappa_1(A) = \|A\|_1\|A^{-1}\|_1$. From Theorem 2.2.19 we know that it is easy to compute $\|A\|_1$; the cost is about n^2 additions. What is harder is to get an estimate of $\|A^{-1}\|_1$, but if we have already solved the system $Ax = b$ by Gaussian elimination, an a posteriori estimate can be obtained at reasonable cost. Note first that for any nonzero $w \in \mathbb{R}^n$

$$\frac{\|A^{-1}w\|_1}{\|w\|_1} \le \max_{y \ne 0} \frac{\|A^{-1}y\|_1}{\|y\|_1} = \|A^{-1}\|_1$$

Thus, taking $w = b$, we have $A^{-1}w = x$,

$$\frac{\|x\|_1}{\|b\|_1} \le \|A^{-1}\|_1 \qquad \text{and} \qquad \kappa_1(A) \ge \frac{\|A\|_1\|x\|_1}{\|b\|_1}$$

The cost of calculating the vector norms is $O(n)$, so this gives an inexpensive lower bound for $\kappa_1(A)$. In many cases it will also be a good order-of-magnitude estimate of $\kappa_1(A)$. More generally, for any nonzero $w \in \mathbb{R}^n$,

$$\kappa_1(A) \ge \frac{\|A\|_1\|A^{-1}w\|_1}{\|w\|_1}$$

Since we already have an LU decomposition of A at hand, we can calculate $A^{-1}w$ by solving $Ac = w$ at a cost of only n^2 flops. If w is chosen roughly in the direction of maximum magnification of A^{-1}, the estimate

$$\kappa_1(A) \approx \frac{\|A\|_1\|A^{-1}w\|_1}{\|w\|_1} \tag{2.4.1}$$

will be quite good. Actually any w chosen at random is likely to have a significant component in the direction of maximum magnification by A^{-1} and therefore to give a reasonable estimate in (2.4.1). Since a random w will occasionally give a severe underestimate of $\kappa_1(A)$, the cautious operator might like to try several different choices of w. Another approach is to use the information at hand (i.e., the LU decomposition) to calculate a special w that is almost certain to have a large component in the direction of maximum magnification by A^{-1}. This approach was followed by Cline et al. (1979) and led to the successful condition number estimator used in LINPACK. We will not go into the details of this estimator, which is documented in [LUG] and [MC], as well as Cline et al. (1979).

Another approach, due to Hager (1984), uses ideas from convex optimization to try to find a w that maximizes $\|Aw\|_1/\|w\|_1$. A survey paper by Higham (1987) compared Hager's approach with the LINPACK approach and concluded that

Hager's approach generally gives better estimates at comparable cost. A modified version of Hager's approach is used in LAPACK. See Higham (1988) and (1990) for details.

2.5
ANALYSIS OF ROUNDOFF ERRORS

This section begins with a discussion of floating-point arithmetic and the effects of roundoff errors. A general approach to roundoff-error analysis is developed and used to analyze the effects of roundoff errors in Gaussian elimination. The findings are that Gaussian elimination with partial pivoting, although not unconditionally stable, is stable in practice; Gaussian elimination with complete pivoting is unconditionally stable; and Cholesky's method (without pivoting) for positive definite systems is unconditionally stable. This means that if these methods are used, the total effect of the roundoff errors is the same as that of a small perturbation of the system. Therefore if A is well conditioned, these methods will solve the system $Ax = b$ accurately. The following rule of thumb is developed: If the entries of A and b are accurate to about s decimal places and $\kappa(A) \approx 10^t$, then the computed solution will be accurate to about $s - t$ decimal places.

Most scientific computations are performed on computers using *floating-point arithmetic*. We will not define this term but instead give some examples. The number $.123456 \times 10^7$ is a typical six-digit decimal floating-point number. It has a *mantissa* $.123456$ and an *exponent* 7. It is called a decimal number because the number base is 10 (and of course the mantissa is interpreted as a base-10 fraction). Because the number has an exponent, the decimal point can "float" rather than remain in a fixed position. For example, $.123456 \times 10^3 = 123.456$, $.123456 \times 10^8 = 12345600.$, and $.123456 \times 10^{-2} = .00123456$. The advantage of the floating-point representation is that it allows very large and very small numbers to be represented accurately. Some other examples of floating-point numbers are $.6542 \times 10^{36}$, a large four-digit decimal number, and $-.71236 \times 10^{-42}$, a small five-digit decimal number. A floating-point number is said to be *normalized* if the first digit of its mantissa is nonzero. Thus the three examples we have looked at so far are all normalized, whereas $.0987 \times 10^6$ and $-.0012346 \times 10^{-4}$ are not normalized. With few exceptions, nonzero floating-point numbers are stored in normalized form.

Each computer's floating-point number system has its own characteristics. Small calculators aside, most computers do not use a base-10 representation. A power of 2 is more convenient architecturally. The most commonly used base is 2 itself, although 16 and 8 have also been used. The number of digits in the mantissa varies from one computer to the next. For example, the IBM 3033, the IBM 3090, and similar machines have, in single-precision format, a mantissa of 6 hexadecimal (base 16) digits. This gives about the same accuracy as 7 decimal digits. In double precision the mantissa has 14 hexadecimal digits, which is equivalent to almost 17 decimal digits. The Cray computers have a 48 binary (base 2) digit mantissa, which is equivalent to just over 14 decimal digits. Many computers conform to the IEEE floating-point standard (ANSI/IEEE Standard 754-1985), in which a single-precision mantissa has 24 binary digits ($\approx$ 7 decimal digits) and a double-precision mantissa has 53 binary digits ($\approx$ 16 decimal digits).

The range of allowable exponents also varies. On the IBM 3033 and 3090, the largest and smallest positive numbers that can be represented are 16^{63} ($\approx 7.2 \times 10^{75}$) and 16^{-64} ($\approx 8.6 \times 10^{-78}$), respectively. In the IEEE floating-point standard, the range of allowable numbers depends on the precision. In single precision, the largest and smallest normalized positive numbers that can be represented are $\approx 2^{128}$ ($\approx 3.4 \times 10^{38}$) and 2^{-126} ($\approx 1.2 \times 10^{-38}$), respectively; in double precision, the corresponding numbers are $\approx 2^{1024}$ ($\approx 1.8 \times 10^{308}$) and 2^{-1022} ($\approx 2.2 \times 10^{-308}$).[§] The Cray computers have an even greater range. If a computation that results in a number too large to be represented by the machine is made, an *overflow* condition results. For example, if we try to multiply 16^{40} by 16^{30} on an IBM 3033, we will get an overflow, as the product 16^{70} is too large to be represented by the machine's floating-point number system. In most systems, an overflow will cause the program to halt. If a number that is nonzero but too small to be represented by the machine is computed, an *underflow* results. For example, this happens when 16^{-40} is multiplied by 16^{-30} on the IBM 3033. The usual remedy for underflow is to set the result of the computation to zero. With one or two exceptions, we will ignore the possibility of overflow or underflow.

Our analysis will not be based on any specific computer. Instead we will follow the common practice of considering an ideal computer that performs all arithmetic operations on floating-point numbers exactly, except that the mantissa of each result must be either chopped or rounded off to the nearest floating-point number before it can be stored.[‖] For example, on an ideal four-digit decimal machine, the product of $.1111 \times 10^2$ and $.1111 \times 10^1$ would be computed as follows. The mantissas are multiplied and the exponents are added to give the result $.01234321 \times 10^3$. The result is then normalized to $.1234321 \times 10^2$. Finally the number is shortened to four digits by either chopping or rounding. This is where the error occurs. In this example the final result is $.1234 \times 10^2$, regardless of whether chopping or rounding is used. Let us consider a second example, in which $.1937 \times 10^3$ is subtracted from $.3426 \times 10^5$. Our ideal machine calculates the exact difference $.340663 \times 10^5$ and then either chops or rounds to get $.3406 \times 10^5$ or $.3407 \times 10^5$, respectively.

Our task is to assess the cumulative effect of roundoff errors on our calculations. To this end we introduce the notation $\mathrm{fl}(C)$ to denote the floating point result of some computation C. For example, if we multiply x by y, the result calculated by the computer is denoted $\mathrm{fl}(xy)$. We can apply this notation to more complicated calculations as well, as long as the order in which the calculations are to be performed is clear. For example $\mathrm{fl}\left(\sum_{i=1}^{n} x_i y_i\right)$ is a perfectly acceptable expression, provided that we have agreed on the order in which the terms are to be added.

Denoting the exact result of the computation C also by the letter C, we have $\mathrm{fl}(C) = C + e$, where e is the *absolute error* of the computation. A more useful measure is the *relative error* $\epsilon = e/C$, provided that $C \neq 0$. You can easily verify that the relative error satisfies the equation

$$\mathrm{fl}(C) = C(1 + \epsilon) \tag{2.5.1}$$

[§] The IEEE standard allows unnormalized numbers smaller than 2^{-126} (or 2^{-1022} in double precision), so that underflows can occur gradually rather than abruptly.

[‖] The IEEE floating-point standard conforms to this model.

Example 2.5.2 One example suffices to show that the relative error is more meaningful than the absolute error. Suppose we perform a computation on a seven-digit decimal machine and get the result $fl(C) = .9876572 \times 10^{17}$, whereas the correct value is $C = .98765432 \times 10^{17}$. The computed value is clearly a good approximation to the true value, but the absolute error is $e = fl(C) - C = .288 \times 10^{12}$, which looks like a large number unless it is compared with C. In the relative error the magnitude of C is automatically taken into account: $\epsilon = e/C = .291 \times 10^{-5}$. Now consider a different computation in which $fl(C) = .9876572 \times 10^{-15}$ and $C = .98765432 \times 10^{-15}$. The mantissas are the same as before, but the exponent has been changed. Now $e = .288 \times 10^{-20}$, which appears extremely small until it is compared with C. By contrast $\epsilon = .291 \times 10^{-5}$, the same as before. That the relative error is approximately 10^{-5} is reflected in the fact that C and $fl(C)$ agree in their first five decimal places. Because of this agreement the difference between $fl(C)$ and C is about five powers of 10 smaller than C. That is, the relative error is approximately 10^{-5}.

Another point worth mentioning is that the absolute error has the same units as C. Thus if C is measured in volts (seconds, meters), then e is also measured in volts (seconds, meters). By contrast the relative error is a dimensionless number; that is, it has no units.

In this text we will always measure errors in relative terms. The relative error appears in various guises. For example, the expressions $\|\delta x\|/\|x\|$, $\|\delta b\|/\|b\|$, and $\|\delta A\|/\|A\|$, with which we worked in Section 2.3, are expressions of relative error. Also, we have already observed that statements about the number of correct digits are actually vague statements about the relative error. Finally, statements about the percent error are also statements about the relative error: percent error = |relative error| × 100.

Our ideal computer does each arithmetic operation exactly and then rounds or chops the result. This implies that the relative error in each individual computation is small. Using the form (2.5.1) to express relative error we have, for any two normalized floating-point numbers x and y,

$$
\begin{aligned}
fl(x \pm y) &= (x \pm y)(1 + \epsilon_1) & |\epsilon_1| &\leq u \\
fl(xy) &= (xy)(1 + \epsilon_2) & |\epsilon_2| &\leq u \qquad (2.5.3) \\
fl(x/y) &= (x/y)(1 + \epsilon_3) & |\epsilon_3| &\leq u
\end{aligned}
$$

where u is the *unit roundoff*, defined as the largest relative error that can occur in a rounding or chopping operation. Obviously the value of u depends on the computer. On a machine with a mantissa of s decimal digits the value of u will be approximately 10^{-s}, since the computed value and the exact value always agree to s decimal places. In fact it is not hard to prove the following more precise statement: If a computer has a mantissa of s digits in base b, then $u = \frac{1}{2}b^{1-s}$ if rounding is used, and $u = b^{1-s}$ if chopping is used. For example, an IBM 3090 with single-precision, chopped arithmetic has $u = 16^{-5} \approx 10^{-6}$, and a computer with double-precision, rounded, IEEE standard floating-point arithmetic has $u = 2^{-53} \approx 10^{-16}$.

The results (2.5.3) can give a false sense of security. We might think that since the error made in each individual operation is small, a great many operations would have to be made before a significant error could accumulate. Unfortunately, this turns out not to be the case. In order to get a realistic idea of what can happen, we need to take account of the fact that the operands x and y usually have some error in them already. Instead of the correct values x and y, the computer works with perturbed values $\hat{x} = x(1 + \epsilon_1)$ and $\hat{y} = y(1 + \epsilon_2)$. Instead of calculating xy or $\mathrm{fl}(xy)$, for example, the computer calculates $\mathrm{fl}(\hat{x}\hat{y})$. We need to compare $\mathrm{fl}(\hat{x}\hat{y})$ with xy. We would like to be able to say that if $|\epsilon_1| \ll 1$ and $|\epsilon_2| \ll 1$, then $\mathrm{fl}(\hat{x}\hat{y}) = xy(1 + \epsilon)$, where $|\epsilon| \ll 1$. It turns out that such a result does hold for multiplication, and there is an analogous result for division. Unfortunately, addition and subtraction do not always behave so well.

Let us begin with the well-behaved operations. The computer multiplies $\hat{x}$ by $\hat{y}$ to get $\mathrm{fl}(\hat{x}\hat{y}) = \hat{x}\hat{y}(1 + \epsilon_3)$, where $|\epsilon_3| \leq u \ll 1$. Thus

$$\mathrm{fl}(\hat{x}\hat{y}) = x(1 + \epsilon_1)y(1 + \epsilon_2)(1 + \epsilon_3)$$

$$= xy(1 + \epsilon_1 + \epsilon_2 + \epsilon_3 + \epsilon_1\epsilon_2 + \epsilon_1\epsilon_3 + \epsilon_2\epsilon_3 + \epsilon_1\epsilon_2\epsilon_3)$$

$$= xy(1 + \epsilon)$$

where $\epsilon = \epsilon_1 + \epsilon_2 + \epsilon_3 + \epsilon_1\epsilon_2 + \epsilon_1\epsilon_3 + \epsilon_2\epsilon_3 + \epsilon_1\epsilon_2\epsilon_3$. The terms involving products of two or three ϵ_i are negligible because all ϵ_i are small. Thus $\epsilon \approx \epsilon_1 + \epsilon_2 + \epsilon_3$. Since $|\epsilon_1| \ll 1$, $|\epsilon_2| \ll 1$, and $|\epsilon_3| \ll 1$, it also holds that $|\epsilon| \ll 1$. We conclude that multiplication is well behaved in the presence of errors in the operands.

In order to analyze division, we begin by recalling from the theory of geometric series that $1/(1 + \epsilon_2) = 1 - \epsilon_2 + \epsilon_2^2 - \epsilon_2^3 + \cdots$. Since $|\epsilon_2| \ll 1$, the approximation $1/(1 + \epsilon_2) \approx 1 - \epsilon_2$, obtained by ignoring quadratic and higher terms, is good. Thus

$$\mathrm{fl}\left(\frac{\hat{x}}{\hat{y}}\right) = \frac{x(1 + \epsilon_1)}{y(1 + \epsilon_2)}(1 + \epsilon_3)$$

$$\approx \frac{x}{y}(1 + \epsilon_1)(1 - \epsilon_2)(1 + \epsilon_3)$$

$$\approx \frac{x}{y}(1 + \epsilon_1 - \epsilon_2 + \epsilon_3)$$

Therefore $\mathrm{fl}(\hat{x}/\hat{y}) = (x/y)(1 + \epsilon)$, where $|\epsilon| \ll 1$. We conclude that division is well behaved in the presence of errors.

Our analysis of addition will be a little bit different. We know that the difference between $\mathrm{fl}(\hat{x} + \hat{y})$ and $\hat{x} + \hat{y}$ is relatively small, so we will simply compare $\hat{x} + \hat{y}$ with $x + y$. (We could have done the same in our analyses of multiplication and division.) This simplifies the analysis slightly and has the advantage of making it clear that any serious damage that is done is caused not by the roundoff error from the current operation but by the previously existing errors.

$$\hat{x} + \hat{y} = x(1 + \epsilon_1) + y(1 + \epsilon_2)$$

$$= (x + y) + x\epsilon_1 + y\epsilon_2$$

$$= (x + y)\left(1 + \frac{x}{x + y}\epsilon_1 + \frac{y}{x + y}\epsilon_2\right)$$

Thus $\hat{x} + \hat{y} = (x + y)(1 + \epsilon)$, where

$$\epsilon = \frac{x}{x + y}\, \epsilon_1 + \frac{y}{x + y}\, \epsilon_2$$

Given that $|\epsilon_1| \ll 1$ and $|\epsilon_2| \ll 1$, we can say that $|\epsilon| \ll 1$ provided that neither x nor y is large compared with $x + y$. However if x or y is large compared with $x + y$, then ϵ can and probably will be large. That is, the computed result is probably inaccurate. This occurs when (and only when) x and y are virtually equal in magnitude and of opposite sign, so that they nearly cancel one another out when they are added.

An identical analysis holds for subtraction.

Exercise 2.5.1 Show that if $\hat{x} = x(1 + \epsilon_1)$ and $\hat{y} = y(1 + \epsilon_2)$, where $|\epsilon_1| \ll 1$ and $|\epsilon_2| \ll 1$, then $\hat{x} - \hat{y} = (x - y)(1 + \epsilon)$, where ϵ is small unless x or y is much larger than $x - y$. ☐

If x and y are nearly equal, so that $x - y$ is much smaller than both x and y, then the computed result $\hat{x} - \hat{y}$ can and probably will be inaccurate.

We conclude that both addition and subtraction are well behaved in the presence of errors, except when the operands nearly cancel one another out. Because cancellation generally signals a sudden loss of accuracy, it is usually called *catastrophic cancellation*.

Example 2.5.4 It is easy to see intuitively how cancellation leads to inaccurate results. Suppose an eight-digit decimal computer is to calculate $x - y$, where $x = .31415927 \cdots \times 10^1$ and $y = .31415916 \cdots \times 10^1$. Due to errors in the computation of x and y, the numbers that are actually stored in the computer's memory are $\hat{x} = .31415929 \times 10^1$ and $\hat{y} = .31415914 \times 10^1$. Clearly these numbers are excellent approximations to x and y, respectively, since they are correct in the first seven decimal places. That is, the relative errors ϵ_1 and ϵ_2 are of magnitude about 10^{-7}. Since $\hat{x}$ and $\hat{y}$ are virtually equal, all but one of the seven accurate digits are canceled off when $\hat{x} - \hat{y}$ is formed: $\hat{x} - \hat{y} = .00000015 \times 10^1 = .15000000 \times 10^{-5}$. In the normalized result only the first digit is correct. The second digit is inaccurate, as are all the zeros that follow it. Thus the computed result is a poor approximation of the true result $x - y = .11 \cdots \times 10^{-5}$. The relative error is about .36 (36 percent).

This example demonstrates a relatively severe case of catastrophic cancellation. In fact a whole range of severities is possible. Suppose for example two numbers are accurate to seven decimal places and they agree with each other in the first three decimal places. Then when their difference is taken, three accurate digits will be lost, and the result will be accurate to four decimal places.

We have demonstrated not only that catastrophic cancellation is dangerous, but also that it is the only mechanism by which a sudden large loss of accuracy can occur. The only other way an inaccurate result can occur is by the gradual accumulation of small errors over a large number of arithmetic operations. Although it is possible to concoct examples where this happens, it is seldom a problem in practice. The small errors that occur are just as likely to cancel one another out, at least in part, as they are to reinforce one another, so they tend to accumulate very slowly. Thus as a practical matter we can say that the only way a computation can go bad is through catastrophic cancellation. In other words, if no cancellations occur during a computation (and the original operands were accurate), the result will be accurate.

Unfortunately it is usually difficult to verify that no cancellation will occur in a given computation, and this makes it hard to prove that the roundoff errors will not spoil the computation. The first attempts at error analysis took the *forward* or *direct* approach, in which one works through the algorithm, attempting to bound the error in each intermediate result. In the end one gets a bound for the error in the final result. This approach usually fails because each time an addition or subtraction must be performed, one must somehow prove either that catastrophic cancellation cannot occur or that cancellation at that step will not destroy the outcome of subsequent computations. This is usually not possible.

Because of the threat of catastrophic cancellation, most of the pioneers in scientific computing were very pessimistic about the possible effects of roundoff errors on their computations. It was feared that any attempt to solve, say, a system of 50 equations in 50 unknowns would produce an inaccurate result. The early attempts to solve systems of linear equations on computers turned out generally better than expected, although disasters did sometimes occur. The issues were not well understood until a new approach to error analysis was developed, mainly by J. H. Wilkinson, in the late fifties. The new approach, called *inverse* or *backward* error analysis, does not attempt to bound the error in the result directly. Instead it pushes the effect of the errors back onto the operands.

Suppose, for example, we are given three floating-point numbers x, y, and z, and we wish to calculate $C = (x + y) + z$. The computer actually calculates $\hat{C} = \text{fl}(\text{fl}(x + y) + z)$. Even if the operands are exact, we cannot assert that the relative error in $\hat{C}$ is small: $\text{fl}(x + y)$ is (probably) slightly in error, so there can be a large relative error in $\hat{C}$ if cancellation takes place when $\text{fl}(x + y)$ and z are added. However, there is something else we can do. We have $\hat{C} = [(x + y)(1 + \epsilon_1) + z](1 + \epsilon_2)$, where $|\epsilon_1|, |\epsilon_2| \le u \ll 1$. Define ϵ_3 by $(1 + \epsilon_3) = (1 + \epsilon_1)(1 + \epsilon_2)$, so that $|\epsilon_3| \approx |\epsilon_1 + \epsilon_2| \ll 1$. Then $\hat{C} = (x + y)(1 + \epsilon_3) + z(1 + \epsilon_2)$. Defining $\overline{x} = x(1 + \epsilon_3)$, $\overline{y} = y(1 + \epsilon_3)$, and $\overline{z} = z(1 + \epsilon_2)$, we have

$$\hat{C} = (\overline{x} + \overline{y}) + \overline{z}$$

This shows that $\hat{C}$ is the *exact* result of performing the calculation $(x + y) + z$ with the slightly inaccurate data $\overline{x}$, $\overline{y}$, and $\overline{z}$. The errors have been shoved back onto the operands. The same can be done with subtraction, multiplication, division, and (with some ingenuity) longer computations.

In general suppose we wish to analyze some long computation $C(z_1, \ldots, z_m)$ involving m operands or input data $z_1, \ldots, z_m$. Instead of trying to show directly that $\text{fl}(C(z_1, \ldots, z_m))$ is close to $C(z_1, \ldots, z_m)$, we show that $\text{fl}(C(z_1, \ldots, z_m))$ is the exact result of operating with slightly perturbed input data; that is,

$$\text{fl}(C(z_1, \ldots, z_m)) = C(\overline{z}_1, \ldots, \overline{z}_m)$$

where $\overline{z}_1, \ldots, \overline{z}_m$ are close to $z_1, \ldots, z_m$. Of course the analysis does not end here. The backward error analysis has to be combined with a *sensitivity analysis* of the problem. If we can show that small perturbations in the operands lead to small perturbations in the (exact) results, then we can conclude that our computed result is accurate.

In Section 2.3 we carried out a sensitivity analysis of the linear system $Ax = b$. The inputs (operands) are A and b, and the result is x. Theorem 2.3.20 shows that if A is well conditioned, then small perturbations in the input data lead to small perturbations in the results. In order to show that Gaussian elimination (or any other algorithm) solves *well-conditioned* systems accurately, we need only carry out a backward error analysis of the algorithm and show that the computed result is the exact solution of a slightly perturbed system. Wilkinson developed this approach and carried it to completion by performing backward error analyses for various forms of Gaussian elimination.

We will examine a theorem of the type proved by Wilkinson, but first some general remarks are in order. The inverse approach to error analysis separates clearly the properties of the problem from the properties of the algorithm. The sensitivity analysis pertains to the problem, and the backward error analysis pertains to the algorithm. We say that the *problem* is *well conditioned* if small changes in the input parameters lead to small changes in the results. Otherwise the problem is said to be *ill conditioned*. We will adopt the attitude that it is unreasonable to expect any algorithm to be able to solve ill-conditioned problems accurately. Therefore we will judge an *algorithm* to be satisfactory if it admits a successful backward error analysis. That is, we make the following (somewhat vague) definition. Let $fl(C(z_1, \ldots, z_m))$ denote the result of performing the calculation $C(z_1, \ldots, z_m)$ in floating-point arithmetic by some specified algorithm. The algorithm is said to be *stable* (for inputs in some specified domain) if for all allowable inputs $z_1, \ldots, z_m$ there exist $\bar{z}_1, \ldots, \bar{z}_m$ that are close to $z_1, \ldots, z_m$, such that

$$fl(C(z_1, \ldots, z_m)) = C(\bar{z}_1, \ldots, \bar{z}_m)$$

A backward error analysis can determine whether an algorithm is stable. Once the algorithm has been found to be stable, it does not follow that it will always return an accurate result. A good result is guaranteed only when the algorithm is applied to a well-conditioned problem. If an algorithm fails to produce an accurate solution to a problem that is ill conditioned, this should not be held against the algorithm, rather it should be attributed to the nature of the problem.

The inverse approach to error analysis is successful because it is much less ambitious than the direct approach. The direct approach attempts to prove that a given algorithm always produces an accurate result, regardless of the sensitivity of the problem. This is usually impossible.

A backward error analysis of the Gaussian elimination algorithm for the linear system $Ax = b$ leads to results such as the following.

THEOREM 2.5.5 Let $\hat{L}$ and $\hat{U}$ be the computed LU factors obtained from $A \in \mathbb{R}^{n \times n}$ by Gaussian elimination without row or column interchanges. Suppose $\hat{U}$ is nonsingular and $\hat{x}$ is the computed solution of the system $Ax = b$ obtained by performing forward substitution with $\hat{L}$ and back substitution with $\hat{U}$. All operations are performed in floating-point arithmetic on an ideal computer with unit roundoff u. Then $\hat{x}$ is the exact solution of a perturbed system $(A + \delta A)\hat{x} = b$, where

$$\frac{\|\delta A\|_\infty}{\|A\|_\infty} \le nu\left(3 + 5\frac{\|\hat{L}\|_\infty\|\hat{U}\|_\infty}{\|A\|_\infty}\right) + O(u^2)$$

The term $O(u^2)$ means terms of order u^2, that is, terms that can be neglected. The theorem is valid for all formulations of Gaussian elimination, because all formulations perform exactly the same arithmetic. The result also holds for the matrix 1-norm and the Frobenius norm, and similar results hold for the matrix 2-norm and other matrix norms.

Theorem 2.5.5 is an easy consequence of Theorem 3.3.2 of [MC, p. 106]. We omit the proof because the actual details of backward error analysis are quite tedious and best skipped on a first reading. The curious reader can consult [MC], [CSLAS], or [AEP], for example. The main feature of Theorem 2.5.5 is the factor $\|\hat{L}\|_\infty\|\hat{U}\|_\infty/\|A\|_\infty$. If this number is not too large, then $\|\delta A\|_\infty/\|A\|_\infty$ is small. Unfortunately it sometimes can be large. If at some step a small pivot is chosen, resulting in large multipliers, then $\|\hat{L}\|_\infty$ and $\|\hat{U}\|_\infty$ will both be large. Consequently Gaussian elimination without pivoting is not a stable algorithm and is seldom used. This is not to say that it is never used or that it cannot be used. It *can* be used, provided that the size of the elements of $\hat{L}$ and $\hat{U}$ is monitored. If the entries of $\hat{L}$ and $\hat{U}$ are not too large and A is well conditioned, then it can be concluded with certainty (by Theorems 2.5.5 and 2.3.18) that the solution is accurate.

Gaussian elimination with row and column interchanges is equivalent to Gaussian elimination without interchanges, applied to a matrix obtained by making the row and column interchanges in advance. Therefore Theorem 2.5.5 is also applicable to Gaussian elimination with partial pivoting, complete pivoting, or any other pivoting strategy. Both partial and complete pivoting guarantee that all entries of $\hat{L}$ have absolute values not exceeding 1, so $\|\hat{L}\|_\infty \le n$. Thus, to guarantee the stability of each of these algorithms, we need only show that the elements of U are not too large. Unfortunately, in the case of partial pivoting there exist matrices for which the resulting $\hat{U}$ is quite large: $\|\hat{U}\|_\infty/\|A\|_\infty \approx 2^{n-1}$. Since 2^{n-1} grows very rapidly with n, we cannot say that Gaussian elimination with partial pivoting is unconditionally stable, except when n is very small. Consider the following example.

Exercise 2.5.2   Let A_n be the n-by-n matrix

$$A_n = \begin{bmatrix} 1 & 0 & 0 & \cdots & 0 & 1 \\ -1 & 1 & 0 & \cdots & 0 & 1 \\ -1 & -1 & 1 & \ddots & \vdots & \vdots \\ \vdots & \vdots & \ddots & 1 & 0 & 1 \\ -1 & -1 & \cdots & -1 & 1 & 1 \\ -1 & -1 & \cdots & -1 & -1 & 1 \end{bmatrix}$$

Show that if Gaussian elimination with partial pivoting is used, then A_n can be reduced to upper-triangular form without row interchanges, and the resulting matrix U satisfies $u_{nn} = 2^{n-1}$ and $\|U\|_\infty/\|A_n\|_\infty = 2^{n-1}/n$. $\square$

Despite the bad news from Exercise 2.5.2, Gaussian elimination with partial pivoting is now and will continue to be widely used. Years of testing and experience have shown that the type of element growth exhibited by the matrix in Exercise 2.5.2 is extremely rare in practical problems. Much more commonly

$$\frac{\|\hat{U}\|_\infty}{\|A\|_\infty} \approx 1$$

Hence, for practical purposes, Gaussian elimination with partial pivoting is considered to be a stable algorithm and is used with confidence. The skeptic has the inexpensive option of performing an a posteriori error analysis, as described at the end of Section 2.3, as a precaution.[§]

For complete pivoting a much more satisfactory rigorous bound on $\|\hat{U}\|_\infty / \|A\|_\infty$ can be obtained. Wilkinson (1961) has shown that the very worst that could possibly happen is

$$\frac{\|\hat{U}\|_\infty}{\|A\|_\infty} \approx n^{1/2} \left(2^1 3^{1/2} 4^{1/3} \cdots n^{1/n-1} \right)^{1/2} \tag{2.5.6}$$

While this number grows rather slowly with n (much more slowly than 2^{n-1}), it is also highly pessimistic. In fact no one has ever found a matrix for which (2.5.6) is approached. Again a more typical ratio is $\|\hat{U}\|_\infty / \|A\|_\infty \approx 1$. Thus Gaussian elimination with complete pivoting is stable.[‖]

In spite of the theoretical superiority of complete pivoting over partial pivoting, as evidenced by (2.5.6), partial pivoting is much more widely used. The reason is simple: partial pivoting works well in practice, and it is less expensive.

Theorem 2.5.5 is also important to designers of Gaussian elimination codes for sparse matrices. In sparse problems one wishes not only to perform the elimination in a stable manner, but also to keep fill-in as small as possible. Therefore the pivots that are selected are not always the largest ones. The stability of the decomposition can be monitored by checking the size of the entries of $\hat{L}$ and $\hat{U}$ as they are produced.

For positive definite systems, Theorem 2.5.5 is also valid for Cholesky's method when $\hat{L}$ and $\hat{U}$ are replaced by $\hat{G}$ and $\hat{G}^T$, respectively. As the following exercise shows, the elements of $\hat{G}$ cannot be large compared with those of A. Thus Cholesky's method is unconditionally stable.

Exercise 2.5.3 Suppose $A = GG^T$, where G is lower triangular. (a) (review) Show that $a_{ii} = \sum_{j=1}^{i} g_{ij}^2$. (b) Prove that $\|G\|_F \|G^T\|_F = \|G\|_F^2 = \sum_{i=1}^{n} a_{ii}$. ☐

While this exercise pertains to the theoretical Cholesky factor G, it is not hard to show that the computed Cholesky factor $\hat{G}$ satisfies a similar bound.

[§] Trefethen and Schreiber (1990) have proposed a statistical explanation for the observed lack of element growth in Gaussian elimination with partial pivoting: correlations in the signs of the elements and multipliers have the effect of suppressing growth.

[‖] Recently Higham and Higham (1989) exhibited several families of matrices that arise naturally in applications, for which $\|U\|_\infty / \|A\|_\infty = O(n)$ regardless of which pivoting strategy is used.

The details in the bound in Theorem 2.5.5 should not be taken too seriously. Bounds of this type are always extremely pessimistic because they must take into account the worst possible case, in which the roundoff errors are always as large as possible and always compound one another. The factor n appears in the bound simply because up to $2(n-1)$ roundoff errors are made in the process of calculating each entry of $\hat{L}$ and $\hat{U}$. The fact that these errors usually cancel each other out for the most part means that in practice we can ignore the n. For similar reasons we can ignore the n in the bound $\|\hat{L}\|_\infty \le n$, which holds if partial or complete pivoting is used. In other words we can ignore the factor $\|\hat{L}\|_\infty$. While we are at it, we might as well ignore the 3 and the 5 also. Then if $\|\hat{U}\|_\infty/\|A\|_\infty \approx 1$, as is usually the case in Gaussian elimination with partial or complete pivoting, we obtain the (very) rough estimate $\|\delta A\|_\infty/\|A\|_\infty \approx u$. That is, the effect of roundoff errors is such that the computed solution $\hat{x}$ is the exact solution of a perturbed problem $(A + \delta A)\hat{x} = b$, in which the magnitude of the perturbation is (usually) roughly the same as that of the initial rounding errors in the representation of A. This error is usually much smaller than the original measurement error in A and b. These considerations lead to a useful rule of thumb.

RULE OF THUMB 2.5.7 Suppose the linear system $Ax = b$ is solved by Gaussian elimination with partial or complete pivoting (or by Cholesky's method in the positive definite case). If the entries of A and b are accurate to about s decimal places and $\kappa(A) \approx 10^t$, where $t < s$, then the entries of the computed solution are accurate to about $s - t$ decimal places.

"Proof" We intended to solve $Ax = b$, but our computed solution $\hat{x}$ satisfies a perturbed equation $(A + \delta A)\hat{x} = b + \delta b$, where δA is the sum of measurement error, initial rounding error, and the effect of roundoff errors made during the computation and δb is the sum of measurement error and initial rounding error. Since the entries of A and b are accurate to about s decimal places,

$$\frac{\|\delta A\|}{\|A\|} \approx 10^{-s} \qquad \text{and} \qquad \frac{\|\delta b\|}{\|b\|} \approx 10^{-s}$$

Preparing to apply Theorem 2.3.18, we note that

$$\kappa(A)\frac{\|\delta A\|}{\|A\|} \approx 10^{t-s} \ll 1 \qquad \text{so} \qquad 1 - \kappa(A)\frac{\|\delta A\|}{\|A\|} \approx 1$$

Thus Theorem 2.3.18 gives roughly

$$\frac{\|\delta x\|}{\|x\|} \approx \kappa(A)\left(\frac{\|\delta A\|}{\|A\|} + \frac{\|\delta b\|}{\|b\|}\right) \approx 10^{t-s}$$

where $\hat{x} = x + \delta x$. That is, the entries of x are accurate to about $s - t$ decimal places.   □

Note that the approximation $\| \delta x \| / \| x \| \approx 10^{t-s}$ does not really imply that all components of x will be accurate to about $s - t$ decimal places. Any component that is a power of 10 or more smaller than the largest component may have fewer than $s - t$ correct digits.

2.6
GAUSSIAN ELIMINATION WITH ILL-CONDITIONED MATRICES

The material in Section 2.3 gives us ample reason to believe that there is no point in trying to solve a severely ill-conditioned system. Further insight can be gained by taking a heuristic look at what happens when one tries to solve an ill-conditioned system by Gaussian elimination. We will assume that the rows and columns are not out of scale.

When we do row operations, we take linear combinations of rows in such a way that zeros are deliberately created. Since the rows of an ill-conditioned matrix are nearly linearly dependent, there is the possibility of rows being made almost exactly zero by row operations. This possibility is encouraged by the progressive introduction of zeros into the array. Let us examine the phenomenon in some detail. A row will be called *bad* if it is nearly a linear combination of previous rows. Suppose the first $k - 1$ steps of the elimination have gone smoothly, but the kth row is bad. We have subtracted multiples of the first $k - 1$ rows from the kth row in such a way that there are now zeros in the first $k - 1$ positions. If the kth row were exactly a linear combination of the previous rows (and exact arithmetic were used), the entire kth row would now have to be zero. (Why?) Since it is only approximately a linear combination of the previous rows, it can still contain nonzero entries, but these entries will typically be small. They are not only small but also inaccurate because they became small through cancellation, as multiples of the earlier rows were subtracted from row k.

One of these small, inaccurate entries is the potential pivot in the (k, k) position. Because it is small, the kth row will be interchanged with a lower row that has a larger entry in its kth position, if such a row exists. In this way the bad rows get shifted downward. Eventually a step will be reached at which only bad rows remain. At this point all choices of pivot are small and inaccurate.[§] Although the presence of small, inaccurate numbers is not necessarily disastrous to the computation, the use of one as a pivot must be avoided if possible. In the present scenario we must choose a small, inaccurate pivot. This is used as the divisor in the computation of not-so-small, inaccurate multipliers, whose error pollutes all subsequent rows. The pivots are also used as divisors in the last step of the back-substitution process. Each component of the computed solution is a quotient whose

---

[§] We have described what typically happens when Gaussian elimination is applied to an ill-conditioned matrix. It is not claimed that the argument is rigorous. In fact the ill-conditioned upper-triangular matrix on page 81 of [MC] is a counterexample.

divisor is a pivot. We cannot expect these quotients to be accurate if the divisors are inaccurate. These observations apply to all variants of Gaussian elimination, since they all perform essentially the same operations.

Example 2.6.1 Consider the ill-conditioned matrix

$$A = \begin{bmatrix} 1000 & 999 \\ 999 & 998 \end{bmatrix}$$

which we examined previously in Section 2.3. Since the rows are nearly linearly dependent, when a zero is created in the $(2, 1)$ position, the entry in the $(2, 2)$ position should become nearly zero as well. In fact the multiplier is $l_{21} = .999$, and the $(2, 2)$ entry becomes

$$998 - (.999)(999) = 998 - 998.001 = -.001$$

This is indeed small, and the result was obtained by severe cancellation. There is no error in the result because it was computed by exact arithmetic. Consider what happens when five-digit decimal floating-point arithmetic is used. The computation yields

$$998.00 - (.99900)(999.00) = 998.00 - 998.00 = 0$$

The matrix appears to be singular!

This example may remind you of a remark that was made in Chapter 1. Not only can a nonsingular matrix appear singular (as just happened); the reverse can occur as well. We already remarked in Chapter 1 that if a Gaussian elimination program attempts to calculate the LU decomposition of a singular matrix, it probably will not recognize that the matrix is singular, because certain numbers that should have been zero are made nonzero by roundoff errors. Thus in numerical practice it is impossible to distinguish between ill-conditioned matrices and singular matrices. In contrast to the theoretical situation, in which there is a clear distinction between singular and nonsingular, we have instead a whole spectrum of condition numbers. (Exceptions to this rule are certain matrices that are obviously singular, such as a matrix with a row or column of zeros or two equal rows.)

The next example shows that the distinction between good and bad rows is not always clear. It can happen that the accuracy of a computation deteriorates gradually over a number of steps.

Example 2.6.2 Consider the 4-by-4 *Hilbert matrix*

$$H_4 = \begin{bmatrix} 1 & 1/2 & 1/3 & 1/4 \\ 1/2 & 1/3 & 1/4 & 1/5 \\ 1/3 & 1/4 & 1/5 & 1/6 \\ 1/4 & 1/5 & 1/6 & 1/7 \end{bmatrix}$$

The rows look very much alike, suggesting that this matrix is ill conditioned. In fact it is. Let us see how the ill conditioning manifests itself during Gaussian elimination. None of the entries in the first column of H_4 is small, so there is no need to pick a small pivot. Since the largest entry is already in the pivotal position, we do not make a row interchange. You can easily check that after one step the transformed array is

$$\begin{bmatrix} 1 & 1/2 & 1/3 & 1/4 \\ 1/2 & 1/12 & 1/12 & 3/40 \\ 1/3 & 1/12 & 4/45 & 1/12 \\ 1/4 & 3/40 & 1/12 & 9/112 \end{bmatrix}$$

The second step operates on the submatrix

$$\begin{bmatrix} 1/12 & 1/12 & 3/40 \\ 1/12 & 4/45 & 1/12 \\ 3/40 & 1/12 & 9/112 \end{bmatrix}$$

all of whose entries are smaller than the original matrix entries. Thus each entry has suffered a small amount of cancellation. Of course these numbers are perfectly accurate because we calculated them by exact arithmetic. If we had used floating-point arithmetic, they would have suffered a slight loss of accuracy because of the cancellation. Notice that all of the entries of this submatrix are quite close to $1/12$; the rows are almost equal.

The potential pivots for the second step are smaller than those for the first step. Again there is no need for a row interchange, and after the step the transformed submatrix is

$$\begin{bmatrix} 1/12 & 1/12 & 3/40 \\ 1 & 1/180 & 1/120 \\ 9/10 & 1/120 & 9/700 \end{bmatrix}$$

The entries of the submatrix

$$\begin{bmatrix} 1/180 & 1/120 \\ 1/120 & 9/700 \end{bmatrix}$$

are even smaller than before; cancellation has taken place once again.

The potential pivots for the third step, $1/180$ and $1/120$, are both quite small. Since the latter is larger, we interchange the rows. After the step we have

$$\begin{bmatrix} 1/120 & 9/700 \\ 2/3 & -1/4200 \end{bmatrix}$$

The final pivot is $-1/4200$, which is very small indeed.

Now let us see what happens when the same operations are performed in rounded three-digit floating-point arithmetic. The original array is

$$\begin{bmatrix} 1.00 & .500 & .333 & .250 \\ .500 & .333 & .250 & .200 \\ .333 & .250 & .200 & .167 \\ .250 & .200 & .167 & .143 \end{bmatrix}$$

Some of the entries are already slightly in error. On the first step the $(4, 3)$ entry (for example) is modified as follows:

$$.167 - (.250)(.333) = .167 - .833 \times 10^{-1} = .837 \times 10^{-1}$$

This is not an exact equation; it is a floating-point result. Comparing it with the correct value $1/12 \approx .833 \times 10^{-1}$, we see that there is a significant error in the third digit. The complete result of the first step is

$$\left[\begin{array}{c|ccc} 1.000 & .500 & .333 & .250 \\ \hline .500 & .830 \times 10^{-1} & .830 \times 10^{-1} & .750 \times 10^{-1} \\ .333 & .830 \times 10^{-1} & .890 \times 10^{-1} & .837 \times 10^{-1} \\ .250 & .750 \times 10^{-1} & .837 \times 10^{-1} & .805 \times 10^{-1} \end{array}\right]$$

The result of the second step is (ignoring the first row and column)

$$\left[\begin{array}{c|cc} .830 \times 10^{-1} & .830 \times 10^{-1} & .750 \times 10^{-1} \\ \hline 1.00 & .600 \times 10^{-2} & .870 \times 10^{-2} \\ .904 & .870 \times 10^{-2} & .127 \times 10^{-1} \end{array}\right]$$

Significant cancellations have now taken place; most of these numbers have only one correct digit. For example, the following computation produced the $(4, 3)$ element:

$$.837 \times 10^{-1} - (.904)(.830 \times 10^{-1}) = (.837 - .750) \times 10^{-1}$$
$$= .870 \times 10^{-2}$$

Comparing this result with the correct value $1/120 \approx .833 \times 10^{-2}$, we see that it has only one correct digit. The result of the third and final step is

$$\left[\begin{array}{c|c} .870 \times 10^{-2} & .127 \times 10^{-1} \\ \hline .690 & -.600 \times 10^{-4} \end{array}\right]$$

The $(4, 4)$ entry $-.600 \times 10^{-4}$ is not even close to the correct value $-1/4200 \approx -.238 \times 10^{-3}$.

Exercise 2.6.1   Work through the computations in Example 2.6.2, observing the cancellations and the accompanying loss of accuracy. Remember that if you wish to
use your calculator to simulate a three-digit machine, it does not suffice simply
to set the display to show three digits. Although only three digits are displayed,
nine or more digits are stored internally. Correct simulation of a three-digit machine requires that each intermediate result be rounded off to three decimal places
before it is used in the next computation. The simplest way to do this is to write
down each intermediate result and then enter it back into the calculator when it is
needed.   ☐

Exercise 2.6.2   Work Example 2.6.2 using four-digit arithmetic instead of three. You
will see that the outcome is not nearly so bad.   ☐

The Hilbert matrix H_4 is just one of an infinite family of Hilbert matrices.
In general H_n is the n-by-n matrix whose (i, j) entry is $1/(1 + j - 1)$. These are
easily the most famous examples of ill-conditioned matrices; the condition number
increases rapidly as n increases. (For more on Hilbert matrices see [CSLAS].)

Exercise 2.6.3   Use a computer to solve the systems

$$H_n x = b_n \qquad n = 3, 4, 5, \ldots ?$$

where b_n is the vector whose ith component is

$$\sum_{j=1}^{n} \frac{1}{i + j - 1}$$

With this choice of b_n the exact solution is $[1, 1, \ldots, 1]^T$. The Hilbert matrices
are positive definite, so you can use either a general Gaussian elimination program
or a program for positive definite systems. Notice that as n increases, the accuracy
of the computed solution decreases. Somewhere around $n = 10$, you will find
that the computed solution is complete garbage; the exact value of n at which this
occurs depends on the precision of the arithmetic being used. If you are using a
package such as LINPACK or MATLAB, you can also estimate or calculate the
condition numbers of the matrices.   ☐

2.7
WHY SMALL PIVOTS SHOULD BE AVOIDED

In Chapter 1 we introduced the partial-pivoting strategy, in which the pivot for the
kth step is chosen to be the largest in magnitude of the potential pivots in the kth
column. The justification given at that time was that we wanted to avoid using
as a pivot a small number that would have been zero except for roundoff errors
made in previous steps. Since then we have studied catastrophic cancellation and
ill-conditioned matrices and can make a more general statement: We wish to avoid

using a small pivot because it may have become small as a result of cancellation in a previous step, in which case it could be very inaccurate. The dangers of using an inaccurate pivot were stated in the previous section in connection with ill conditioning. The only difference between that scenario and the present one is that now we wish to consider what happens when one chooses a small pivot even though a large pivot is available. In this case the resulting inaccurate multipliers are not merely not so small; some of them will be quite large and therefore do even greater damage.

However, this is not the only reason for avoiding small pivots. Further justification is given by Theorem 2.5.5. During the discussion of that theorem, we remarked that small pivots lead to large multipliers; this causes both the computed L and U to have large entries and thus makes it impossible to infer stability from Theorem 2.5.5. Whereas this gives us further incentive for avoiding small pivots, it does not exhibit the mechanism by which small pivots can ruin a computation simply by being small. For this it is very helpful to look at an example.

Consider the linear system

$$\begin{bmatrix} .002 & 1.231 & 2.471 \\ 1.196 & 3.165 & 2.543 \\ 1.475 & 4.271 & 2.142 \end{bmatrix} \begin{bmatrix} x_1 \\ x_2 \\ x_3 \end{bmatrix} = \begin{bmatrix} 3.704 \\ 6.904 \\ 7.888 \end{bmatrix} \qquad (2.7.1)$$

This system is well conditioned, and we will see that it can be solved accurately by Gaussian elimination with partial pivoting. But first let us observe what happens when we use Gaussian elimination without row interchanges. We will see that the use of the exceptionally small $(1, 1)$ entry as a pivot destroys the computation. The computations will be done in rounded four-digit decimal floating-point arithmetic. You should think of the numbers in (2.7.1) as being exact; you can easily check that the exact solution is $x = [1, 1, 1]^T$. The roundoff error effects that you are about to observe are caused not because the small pivot is inaccurate (it is not!), but simply because it is small.

The multipliers for the first step are

$$l_{21} = \frac{1.196}{.002} = 598.0 \qquad l_{31} = \frac{1.475}{.002} = 737.5 \qquad (2.7.2)$$

These are multiplied by the first row and subtracted from the second and third rows, respectively. For example the $(2, 2)$ entry is altered as follows: 3.165 is replaced by

$$3.165 - (598.0)(1.231) = 3.165 - 736.1 = -732.9 \qquad (2.7.3)$$

These equations are not exact; rounded four-digit arithmetic has been used. Notice that the resulting entry is much larger than the original entry and that the last two digits of 3.165 were lost when a large number was subtracted from it. This type of information loss is called *swamping*. The small number was *swamped* by the large one. You can (and should) check that swamping also occurs when the $(2, 3)$, $(3, 2)$, and $(3, 3)$ entries are modified. In fact three digits are swamped in the $(2, 3)$ and

$(3, 3)$ positions. At the end of the first step the modified coefficient matrix looks like

$$\begin{bmatrix} .002 & 1.231 & 2.471 \\ \overline{598.0} & -732.9 & -1475. \\ 737.5 & -903.6 & -1820. \end{bmatrix}$$

The second step works with the submatrix

$$\tilde{A} = \begin{bmatrix} -732.9 & -1475. \\ -903.6 & -1820. \end{bmatrix}$$

Since this matrix was obtained by subtracting very large multiples of the row $[1.231, 2.471]$ from the much smaller numbers that originally occupied these rows, the two rows of $\tilde{A}$ are almost exact multiples of $[1.231, 2.471]$. Thus the rows of $\tilde{A}$ are nearly linearly dependent; that is, $\tilde{A}$ is ill conditioned. You can easily check that $\kappa_\infty(\tilde{A}) \approx 8400$, which is very large, considering that four-digit arithmetic is being used.

The multiplier for the second step is $l_{32} = (-903.6)/(-732.9) = 1.233$. It is used only to modify the $(3, 3)$ entry as follows:

$$-1820. - (1.233)(-1475.) = -1820. + 1819. = -1.000$$

Notice the severe cancellation that occurs here. This is just an attempt to recover the information that was lost through swamping in the previous step. Unfortunately that information is gone, and consequently the number -1.000 is inaccurate. At the end of the second step the LU decomposition is complete:

$$\begin{bmatrix} .002 & 1.231 & 2.471 \\ \overline{598.0} & -732.9 & -1475. \\ 737.5 & 1.233 & -1.000 \end{bmatrix}$$

The forward substitution step gives

$$y_1 = 3.704$$
$$y_2 = 6.904 - (598.0)(3.704) = 6.904 - 2215. = -2208.$$
$$y_3 = 7.888 - (737.5)(3.704) - (1.233)(-2208.)$$
$$= 7.888 - 2732. + 2722. = -2724. + 2722. = -2.000$$

Notice that the last three digits of 6.904 were swamped in the computation of y_2. Notice also that severe cancellation occurred in the calculation of y_3. Thus y_3 is inaccurate.

The first step of back substitution is

$$x_3 = \frac{y_3}{u_{33}} = \frac{-2.000}{-1.000} = 2.000$$

The computed x_3, being the quotient of two very inaccurate numbers, is also inaccurate. Recall that the correct value is 1.000. You can carry out the rest of the back substitution process and find that the computed solution is $[4.000, -1.012, 2.000]^T$, which is nothing like the true solution.

Let us summarize what went wrong, speaking in general terms (and heuristically!). When a pivot is used that is much smaller than the other potential pivots in the same column, large multipliers will result. Thus very large multiples of the pivotal row will be subtracted from the remaining rows. In the process the numbers that occupied these rows will be swamped. The resulting submatrix (the matrix that will be operated on in the next step) will be ill conditioned because each of its rows is almost exactly a multiple of the pivotal row. Because of this ill conditioning, there will be cancellations in later steps. These cancellations are actually just an attempt to uncover the information that was lost due to swamping, but that information is gone.

Now let us see what happens when we solve (2.7.1) using partial pivoting. Interchanging rows 1 and 3, we obtain the system

$$\begin{bmatrix} 1.475 & 4.271 & 2.142 \\ 1.196 & 3.165 & 2.543 \\ .002 & 1.231 & 2.471 \end{bmatrix} \begin{bmatrix} x_1 \\ x_2 \\ x_3 \end{bmatrix} = \begin{bmatrix} 7.888 \\ 6.904 \\ 3.704 \end{bmatrix}$$

After one step the partially reduced matrix has the form

$$\begin{bmatrix} 1.475 & 4.271 & 2.142 \\ .8108 & -.2980 & .8060 \\ 1.356 \times 10^{-3} & 1.225 & 2.468 \end{bmatrix}$$

You should carry out this computation and see for yourself that the information in the $(2, 2)$, $(2, 3)$, $(3, 2)$, and $(3, 3)$ positions is not swamped. There is, however, a slight cancellation in the $(2, 2)$ and $(2, 3)$ positions. The partial-pivoting strategy dictates that we interchange rows 2 and 3. In this way we avoid using the slightly inaccurate number $-.2980$ as a pivot. After step 2 the LU decomposition is complete:

$$\begin{bmatrix} 1.475 & 4.271 & 2.142 \\ 1.356 \times 10^{-3} & 1.225 & 2.468 \\ .8108 & -.2433 & 1.407 \end{bmatrix}$$

Forward substitution yields $y = [7.888, 3.693, 1.407]^T$, and back substitution gives the computed result

$$x = [1.000, 1.000, 1.000]^T$$

which is exactly equal to the true solution. You should work through the computations and see for yourself that no serious cancellations occurred and no information was swamped. Although it is a matter of luck that the computed solution exactly equals the true solution, it is not luck that the computation yielded an accurate result. Accuracy is guaranteed by the well-conditioned coefficient matrix, together with Theorem 2.5.5.

Exercise 2.7.1   Work through the details of the computations performed in this section.

□

2.8
LIMITATION OF THE NORM APPROACH

We have adopted the standard approach to sensitivity analysis, in which everything is measured by norms. The advantage of this approach is its simplicity: the error in a computed solution is expressed by a single number $\| \delta x \|/\| x \|$, and a single number $\kappa(A)$ summarizes the sensitivity of a linear system. As you may have noticed, this approach also has its limitations. A single number cannot possibly describe completely the error in a vector because the vector has n components. Consequently the statement "$\| \delta x \|/\| x \|$ is small" cannot guarantee that the error in each component is small.

Example 2.8.1   Suppose

$$x = \begin{bmatrix} 1.04 \\ 2.35 \\ 4.26 \times 10^{-6} \end{bmatrix} \quad \text{and} \quad \delta x = \begin{bmatrix} 1.32 \times 10^{-5} \\ 5.46 \times 10^{-6} \\ 1.02 \times 10^{-5} \end{bmatrix}$$

Then

$$x + \delta x = \begin{bmatrix} 1.0400132 \\ 2.35000546 \\ 1.446 \times 10^{-5} \end{bmatrix}$$

Even though $\| \delta x \|_\infty / \| x \|_\infty < 10^{-5}$, the relative perturbation in the third component is large. Obviously this danger exists whenever x has one or more components that are much smaller in magnitude than the largest component.

The most important thing to be said about the shortcoming illustrated by this example is that in many contexts it is not really anything to worry about. Consider for example an electrical circuit problem in which the components of the solution represent the voltages at the nodes of the circuit. Suppose the actual voltages are $E_1 = 1.04$, $E_2 = 2.35$, and $E_3 = 4.26 \times 10^{-6}$ and the computed voltages are $\hat{E}_1 = 1.04$, $\hat{E}_2 = 2.35$, and $\hat{E}_3 = 1.45 \times 10^{-5}$. It does not matter that the relative error in E_3 is large. What matters is that the error is small relative to the voltage *differences* in the system. Many problems are of this nature. It is easy to imagine examples involving temperature, location, time, and so on.

If you have a problem in which one of the components of the solution is small and you really do need to compute that component with a low relative error, you might be able to do it by rescaling the problem [see Skeel (1979)] or by iterative refinement [see Skeel (1980)]. Scaling and iterative refinement will be discussed briefly in the remaining two sections of the chapter.

2.9
SCALING

In a linear system $Ax = b$, any equation can be multiplied by any nonzero constant without changing the solution of the system. Such an operation is called a *row scaling* operation. A similar operation can be applied to the columns of A. In contrast to row scaling operations, *column scaling* operations do change the solution.

Exercise 2.9.1   Show that if the nonsingular linear system $Ax = b$ is altered by multiplication of its jth column by $c \neq 0$, then the solution is altered only in the jth component, which is multiplied by $1/c$.   ☐

Scaling operations can be viewed as changes of measurement units. Suppose the entries of the jth column of A are masses expressed in grams, and x_j is an acceleration measured in meters/second^2. Multiplication of the jth column by $1/1000$ is the same as changing the units of its entries from grams to kilograms. At the same time x_j is multiplied by 1000, which is the same as changing its units from meters/second^2 to millimeters/second^2. Similarly a rescaling of the ith row means a change of units for the entries of the ith row and the ith component of the right-hand side.

A discussion of scaling operations is made necessary by the fact that these operations affect the numerical properties of a system. This discussion has been placed toward the end of the chapter because in most cases rescaling is unnecessary and undesirable; usually an appropriate scaling is determined by the physical units of the problem. Consider for example the electrical circuit problem (1.1.1), with which this book began. In that system all entries of the coefficient matrix have the same units (1/ohm), all components of the solutions have the same units (volts), and all entries of the right-hand side have the same units (amperes). One could rescale this system so that, for example, one of the unknowns is expressed in millivolts while the others remain in volts, but this should not be done without a good reason. In most cases it will be best not to rescale.

Let us look at a few examples that illustrate some of the effects of scaling.

Example 2.9.1   The first example shows that a small pivot cannot be "cured" by multiplying its row by a large number. Consider the system

$$\begin{bmatrix} 2.000 & 1231. & 2471. \\ 1.196 & 3.165 & 2.543 \\ 1.475 & 4.271 & 2.142 \end{bmatrix} \begin{bmatrix} x_1 \\ x_2 \\ x_3 \end{bmatrix} = \begin{bmatrix} 3704. \\ 6.904 \\ 7.888 \end{bmatrix}$$

which was obtained from the system (2.7.1) by multiplying the first row by 1000. We used (2.7.1) to illustrate the damaging effects of using a small number as a pivot. Now that the first row has been multiplied by 1000, the (1, 1) entry is no longer small. In fact, it is now the largest entry in the first column. Let us see what happens when it is used as a pivot.

Using four-digit decimal arithmetic, the multipliers for the first step are

$$l_{21} = \frac{1.196}{2.000} = .5980 \qquad l_{31} = \frac{1.475}{2.000} = .7375$$

Comparing these with (2.7.2), we see that they are 1000 times smaller than before. In step 1 the $(2, 2)$ entry is altered as follows:

$$3.165 - (.5980)(1231.) = 3.165 - 736.1 = -732.9$$

Comparing this with (2.7.3), we see that in spite of the smaller pivot, the outcome is the same as before. This time the number 3.165 is swamped by the large entry 1231. Notice that this computation is essentially identical with (2.7.3). The result is *exactly* the same, including the roundoff errors. You can check that when the $(2, 3)$, $(3, 2)$, and $(3, 3)$ entries are modified, swamping occurs just as before, and indeed the computations are essentially the same as before and yield exactly the same result. Thus, after the first step, the modified coefficient matrix is

$$\begin{bmatrix} 2.000 & 1231. & 2471. \\ .5980 & -732.9 & -1475. \\ .7375 & -903.6 & -1820. \end{bmatrix}$$

The submatrix for the second step is

$$\begin{bmatrix} -732.9 & -1475. \\ -903.6 & -1820. \end{bmatrix}$$

which is exactly the same as before. If we continue the computation, we will have the same disastrous outcome. This time the swamping occurred not because large multiples of the first row were subtracted from the other rows, but because the first row itself is large.

How could this disaster have been predicted? Looking at the coefficient matrix, we see that it is ill conditioned: the rows (and the columns) are out of scale. It is interesting that we have two different explanations for the same numerical disaster. With the original system we blamed a small pivot; with the rescaled system we blame ill conditioning.

This example also illustrates an interesting theorem of Bauer. Suppose we solve the system $Ax = b$ by Gaussian elimination, using some specified sequence of row and column interchanges, on a computer that uses base β floating-point arithmetic. If the system is then rescaled by multiplying the rows and columns by powers of β and solved again using the same sequence of row and column interchanges, the result will be exactly the same as before, including roundoff errors. All roundoff errors at all steps are the same as before. In our present example $\beta = 10$. Multiplying the first row by 10^{-3} has no effect on the arithmetic. It is not hard to prove Bauer's theorem; you might like to do so as an exercise. At the

very least you should convince yourself that it is true. The examples that follow
should help.

Bauer's theorem has an interesting consequence: If the scaling factors are
always chosen to be powers of β, then the only way rescaling affects the numerical
properties of Gaussian elimination is by changing the choices of pivot. If scaling
factors that are not powers of β are used, there will be additional roundoff errors
associated with the rescaling, but it remains true that the principal effect of rescaling
is to alter the pivoting strategy.

Example 2.9.2 Let us solve the system

$$\begin{bmatrix} .003 & .217 \\ .277 & .138 \end{bmatrix} \begin{bmatrix} x_1 \\ x_2 \end{bmatrix} = \begin{bmatrix} .437 \\ .553 \end{bmatrix} \tag{2.9.3}$$

using rounded three-digit decimal floating-point arithmetic without row or column interchanges. The exact solution is $x = [1 \; 2]^T$. The multiplier is $l_{21} =$
$.277/.003 = 92.3$, and $u_{22} = .138 - (92.3)(.217) = .138 - 20.0 = -19.9$, so
the computed LU decomposition is

$$\begin{bmatrix} 1 & 0 \\ 92.3 & 1 \end{bmatrix} \begin{bmatrix} .003 & .217 \\ 0 & -19.9 \end{bmatrix}$$

The forward substitution gives $y_1 = .437$ and $y_2 = .553 - (92.3)(.437) = .553 -$
$40.3 = -39.7$. Finally the back substitution gives $x_2 = (-39.7)/(-19.9) =$
1.99 and $x_1 = (.003)^{-1}[.437 - (.217)(1.99)] = (.003)^{-1}(.437 - .432) =$
$.00500/.003 = 1.67$. Thus the computed solution is $\hat{x} = [1.67 \; 1.99]^T$, whose
first component is inaccurate.

Exercise 2.9.2 (a) Calculate $\kappa_\infty(A)$, where A is the coefficient matrix of (2.9.3). Conclude that A is well conditioned. (b) Perform Gaussian elimination on (2.9.3) with
the rows interchanged, using rounded 3-digit decimal floating-point arithmetic, and
note that an accurate solution is obtained. (Remember to round each intermediate
result to three decimal places before using it in the next calculation.) ☐

Example 2.9.4 Now let us solve

$$\begin{bmatrix} .300 & 21.7 \\ .277 & .138 \end{bmatrix} \begin{bmatrix} x_1 \\ x_2 \end{bmatrix} = \begin{bmatrix} 43.7 \\ .553 \end{bmatrix} \tag{2.9.5}$$

which was obtained by multiplying the first row of (2.9.3) by 10^2. By Bauer's
theorem the outcome should be the same as in Example 2.9.2. Let us check that it
is. The multiplier is $l_{21} = .277/.300 = .923$, and $u_{22} = .138 - (.923)(21.7) =$
$.138 - 20.0 = -19.9$, so the computed LU decomposition is

$$\begin{bmatrix} 1 & 0 \\ .923 & 1 \end{bmatrix} \begin{bmatrix} .300 & 21.7 \\ 0 & -19.9 \end{bmatrix}$$

The forward substitution yields $y_1 = 43.7$ and $y_2 = .553 - (.923)(43.7) = .553 - 40.3 = -39.7$. Finally the back substitution yields $x_2 = (-39.7)/(-19.9) = 1.99$ and $x_1 = (.300)^{-1}[43.7 - (21.7)(1.99)] = (.300)^{-1}(43.7 - 43.2) = .500/.300 = 1.67$. Thus the computed solution is again $\hat{x} = [1.67 \ 1.99]^T$. All intermediate results are identical to those in Example 2.9.2, except for powers of 10.

Exercise 2.9.3 (a) Calculate $\kappa_\infty(A)$, where A is the coefficient matrix of (2.9.5). A is ill conditioned because its rows (and columns) are out of scale. (b) Perform Gaussian elimination on (2.9.5) with the rows interchanged, using rounded three-digit decimal arithmetic. Note that, as guaranteed by Bauer's theorem, the computations and outcome are identical to those in Exercise 2.9.2, part (b). Thus an ill-conditioned coefficient matrix does not absolutely guarantee an inaccurate result. (However, if the partial-pivoting strategy is used, the row interchange will not be made.) □

Exercise 2.9.4 Solve (2.9.5) by Gaussian elimination with the columns interchanged, using rounded three-digit decimal arithmetic. (This is the complete-pivoting strategy. Note that an accurate result is obtained.) □

It is reasonable to expect that complete pivoting will perform much better than partial pivoting on problems that are ill conditioned simply because they are out of scale, since complete pivoting guarantees that the swamping phenomenon observed in Section 2.7 and Examples 2.9.1, 2.9.2, and 2.9.4 will not occur. The empirical fact that partial pivoting works well enough in practice suggests that badly scaled problems occur rarely in practice.

Example 2.9.6 Now let us solve

$$\begin{bmatrix} .300 & .217 \\ .277 & .00138 \end{bmatrix} \begin{bmatrix} x_1 \\ x_2 \end{bmatrix} = \begin{bmatrix} 43.7 \\ .553 \end{bmatrix} \tag{2.9.7}$$

which was obtained from (2.9.5) by multiplying the second column by 1/100. The exact solution is therefore $x = [1 \ 200]^T$. By Bauer's theorem the outcome should be the same as in Examples 2.9.2 and 2.9.4. The multiplier is $l_{21} = .277/.300 = .923$, and $u_{22} = .00138 - (.923)(.217) = .00138 - .200 = -.199$, so the computed LU decomposition is

$$\begin{bmatrix} 1 & 0 \\ .923 & 1 \end{bmatrix} \begin{bmatrix} .300 & .217 \\ 0 & -.199 \end{bmatrix}$$

The forward substitution gives $y_1 = 43.7$ and $y_2 = .553 - (.923)(43.7) = .553 - 40.3 = -39.7$. Finally the back substitution gives $x_2 = (-39.7)/(-.199) = 199.$, and $x_1 = (.300)^{-1}[43.7 - (.217)(199.)] = (.300)^{-1}(43.7 - 43.2) = .500/.300 = 1.67$. Thus the computed solution is $\hat{x} = [1.67 \ 199.]^T$. All computations were identical to those in Examples 2.9.2 and 2.9.4.

Although the computed solution $\hat{x} = [1.67 \ 199.]^T$ has an inaccurate first component, it should not necessarily be viewed as a bad result. The inaccurate component is much smaller than the accurate component, and in fact $\|\delta x\|_{\infty}/\|x\|_{\infty} = .005$ where $\hat{x} = x + \delta x$. The result that $\|\delta x\|_{\infty}/\|x\|_{\infty}$ is small is guaranteed by the fact that the coefficient matrix is well conditioned, together with the fact that the computed L and U are not large. Notice, by the way, that the pivot .300 is the one that would have been chosen by both the partial- and complete-pivoting strategies. It is easy to imagine situations in which the computed result of Example 2.9.6 is acceptable. Suppose, for example, that x_1 and x_2 represent voltages expressed in the same units. If all that matters is the voltage difference, then the result is okay, since the computed difference $\hat{x}_2 - \hat{x}_1 = 197.33$ differs from the correct difference 199 by only about 1 percent.

Exercise 2.9.5

(a) Calculate $\kappa_{\infty}(A)$, where A is the coefficient matrix of (2.9.7).

(b) Perform Gaussian elimination on (2.9.7) with the rows interchanged, using rounded, three-digit decimal arithmetic.

(c) Perform Gaussian elimination on (2.9.7) with the columns interchanged, using rounded, three-digit decimal arithmetic.   □

2.10
ITERATIVE REFINEMENT

Suppose the matrix A is nonsingular, so that the system $Ax = b$ has a unique solution. Let $\hat{x}$ denote an approximation to the solution, and let r denote the residual associated with $\hat{x}$; that is, $r = A\hat{x} - b$. The approximation $\hat{x}$ may have been obtained by Gaussian elimination or by some other means. If we could solve the residual system $Az = r$ exactly, then the vector $x = \hat{x} - z$ would be the exact solution of $Ax = b$, as you can easily check. This simple observation has been used as the basis for a scheme for improving the accuracy of solutions to ill-conditioned systems. Suppose $\hat{x}$ has been obtained by Gaussian elimination. Then an LU decomposition of A is available, so the system $Az = r$ can be solved in only n^2 flops using this decomposition. Of course the computed solution $\hat{z}$ is not exact, so $\hat{x} - \hat{z}$ is not the exact solution of $Ax = b$. Nevertheless it is not unreasonable to hope that $\hat{x} - \hat{z}$ will be an improvement over $\hat{x}$. If this is really the case, then perhaps we can improve the solution even more by calculating the residual associated with $\hat{x} - \hat{z}$ and repeating the process. In fact we can repeat it as many times as we wish. This gives the following *iterative refinement* algorithm.

$$
\begin{aligned}
&\text{for } i = 1, \ldots, m \\
&\quad \left\lceil r \leftarrow A\hat{x} - b \right. \\
&\quad \left| \text{Calculate } \hat{z}, \text{ an approximate solution of } Az = r. \right. \\
&\quad \left| \quad \text{(Use the } LU \text{ decomposition that was computed previously.)} \right. \\
&\quad \left| \hat{x} \leftarrow \hat{x} - \hat{z} \right. \\
&\quad \left\lfloor \text{if } (\|\hat{z}\|/\|\hat{x}\| \text{ is sufficiently small) exit}\quad \text{(successful completion)} \right. \\
&\text{set flag indicating failure} \\
&\text{exit}
\end{aligned}
$$

(2.10.1)

This is called an *iterative* algorithm because the number of steps or *iterations* to be performed is not known in advance. The iterations are terminated as soon as the corrections become sufficiently small. Any iterative algorithm should have an upper bound on the number of iterations it is willing to attempt before abandoning the process as a failure. In (2.10.1) that number is denoted by m. Notice that if we wish to carry out this procedure, we must save copies of A and b to use in the computation of the residuals. Furthermore, if we are to have any hope at all for success, the residuals must be calculated in extended-precision arithmetic. This means that if we are performing our calculations in single precision, the step $r \leftarrow A\hat{x} - b$ should be done in double precision. It is not surprising that this should be necessary. As $\hat{x}$ gets close to the true solution, the residual becomes small. In the calculation of r, severe cancellation occurs when b is subtracted from $A\hat{x}$. Unless the computation is performed in double precision, r will have few or no correct significant digits.

It turns out that if A is not too badly conditioned [say $\kappa(A) \ll 1/u$] and the residuals are calculated with extended-precision arithmetic, (2.10.1) actually converges to the true solution of $Ax = b$. Thus iterative refinement can be used to solve $Ax = b$ to the full precision of the computation. That is, if the computations (other than the residual calculation) are performed in arithmetic that is accurate to seven decimal places, then it is possible to determine x to seven-decimal-places accuracy. See [CSLAS] for details.

The preceding paragraph seems to contradict the earlier assertion that it is futile to try to solve an ill-conditioned system. However, iterative refinement has the following important weakness: The system that is solved so accurately is the one whose coefficient matrix A and right-hand-side vector b are exactly what is stored in the computer. Because of measurement and representation errors, A and b are mere approximations to the true data for the physical problem that we are trying to solve. If the problem is ill conditioned, the exact solution of $Ax = b$ can be a very unsatisfactory approximation to the solution of the problem we really wish to solve.

If the improvement of solutions of ill-conditioned problems were the only application of iterative refinement, we would not have bothered to consider it in this book. But there is another interesting application that has emerged more recently. As we saw in Section 2.5, Gaussian elimination with partial pivoting is not unconditionally stable. However Jankowski and Woźniakowski (1977) demonstrated that this and other algorithms for solving $Ax = b$ can be made stable by combining them with iterative refinement. For this application it is not required that the residuals be calculated with extended-precision arithmetic. Skeel (1980) showed that just one step of iterative refinement suffices to make Gaussian elimination with partial pivoting stable. Again extended precision is not necessary. Thus we have an inexpensive way of guaranteeing stability. The only inconvenience is that copies of A and b must be saved for the computation of r. It is important to realize that the solution computed this way is not necessarily accurate if A is ill conditioned. All that is guaranteed is that the computed solution is the exact solution of $(A + \delta A)\hat{x} = b$, where δA is some small perturbation of A. See the works cited above for details.

3

Orthogonal Matrices and the Least-Squares Problem

3.1
THE DISCRETE LEAST-SQUARES PROBLEM

A task that occurs not infrequently in scientific investigations is that of finding a straight line that "fits" some set of data points. Typically we have a fairly large number of points (t_i, y_i), $i = 1, \ldots, n$, collected from some scientific experiment, and often we have some theoretical reason to believe that these points should lie on a straight line. Thus we seek a linear function $p(t) = a_0 + a_1 t$ such that $p(t_i) = y_i$, $i = 1, \ldots, n$. In practice of course the points will deviate somewhat from a straight line, so it is impossible to find a linear $p(t)$ that passes through all of them. Instead we settle for a line that fits the data well, in the sense that the errors

$$|y_i - p(t_i)| \qquad i = 1, \ldots, n$$

are made as small as possible.

It is generally impossible to find a p for which all the numbers $|y_i - p(t_i)|$, $i = 1, \ldots, n$, are simultaneously minimized. Therefore we seek a p that strikes a good compromise. Specifically, let $r = [r_1, \ldots, r_n]^T$ denote the vector of residuals $r_i = y_i - p(t_i)$. We can solve our problem by choosing a vector norm $\| \cdot \|$ and taking our compromise function to be that p for which $\|r\|$ is made as small as possible. Of course the solution depends on the choice of norm. For example, if we choose the Euclidean norm, we minimize the quantity

$$\|r\|_2 = \left(\sum_{i=1}^{n} |y_i - p(t_i)|^2 \right)^{1/2}$$

whereas if we choose the 1-norm or the ∞-norm, we minimize respectively the

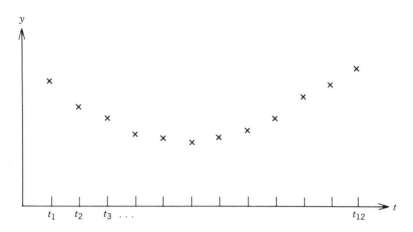

Figure 3.1

quantities

$$\|r\|_1 = \sum_{i=1}^{n} |y_i - p(t_i)|$$

$$\|r\|_\infty = \max_{1 \le i \le n} |y_i - p(t_i)|$$

The problem has been studied for a variety of norms, including the 1-, 2-, and ∞-norms. By far the nicest theory is that based on the 2-norm, and it is that theory that we will study in this chapter. To minimize $\|r\|_2$ is the same as to minimize

$$\|r\|_2^2 = \sum_{i=1}^{n} |y_i - p(t_i)|^2$$

Thus we are minimizing the sum of the squares of the residuals. For this reason the problem of minimizing $\|r\|_2$ is called the *least-squares problem*.[§]

A straight line is the graph of a polynomial of first degree. Sometimes it is desirable to fit a data set with a polynomial of higher degree. For example, it might be useful to approximate the data of Figure 3.1 by a polynomial of degree 2. If we decide to approximate our data by a polynomial of degree $\le m - 1$, then the task is to seek $p(t) = a_0 + a_1 t + a_2 t^2 + \cdots + a_{m-1} t^{m-1}$ such that

$$p(t_i) = y_i \qquad i = 1, \ldots, n \qquad (3.1.1)$$

Since the number of data points is typically large and the degree of the polynomial fairly low, it will usually be the case that $n \gg m$. In this case it is too much to ask for a p that satisfies all the equations (3.1.1) exactly, but for the moment let us act as if that were our goal. The set of polynomials of degree $\le m - 1$ is a

[§] It is called the *discrete* least-squares problem because a finite (discrete) data set $(t_i, y_i), i = 1, \ldots, n$, is being approximated. The *continuous* least-squares problem, in which a continuum of data points is approximated, will be discussed briefly in Section 3.5.

vector space of dimension m. If $\phi_1, \phi_2, \ldots, \phi_m$ is a basis of this space, then each polynomial p in the space can be expressed in the form

$$p(t) = \sum_{j=1}^{m} x_j \phi_j(t) \qquad (3.1.2)$$

for some unique choice of coefficients $x_1, x_2, \ldots, x_m$. The obvious basis is $\phi_1(t) = 1$, $\phi_2(t) = t$, $\phi_3(t) = t^2$, $\ldots$, $\phi_m(t) = t^{m-1}$, but there are many others, some of which may be better from a computational standpoint.

Substituting the expression (3.1.2) into the equations (3.1.1), we get a set of n linear equations in the m unknowns $x_1, \ldots, x_m$:

$$\sum_{j=1}^{m} x_j \phi_j(t_i) = y_i \qquad i = 1, \ldots, n$$

which can be written in matrix form as

$$\begin{bmatrix} \phi_1(t_1) & \phi_2(t_1) & \cdots & \phi_m(t_1) \\ \phi_1(t_2) & \phi_2(t_2) & \cdots & \phi_m(t_2) \\ \vdots & \vdots & & \vdots \\ \phi_1(t_n) & \phi_2(t_n) & \cdots & \phi_m(t_n) \end{bmatrix} \begin{bmatrix} x_1 \\ x_2 \\ \vdots \\ x_m \end{bmatrix} = \begin{bmatrix} y_1 \\ y_2 \\ \vdots \\ y_n \end{bmatrix} \qquad (3.1.3)$$

If $n > m$, as is usually the case, this is an *overdetermined system*; that is, it has more equations than unknowns. Thus we cannot expect to find an x that satisfies (3.1.3) exactly. Instead we might seek an x for which the sum of the squares of the residuals is minimized.

It is easy to imagine further generalizations of this problem. For example, the functions $\phi_1, \ldots, \phi_m$ could be taken to be trigonometric or exponential or some other kind of nonpolynomial functions. Even more generally we can consider the overdetermined system

$$Ax = b \qquad (3.1.4)$$

where $A \in \mathbb{R}^{n \times m}$, $n > m$, and $b \in \mathbb{R}^n$. The *least-squares problem* for the system (3.1.4) is to find $x \in \mathbb{R}^m$ for which $\| r \|_2$ is minimized, where $r = b - Ax$ is the vector of residuals. With the help of the orthogonal matrices introduced in the next section, we will develop an algorithm to solve the least-squares problem for the overdetermined system (3.1.4), which includes (3.1.3) as a special case.

3.2
ORTHOGONAL MATRICES, ROTATORS, AND REFLECTORS

In this section we will develop powerful tools for solving the least-squares problem. As these tools are also of fundamental importance to the chapters that follow, this is one of the most important sections of the book.

As in the first two chapters we will restrict our attention to real vectors and matrices. However, the entire theory can be carried over to the complex case, and this is done in part in exercises at the end of this section. You should work through those exercises because the results will be needed in Chapter 4.

We begin by introducing an inner product in $\mathbb{R}^n$. Given two vectors $x = (x_1, \ldots, x_n)^T$ and $y = (y_1, \ldots, y_n)^T$ in $\mathbb{R}^n$, we define the *inner product* of x and y, denoted (x, y), by

$$(x, y) = \sum_{i=1}^{n} x_i y_i$$

Although the inner product of x and y is a real number, it can also be expressed as a matrix product: $(x, y) = y^T x = x^T y$. The inner product has the following properties, which you can easily verify:

$$(x, y) = (y, x)$$
$$(\alpha_1 x_1 + \alpha_2 x_2, y) = \alpha_1(x_1, y) + \alpha_2(x_2, y)$$
$$(x, \alpha_1 y_1 + \alpha_2 y_2) = \alpha_1(x, y_1) + \alpha_2(x, y_2)$$
$$(x, x) \geq 0 \qquad \text{with equality if and only if } x = 0$$

for all $x, x_1, x_2, y, y_1, y_2 \in \mathbb{R}^n$ and $\alpha_1, \alpha_2 \in \mathbb{R}$.

Note also the close relationship between the inner product and the Euclidean norm: $\|x\|_2 = \sqrt{(x, x)}$. The Cauchy–Schwarz inequality (Theorem 2.2.5) can be stated much more concisely in terms of the inner product and the Euclidean norm: For every $x, y \in \mathbb{R}^n$,

$$|(x, y)| \leq \|x\|_2 \|y\|_2 \tag{3.2.1}$$

When $n = 2$ (or 3) the inner product coincides with the familiar dot product from analytic geometry. Recall that if x and y are two nonzero vectors in a plane and θ is the angle between them, then

$$\cos \theta = \frac{(x, y)}{\|x\|_2 \|y\|_2}$$

It is not unreasonable to employ this formula to *define* the angle between two vectors in $\mathbb{R}^n$. Note first that (3.2.1) guarantees that $(x, y)/\|x\|_2 \|y\|_2$ always lies between -1 and 1, so it is the cosine of some angle. We now define the *angle*[§] between (nonzero) x and $y \in \mathbb{R}^n$ to be

$$\theta = \arccos \frac{(x, y)}{\|x\|_2 \|y\|_2} \tag{3.2.2}$$

[§] The angle between two vectors in, say, $\mathbb{R}^{100}$ is just as real as the angle between two vectors in a plane. In fact x and y span a two-dimensional subspace of $\mathbb{R}^{100}$. This subspace is nothing but a copy of $\mathbb{R}^2$, that is, a plane. Viewed in this plane, x and y have an angle θ between them. It is this angle that is produced by formula (3.2.2).

If $x = 0$ or $y = 0$, we define $\theta = \pi/2$. Two vectors x and y are said to be *orthogonal* if the angle between them is $\pi/2$ radians. Clearly this is the case if and only if $(x, y) = 0$.

A matrix $Q \in \mathbb{R}^{n \times n}$ is said to be *orthogonal* if $QQ^T = I$. This equation says that Q has an inverse, and $Q^{-1} = Q^T$. Since a matrix always commutes with its inverse, we have $Q^T Q = I$ as well. For square matrices the equations

$$QQ^T = I \qquad Q^T Q = I \qquad Q^T = Q^{-1}$$

are equivalent, and any one of them could be taken as the definition of an orthogonal matrix.

Exercise 3.2.1   (a) Show that if Q is orthogonal, then Q^{-1} is orthogonal. (b) Show that if Q_1 and Q_2 are orthogonal, then $Q_1 Q_2$ is orthogonal.   □

Exercise 3.2.2   If Q is orthogonal, then $\det(Q) = \pm 1$.   □

THEOREM 3.2.3   If $Q \in \mathbb{R}^{n \times n}$ is orthogonal, then for all $x, y \in \mathbb{R}^n$,

   (a) $(Qx, Qy) = (x, y)$
   (b) $\| Qx \|_2 = \| x \|_2$

Proof

   (a) $(Qx, Qy) = (Qy)^T Qx = y^T Q^T Qx = y^T x = (x, y)$.
   (b)   Set $x = y$ in part (a) and take square roots.   □

Part (b) of the theorem says Qx and x have the same length. Thus *orthogonal transformations preserve lengths*. Combining parts (a) and (b), we find easily that

$$\arccos \frac{(Qx, Qy)}{\| Qx \|_2 \| Qy \|_2} = \arccos \frac{(x, y)}{\| x \|_2 \| y \|_2}$$

Thus the angle between Qx and Qy is the same as the angle between x and y. We conclude that *orthogonal transformations preserve angles*.

In the least-squares problem we wish to find x such that $\| b - Ax \|_2$ is minimized. Theorem 3.2.3, part (b), shows that for every orthogonal matrix Q, $\| b - Ax \|_2 = \| Qb - QAx \|_2$. Therefore the solution of a least-squares problem is unchanged when A and b are replaced by QA and Qb, respectively. We will eventually solve the least-squares problem by finding a Q for which QA has a very simple form, from which the solution of the least-squares problem will be easy to determine.

Exercise 3.2.3   Prove the converse of Theorem 3.2.3, part (a): If Q satisfies $(Qx, Qy) = (x, y)$ for all $x, y \in \mathbb{R}^n$, then Q is orthogonal. [The condition of part (b) also implies that Q is orthogonal, but this is harder to prove.]   □

Exercise 3.2.4 Show that if Q is orthogonal, then $\|Q\|_2 = 1$, $\|Q^{-1}\|_2 = 1$, and $\kappa_2(Q) = 1$. Thus Q is perfectly conditioned with respect to the 2-condition number. This suggests that orthogonal matrices will have good computational properties. $\square$

There are two types of orthogonal transformations that are widely used in matrix computations: rotators and reflectors. We will discuss rotators first because they are simpler.

Rotators

Consider the plane $\mathbb{R}^2$. The operator that rotates each vector through a fixed angle θ is a linear transformation, so it can be represented by a matrix. Let

$$Q = \begin{bmatrix} q_{11} & q_{12} \\ q_{21} & q_{22} \end{bmatrix}$$

be this matrix. Then Q is completely determined by its action on the two vectors $\begin{bmatrix} 1 \\ 0 \end{bmatrix}$ and $\begin{bmatrix} 0 \\ 1 \end{bmatrix}$, because

$$\begin{bmatrix} q_{11} & q_{12} \\ q_{21} & q_{22} \end{bmatrix}\begin{bmatrix} 1 \\ 0 \end{bmatrix} = \begin{bmatrix} q_{11} \\ q_{21} \end{bmatrix} \quad \text{and} \quad \begin{bmatrix} q_{11} & q_{12} \\ q_{21} & q_{22} \end{bmatrix}\begin{bmatrix} 0 \\ 1 \end{bmatrix} = \begin{bmatrix} q_{12} \\ q_{22} \end{bmatrix}$$

Since the action of Q is to rotate each vector through the angle θ, clearly $Q\begin{bmatrix} 1 \\ 0 \end{bmatrix} = \begin{bmatrix} \cos\theta \\ \sin\theta \end{bmatrix}$ and $Q\begin{bmatrix} 0 \\ 1 \end{bmatrix} = \begin{bmatrix} -\sin\theta \\ \cos\theta \end{bmatrix}$ (see Figure 3.2). Thus $\begin{bmatrix} q_{11} \\ q_{21} \end{bmatrix} = \begin{bmatrix} \cos\theta \\ \sin\theta \end{bmatrix}$ and $\begin{bmatrix} q_{12} \\ q_{22} \end{bmatrix} = \begin{bmatrix} -\sin\theta \\ \cos\theta \end{bmatrix}$; that is,

$$Q = \begin{bmatrix} \cos\theta & -\sin\theta \\ \sin\theta & \cos\theta \end{bmatrix}$$

A matrix of this form is called a *rotator* (or *rotation*).

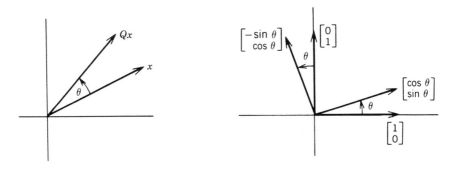

Figure 3.2 Rotation through an angle θ.

Exercise 3.2.5 Verify that every rotator is an orthogonal matrix with determinant 1. What does the inverse of a rotator look like? What transformation does it represent? □

Rotators can be used to create zeros in a vector or matrix. For example, if $x = \begin{bmatrix} x_1 \\ x_2 \end{bmatrix}$ is a vector with $x_2 \neq 0$, let us see how to find a rotator

$$Q = \begin{bmatrix} \cos\theta & -\sin\theta \\ \sin\theta & \cos\theta \end{bmatrix}$$

such that $Q^T x$ has a zero as its second component: $Q^T x = \begin{bmatrix} y \\ 0 \end{bmatrix}$ for some y. Now

$$Q^T x = \begin{bmatrix} \cos\theta & \sin\theta \\ -\sin\theta & \cos\theta \end{bmatrix} \begin{bmatrix} x_1 \\ x_2 \end{bmatrix} = \begin{bmatrix} (\cos\theta)x_1 + (\sin\theta)x_2 \\ -(\sin\theta)x_1 + (\cos\theta)x_2 \end{bmatrix}$$

which has the form $\begin{bmatrix} y \\ 0 \end{bmatrix}$ if and only if

$$x_1 \sin\theta = x_2 \cos\theta \qquad (3.2.4)$$

Thus θ can be taken to be $\arctan(x_2/x_1)$ or any other angle satisfying $\tan\theta = x_2/x_1$. But notice that we can determine Q without calculating θ itself: we need only $\cos\theta$ and $\sin\theta$. Clearly the choice $\cos\theta = x_1$ and $\sin\theta = x_2$ would satisfy (3.2.4). However, this choice violates the basic trigonometric identity $\cos^2\theta + \sin^2\theta = 1$, unless $x_1^2 + x_2^2 = 1$. Therefore we take instead

$$\cos\theta = \frac{x_1}{\sqrt{x_1^2 + x_2^2}} \qquad \text{and} \qquad \sin\theta = \frac{x_2}{\sqrt{x_1^2 + x_2^2}} \qquad (3.2.5)$$

obtained by dividing our original choice by $\|x\|_2 = \sqrt{x_1^2 + x_2^2}$. Now the basic trigonometric identity is satisfied, and there is a unique $\theta \in [0, 2\pi)$ for which (3.2.5) holds.

Exercise 3.2.6 We have just shown that for every $x \in \mathbb{R}^2$, there exists a rotator Q such that $Q^T x = \begin{bmatrix} y \\ 0 \end{bmatrix}$ for some y. (a) Give the geometric interpretation of this fact. (b) Show that if $\cos\theta$ and $\sin\theta$ are given by (3.2.5), then $y = \sqrt{x_1^2 + x_2^2} = \|x\|_2$. □

Now let us see how we can use a rotator to simplify a matrix

$$A = \begin{bmatrix} a_{11} & a_{12} \\ a_{21} & a_{22} \end{bmatrix}$$

We have just seen that there is a rotator Q such that

$$Q^T \begin{bmatrix} a_{11} \\ a_{21} \end{bmatrix} = \begin{bmatrix} r_{11} \\ 0 \end{bmatrix}$$

where $r_{11} = \sqrt{a_{11}^2 + a_{21}^2}$. Define r_{12} and r_{22} by

$$\begin{bmatrix} r_{12} \\ r_{22} \end{bmatrix} = Q^T \begin{bmatrix} a_{12} \\ a_{22} \end{bmatrix}$$

and let

$$R = \begin{bmatrix} r_{11} & r_{12} \\ 0 & r_{22} \end{bmatrix}$$

Then $Q^T A = R$. This shows that we can transform A to an upper triangular matrix by multiplying it by the orthogonal matrix Q^T. As we shall soon see, it is possible to carry out such a transformation not just for 2-by-2 matrices, but for all $A \in \mathbb{R}^{n \times n}$. That is, for every $A \in \mathbb{R}^{n \times n}$ there is an orthogonal matrix $Q \in \mathbb{R}^{n \times n}$ and an upper triangular matrix $R \in \mathbb{R}^{n \times n}$ such that $Q^T A = R$.

A transformation of this type can be used to solve a system of linear equations $Ax = b$. Multiplying on the left by Q^T, we transform the system $Ax = b$ to the equivalent system $Q^T A x = Q^T b$, or $Rx = c$, where $c = Q^T b$. It is a simple matter to calculate c, given b. Then x can be obtained from the upper triangular system $Rx = c$ by back substitution.

Another useful viewpoint is reached by rewriting the equation $Q^T A = R$ as

$$A = QR$$

This shows that A can be expressed as a product QR, where Q is orthogonal and R is upper triangular. The *QR decomposition*, as it is called, can be used in much the same way as an *LU* decomposition to solve a linear system $Ax = b$. If we have a QR decomposition of A, we can rewrite the system as $QRx = b$. Letting $c = Rx$, we can find c by solving $Qc = b$. This doesn't cost much; Q is orthogonal, so $c = Q^T b$. We can then obtain x by solving $Rx = c$ by back substitution. Notice that the computations required here are no different than those required by the method outlined in the previous paragraph. Thus, we have actually derived a single method from two different points of view.

Example 3.2.6   We will use a QR decomposition to solve the system

$$\begin{bmatrix} 1 & 2 \\ 1 & 3 \end{bmatrix} \begin{bmatrix} z_1 \\ z_2 \end{bmatrix} = \begin{bmatrix} 1 \\ 2 \end{bmatrix}$$

First we require Q such that $Q^T \begin{bmatrix} 1 \\ 1 \end{bmatrix} = \begin{bmatrix} * \\ 0 \end{bmatrix}$. We use (3.2.5) with $x_1 = 1$ and $x_2 = 1$ to get $c = \cos \theta = 1/\sqrt{2}$ and $s = \sin \theta = 1/\sqrt{2}$. Then

$$Q = \begin{bmatrix} c & -s \\ s & c \end{bmatrix} = \frac{1}{\sqrt{2}} \begin{bmatrix} 1 & -1 \\ 1 & 1 \end{bmatrix}$$

and

$$R = Q^T A = \frac{1}{\sqrt{2}} \begin{bmatrix} 1 & 1 \\ -1 & 1 \end{bmatrix} \begin{bmatrix} 1 & 2 \\ 1 & 3 \end{bmatrix} = \frac{1}{\sqrt{2}} \begin{bmatrix} 2 & 5 \\ 0 & 1 \end{bmatrix}$$

Solving $Qc = b$, we have $c = Q^T b = \frac{1}{\sqrt{2}} \begin{bmatrix} 1 & 1 \\ -1 & 1 \end{bmatrix} \begin{bmatrix} 1 \\ 2 \end{bmatrix} = \frac{1}{\sqrt{2}} \begin{bmatrix} 3 \\ 1 \end{bmatrix}$. Finally
we solve $Rz = c$ by back substitution and get $z_2 = 1$ and $z_1 = -1$.

Exercise 3.2.7 Use a QR decomposition to solve the linear system

$$\begin{bmatrix} 2 & 3 \\ 5 & 7 \end{bmatrix} \begin{bmatrix} x_1 \\ x_2 \end{bmatrix} = \begin{bmatrix} 12 \\ 29 \end{bmatrix} \qquad \square$$

We now turn our attention to n-by-n matrices. A *plane rotator* is a matrix of
the form

$$Q = \begin{bmatrix} 1 & & & & & & & & & \\ & 1 & & & & & & & & \\ & & \ddots & & & & & & & \\ & & & 1 & & & & & & \\ & & & & c & & -s & & & \\ & & & & & 1 & & & & \\ & & & & & & \ddots & & & \\ & & & & & & & 1 & & \\ & & & & s & & c & & & \\ & & & & & & & & 1 & \\ & & & & & & & & & \ddots \\ & & & & & & & & & & 1 \end{bmatrix} \begin{matrix} \\ \\ \\ \\ \leftarrow i \\ \\ \\ \\ \leftarrow j \\ \\ \\ \end{matrix} \qquad \begin{matrix} c = \cos\theta \\ s = \sin\theta \end{matrix}$$

$$\underset{i}{\uparrow} \qquad\qquad \underset{j}{\uparrow}$$

$$(3.2.7)$$

All the entries that have not been filled in are zeros. Thus a plane rotator looks like
an identity matrix, except that one pair of rows and columns contains a rotator.
Plane rotators are used extensively in matrix computations and are called *Givens
transformations* or *Givens rotations* in some contexts and *Jacobi transformations*
in others. We will usually drop the adjective "plane" and refer simply to *rotators*.

Exercise 3.2.8 Prove that every plane rotator is orthogonal and has determinant 1. $\square$

Exercise 3.2.9 Let Q be the plane rotator (3.2.7). Show that the transformations $x \rightarrow$
Qx and $x \rightarrow Q^T x$ alter only the ith and jth entries of x and that the effect
on these entries is the same as that of the 2-by-2 rotators $Q = \begin{bmatrix} c & -s \\ s & c \end{bmatrix}$ and
$Q^T = \begin{bmatrix} c & s \\ -s & c \end{bmatrix}$ on the vector $\begin{bmatrix} x_i \\ x_j \end{bmatrix}$. $\square$

From Exercise 3.2.9 and (3.2.5) we see that we can transform any vector x to one whose jth entry is zero by applying the plane rotator Q^T, where

$$c = \frac{x_i}{\sqrt{x_i^2 + x_j^2}} \qquad s = \frac{x_j}{\sqrt{x_i^2 + x_j^2}} \qquad (3.2.8)$$

(If $x_i = x_j = 0$, take $c = 1$ and $s = 0$.)

It is also important to note the effect of a rotator on a matrix. Let $A \in \mathbb{R}^{n \times m}$, and consider the transformations $A \to QA$ and $A \to Q^TA$, where Q is as in (3.2.7). It follows easily from Exercise 3.2.9 that these transformations alter only the ith and jth rows of A. Transposing these results, we see that for $B \in \mathbb{R}^{m \times n}$, the transformations $B \to BQ$ and $B \to BQ^T$ alter only the ith and jth columns of B.

Exercise 3.2.10

    (a) Show that the ith and jth rows of QA are linear combinations of the ith and jth rows of A.

    (b) Show that the ith and jth columns of BQ are linear combinations of the ith and jth columns of B.   □

The geometric interpretation of the action of a plane rotator is clear. All vectors lying in the $x_i x_j$ plane are rotated through an angle θ. All vectors orthogonal to the $x_i x_j$ plane are left fixed. A typical vector x is neither in the $x_i x_j$ plane nor orthogonal to it but can be expressed (uniquely) as a sum $x = p + p^\perp$, where p is in the $x_i x_j$ plane and $p^\perp$ is orthogonal to it. The plane rotator rotates p through the angle θ and leaves $p^\perp$ fixed.

THEOREM 3.2.9   Let $A \in \mathbb{R}^{n \times n}$. Then there exists an orthogonal matrix Q and an upper-triangular matrix R such that $A = QR$.

Proof   We will sketch a proof in which Q is taken to be a product of rotators. A more detailed proof using reflectors will be given later. Let Q_{21} be a rotator acting in the $x_1 x_2$ plane, such that Q_{21}^T makes the transformation

$$\begin{bmatrix} a_{11} \\ a_{21} \\ a_{31} \\ a_{41} \\ \vdots \\ a_{n1} \end{bmatrix} \to \begin{bmatrix} * \\ 0 \\ a_{31} \\ a_{41} \\ \vdots \\ a_{n1} \end{bmatrix}$$

Then $Q_{21}^T A$ has a zero in the $(2, 1)$ position. Similarly we can find a plane rotator Q_{31}, acting in the $x_1 x_3$ plane, such that $Q_{31}^T(Q_{21}^T A)$ has a zero in the $(3, 1)$ position. This rotator does not disturb the zero in the $(2, 1)$ position because Q_{31}^T leaves the second row of $Q_{21}^T A$ unchanged. Continuing in this manner, we create rotators $Q_{41}, Q_{51}, \ldots, Q_{n1}$ such that $Q_{n1}^T Q_{n-1,1}^T \cdots Q_{21}^T A$ has zeros in the entire first column, except for the $(1, 1)$ position.

Now we go to work on the second column. Let Q_{32} be a plane rotator acting in the x_2x_3 plane such that the $(3, 2)$ entry of $Q_{32}^T(Q_{n1}^T \cdots Q_{21}^T A)$ is zero. This rotator does not disturb the zeros that were created previously in the first column. (Why?) Let $Q_{42}, Q_{52}, \ldots, Q_{n2}$ be rotators such that $Q_{n2}^T \cdots Q_{32}^T Q_{n1}^T \cdots Q_{21}^T A$ has zeros in columns 1 and 2 below the main diagonal.

Next we take care of the third column, fourth column, and so on. In all we create rotators $Q_{21}, Q_{31}, \ldots, Q_{n1}, Q_{32}, \ldots, Q_{n,n-1}$ such that

$$R = Q_{n,n-1}^T Q_{n,n-2}^T \cdots Q_{21}^T A$$

is upper triangular. Let $Q = Q_{21}Q_{31}\cdots Q_{n,n-1}$. Then Q, being a product of orthogonal matrices, is itself orthogonal, and $R = Q^T A$; that is, $A = QR$. ☐

Exercise 3.2.11 A mathematically precise proof of Theorem 3.2.9 would have used induction on n. Sketch an inductive proof of Theorem 3.2.9. ☐

The proof of Theorem 3.2.9 is constructive; it gives us a recipe for calculating Q and R. Let us see how much arithmetic this construction requires. First consider the cost of multiplying a simple rotator by a vector:

$$\begin{bmatrix} c & s \\ -s & c \end{bmatrix}\begin{bmatrix} x_1 \\ x_2 \end{bmatrix} = \begin{bmatrix} cx_1 + sx_2 \\ -sx_1 + cx_2 \end{bmatrix} \tag{3.2.10}$$

Clearly four multiplications and two additions are required. Now what does it cost to convert A to $Q_{21}^T A$? Q_{21}^T operates only on rows 1 and 2, and within those rows it converts $\begin{bmatrix} a_{1j} \\ a_{2j} \end{bmatrix}$ to $\begin{bmatrix} ca_{1j} + sa_{2j} \\ -sa_{1j} + ca_{2j} \end{bmatrix}$, $j = 1, \ldots, n$, as in (3.2.10). The total cost is $4n$ multiplications and $2n$ additions. The cost of determining c and s is 2 multiplications, one square root, and two divisions [cf., (3.2.8) with x_i and x_j replaced by a_{11} and a_{21}, respectively], which is negligible in comparison with $4n$. The cost of applying each of $Q_{31}^T, Q_{41}^T, \ldots, Q_{n1}^T$ is also $4n$ multiplications and $2n$ additions. Since there are $n - 1$ of these rotators in all, the total cost of creating zeros in the first column of A is about $4n^2$ multiplications and $2n^2$ additions.

The cost of creating zeros in the second column is slightly less because there is no need to operate on the zeros in the first column and there is one less rotator. Indeed the task of creating zeros in the second column is the same as that of creating zeros in the first column, except that the operations are performed on the $(n - 1)$-by-$(n - 1)$ submatrix obtained by deleting the first row and first column. Thus the cost is about $4(n - 1)^2$ multiplications and $2(n - 1)^2$ additions. Similarly the task of creating zeros in the third column is about $4(n - 2)^2$ multiplications and $2(n - 2)^2$ additions, and so on.

Thus the total cost of transforming A to upper-triangular form is about $4n^2 + 4(n - 1)^2 + 4(n - 2)^2 + \cdots \approx 4n^3/3$ multiplications and $2n^3/3$ additions. This is roughly three times as expensive as an LU decomposition, which requires $n^3/3$ multiplications and $n^3/3$ additions. Not included here is the cost of calculating Q by accumulating the rotators Q_{ij}, but then this is not usually necessary. For example, if one wishes to solve the linear system $Ax = b$ by a QR decomposition, the computation $c = Q^T b$ can be carried out by applying the rotators $Q_{21}^T, Q_{31}^T, \ldots, Q_{n,n-1}^T$ to b successively: $c = Q_{n,n-1}^T \cdots Q_{21}^T b$.

Exercise 3.2.12   Show that the computation $c = Q_{n,n-1}^T \cdots Q_{31}^T Q_{21}^T b$ requires about $2n^2$ multiplications and n^2 additions. $\qquad\square$

As we will soon see, the *QR* decomposition can be carried out using reflectors with only $2n^3/3$ multiplications and $2n^3/3$ additions; that is, $2n^3/3$ flops. This is still twice as expensive as an *LU* decomposition.

It turns out that there is a clever way to implement rotators so that the number of multiplications required is halved. Thus one can compute with rotators at roughly the same cost as with reflectors. Rotators implemented in this special, somewhat complicated, way are called *fast Givens rotations*, *fast scaled rotators*, or simply *fast rotators*. Fast rotators are explained in Appendix A.

Reflectors

We begin with the case $n = 2$, just as we did for rotators. Let $\mathcal{L}$ be any line in $\mathbb{R}^2$ that passes through the origin. The operator that reflects each vector in $\mathbb{R}^2$ through the line $\mathcal{L}$ is a linear transformation, so it can be represented by a matrix. Let us determine that matrix. Let v be a nonzero vector lying in $\mathcal{L}$. Then every vector that lies in $\mathcal{L}$ is a multiple of v. Let u be a nonzero vector orthogonal to $\mathcal{L}$. Then $\{u, v\}$ is a basis of $\mathbb{R}^2$, so every $x \in \mathbb{R}^2$ can be expressed as a linear combination of u and v: $x = \alpha u + \beta v$. The reflection of x through $\mathcal{L}$ is $-\alpha u + \beta v$ (Figure 3.3), so the matrix Q of the reflection must satisfy $Q(\alpha u + \beta v) = -\alpha u + \beta v$ for all α and β. For this it is necessary and sufficient that

$$Qu = -u \qquad \text{and} \qquad Qv = v$$

Without loss of generality we can assume that u was chosen to be a unit vector: $\|u\|_2 = 1$.

Consider the matrix $P = uu^T \in \mathbb{R}^{2 \times 2}$. P does not have the desired properties; indeed $Pu = (uu^T)u = u(u^Tu) = u\|u\|_2^2 = u$, and $Pv = (uu^T)v = u(u^Tv) = u(u, v) = 0$ because u and v are orthogonal. With a little thought we can produce a Q with the desired properties by combining P and the identity matrix. Indeed

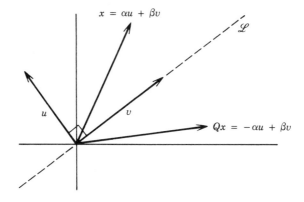

Figure 3.3   Reflection through a line.

let $Q = I - 2P$. Then $Qu = u - 2Pu = -u$ and $Qv = v - 2Pv = v$. To summarize: The matrix $Q \in \mathbb{R}^{2 \times 2}$ that reflects vectors through the line $\mathscr{L}$ is given by $Q = I - 2uu^T$, where u is a unit vector orthogonal to the line.

Now we are ready to consider general n.

THEOREM 3.2.11   Let $u \in \mathbb{R}^n$ such that $\|u\|_2 = 1$, and define $P \in \mathbb{R}^{n \times n}$ by $P = uu^T$. Then:[§]

(a) $Pu = u$.   **(c)** $P^2 = P$.

(b) $Pv = 0$ if $(u, v) = 0$.   **(d)** $P^T = P$.

Exercise 3.2.13   Prove Theorem 3.2.11.   □

THEOREM 3.2.12   Let $u \in \mathbb{R}^n$ be such that $\|u\|_2 = 1$, and define $Q \in \mathbb{R}^{n \times n}$ by $Q = I - 2uu^T = I - 2P$. Then:

(a) $Qu = -u$.

(b) $Qv = v$ if $(u, v) = 0$.

(c) $Q = Q^T$ (Q is symmetric).

(d) $Q^T = Q^{-1}$ (Q is orthogonal).

(e) $Q^{-1} = Q$ (Q is an involution).

Exercise 3.2.14   Prove Theorem 3.2.12. [Notice that any matrix that satisfies any two of the properties (c), (d), and (e) automatically satisfies the third.]   □

Part (a) implies that $Qx = -x$ for any x that is a multiple of u.

Matrices Q of the form of Theorem 3.2.12 are called *reflectors* or *Householder transformations*, after A. S. Householder, who first used them in matrix computations. The set $H = \{v \in \mathbb{R}^n \,|\, (u, v) = 0\}$ is an $(n - 1)$-dimensional subspace of $\mathbb{R}^n$ known as a *hyperplane*. The matrix Q maps each $x \in \mathbb{R}^n$ to its reflection through the hyperplane H. This can be visualized by thinking of the case $n = 3$, in which H is just an ordinary plane through the origin.

In Theorems 3.2.11 and 3.2.12, u was taken to be a unit vector. This is a convenient choice for the statements and proofs of those theorems, but in computation it is usually more convenient not to normalize the vector. The following theorem makes it possible to skip the normalization.

THEOREM 3.2.13   Let u be a nonzero vector in $\mathbb{R}^n$, and define $\gamma = 2/\|u\|_2^2$ and $Q = I - \gamma uu^T$. Then Q is a reflector.

Proof   Let $\hat{u} = u/\|u\|_2$. Then $\|\hat{u}\|_2 = 1$, and you can easily verify that $Q = I - 2\hat{u}\hat{u}^T$.   □

[§] A matrix satisfying $P^2 = P$ is called a *projector* or *idempotent*. A projector that is also symmetric ($P^T = P$) is called an *orthoprojector* or (more frequently) *orthogonal projection*. We prefer to avoid the latter term because orthoprojectors are not orthogonal matrices. The matrix $P = uu^T$ has rank 1, since its range consists of multiples of u. Thus the properties of P can be summarized by saying that P is a rank-1 orthoprojector.

THEOREM 3.2.14 Let $x, y \in \mathbb{R}^n$ such that $x \neq y$ but $\| x \|_2 = \| y \|_2$. Then there is a unique reflector Q such that $Qx = y$.

Proof We will not prove the uniqueness part because it is not important to our development (but see Exercise 3.2.24). In order to establish the existence of Q, we must find u such that $(I - \gamma u u^T)x = y$, where $\gamma = 2/\| u \|_2^2$. To see how to proceed consider the case $n = 2$, which is depicted in Figure 3.4a. Let $\mathcal{L}$ denote the line that bisects the angle between x and y. The reflection of x through this (and only this) line is y. Thus we require a vector u that is orthogonal to $\mathcal{L}$. It appears that $u = x - y$, or any multiple thereof, is the right choice.

Let $u = (x - y)/\| x - y \|_2$ and $Q = I - 2uu^T$. To prove that $Qx = y$, we first decompose x into a sum

$$x = \tfrac{1}{2}(x - y) + \tfrac{1}{2}(x + y)$$

Clearly $\tfrac{1}{2}(x - y)$ is a multiple of u. Therefore by Theorem 3.2.12, part (a), $Q(\tfrac{1}{2}(x - y)) = -\tfrac{1}{2}(x - y)$. Figure 3.4b suggests that $\tfrac{1}{2}(x + y)$ is orthogonal to u. To see that this is so, we simply compute the inner product: $(\tfrac{1}{2}(x + y), u) = \tfrac{1}{2}\| x - y \|_2^{-1}(x + y, x - y)$. Now $(x + y, x - y) = (x, x) + (y, x) - (x, y) - (y, y) = \| x \|_2^2 - \| y \|_2^2 = 0$, because $\| x \|_2 = \| y \|_2$. It now follows from Theorem

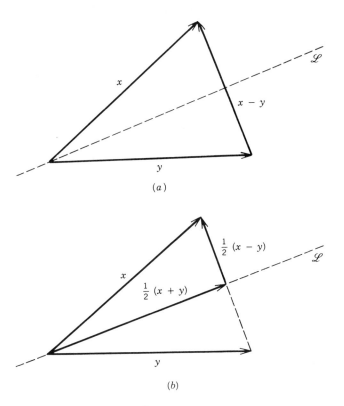

(a)

(b)

Figure 3.4

3.2.12, part (b), that $Q(\frac{1}{2}(x + y)) = \frac{1}{2}(x + y)$. Finally,

$$Qx = Q(\frac{1}{2}(x - y)) + Q(\frac{1}{2}(x + y))$$
$$= -\frac{1}{2}(x - y) + \frac{1}{2}(x + y) = y \qquad \square$$

It follows easily from this theorem that reflectors can be used to create zeros in vectors and matrices: Let $x = [x_1, x_2, \ldots, x_n]^T$ be a vector such that not all of the entries $x_2, x_3, \ldots, x_n$ are zero, and suppose we want to transform x to a vector $y = [y_1, y_2, \ldots, y_n]^T$ such that $y_2 = y_3 = \cdots = y_n = 0$. Let $\sigma = \pm\|x\|_2$, and define $y = [-\sigma, 0, \ldots, 0]^T$. Then obviously $\|x\|_2 = \|y\|_2$ and $y \neq x$, so by Theorem 3.2.14 there exists a reflector Q such that $Qx = y$. From the proof of Theorem 3.2.14 we know that $Q = I - \gamma u u^T$, where $u = x - y = [x_1 + \sigma, x_2, x_3, \ldots, x_n]^T$ and $\gamma = 2/\|u\|_2^2$. It is a simple matter to calculate u and γ. We specified that $\sigma = \pm\|x\|_2$, but we did not specify the sign. In theory either choice works, but in practice σ should be chosen so that its sign is the same as that of x_1. This ensures that cancellation cannot occur in the calculation of $x_1 + \sigma$. The calculation of γ is even easier than it looks. Notice that

$$\|u\|_2^2 = (x_1 + \sigma)^2 + x_2^2 + \cdots + x_n^2$$
$$= x_1^2 + 2x_1\sigma + \sigma^2 + x_2^2 + \cdots + x_n^2$$
$$= 2x_1\sigma + \sigma^2 + \|x\|_2^2$$

But $\|x\|_2^2 = \sigma^2$, regardless of which sign is chosen for σ, so $\|u\|_2^2 = 2x_1\sigma + 2\sigma^2 = 2\sigma(x_1 + \sigma) = 2\sigma u_1$. Thus $\gamma = 1/\sigma u_1$.

Before we summarize this computation, let us point out one other potential danger. In order to calculate σ, we must calculate $\|x\|_2 = (x_1^2 + x_2^2 + \cdots + x_n^2)^{1/2}$. Since squaring doubles the exponent of a number, an overflow can occur if some of the entries of x are very large and an underflow can occur if some of the entries are very small. An overflow will stop the computation. An underflow may or may not stop the computation, depending on which compiler options are in effect. If the computation is not stopped, the underflow will be set to zero, which can sometimes be dangerous.

Example 3.2.15 Consider the calculation of $\|x\|_2$, where $x = [10^{-49}, 10^{-50}, 10^{-50}, \ldots, 10^{-50}] \in \mathbb{R}^{11}$, on a machine that sets to zero any number whose absolute value is less than 10^{-99}. Now $x_1^2 = 10^{-98}$ and $x_2^2 = x_3^2 = \cdots = x_{11}^2 = 10^{-100}$, each of which is set to zero by the machine. Thus the computed norm is $\sqrt{x_1^2} = 10^{-49}$. This is in error by almost 5 percent, as the true value of the norm is $\sqrt{1.1 \times 10^{-98}} \approx 1.0488 \times 10^{-49}$.

Problems with overflows and underflows in the calculation of $\|x\|_2$ can be avoided by the following simple procedure: Let $m = \|x\|_\infty = \max_{1 \leq i \leq n}|x_i|$. If $m = 0$, then $\|x\|_2 = 0$. Otherwise let $\hat{x} = (1/m)x$. Then $\|x\|_2 = m\|\hat{x}\|_2 = m\sqrt{\hat{x}_1^2 + \cdots + \hat{x}_n^2}$. This scaling procedure eliminates the possibility of overflow because $|\hat{x}_i| \leq 1$ for all i. Underflows are still possible, but they occur only when some terms are so much smaller than the largest term ($=1$) that the error incurred by ignoring them is well below the machine's unit roundoff. Thus these underflows are harmless and can safely be set to zero.

Exercise 3.2.15 Use the scaling procedure outlined above to calculate the norm of the vector x of Example 3.2.15. □

Exercise 3.2.16 The danger of underflows and overflows exists not only in the calculation of reflectors, but in the calculation of rotators as well. Modify the computation (3.2.8) so that the danger of overflows and harmful underflows is eliminated. □

Now let us summarize the procedure for calculating Q such that $Qx = y$, where $y = [-\sigma \ 0 \ \cdots \ 0]^T$, incorporating the scaling procedure to calculate $\| x \|_2$. In this case it is not necessary to multiply the scaling factor m back into x or u afterwards because, as you can easily check, x and its scaled multiple $\hat{x}$ both determine the same matrix Q. Thus we can simply work with $\hat{x}$ instead of x. Here is the procedure: Calculate

$$m = \max_{1 \le i \le n} |x_i| \qquad\qquad \hat{x}_i = x_i/m \qquad i = 1, \ldots, n$$

$$\hat{\sigma} = \text{sign}(\hat{x}_1) \sqrt{\hat{x}_1^2 + \cdots + \hat{x}_n^2} \qquad u_1 = \hat{x}_1 + \hat{\sigma}$$

$$u_i = \hat{x}_i \qquad i = 2, \ldots, n \qquad\qquad \gamma = \frac{1}{\hat{\sigma} u_1} \qquad \sigma = \hat{\sigma} m$$

The procedure could be implemented as follows.

Given $x \in \mathbb{R}^n$, this algorithm calculates σ, γ, and u such that $Q = I - \gamma u u^T$ is a reflector for which $Qx = [-\sigma \ 0 \ \cdots \ 0]^T$. u is stored over x. If $x = 0$, γ is set to zero, giving $Q = I$.

$$
\begin{aligned}
&m \leftarrow \max_{1 \le i \le n} |x_i| \\
&\text{if } (m = 0) \\
&\text{then} \\
&\quad \left[\gamma \leftarrow 0 \right. \\
&\text{else} \qquad\qquad\qquad\qquad\qquad\qquad\qquad\qquad\qquad (3.2.16) \\
&\quad \left[\begin{array}{l} \text{for } i = 1, \ldots, n \\ \quad \left[x_i \leftarrow x_i/m \right. \\ \sigma \leftarrow \sqrt{x_1^2 + x_2^2 + \cdots + x_n^2} \\ \text{if } (x_1 < 0) \ \sigma \leftarrow -\sigma \\ x_1 \leftarrow x_1 + \sigma \\ \gamma \leftarrow 1/\sigma x_1 \\ \sigma \leftarrow \sigma m \end{array} \right.
\end{aligned}
$$

You can readily see that cancellation cannot occur in this algorithm: The sum $x_1^2 + x_2^2 + \cdots + x_n^2$ involves positive numbers only, and we have arranged the computation so that cancellation does not occur in the sum $x_1 + \sigma$. Therefore the computed σ, γ, and u are accurate.

The algorithm requires $n + 2$ multiplications, $n + 1$ divisions, n additions, n comparisons (mostly to determine $\max|x_i|$), and one square root. Thus the cost is $O(n)$.

The algorithm does not calculate Q explicitly, but then Q is seldom needed; it suffices to save γ and u. Suppose, for example, we need the product Qa, where $a = [a_1 \cdots a_n]^T \in \mathbb{R}^n$. Then $Qa = (I - \gamma u u^T)a = a - \gamma u(u^T a)$. Recognizing $u^T a$ as a scalar, we see that $Qa = a - (\gamma u^T a)u$. Thus Qa is obtained by subtracting a scalar multiple of u from a. This can be done by the following algorithm.

Algorithm to calculate Qa and store it over a, where $a \in \mathbb{R}^n$ and $Q = (I - \gamma u u^T)$ is a reflector

$$
\begin{aligned}
&\tau \leftarrow \sum_{i=1}^{n} u_i a_i \\
&\tau \leftarrow \gamma\tau \\
&\text{for } i = 1, \dots, n \\
&\quad \lfloor\, a_i \leftarrow a_i - \tau u_i
\end{aligned}
\tag{3.2.17}
$$

The algorithm requires $2n + 1$ multiplications and $2n - 1$ additions, that is, about $2n$ flops. This is dramatically less expensive than assembling Q explicitly and multiplying Q by a in the conventional manner. You will recall that that computation normally requires n^2 flops.

Exercise 3.2.17

(a) Find a reflector Q that maps the vector $x = [3\ 4\ 1\ 3\ 1]^T$ to a vector of the form $[-\sigma\ 0\ 0\ 0\ 0]^T$. You need not rescale x. Write Q two ways: (i) in the form $I - \gamma u u^T$ and (ii) as a single (completely assembled) matrix.

(b) Let $a = [0\ 2\ 1\ -1\ 0]^T$. Calculate Qa two different ways: (i) the efficient way, using $(I - \gamma u u^T)$, and (ii) using the assembled matrix Q. □

Now suppose we wish to calculate QA, where A is an n-by-m matrix. Letting $a_1, \dots, a_m$ denote the columns of A, we have $QA = Q[a_1\ a_2\ \cdots\ a_m] = [Qa_1\ Qa_2\ \cdots\ Qa_m]$, so the whole operation amounts to multiplying Q by each of the vectors $a_1, \dots, a_m$. The total flop count using algorithm (3.2.17) is $2nm$. This compares with $n^2 m$ flops if QA is calculated in the usual manner.

Theorem 3.2.9 states that every $A \in \mathbb{R}^{n \times n}$ can be expressed as a product $A = QR$, where Q is orthogonal and R is upper triangular. We used rotators to prove Theorem 3.2.9, but promised a second proof based on reflectors. We are now ready for that proof.

Proof of Theorem 3.2.9 Using Reflectors   The proof is by induction on n. When $n = 1$, take $Q = [1]$ and $R = [a_{11}]$ to get $A = QR$. Now we will assume the result holds for $(n - 1)$-by-$(n - 1)$ matrices and show that it follows for n-by-n matrices. Let $Q_1 \in \mathbb{R}^{n \times n}$ be a reflector that creates zeros in the first column of A:

$$Q_1 \begin{bmatrix} a_{11} \\ a_{21} \\ \vdots \\ a_{n1} \end{bmatrix} = \begin{bmatrix} -\sigma_1 \\ 0 \\ \vdots \\ 0 \end{bmatrix} \tag{3.2.18}$$

Recalling that Q_1 is symmetric, we see that $Q_1^T A$ has the form

$$Q_1^T A = Q_1 A = \begin{bmatrix} -\sigma_1 & \hat{a}_{12} \cdots \hat{a}_{1n} \\ 0 & \\ \vdots & \hat{A}_2 \\ 0 & \end{bmatrix} \tag{3.2.19}$$

By the induction hypothesis $\hat{A}_2$ has a QR decomposition: $\hat{A}_2 = \hat{Q}_2 \hat{R}_2$, where $\hat{Q}_2$ is orthogonal and $\hat{R}_2$ is upper triangular. Define $Q_2 \in \mathbb{R}^{n \times n}$ by

$$Q_2 = \begin{bmatrix} 1 & 0 \cdots 0 \\ 0 & \\ \vdots & \hat{Q}_2 \\ 0 & \end{bmatrix}$$

Then obviously Q_2 is orthogonal, and

$$Q_2^T Q_1^T A = \begin{bmatrix} 1 & 0 \cdots 0 \\ 0 & \\ \vdots & \hat{Q}_2^T \\ 0 & \end{bmatrix} \begin{bmatrix} -\sigma_1 & \hat{a}_{12} \cdots \hat{a}_{1n} \\ 0 & \\ \vdots & \hat{A}_2 \\ 0 & \end{bmatrix} = \begin{bmatrix} -\sigma_1 & \hat{a}_{12} \cdots \hat{a}_{1n} \\ 0 & \\ \vdots & \hat{R}_2 \\ 0 & \end{bmatrix} \tag{3.2.20}$$

This matrix is upper triangular; let us call it R. Thus $Q_2^T Q_1^T A = R$. Let $Q = Q_1 Q_2$. Then Q is orthogonal and $Q^T A = R$; therefore $A = QR$.   □

This proof can be turned into an algorithm for constructing Q and R. Step 1 calculates a reflector $Q_1 = (I - \gamma_1 u^{(1)} u^{(1)T})$ to perform the task (3.2.18) and then uses Q_1 to make the transformation (3.2.19). In practice we do not form Q_1 explicitly; we just save $-\sigma_1$, γ_1, and $u^{(1)}$, which were calculated by (3.2.16) with x taken to be $[a_{11} \; a_{21} \; \cdots \; a_{n1}]^T$. There is no need to transform the first column of A to the form $[-\sigma_1 \; 0 \; \cdots \; 0]^T$ as shown in (3.2.19); the zeros need not be stored, and $-\sigma_1$ can be stored elsewhere. Instead the first column of A can be used to store $u^{(1)}$, which is just $[a_{11} + \sigma_1, a_{21}, \ldots, a_{n1}]^T$ anyway. Columns 2 through n can be transformed as indicated in (3.2.19). This is done by multiplying each column by Q_1, using (3.2.17), and storing the result in that column. Each of these multiplications can be done in $2n$ flops. Since $n - 1$ columns are processed, the total flop count is about $2n^2$. Since the cost of determining $-\sigma_1$, γ_1, and $u^{(1)}$ is $O(n)$, we ignore it.

The rest of the algorithm consists of reducing $\hat{A}_2$ to upper-triangular form. Comparing (3.2.19) with (3.2.20), we see that the first row and column remain unchanged after step 1, so they can be ignored. Step 2 is identical to step 1, except that it operates on $\hat{A}_2$ to convert it to the form

$$
\left[
\begin{array}{c|ccc}
-\sigma_2 & \hat{a}_{23} \cdots \hat{a}_{2n} \\
\hline
0 & \\
\vdots & & \hat{A}_3 \\
0 &
\end{array}
\right]
$$

In practice we store $-\sigma_2$ elsewhere and use the first column to store $u^{(2)}$, the vector associated with the reflector for step 2. The cost of step 2 is about $2(n-1)^2$ flops, since it is identical to step 1 except that it acts on an $(n-1)$-by-$(n-1)$ matrix. Step 3 is identical to the previous steps except that it acts on the smaller submatrix $\hat{A}_3$. The cost of the third step is about $(n-2)^2$ flops.

After $n-1$ steps, the matrix has been reduced to upper-triangular form. The array that originally held A now looks like

$$
\left[
\begin{array}{c|c|c|c|c|c}
 & r_{12} & r_{13} & \cdots & r_{1,n-1} & r_{1n} \\
 & & r_{23} & \cdots & r_{2,n-1} & r_{2n} \\
u^{(1)} & u^{(2)} & u^{(3)} & \cdots & & \vdots \\
 & & & & u^{(n-1)} & \begin{array}{c} r_{n-1,n} \\ \hline r_{nn} \end{array}
\end{array}
\right]
$$

The main diagonal elements $r_{11}, r_{22}, \ldots, r_{n-1,n-1}$ are precisely $-\sigma_1, -\sigma_2, \ldots, -\sigma_{n-1}$, and are stored elsewhere. For uniformity we define $-\sigma_n = r_{nn}$ and store $-\sigma_n$ in the same array with $-\sigma_1, \ldots, -\sigma_{n-1}$. Another array holds $\gamma_1, \gamma_2, \ldots, \gamma_{n-1}$. The information to construct Q is contained in $u^{(1)}, \ldots, u^{(n-1)}$ and $\gamma_1, \ldots, \gamma_{n-1}$. Clearly $R = Q_{n-1} \cdots Q_2 Q_1 A$, where $Q_1 = I - \gamma_1 u^{(1)} u^{(1)T}$,

$$
Q_2 = \left[
\begin{array}{c|ccc}
1 & 0 & \cdots & 0 \\
\hline
0 & \\
\vdots & & I - \gamma_2 u^{(2)} u^{(2)T} \\
0 &
\end{array}
\right]
$$

and in general

$$
Q_i = \left[
\begin{array}{c|c}
\overset{i-1}{I} & \overset{n-i+1}{0} \\
\hline
0 & I - \gamma_i u^{(i)} u^{(i)T}
\end{array}
\right]
\begin{array}{c}
i-1 \\
n-i+1
\end{array}
$$

Letting $Q = Q_1 Q_2 \cdots Q_{n-1}$, we have $Q^T = Q_{n-1}^T Q_{n-2}^T \cdots Q_1^T = Q_{n-1} \cdots Q_1$, $R = Q^T A$, and finally $A = QR$. As we shall see, there is no need to assemble Q explicitly. The algorithm can be summarized as follows:

Algorithm to calculate the QR decomposition of $A \in \mathbb{R}^{n \times n}$ using reflectors

$$
\begin{aligned}
&\text{for } k = 1, \ldots, n - 1 \\
&\quad\left[\begin{array}{l}
\text{Determine a reflector } Q_k = I - \gamma_k u^{(k)} u^{(k)T} \text{ such that} \\
Q_k[a_{kk} \ \cdots \ a_{nk}]^T = [-\sigma_k \ 0 \ \cdots \ 0]^T. \\
\text{Store } u^{(k)} \text{ over } [a_{kk} \ \cdots \ a_{nk}]^T, \text{ and save } \gamma_k \text{ and } -\sigma_k \text{ (see 3.2.16).} \\
\text{for } j = k + 1, \ldots, n \\
\quad\left[\begin{bmatrix} a_{kj} \\ \vdots \\ a_{nj} \end{bmatrix} \leftarrow Q_k \begin{bmatrix} a_{kj} \\ \vdots \\ a_{nj} \end{bmatrix} \quad \text{(see 3.2.17)}\right.
\end{array}\right. \\
&-\sigma_n \leftarrow a_{nn} \\
&\gamma_n \leftarrow a_{nn}
\end{aligned}
\tag{3.2.21}
$$

γ_n functions as a flag. If γ_n or any other γ_i is zero, A is singular. (Why?)

Recalling that the flop count for step one is $2n^2$, for step two $2(n - 1)^2$, and so on, we see that the total flop count for (3.2.21) is $2n^2 + 2(n - 1)^2 + \cdots \approx 2n^3/3$, twice the cost of an LU decomposition.

In Chapter 1 we observed that there are numerous ways to organize the operations in an LU decomposition. With a bit of ingenuity the same can be done with (3.2.21). A couple of possibilities are explored in Exercises 3.2.18 and 3.2.19. Which version is best depends on the computer and compiler being used.

Exercise 3.2.18   (a) Write out the j-loop of (3.2.21) in detail. This is column-oriented code. (b) Carefully reorganize the code from part (a) so that the data is accessed by rows instead of columns. You will need one extra vector for temporary storage of inner products.   □

Exercise 3.2.19   Write a version of (3.2.21) in which all updates of column k are delayed until step k.   □

Let us see how the QR decomposition, as calculated using reflectors, can be used to solve the system of linear equations $Ax = b$. This is admittedly an expensive way to solve a linear system, but it is a good illustration because the operations are similar to those used to solve the least-squares problem. Since $A = QR$, our task is to solve $QRx = b$. Letting $Rx = c$, we must first solve $Qc = b$ for c, then we can solve $Rx = c$ for x. First of all $c = Q^T b = Q_{n-1} \cdots Q_2 Q_1 b$, so we can apply the reflectors $Q_1, Q_2, \ldots, Q_{n-1}$ to b successively to obtain c. Naturally we use (3.2.17). Applying Q_1 costs $2n$ flops. Applying Q_2 costs only $2(n - 1)$ flops because it is really only an $(n - 1)$-by-$(n - 1)$ reflector; it does not touch the first entry of $Q_1 b$. Similarly Q_3 does not touch the first two entries of $Q_2 Q_1 b$, and the cost of applying it is $2(n - 2)$ flops. The total cost of calculating c is therefore $2n + 2(n - 1) + 2(n - 2) + \cdots \approx n^2$ flops. This is the same as it would have cost to calculate $c = Q^T b$ if Q had been available explicitly. Finally $Rx = c$ can be solved for x by back substitution at a cost of $n^2/2$ flops. Remember that the pivots are stored as $-\sigma_1, \ldots, -\sigma_n$. The algorithm can be summarized as follows:

Algorithm to solve $Ax = b$ for $A \in \mathbb{R}^{n \times n}$ using the QR decomposition calculated by (3.2.21), x stored over b

$$
\begin{aligned}
&\text{for } i = 1, \ldots, n \\
&\quad \Big[\text{ if } (\gamma_i = 0) \text{ set flag, exit (A is singular)} \\
&\text{for } i = 1, \ldots, n - 1 \\
&\quad \Big[\; [b_i \;\; \cdots \;\; b_n]^T \leftarrow (I - \gamma_i u^{(i)} u^{(i)T})[b_i \;\; \cdots \;\; b_n]^T \quad \text{(see 3.2.17)} \quad\quad (3.2.22) \\
&(c \text{ has been stored over } b. \text{ Now solve } Rx = c) \\
&\text{for } j = n, n - 1, \ldots, 1 \\
&\quad \left[\begin{array}{l} b_j \leftarrow -b_j/\sigma_j \\ \quad \text{for } i = 1, \ldots, j - 1 \text{ (skip when } j = 1) \\ \quad \big[b_i \leftarrow b_i - a_{ij} b_j \end{array} \right\} \quad \begin{array}{l} \text{back substitution} \\ \text{(column-oriented)} \end{array}
\end{aligned}
$$

Exercise 3.2.20   Write Fortran subroutines to solve the linear system $Ax = b$ via a QR decomposition. (In the next section you will find out how to make minor modifications in your routines so that they can solve the least-squares problem.)

$\square$

For nonsingular matrices the QR decomposition theorem can be strengthened to include the uniqueness of Q and R, a fact that is of considerable theoretical importance.

THEOREM 3.2.23   Let $A \in \mathbb{R}^{n \times n}$ be nonsingular. Then there exist unique $Q, R \in \mathbb{R}^{n \times n}$ such that Q is orthogonal, R is upper triangular with positive main-diagonal entries, and $A = QR$.

Proof   By Theorem 3.2.9 $A = \hat{Q}\hat{R}$, where $\hat{Q}$ is orthogonal and $\hat{R}$ is upper triangular, but $\hat{R}$ does not necessarily have positive main-diagonal entries. Since A is nonsingular, $\hat{R}$ must also be nonsingular, so its main-diagonal entries must be nonzero. Let D be the diagonal matrix given by

$$
d_{ii} = \begin{cases} 1 & \text{if } \hat{r}_{ii} > 0 \\ -1 & \text{if } \hat{r}_{ii} < 0 \end{cases}
$$

Then $D = D^T = D^{-1}$ is orthogonal. Let $Q = \hat{Q}D^{-1}$ and $R = D\hat{R}$. Then Q is orthogonal, R is upper triangular with $r_{ii} = d_{ii}\hat{r}_{ii} > 0$, and $A = QR$. This establishes existence.

Uniqueness can be established using the same ideas (see Exercise 3.2.21). For variety we will consider another proof, which is based on an interesting connection between the QR and Cholesky decompositions. Suppose $A = Q_1 R_1 = Q_2 R_2$, where Q_1 and Q_2 are orthogonal and R_1 and R_2 are upper triangular with positive main-diagonal entries. $A^T A$ is a positive definite matrix, and

$$
A^T A = R_1^T R_1 = R_2^T R_2
$$

since $Q_1^T Q_1 = Q_2^T Q_2 = I$. R_1^T is lower triangular with positive main-diagonal entries, so it is the Cholesky factor of $A^T A$. But R_2^T is also the Cholesky factor of $A^T A$. By uniqueness of the Cholesky factor, $R_1 = R_2$. Finally $Q_1 = AR_1^{-1} = AR_2^{-1} = Q_2$.

$\square$

Exercise 3.2.21 This leads to another uniqueness proof for Theorem 3.2.23. (a) Suppose $B \in \mathbb{R}^{n \times n}$ is both orthogonal and upper triangular. Prove that B must be a diagonal matrix whose main-diagonal entries are ± 1. (b) Suppose $Q_1 R_1 = Q_2 R_2$, where Q_1 and Q_2 are orthogonal and R_1 and R_2 are upper triangular and nonsingular. Show that there is a diagonal matrix D with main-diagonal entries ± 1, such that $R_2 = D R_1$ and $Q_1 = Q_2 D$. $\qquad\square$

Exercise 3.2.22

(a) Let $A = \begin{bmatrix} 1 & 2 \\ 1 & 3 \end{bmatrix}$. Find a reflector $\hat{Q}$ and an upper-triangular matrix $\hat{R}$ such that $\hat{A} = \hat{Q}\hat{R}$. Assemble $\hat{Q}$ and simplify it.

$$\left(\text{Solution}: \hat{Q} = \frac{1}{\sqrt{2}} \begin{bmatrix} -1 & -1 \\ -1 & 1 \end{bmatrix}, \quad \hat{R} = \frac{1}{\sqrt{2}} \begin{bmatrix} -2 & -5 \\ 0 & 1 \end{bmatrix}. \right)$$

(b) Compare your solution from part (a) with the QR decomposition of the same matrix (obtained by using a rotator) in Example 3.2.6. Find a diagonal matrix D whose main-diagonal entries are ± 1, such that $Q = \hat{Q}D$ and $\hat{R} = DR$. $\qquad\square$

Exercise 3.2.23 (a) Show that every 2-by-2 reflector has the form $\begin{bmatrix} c & s \\ s & -c \end{bmatrix}$, where $c^2 + s^2 = 1$. (b) How does the reflector $Q = \begin{bmatrix} c & s \\ s & -c \end{bmatrix}$ act on the unit vectors $\begin{bmatrix} 1 \\ 0 \end{bmatrix}$ and $\begin{bmatrix} 0 \\ 1 \end{bmatrix}$? Contrast the action of Q with that of the rotator $\tilde{Q} = \begin{bmatrix} c & -s \\ s & c \end{bmatrix}$. Draw a picture. By examining your picture, determine the line through which vectors are reflected by Q. Assuming $c > 0$ and $s > 0$, show that this line has slope $\tan(\theta/2)$, where $\theta = \arccos c = \arcsin s$. $\qquad\square$

Exercise 3.2.24 Show that if $Qx = y$, where $Q = I - \gamma u u^T \in \mathbb{R}^{n \times n}$, then u must be a multiple of $x - y$. (This establishes the uniqueness part of Theorem 3.2.14.) $\qquad\square$

Stability of Computations with Reflectors and Rotators

The numerical properties of both rotators and reflectors are excellent. A detailed analysis is carried out by Wilkinson [AEP], pages 126 to 162. We will content ourselves with a brief, informal summary. Suppose we wish to convert A to QA, where Q is either a rotator determined by (3.2.8) or a reflector determined by (3.2.16). In the case of the reflector, (3.2.17) would be used to calculate QA. In both cases Wilkinson has shown that

$$\text{fl}\,(QA) = Q(A + E)$$

where $\| E \|_2 / \| A \|_2$ is extremely small. In other words, the computed result is the

exact result of applying Q to a matrix that is very close to A. That is, the computation of QA is stable (cf., Section 2.5).

The stability is preserved under the repeated application of rotators and/or
reflectors. Consider for example the computation of $Q_2 Q_1 A$. First of all, $fl(Q_1 A) = Q_1(A + E_1)$. Thus $fl(Q_2 Q_1 A) = fl(Q_2(Q_1(A + E_1))) = Q_2(Q_1(A + E_1) + E_2) = Q_2 Q_1(A + E_1 + Q_1^T E_2) = Q_2 Q_1(A + E)$, where $E = E_1 + Q_1^T E_2$. Since $\|E\|_2 \le \|E_1\|_2 + \|E_2\|_2$, we can assert that $\|E\|_2 / \|A\|_2$ is very small.

Many of the algorithms in this book consist of repeated modifications of a
matrix by reflectors or rotators. The two QR decomposition algorithms discussed
in this section are of this type. The preceding remarks imply that all algorithms of
this type are stable.

The Complex Case

The following exercises develop the complex analogs of the results obtained in this
section. The *inner product* on $\mathbb{C}^n$ is defined by

$$(x, y) = \sum_{i=1}^{n} x_i \bar{y}_i \qquad \left\{ x = [x_1, \ldots, x_n]^T, y = [y_1, \ldots, y_n]^T \right\}$$

where the bar denotes complex conjugation. Clearly $(x, y) = y^* x = \overline{x^* y}$, where
y^* denotes the conjugate transpose of y.

Exercise 3.2.25   Show that the inner product on $\mathbb{C}^n$ satisfies the following properties:

(a) $(x, y) = \overline{(y, x)}$.

(b) $(\alpha_1 x_1 + \alpha_2 x_2, y) = \alpha_1(x_1, y) + \alpha_2(x_2, y)$.

(c) $(x, \alpha_1 y_1 + \alpha_2 y_2) = \bar{\alpha}_1(x, y_1) + \bar{\alpha}_2(x, y_2)$.

(d) (x, x) is real, $(x, x) \ge 0$, and $(x, x) = 0$ if and only if $x = 0$.

(e) $\sqrt{(x, x)} = \|x\|_2$.   ☐

The complex analogs of orthogonal matrices are unitary matrices. A matrix
$U \in \mathbb{C}^{n \times n}$ is *unitary* if $UU^* = I$, where U^* is the conjugate transpose of U.
Equivalent statements are $U^* = U^{-1}$ and $U^* U = I$. Note that the orthogonal
matrices are just the real unitary matrices.

Exercise 3.2.26   (a) Show that if U is unitary, then U^{-1} is unitary. (b) Show that if U_1
and U_2 are unitary, then $U_1 U_2$ is unitary.   ☐

Exercise 3.2.27   Show that if U is unitary then $|\det(U)| = 1$.   ☐

Exercise 3.2.28   Show that if $U \in \mathbb{C}^{n \times n}$ is unitary and $x, y \in \mathbb{C}^n$, then (a)
$(Ux, Uy) = (x, y)$, (b) $\|Ux\|_2 = \|x\|_2$, (c) $\|U\|_2 = \|U^{-1}\|_2 = \kappa_2(U) = 1$. ☐

Thus unitary matrices preserve the 2-norm and the inner product.

Exercise 3.2.29 (Complex Rotators)   Given a nonzero $\begin{bmatrix} a \\ b \end{bmatrix} \in \mathbb{C}^2$ define $U \in \mathbb{C}^{2 \times 2}$

by

$$U = \frac{1}{m} \begin{bmatrix} a & -\bar{b} \\ b & \bar{a} \end{bmatrix} \qquad m = \sqrt{|a|^2 + |b|^2}$$

Verify that:

(a) U is unitary.

(b) $\det(U) = 1$.

(c) $U^* \begin{bmatrix} a \\ b \end{bmatrix} = \begin{bmatrix} m \\ 0 \end{bmatrix}$.   □

The extension to complex plane rotators in $\mathbb{C}^{n \times n}$ is obvious.

Exercise 3.2.30 (Complex Reflectors)   Let $u \in \mathbb{C}^n$ such that $\| u \|_2 = 1$, and define $Q \in \mathbb{C}^{n \times n}$ by $Q = I - 2uu^*$. Verify that:

(a) $Qu = -u$.

(b) $Qv = v$ if $(u, v) = 0$.

(c) $Q = Q^*$ (Q is Hermitian).

(d) $Q^* = Q^{-1}$ (Q is unitary).

(e) $Q^{-1} = Q$ (Q is an involution).   □

Exercise 3.2.31   (a) Prove that if $x, y \in \mathbb{C}^n$, $x \neq y$, $\| x \|_2 = \| y \|_2$, and (x, y) is real, then there exists a complex reflector Q such that $Qx = y$. (b) Let $x \in \mathbb{C}^n$ be a nonzero vector. Express x_1 in polar form as $x_1 = re^{i\theta}$ (with $\theta = 0$ if $x_1 = 0$). Let $\sigma = \| x \|_2 e^{i\theta}$ and $y = [-\sigma \ 0 \ \cdots \ 0]^T$. Show that $\| x \|_2 = \| y \|_2$, $x \neq y$, and (x, y) is real. Write an algorithm that determines a reflector Q such that $Qx = y$. [This is the complex analogue of (3.2.16).]   □

Exercise 3.2.32   Show that for every $A \in \mathbb{C}^{n \times n}$ there exist unitary Q and upper-triangular R such that $A = QR$. Write a constructive proof using either rotators or reflectors.   □

If A is nonsingular, then R can be chosen so that the entries on its main diagonal are (real and) positive, in which case Q and R are unique. This is the complex analogue of Theorem 3.2.23.

THEOREM 3.2.24   Let $A \in \mathbb{C}^{n \times n}$ be nonsingular. Then there exist unique $Q, R \in \mathbb{C}^{n \times n}$ such that Q is unitary, R is upper triangular with real, positive main-diagonal entries, and $A = QR$.

Exercise 3.2.33   Prove Theorem 3.2.24.   □

Exercise 3.2.34   Let $u \in \mathbb{C}^n$ with $\|u\|_2 = 1$. Then $I - \gamma uu^*$ is unitary if $\gamma = 2$. Another choice of γ for which $I - \gamma uu^*$ is unitary is $\gamma = 0$. Find the set of all $\gamma \in \mathbb{C}$ for which $I - \gamma uu^*$ is unitary. $\quad\square$

3.3
SOLUTION OF THE LEAST-SQUARES PROBLEM

Consider an overdetermined system

$$Ax = b \qquad A \in \mathbb{R}^{n \times m}, b \in \mathbb{R}^n, n > m \qquad (3.3.1)$$

Our task is to find $x \in \mathbb{R}^m$ such that $\|r\|_2 = \|b - Ax\|_2$ is minimized. So far we do not know whether this problem has a solution. If it does have a solution, we do not know whether the solution is unique. These fundamental questions will be answered in this section. We will settle the existence question affirmatively by constructing a solution. The solution is unique if and only if A has full rank.

Let $Q \in \mathbb{R}^{n \times n}$ be any orthogonal matrix, and consider the transformed system

$$Q^T Ax = Q^T b \qquad (3.3.2)$$

obtained by multiplying both sides of the original system by Q^T. Let s be the residual for the transformed system. Then $s = Q^T b - Q^T Ax = Q^T(b - Ax) = Q^T r$. Since Q^T is orthogonal, $\|s\|_2 = \|r\|_2$. Thus, for a given x, the residual of the transformed system has the same norm as the residual of the original system. Therefore $x \in \mathbb{R}^m$ minimizes $\|r\|_2$ if and only if it minimizes $\|s\|_2$; that is, the two overdetermined systems have the same least-squares solution(s).

It seems reasonable to try to find an orthogonal $Q \in \mathbb{R}^{n \times n}$ for which the system (3.3.2) has a particularly simple form. In the previous section we learned that given a square matrix A, there exists an orthogonal Q such that $Q^T A = R$, where R is upper triangular. Thus the linear system $Ax = b$ can be transformed to $Q^T Ax = Q^T b$ ($Rx = c$), which can easily be solved by back substitution. The same approach works for the least-squares problem, but first we need a QR decomposition theorem for nonsquare matrices.

THEOREM 3.3.3   Let $A \in \mathbb{R}^{n \times m}, n > m$. Then there exist $Q \in \mathbb{R}^{n \times n}$ and $R \in \mathbb{R}^{n \times m}$, such that Q is orthogonal and $R = \begin{bmatrix} \hat{R} \\ 0 \end{bmatrix}$, where $\hat{R} \in \mathbb{R}^{m \times m}$ is upper triangular, and $A = QR$.

Proof   Let $\bar{A} = [A \ \tilde{A}] \in \mathbb{R}^{n \times n}$, where $\tilde{A} \in \mathbb{R}^{n \times (n-m)}$ is chosen arbitrarily. Then by Theorem 3.2.9 there exist $Q, \bar{R} \in \mathbb{R}^{n \times n}$ such that Q is orthogonal, $\bar{R}$ is upper triangular, and $\bar{A} = Q\bar{R}$. Now partitioning $\bar{R} : \bar{R} = [R \ \tilde{R}]$, where $R \in \mathbb{R}^{n \times m}$, we find that $A = QR$. Since $\bar{R}$ is upper triangular, R has the form $\begin{bmatrix} \hat{R} \\ 0 \end{bmatrix}$, where $\hat{R} \in \mathbb{R}^{m \times m}$ is upper triangular. $\quad\square$

Since this QR decomposition is obtained by retaining a portion of an n-by-n QR decomposition, any algorithm that computes a square QR decomposition can easily be modified to compute an n-by-m QR decomposition. In practice there is no need to choose a matrix $\tilde{A}$ with which to augment A, since that portion of the matrix is thrown away in the end anyway. One simply operates on A alone and quits upon running out of columns, that is, after m steps. For example, algorithm (3.2.21) can be adapted to n-by-m matrices by changing the upper limit on the loop indices to m and deleting the last two instructions, which are irrelevant in this case. Q is produced implicitly as a product of m reflectors: $Q = Q_1 Q_2 \cdots Q_m$.

Exercise 3.3.1 Show that the flop count for a QR decomposition of an n-by-m matrix using reflectors is approximately $nm^2 - m^3/3$. (Thus if $n \gg m$, the flop count is about nm^2.) □

The QR decomposition of Theorem 3.3.3 may or may not be useful for solving the least-squares problem, depending on whether or not A has full rank. Recall that the *rank* of a matrix is the number of linearly independent columns, which is the same as the number of linearly independent rows. The matrix $A \in \mathbb{R}^{n \times m}$ ($n \geq m$) has *full rank* if its rank is m; that is, if its columns are linearly independent. In Theorem 3.3.3 the equation $A = QR$ implies that rank$(A) \leq$ rank(R).[§] On the other hand the equation $R = Q^T A$ implies that rank$(R) \leq$ rank(A). Thus rank$(A) =$ rank(R), and R has full rank if and only if A does. Clearly rank$(R) =$ rank$(\hat{R})$, and $\hat{R}$ has full rank if and only if it is nonsingular. Thus $\hat{R}$ is nonsingular if and only if A has full rank.

Full-Rank Case

Now consider an overdetermined system $Ax = b$ for which A has full rank. Using the QR decomposition, we can transform the system to the form $Q^T A x = Q^T b$, or $Rx = c$, where $c = Q^T b$. Writing $c = \begin{bmatrix} \hat{c} \\ d \end{bmatrix}$, where $\hat{c} \in \mathbb{R}^m$, we can express the residual $s = c - Rx$ as

$$s = \begin{bmatrix} \hat{c} \\ d \end{bmatrix} - \begin{bmatrix} \hat{R} \\ 0 \end{bmatrix} x = \begin{bmatrix} \hat{c} - \hat{R}x \\ d \end{bmatrix}$$

Thus

$$\| s \|_2^2 = \sum_{i=1}^{n} |s_i|^2 = \| \hat{c} - \hat{R}x \|_2^2 + \| d \|_2^2 \qquad (3.3.4)$$

Since the term $\| d \|_2^2$ is independent of x, $\| s \|_2$ is minimized exactly when $\| \hat{c} - \hat{R}x \|_2^2$ is minimized. Obviously $\| \hat{c} - \hat{R}x \|_2^2 \geq 0$, with equality if and only if $\hat{R}x = \hat{c}$. Since A has full rank, $\hat{R}$ is nonsingular. Thus the system $\hat{R}x = \hat{c}$ has a unique

[§] We recall from elementary linear algebra that rank$(ST) \leq$ rank(T) for any matrices S and T for which the product ST is defined. You might like to prove this result yourself.

solution, which is then the unique minimizer of $\| s \|_2$. We summarize these findings as a theorem.

THEOREM 3.3.5 Let $A \in \mathbb{R}^{n \times m}$ and $b \in \mathbb{R}^n$, $n > m$, and suppose A has full rank. Then the least-squares problem for the overdetermined system $Ax = b$ has a unique solution, which can be found by solving the nonsingular system $\hat{R}x = \hat{c}$, where $\hat{c} \in \mathbb{R}^m$, $\begin{bmatrix} \hat{c} \\ d \end{bmatrix} = Q^T b$, and Q and $\hat{R}$ are as in Theorem 3.3.3.

Assuming that the QR decomposition is calculated by reflectors, Q is stored as a product of reflectors $Q_1 Q_2 \cdots Q_m$. Thus the computation $c = Q^T b = Q_m Q_{m-1} \cdots Q_1 b$ is accomplished by applying reflectors to b using (3.2.17). Recalling that each reflector does a bit less work than the previous one, we find that the total flop count for this step is about $2n + 2(n-1) + 2(n-2) + \cdots + 2(n-m+1) \approx 2nm - m^2$. The equation $\hat{R}x = \hat{c}$ can be solved by back substitution at a cost of $m^2/2$ flops. It is also advisable to calculate $\| d \|_2$ ($n - m$ flops), as this is the norm of the residual associated with the least-squares solution. [See (3.3.4).] Therefore $\| d \|_2$ is a measure of goodness of fit. Comparing these flop counts with the result of Exercise 3.3.1, we see that the most expensive step is the computation of the QR decomposition.

Example 3.3.6 We will use a QR decomposition to solve the least-squares problem for the overdetermined system

$$\begin{bmatrix} 3 & -2 \\ 0 & 3 \\ 4 & 4 \end{bmatrix} \begin{bmatrix} x_1 \\ x_2 \end{bmatrix} = \begin{bmatrix} 1 \\ 2 \\ 4 \end{bmatrix}$$

Clearly the coefficient matrix has full rank. The first step in the QR decomposition is to find a reflector Q_1 such that $Q_1 [3 \ 0 \ 4]^T = [-\sigma_1 \ 0 \ 0]^T$. $Q_1 = (I - \gamma_1 u^{(1)} u^{(1)T})$, where $\sigma_1 = +\sqrt{3^2 + 0^2 + 4^2} = 5$, $u^{(1)T} = [3 + \sigma_1 \ 0 \ 4] = [8 \ 0 \ 4]$ and $\gamma_1 = 1/\sigma_1(3 + \sigma_1) = 1/40$. Since

$$Q_1 \begin{bmatrix} -2 \\ 3 \\ 4 \end{bmatrix} = (I - \gamma_1 u^{(1)} u^{(1)T}) \begin{bmatrix} -2 \\ 3 \\ 4 \end{bmatrix} = \begin{bmatrix} -2 \\ 3 \\ 4 \end{bmatrix} - \left\{ \frac{1}{40} [8 \ 0 \ 4] \begin{bmatrix} -2 \\ 3 \\ 4 \end{bmatrix} \right\} \begin{bmatrix} 8 \\ 0 \\ 4 \end{bmatrix}$$

$$= \begin{bmatrix} -2 \\ 3 \\ 4 \end{bmatrix} - 0 \begin{bmatrix} 8 \\ 0 \\ 4 \end{bmatrix} = \begin{bmatrix} -2 \\ 3 \\ 4 \end{bmatrix}$$

$$Q_1 A = \begin{bmatrix} -5 & -2 \\ 0 & 3 \\ 0 & 4 \end{bmatrix}$$

The second and final step of the QR decomposition is to find a reflector $\hat{Q}_2 \in \mathbb{R}^{2 \times 2}$ such that $\hat{Q}_2 \begin{bmatrix} 3 \\ 4 \end{bmatrix} = \begin{bmatrix} -\sigma_2 \\ 0 \end{bmatrix}$. $\hat{Q}_2 = I - \sigma_2 u^{(2)} u^{(2)T}$, where $\sigma_2 = +\sqrt{3^2 + 4^2} =$

5, $u^{(2)} = [3 + \sigma_2 \; 4]^T = [8 \; 4]^T$, and $\gamma_2 = 1/\sigma_2(3 + \sigma_2) = 1/40$. Letting

$$Q_2 = \begin{bmatrix} 1 & 0 & 0 \\ \hline 0 & & \\ 0 & & \hat{Q}_2 \end{bmatrix}$$

we have

$$R = Q_2 Q_1 A = \begin{bmatrix} -5 & -2 \\ 0 & -5 \\ 0 & 0 \end{bmatrix}$$

and

$$\hat{R} = \begin{bmatrix} -5 & -2 \\ 0 & -5 \end{bmatrix}$$

Next we must compute $c = Q^T b = Q_2 Q_1 b$.

$$Q_1 b = (I - \gamma_1 u^{(1)} u^{(1)T}) b = \begin{bmatrix} 1 \\ 2 \\ 4 \end{bmatrix} - \frac{1}{40}(24) \begin{bmatrix} 8 \\ 0 \\ 4 \end{bmatrix} = \begin{bmatrix} -19/5 \\ 2 \\ 8/5 \end{bmatrix}$$

$$\hat{Q}_2 \begin{bmatrix} 2 \\ 8/5 \end{bmatrix} = (I - \gamma_2 u^{(2)} u^{(2)T}) \begin{bmatrix} 2 \\ 8/5 \end{bmatrix} = \begin{bmatrix} 2 \\ 8/5 \end{bmatrix} - \frac{1}{40}\left(\frac{112}{5}\right) \begin{bmatrix} 8 \\ 4 \end{bmatrix} = \begin{bmatrix} -62/25 \\ -16/25 \end{bmatrix}$$

Thus

$$c = Q_2 Q_1 b = \begin{bmatrix} -19/5 \\ -62/25 \\ -16/25 \end{bmatrix}$$

Letting $c = \begin{bmatrix} \hat{c} \\ d \end{bmatrix}$, $\hat{c} \in \mathbb{R}^2$, we have

$$\hat{c} = \begin{bmatrix} -19/5 \\ -62/25 \end{bmatrix} \qquad d = [-16/25]$$

The least-squares solution is found by solving $\hat{R}x = \hat{c}$ by back substitution:

$$\begin{bmatrix} -5 & -2 \\ 0 & -5 \end{bmatrix} \begin{bmatrix} x_1 \\ x_2 \end{bmatrix} = \begin{bmatrix} -19/25 \\ -62/25 \end{bmatrix}$$

Thus $x_2 = 62/125$ and $x_1 = [-19/5 - (-2)x_2]/(-5) = 351/625$.

$$x = \begin{bmatrix} 351/625 \\ 62/125 \end{bmatrix}$$

Finally $\| d \|_2 = 16/25$, which shows that the fit is not particularly good.

Exercise 3.3.2   As a check of the work in the previous example, calculate Ax, where x is the computed solution. Compare Ax with b, calculate $b - Ax$, and verify that $\|b - Ax\|_2 = 16/25 = \|d\|_2$.   ☐

Exercise 3.3.3   It is a well-known fact from algebra that every nonzero polynomial of degree $\leq m - 1$ is zero at at most $m - 1$ distinct points. Use this fact to prove that the matrix of Eq. 3.1.3 has full rank.   ☐

Exercise 3.3.3 demonstrates that the least-squares problem for data fitting by a polynomial of degree $\leq m - 1$, with $m < n$, always has a unique solution that can be computed by the QR decomposition method. The result is also valid when $n = m$, in which case the matrix of Eq. 3.2.4 is square and, being of full rank, nonsingular. The system is not overdetermined; it has a unique solution. This proves that for any m data points with distinct abscissas, there is a unique polynomial of degree $\leq m - 1$ that passes through the data points.

Exercise 3.3.4   Write a pair of Fortran subroutines to solve the least-squares problem for the overdetermined system $Ax = b$, where $A \in \mathbb{R}^{n \times m}$, $n > m$, and A has full rank. Most of the code for these subroutines can be obtained by making minor modifications of algorithms (3.2.21) and (3.2.22).

The first subroutine uses reflectors to carry out a QR decomposition, $A = QR$, where $Q \in \mathbb{R}^{n \times n}$ is a product of m reflectors: $Q = Q_1 Q_2 \cdots Q_m$ and $R = \begin{bmatrix} \hat{R} \\ 0 \end{bmatrix} \in \mathbb{R}^{n \times m}$ where $\hat{R}$ is m by m and upper triangular. All of these operations should be carried out in a single n-by-m array plus two 1-dimensional arrays of length m to hold the parameters $-\sigma_i$ and γ_i associated with the reflectors Q_i. If $\hat{R}$ has any zero pivots ($\sigma_i = 0$), the subroutine should set an error flag to indicate that A does not have full rank.

The second subroutine uses the QR decomposition to find x. First $c = Q^T b = Q_m Q_{m-1} \cdots Q_1 b$ is calculated by applying the reflectors successively. An additional one-dimensional array is needed for b. This array can also be used for c (and intermediate results). The solution x is found by solving $\hat{R}x = \hat{c}$ by back substitution, where $c = \begin{bmatrix} \hat{c} \\ d \end{bmatrix}$. The subroutine should calculate $\|d\|_2$, which is the Euclidean norm of the residual.

Write clear, structured code and document all subroutine parameters clearly. (See the Fortran programming tips at the end of Section 1.6.)

Use your subroutines to solve the following problems.

(a) Find the least-squares quadratic polynomial for the data

t_i	-1	-0.75	-0.5	0	0.25	0.5	0.75
y_i	1.00	0.8125	0.75	1.00	1.3125	1.75	2.3125

(The correct solution is $\phi(t) = 1 + t + t^2$, which fits the data exactly.)

(b) Find the least-squares linear polynomials for the two sets of data

t_i	1000	1050	1060	1080	1110	1130
(a) y_i	6010	6153	6421	6399	6726	6701
(b) y_i	9422	9300	9220	9150	9042	8800

using the basis $\phi_1(t) = 1000$, $\phi_2(t) = t - 1065$. Notice that the QR decomposition only needs to be done once. Plot your solutions and the data points.

(c)
$$
\begin{bmatrix}
1 & 2 & 3 & 4 \\
5 & 6 & 7 & 8 \\
9 & 10 & 11 & 12 \\
1 & 1 & 1 & 1 \\
3 & 2 & 1 & 0
\end{bmatrix}
\begin{bmatrix}
x_1 \\ x_2 \\ x_3 \\ x_4
\end{bmatrix}
=
\begin{bmatrix}
10 \\ 26 \\ 42 \\ 4 \\ 6
\end{bmatrix}
$$

Examine σ_1, σ_2, σ_3, and σ_4. You will find that some of them are close to zero; this suggests that the coefficient matrix might not have full rank. In fact its rank is 2. This problem illustrates the unfortunate fact that your program is unlikely to detect rank deficiency. While two of σ_1, σ_2, σ_3, and σ_4 should equal zero in principle, roundoff errors have made them all nonzero in practice. On the brighter side you will notice that the "solution" that your program returned does fit the equations remarkably well. In fact it is (up to roundoff error) a solution of the least-squares problem. As we shall see below, the least-squares problem for a rank-deficient matrix has infinitely many solutions, of which your program calculated one.   □

Rank-Deficient Case

While most least-squares problems have full rank, it is obviously desirable to be able to handle problems that are not of full rank as well. Most such problems (and full-rank problems also) can be handled by a variant of the QR method called the QR decomposition with column pivoting. We will discuss that method here. A more reliable and more expensive method based on the singular-value decomposition will be discussed in Chapter 7.

If A does not have full rank, the QR method breaks down because $\hat{R}$ is singular; at least one of its main-diagonal entries $-\sigma_1, \ldots, -\sigma_m$ is zero. The QR decomposition with column pivoting makes column interchanges so that the zero pivots are moved to the lower right-hand corner of $\hat{R}$. The resulting decomposition is suitable for solving the rank-deficient least-squares problem.

Our initial development will ignore the effects of roundoff errors. At step 1 the 2-norm of each column of A is computed. If the jth column has the largest norm, then columns 1 and j are interchanged. The rest of step 1 is the same as before: A reflector that transforms the first column to the form $[-\sigma_1 \ 0 \ 0 \ \cdots \ 0]^T$ is determined. This reflector is then applied to columns 2 through m. Since $|\sigma_1|$

equals the 2-norm of the first column, the effect of the column interchange is to make $|\sigma_1|$ as large as possible. In particular $\sigma_1 \neq 0$ unless $A = 0$. The second step operates on the submatrix obtained by ignoring the first row and column. Otherwise it is identical to the first step, except that when the columns are interchanged, the full columns should be swapped, not just the portions that lie in the submatrix. This implies that the effect of the column interchange is the same as that of making the corresponding interchange in the columns of A before the QR decomposition is begun. Each step operates on one less row and column than the previous step, just as before, except that the column interchanges involve the entire columns.

If the matrix has full rank, the algorithm terminates after m steps. The result is a decomposition $\hat{A} = QR$, where $\hat{A}$ is a matrix obtained from A by permuting the columns. $R = \begin{bmatrix} \hat{R} \\ 0 \end{bmatrix}$, where $\hat{R}$ is upper triangular and nonsingular. If A does not have full rank, there will come a step at which we are forced to take $-\sigma_i = 0$. This happens when and only when all entries of the remaining submatrix are zero. Suppose this occurs after r steps have been completed. Letting $Q_i \in \mathbb{R}^{n \times n}$ denote the reflector obtained at step i, we have

$$Q_r Q_{r-1} \cdots Q_1 \hat{A} = \begin{bmatrix} R_{11} & R_{12} \\ 0 & 0 \end{bmatrix} = R$$

where R_{11} is r by r, upper triangular, and nonsingular. Its main-diagonal entries are $-\sigma_1, -\sigma_2, \ldots, -\sigma_r$, which are all nonzero. R_{12} is r by $(m - r)$. Letting $Q = Q_1 Q_2 \cdots Q_r$, we see that $Q^T = Q_r Q_{r-1} \cdots Q_1$, $Q^T \hat{A} = R$, and $\hat{A} = QR$. Since $\text{rank}(A) = \text{rank}(\hat{A}) = \text{rank}(R)$, we have $\text{rank}(A) = r$. We summarize these findings as a theorem.

THEOREM 3.3.7   Let $A \in \mathbb{R}^{n \times m}$ with $\text{rank}(A) = r > 0$. Then there exist matrices $\hat{A}$, Q, and R, such that $\hat{A}$ is obtained from A by permuting its columns, $Q \in \mathbb{R}^{n \times n}$ is orthogonal, $R = \begin{bmatrix} R_{11} & R_{12} \\ 0 & 0 \end{bmatrix} \in \mathbb{R}^{n \times m}$, $R_{11} \in \mathbb{R}^{r \times r}$ is nonsingular and upper triangular, and

$$\hat{A} = QR$$

Let us see how we can use this result to solve the least-squares problem. Given $x \in \mathbb{R}^m$, denote by $\hat{x}$ the vector obtained from x by making the same sequence of interchanges to its entries as were made to the columns of A in transforming A to $\hat{A}$. Then $\hat{A}\hat{x} = Ax$, so the problem of minimizing $\| b - Ax \|_2$ is the same as that of minimizing $\| b - \hat{A}\hat{x} \|_2$. An application of Q^T to the overdetermined system $\hat{A}\hat{x} = b$ transforms it to the form $R\hat{x} = Q^T b = c$, or

$$\begin{bmatrix} R_{11} & R_{12} \\ 0 & 0 \end{bmatrix} \begin{bmatrix} \hat{x}_1 \\ \hat{x}_2 \end{bmatrix} = \begin{bmatrix} \hat{c} \\ d \end{bmatrix}$$

where $\hat{x} = \begin{bmatrix} \hat{x}_1 \\ \hat{x}_2 \end{bmatrix}$, $\hat{x}_1 \in \mathbb{R}^r$, $c = \begin{bmatrix} \hat{c} \\ d \end{bmatrix}$, and $\hat{c} \in \mathbb{R}^r$. The residual for this transformed system is

$$s = \begin{bmatrix} \hat{c} - R_{11}\hat{x}_1 - R_{12}\hat{x}_2 \\ d \end{bmatrix}$$

whose norm is

$$\|s\|_2 = \sqrt{\|\hat{c} - R_{11}\hat{x}_1 - R_{12}\hat{x}_2\|_2^2 + \|d\|_2^2}$$

Clearly there is nothing we can do with the term $\|d\|_2^2$; we minimize $\|s\|_2$ by minimizing $\|\hat{c} - R_{11}\hat{x}_1 - R_{12}\hat{x}_2\|_2^2$. This term can never be negative, but there are many choices of $\hat{x}$ for which it is zero. Each of these $\hat{x}$ is a solution of the least-squares problem for the overdetermined system $\hat{A}\hat{x} = b$.

To see how to compute these $\hat{x}$, recall that R_{11} is nonsingular. Thus for any choice of $\hat{x}_2 \in \mathbb{R}^{m-r}$ there exists a unique $\hat{x}_1 \in \mathbb{R}^r$ such that

$$R_{11}\hat{x}_1 = \hat{c} - R_{12}\hat{x}_2$$

Since R_{11} is upper triangular, $\hat{x}_1$ can be calculated by back substitution. Then $\hat{c} - R_{11}\hat{x}_1 - R_{12}\hat{x}_2 = 0$, and $\hat{x} = \begin{bmatrix} \hat{x}_1 \\ \hat{x}_2 \end{bmatrix}$ is a solution to the least-squares problem for $\hat{A}\hat{x} = b$. We have thus established the following theorem.

THEOREM 3.3.8   Let $A \in \mathbb{R}^{n \times m}$ and $b \in \mathbb{R}^n, n > m$. Then the least-squares problem for the overdetermined system $Ax = b$ always has a solution. If rank$(A) < m$, there are infinitely many solutions.

Not only have we proved this theorem, but we have also established an algorithm that can be used to calculate any solution of the least-squares problem. Thus we have solved the problem in principle.

In practice the situation is complicated by roundoff errors. After r steps of the QR decomposition with column pivoting, A will have been transformed to the form

$$Q_r \cdots Q_1 \hat{A} = \begin{bmatrix} R_{11} & R_{12} \\ 0 & R_{22} \end{bmatrix}$$

where $R_{11} \in \mathbb{R}^{r \times r}$ is nonsingular.

If rank$(A) = r$, then in principle $R_{22} = 0$, and the algorithm terminates. In practice R_{22} will have been contaminated by roundoff errors and will not be exactly zero. Our criterion for determining the rank of A must take this into account. Thus for example we might decide R_{22} is "numerically zero" if the norm of its largest column is less than $\epsilon\|A\|$, where ϵ is some small parameter depending on the machine precision and the accuracy of the data. This approach generally works well. Unfortunately it is not 100 percent reliable. There exist matrices of the form

$$R = \begin{bmatrix} -\sigma_1 & & & * \\ & -\sigma_2 & & \\ & & \ddots & \\ 0 & & & -\sigma_m \end{bmatrix}$$

that are "nearly" rank deficient, for which none of the $|\sigma_i|$ is particularly small. An example due to Kahan is given in [SLS, p. 31] and [MC, p. 245]. The near rank deficiency of these matrices would not be detected by our simple criterion. A more reliable approach to the detection of rank deficiency is to use the singular-value decomposition (Chapter 7). See also Chan (1987).

A few other implementation details need to be mentioned. At each step we need to know the norms of the columns of the remaining submatrix. The cost of calculating these norms can be substantial. You can easily check that if the norm calculations are done in a straightforward manner, their cost will be comparable to that of the rest of the arithmetic in the algorithm. Fortunately the cost can be decreased substantially for steps $2, 3, \ldots, m$ by using information from the previous step rather than recomputing the norms from scratch.

For example, let us see how we can get the norm information for the second step by using the information from the first step. Let $v_1, v_2, \ldots, v_m$ denote the squares of the norms of the columns of A (calculated at a cost of nm flops). We work with the squares for convenience. For notational simplicity let us assume that v_1 is the largest (or that the column interchange for step 1 has already been made). After step 1 we have

$$Q_1 A = \begin{bmatrix} -\sigma_1 & \tilde{a}_{12} & \cdots & \tilde{a}_{1m} \\ 0 & \tilde{a}_{22} & \cdots & \tilde{a}_{2m} \\ \vdots & \vdots & & \vdots \\ 0 & \tilde{a}_{n2} & \cdots & \tilde{a}_{nm} \end{bmatrix}$$

Since Q_1 is orthogonal, it preserves the lengths of vectors. Therefore the norms of the columns of $Q_1 A$ are the same as those of the columns of A. For the second step we need the squares of the norms of the columns of the submatrix

$$\begin{bmatrix} \tilde{a}_{22} & \cdots & \tilde{a}_{2m} \\ \vdots & & \vdots \\ \tilde{a}_{n2} & \cdots & \tilde{a}_{nm} \end{bmatrix}$$

These can clearly be obtained by the operations

$$v_j \leftarrow v_j - \tilde{a}_{1j}^2 \qquad j = 2, \ldots, m$$

This costs a mere $m - 1$ flops instead of the $(n - 1)(m - 1)$ flops that would have been required to calculate the norms from scratch.

Since the calculation of 2-norms is required, there is some danger of overflow or underflow. This danger can be virtually eliminated by rescaling the entire problem in advance. For example, one can calculate $\nu = \max|a_{ij}|$ and divide all entries of A by b and ν. The cost of this scaling operation is $O(nm)$. If this is done, then it is not necessary to perform scaling operations in the calculation of the reflectors (3.2.16).

Finally we make some observations concerning column interchanges. We have already noted that each column swap requires the interchange of the entire columns. Of course the norm information has to be interchanged too. It is also necessary to keep a record of the interchanges, since the computed solution $\hat{x}$ is the least-squares solution of the permuted system $\hat{A}\hat{x} = b$. To get the least-squares solution of the original problem, we must apply the inverse permutation to the entries of $\hat{x}$. This is accomplished by performing the interchanges in the reverse order.

Exercise 3.3.5   Show that $-\sigma_1, -\sigma_2, \ldots, -\sigma_r$, the main-diagonal entries of R_{11}, satisfy $|\sigma_1| \geq |\sigma_2| \geq \cdots \geq |\sigma_r|$.   □

3.4
ORTHONORMAL VECTORS
AND THE GRAM–SCHMIDT METHOD

In this section we introduce the idea of an orthonormal set, a new formulation of the QR decomposition, and the Gram–Schmidt method for orthonormalizing a linearly independent set. The main result is that performing a Gram–Schmidt orthonormalization is equivalent to calculating a QR decomposition. It follows that the Gram–Schmidt method can be used to solve the least-squares problem.

A set of vectors $q_1, q_2, \ldots, q_k \in \mathbb{R}^n$ is said to be *orthonormal* if the vectors are pairwise orthogonal, and each vector has Euclidean norm 1; that is,

$$(q_i, q_j) = \begin{cases} 0 & \text{if } i \neq j \\ 1 & \text{if } i = j \end{cases}$$

Example 3.4.1   Let $e_1, \ldots, e_n$ be the columns of the identity matrix:

$$e_1 = \begin{bmatrix} 1 \\ 0 \\ 0 \\ \vdots \\ 0 \end{bmatrix} \qquad e_2 = \begin{bmatrix} 0 \\ 1 \\ 0 \\ \vdots \\ 0 \end{bmatrix} \qquad \cdots \qquad e_n = \begin{bmatrix} 0 \\ \vdots \\ 0 \\ 0 \\ 1 \end{bmatrix}$$

It is evident that $e_1, e_2, \ldots, e_n$ form an orthonormal set. In fact, they form an *orthonormal basis*, since they are also a basis of $\mathbb{R}^n$. From now on we will call $e_1, e_2, \ldots, e_n$ the *standard basis* of $\mathbb{R}^n$, and the notation $e_1, e_2, \ldots, e_n$ will be reserved for this basis.

THEOREM 3.4.2   Let $Q \in \mathbb{R}^{n \times n}$. Then Q is an orthogonal matrix if and only if its columns (rows) form an orthonormal set.

Proof   Let $q_1, q_2, \ldots, q_n$ denote the columns of Q. Then

$$Q^T Q = \begin{bmatrix} q_1^T \\ q_2^T \\ \vdots \\ q_n^T \end{bmatrix} [q_1 \ q_2 \ \cdots \ q_n] = \begin{bmatrix} q_1^T q_1 & q_1^T q_2 & \cdots & q_1^T q_n \\ q_2^T q_1 & q_2^T q_2 & \cdots & q_2^T q_n \\ \vdots & \vdots & & \vdots \\ q_n^T q_1 & q_n^T q_2 & \cdots & q_n^T q_n \end{bmatrix}$$

Thus the entries of Q^TQ are the inner products (q_i, q_j). Clearly $Q^TQ = I$ if and only if $q_1, q_2, \ldots, q_n$ form an orthonormal set. The analogous theorem for the rows follows from considering the product QQ^T or from the simple observation that Q is orthogonal if and only if Q^T is. $\square$

Exercise 3.4.1 (a) Let $A \in \mathbb{R}^{n \times n}$ and let $e_1, \ldots, e_n$ denote the standard basis of $\mathbb{R}^n$. Verify that the ith column of A is $Ae_i, i = 1, \ldots, n$. Thus $A = [Ae_1 \, Ae_2 \, \cdots \, Ae_n]$. This simple observation will be used repeatedly. (b) Use the observation of part (a) together with the inner-product preserving property of orthogonal matrices to obtain a second proof that the columns of an orthogonal matrix are orthonormal. $\square$

Using Theorem 3.4.2 as a guide, we introduce a new class of nonsquare matrices that possess some of the properties of orthogonal matrices. The matrix $Q \in \mathbb{R}^{n \times m}, n \geq m$, will be called *isometric* (or an *isometry*) if its columns are orthonormal.

Exercise 3.4.2 Prove that $Q \in \mathbb{R}^{n \times m}$ is isometric if and only if $Q^TQ = I \, (\in \mathbb{R}^{m \times m})$.
$\square$

The result of this exercise does *not* imply that Q^T is Q^{-1}. Only square matrices can have inverses. It is also *not* true that QQ^T is the identity matrix.

Exercise 3.4.3 Let $Q \in \mathbb{R}^{n \times m}$ $(n > m)$ be an isometry with columns $q_1, q_2, \ldots, q_m$.

(a) Show that $QQ^Tv = 0$ if v is orthogonal to $q_1, q_2, \ldots, q_m$.

(b) Show that $QQ^Tq_i = q_i, i = 1, \ldots, m$. Thus $QQ^Tv = v$ whenever v is a linear combination of $q_1, q_2, \ldots, q_m$. Therefore QQ^T behaves like the identity matrix on a proper subspace of $\mathbb{R}^n$.

(c) Show that $(QQ^T)^2 = QQ^T$. Thus QQ^T is a projector. In fact it is an orthoprojector. (Cf., footnote to Theorem 3.2.11.) $\square$

Exercise 3.4.4 Show that if $Q \in \mathbb{R}^{n \times m}$ is an isometry, then (a) $(Qx, Qy) = (x, y)$ for all $x, y \in \mathbb{R}^m$, (b) $\|Qx\|_2 = \|x\|_2$ for all $x \in \mathbb{R}^m$. (Note that the norm and inner product on the left-hand side of these equations are the norm and inner product on $\mathbb{R}^m$.) $\square$

Thus isometries preserve inner products, norms, and angles. The converse of both parts of Exercise 3.4.4 holds as well.[§]
The *QR* decomposition theorem for nonsquare matrices can be restated more elegantly in terms of an isometry.

Theorem 3.4.3 Let $A \in \mathbb{R}^{n \times m}, n \geq m$. Then there exist matrices $\hat{Q}$ and $\hat{R}$ such that $\hat{Q} \in \mathbb{R}^{n \times m}$ is an isometry, $\hat{R} \in \mathbb{R}^{m \times m}$ is upper triangular, and

---

[§] The name isometry comes from the norm preserving property: An isometric operator is one that preserves the metric (i.e., the norm in this case).

$$A = \hat{Q}\hat{R}$$

Proof If $n = m$, this is just Theorem 3.2.9. If $n > m$, we know from Theorem 3.3.3 that there exist matrices $Q \in \mathbb{R}^{n \times n}$ and $R \in \mathbb{R}^{n \times m}$ such that Q is orthogonal, $R = \begin{bmatrix} \hat{R} \\ 0 \end{bmatrix}$, $\hat{R} \in \mathbb{R}^{m \times m}$ is upper triangular, and $A = QR$. Let $\hat{Q} \in \mathbb{R}^{n \times m}$ be the matrix consisting of the first m columns of Q. Clearly $\hat{Q}$ is isometric. Let $\tilde{Q} \in \mathbb{R}^{n \times (n-m)}$ be the last $n - m$ columns of Q. Then

$$A = QR = [\hat{Q} \ \tilde{Q}] \begin{bmatrix} \hat{R} \\ 0 \end{bmatrix} = \hat{Q}\hat{R} + \tilde{Q}0$$

that is, $A = \hat{Q}\hat{R}$. Since $\hat{Q}$ and $\hat{R}$ have the desired properties, the proof is complete. $\square$

If A has full rank, Theorem 3.4.3 can be strengthened to include the uniqueness of $\hat{Q}$ and $\hat{R}$.

THEOREM 3.4.4 Let $A \in \mathbb{R}^{n \times m}$, $n \geq m$, and suppose rank$(A) = m$. Then there exist unique $\hat{Q} \in \mathbb{R}^{n \times m}$ and $\hat{R} \in \mathbb{R}^{m \times m}$, such that $\hat{Q}$ is isometric, $\hat{R}$ is upper triangular with positive entries on the main diagonal, and

$$A = \hat{Q}\hat{R}$$

The proof is similar to that of Theorem 3.2.23 and is left as an exercise for you.

Exercise 3.4.5 Prove Theorem 3.4.4. $\square$

In the full-rank case the decomposition of Theorem 3.4.3 or 3.4.4 can be used to solve the least-squares problem for the overdetermined system $Ax = b$. Let $\hat{c} = \hat{Q}^T b \in \mathbb{R}^m$. You can easily check that this vector is the same as the vector $\hat{c}$ of Theorem 3.3.5, and the least-squares problem can be solved by solving $\hat{R}x = \hat{c}$ by back substitution. When we use this decomposition, we do not get the norm of the residual in the form $\| d \|_2$.

Exercise 3.4.6 Let $A \in \mathbb{R}^{n \times m}$, $n \geq m$, and suppose rank$(A) = r$. Let $\hat{A}$ be any matrix obtained from A by permuting the columns in such a way that the first r columns are linearly independent. Prove that there exist unique $\hat{Q} \in \mathbb{R}^{n \times r}$ and $R \in \mathbb{R}^{r \times m}$, such that $\hat{Q}$ is an isometry, $R = [\hat{R} \ \tilde{R}]$, $\hat{R} \in \mathbb{R}^{r \times r}$ is upper triangular with positive entries on the main diagonal, and $\hat{A} = \hat{Q}R$. $\square$

Before we discuss the Gram–Schmidt process, we should review some of the elementary facts about subspaces of $\mathbb{R}^n$. Recall that a *subspace* of $\mathbb{R}^n$ is a nonempty subset $\mathcal{S}$ of $\mathbb{R}^n$ that is closed under addition and scalar multiplication. That is, $\mathcal{S}$ is a subspace of $\mathbb{R}^n$ if and only if whenever $v, w \in \mathcal{S}$ and $c \in \mathbb{R}$, then $v + w \in \mathcal{S}$ and $cv \in \mathcal{S}$. Given vectors $v_1, \ldots, v_m \in \mathbb{R}^n$, a *linear combination* of $v_1, \ldots, v_m$ is a vector of the form $c_1 v_1 + c_2 v_2 + \cdots + c_m v_m$, where

$c_1, c_2, \ldots, c_m \in \mathbb{R}$. The numbers $c_1, c_2, \ldots, c_m$ are called the *coefficients* of the linear combination. In sigma notation a linear combination looks like $\sum_{k=1}^{m} c_k v_k$. The *span* of $v_1, \ldots, v_m$ is the set of all linear combinations of $v_1, \ldots, v_m$. This set is often denoted span$\{v_1, \ldots, v_m\}$, but we will use the more compact notation $\langle v_1, \ldots, v_m \rangle$. Thus in particular $\langle v \rangle$ denotes the set of all scalar multiples of v. It is clear that $\langle v_1, \ldots, v_m \rangle$ is closed under addition and scalar multiplication; that is, it is a subspace of $\mathbb{R}^n$.

Exercise 3.4.7 Show that if $\langle v_1, \ldots, v_m \rangle = \langle w_1, \ldots, w_p \rangle$, then $\langle v_1, \ldots, v_m, u \rangle = \langle w_1, \ldots, w_p, u \rangle$. ☐

We have already used terms such as *linear independence* and *basis*, but let us now take a moment to review their precise meaning. The vectors $v_1, \ldots, v_m$ are *linearly independent* if the equation $c_1 v_1 + c_2 v_2 + \cdots + c_m v_m = 0$ has no solutions other than $c_1 = c_2 = \cdots = c_m = 0$. In other words, the only linear combination of $v_1, \ldots, v_m$ that equals zero is the one whose coefficients are all zero. Every orthonormal set is linearly independent: Let $q_1, \ldots, q_m$ be an orthonormal set, and suppose $c_1 q_1 + \cdots + c_m q_m = 0$. Taking the inner product of both sides of this equation with q_i, we find that

$$c_1(q_1, q_i) + \cdots + c_i(q_i, q_i) + \cdots + c_m(q_m, q_i) = (0, q_i) = 0$$

Since $(q_j, q_i) = 0$ except when $j = i$ and $(q_i, q_i) = 1$, this equation collapses to $c_i = 0$. This is true for all i, which shows that $q_1, \ldots, q_m$ are linearly independent. Of course the converse if false; there are lots of linearly independent sets that are not orthonormal.

Exercise 3.4.8 Give an example of a set of three linearly independent vectors in $\mathbb{R}^3$ that are not orthogonal and do not have norm 1. ☐

Exercise 3.4.9 Prove that the vectors $v_1, \ldots, v_m$ are linearly independent if and only if none of them can be expressed as a linear combination of the others, in symbols, $v_i \notin \langle v_1, \ldots, v_{i-1}, v_{i+1}, \ldots, v_m \rangle$, $i = 1, \ldots, m$. ☐

Let $\mathscr{S}$ be a subspace of $\mathbb{R}^n$, and let $v_1, \ldots, v_m \in \mathscr{S}$. Then clearly $\langle v_1, \ldots, v_m \rangle \subseteq \mathscr{S}$. We say that $v_1, \ldots, v_m$ *span* $\mathscr{S}$ if $\langle v_1, \ldots, v_m \rangle = \mathscr{S}$. This means that every member of $\mathscr{S}$ can be expressed as a linear combination of $v_1, \ldots, v_m$. In this case we say that $v_1, \ldots, v_m$ form a *spanning set* for $\mathscr{S}$. A *basis* of $\mathscr{S}$ is a spanning set that is linearly independent. If $v_1, \ldots, v_m$ are linearly independent, then they form a basis of $\langle v_1, \ldots, v_m \rangle$, since they are a spanning set by definition. Recall from your elementary linear algebra class that every subspace of $\mathbb{R}^n$ has a basis; in fact it has many bases. Any two bases of $\mathscr{S}$ have the same number of elements. This number is called the *dimension* of the subspace. Thus, for example, if $v_1, \ldots, v_m$ are independent, then $\langle v_1, \ldots, v_m \rangle$ has dimension m.

The Gram–Schmidt process is an algorithm that produces orthonormal bases. Let $\mathscr{S}$ be a subspace of $\mathbb{R}^n$, and let $v_1, \ldots, v_m$ be a basis of $\mathscr{S}$. The Gram–Schmidt process uses $v_1, \ldots, v_m$ to produce orthonormal vectors $q_1, \ldots, q_m$ that form a basis of $\mathscr{S}$. Thus $\mathscr{S} = \langle v_1, \ldots, v_m \rangle = \langle q_1, \ldots, q_m \rangle$. In fact more is true: The vectors $q_1, \ldots, q_m$ also satisfy

$$\begin{aligned}
\langle q_1 \rangle &= \langle v_1 \rangle \\
\langle q_1, q_2 \rangle &= \langle v_1, v_2 \rangle \\
\langle q_1, q_2, q_3 \rangle &= \langle v_1, v_2, v_3 \rangle \\
\vdots \qquad\qquad &\qquad\qquad \vdots \\
\langle q_1, q_2, \ldots, q_m \rangle &= \langle v_1, v_2, \ldots, v_m \rangle
\end{aligned} \tag{3.4.5}$$

This feature turns out to be very important.

We are given linearly independent vectors $v_1, \ldots, v_m \in \mathbb{R}^n$, and we seek orthonormal $q_1, \ldots, q_m$ satisfying (3.4.5). In order to satisfy $\langle q_1 \rangle = \langle v_1 \rangle$, we must choose q_1 to be a multiple of v_1. Since we also require $\|q_1\|_2 = 1$, we define $q_1 = (1/r_{11})v_1$, where $r_{11} = \|v_1\|_2$. We know that $r_{11} \neq 0$ because $v_1, \ldots, v_m$ are linearly independent, so $v_1 \neq 0$. The equation $q_1 = (1/r_{11})v_1$ implies that $q_1 \in \langle v_1 \rangle$; hence $\langle q_1 \rangle \subseteq \langle v_1 \rangle$. Conversely, the equation $v_1 = r_{11}q_1$ implies $v_1 \in \langle q_1 \rangle$, and therefore $\langle v_1 \rangle \subseteq \langle q_1 \rangle$. Thus $\langle q_1 \rangle = \langle v_1 \rangle$.

The second step of the algorithm is to find q_2 such that q_2 is orthogonal to q_1, $\|q_2\|_2 = 1$, and $\langle q_1, q_2 \rangle = \langle v_1, v_2 \rangle$. We will use Figure 3.5 as a guide. In this figure the space $\langle v_1, v_2 \rangle$ is represented by the plane of the page. We require q_2 such that q_1, q_2 is an orthonormal set that spans this plane. The figure suggests that we can produce a vector $\tilde{q}_2$ that lies in the plane and is orthogonal to q_1 by subtracting just the right multiple of q_1 (or of v_1) from v_2. We can then obtain q_2 by scaling $\tilde{q}_2$. Thus let

$$\tilde{q}_2 = v_2 - r_{12}q_1 \tag{3.4.6}$$

where the scalar r_{12} is to be determined. We must choose r_{12} so that $(\tilde{q}_2, q_1) = 0$. This equation implies $0 = (v_2 - r_{12}q_1, q_1) = (v_2, q_1) - r_{12}(q_1, q_1)$, and since $(q_1, q_1) = 1$, $r_{12} = (v_2, q_1)$. Conversely, this choice of r_{12} guarantees that $(\tilde{q}_2, q_1) = 0$. It is clear that $\tilde{q}_2 \neq 0$, for if this were not the case, we would have $v_2 = r_{12}q_1 \in \langle q_1 \rangle$. But the linear independence of v_1 and v_2 implies that $v_2 \notin \langle v_1 \rangle = \langle q_1 \rangle$. Let $r_{22} = \|\tilde{q}_2\|_2 \neq 0$, and define $q_2 = (1/r_{22})\tilde{q}_2$. Then clearly

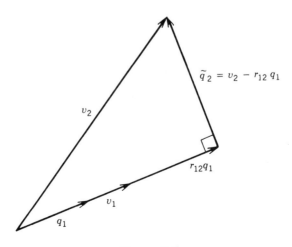

Figure 3.5

$\|q_2\|_2 = 1$ and $(q_1, q_2) = 0$. It is intuitively clear from Figure 3.5 that $\langle q_1, q_2 \rangle = \langle v_1, v_2 \rangle$, and we can easily prove that it is so. First $q_1 \in \langle v_1 \rangle \subseteq \langle v_1, v_2 \rangle$. Also $q_2 \in \langle q_2 \rangle \subseteq \langle q_1, v_2 \rangle = \langle v_1, v_2 \rangle$. Since $q_1, q_2 \in \langle v_1, v_2 \rangle, \langle q_1, q_2 \rangle \subseteq \langle v_1, v_2 \rangle$. Conversely $v_1 \in \langle q_1 \rangle \subseteq \langle q_1, q_2 \rangle$, and $v_2 = r_{12}q_1 + \tilde{q}_2 = r_{12}q_1 + r_{22}q_2 \in \langle q_1, q_2 \rangle$. Since $v_1, v_2 \in \langle q_1, q_2 \rangle, \langle v_1, v_2 \rangle \subseteq \langle q_1, q_2 \rangle$. Thus $\langle q_1, q_2 \rangle = \langle v_1, v_2 \rangle$.

Now suppose we have found orthonormal vectors $q_1, \ldots, q_{k-1}$ such that $\langle q_1, \ldots, q_i \rangle = \langle v_1, \ldots, v_i \rangle$, $i = 1, \ldots, k - 1$. Let us see how to determine q_k. By analogy with (3.4.6) we seek $\tilde{q}_k$ of the form

$$\tilde{q}_k = v_k - \sum_{j=1}^{k-1} r_{jk} q_j \qquad (3.4.7)$$

such that $\tilde{q}_k$ is orthogonal to $q_1, \ldots, q_{k-1}$. (You might find it useful to draw a picture of the case $k = 3$, in which $\tilde{q}_3$ has the form $\tilde{q}_3 = v_3 - r_{13}q_1 - r_{23}q_2$. The scalars r_{13} and r_{23} must be chosen so that q_3 is orthogonal to q_1 and q_2.) The equations $(\tilde{q}_k, q_i) = 0$, $i = 1, \ldots, k - 1$, imply that

$$(v_k, q_i) - \sum_{j=1}^{k-1} r_{jk}(q_j, q_i) = 0 \qquad i = 1, \ldots, k - 1$$

Since $(q_i, q_j) = 0$ when $i \neq j$, and $(q_i, q_i) = 1$, these equations reduce to

$$r_{ik} = (v_k, q_i) \qquad i = 1, \ldots, k - 1 \qquad (3.4.8)$$

Conversely, if r_{ik} are defined by (3.4.8), then $\tilde{q}_k$ is orthogonal to $q_1, \ldots, q_{k-1}$. As in the case $k = 2$, it is easy to show that $\tilde{q}_k \neq 0$. Otherwise we would have $v_k = \sum_{j=1}^{k-1} r_{jk} q_j \in \langle q_1, \ldots, q_{k-1} \rangle$. But the linear independence of $v_1, \ldots, v_k$ implies that $v_k \notin \langle v_1, \ldots, v_{k-1} \rangle = \langle q_1, \ldots, q_{k-1} \rangle$. Let

$$r_{kk} = \|\tilde{q}_k\|_2 \neq 0 \qquad (3.4.9)$$

and define

$$q_k = \frac{1}{r_{kk}} \tilde{q}_k \qquad (3.4.10)$$

Then clearly $\|q_k\|_2 = 1$ and $(q_i, q_k) = 0$, $i = 1, \ldots, k - 1$.

It is also easy to verify that $\langle q_1, \ldots, q_k \rangle = \langle v_1, \ldots, v_k \rangle$. First of all $\langle q_1, \ldots, q_{k-1} \rangle = \langle v_1, \ldots, v_{k-1} \rangle \subseteq \langle v_1, \ldots, v_k \rangle$, and $q_k \in \langle \tilde{q}_k \rangle \subseteq \langle q_1, \ldots, q_{k-1}, v_k \rangle = \langle v_1, \ldots, v_{k-1}, v_k \rangle$ (Exercise 3.4.7). Thus $\langle q_1, \ldots, q_k \rangle \subseteq \langle v_1, \ldots, v_k \rangle$. Conversely $\langle v_1, \ldots, v_{k-1} \rangle = \langle q_1, \ldots, q_{k-1} \rangle \subseteq \langle q_1, \ldots, q_k \rangle$, and combining Eqs. 3.4.7 and 3.4.10 we find that

$$v_k = \sum_{j=1}^{k-1} r_{jk} q_j + r_{kk} q_k \qquad (3.4.11)$$

which implies that $v_k \in \langle q_1, \ldots, q_k \rangle$. Thus $\langle v_1, \ldots, v_k \rangle \subseteq \langle q_1, \ldots, q_k \rangle$. We conclude that $\langle q_1, \ldots, q_k \rangle = \langle v_1, \ldots, v_k \rangle$.

Equations 3.4.8, 3.4.7, 3.4.9, and 3.4.10 express the kth step of the Gram–Schmidt algorithm. Performing this step for $k = 1, \ldots, m$ produces $q_1, \ldots, q_m$. The algorithm is also summarized below.

Classical Gram–Schmidt Algorithm

Given linearly independent $v_1, \ldots, v_m \in \mathbb{R}^n$, this algorithm produces an orthonormal set $q_1, \ldots, q_m$ such that $\langle q_1, \ldots, q_i \rangle = \langle v_1, \ldots, v_i \rangle$, $i = 1, \ldots, m$. $q_1, \ldots, q_m$ are stored over $v_1, \ldots, v_m$.

$$
\begin{aligned}
&\text{for } k = 1, \ldots, m \\
&\quad \left[\begin{aligned}
&\text{for } i = 1, \ldots, k-1 \ (\text{skipped when } k = 1) \\
&\quad \left[\begin{aligned}
r_{ik} &= (v_k, v_i) \\
v_k &\leftarrow v_k - r_{ik} v_i
\end{aligned} \right\} \ (v_i \text{ is really } q_i) \\
&\ (v_k \text{ now contains } \tilde{q}_k.) \\
&\ r_{kk} = \| v_k \|_2 \\
&\ \text{if } (r_{kk} = 0) \text{ set flag } (v_1, \ldots, v_k \text{ are dependent}), \text{ exit} \\
&\ v_k \leftarrow \frac{1}{r_{kk}} v_k
\end{aligned} \right. \\
&\text{exit}
\end{aligned}
\tag{3.4.12}
$$

The promised connection with the QR decomposition follows from (3.4.11). There are actually m such equations, one for each value of k. Writing out these equations, we have

$$
\begin{aligned}
v_1 &= q_1 r_{11} \\
v_2 &= q_1 r_{12} + q_2 r_{22} \\
v_3 &= q_1 r_{13} + q_2 r_{23} + q_3 r_{33} \\
&\ \ \vdots \\
v_m &= q_1 r_{1m} + q_2 r_{2m} + \cdots + q_m r_{mm}
\end{aligned}
$$

These can be packed into a single matrix equation

$$
[v_1 \ v_2 \ v_3 \ \cdots \ v_m] = [q_1 \ q_2 \ q_3 \ \cdots \ q_m]
\begin{bmatrix}
r_{11} & r_{12} & r_{13} & \cdots & r_{1m} \\
0 & r_{22} & r_{23} & \cdots & r_{2m} \\
0 & 0 & r_{33} & \cdots & r_{3m} \\
\vdots & \vdots & \vdots & \ddots & \vdots \\
0 & 0 & 0 & \cdots & r_{mm}
\end{bmatrix}
$$

Defining

$$
V = [v_1 \ v_2 \ \cdots \ v_m] \in \mathbb{R}^{n \times m}
$$

$$
Q = [q_1 \ q_2 \ \cdots \ q_m] \in \mathbb{R}^{n \times m}
$$

$$
R = \begin{bmatrix}
r_{11} & r_{12} & \cdots & r_{1m} \\
0 & r_{22} & \cdots & r_{2m} \\
\vdots & \vdots & \ddots & \vdots \\
0 & 0 & \cdots & r_{mm}
\end{bmatrix} \in \mathbb{R}^{m \times m}
\tag{3.4.13}
$$

we see that V has full rank, Q is isometric, R is upper triangular with positive entries on the main diagonal [cf., (3.4.9)], and

$$V = QR$$

Thus Q and R are the unique factors of V guaranteed by Theorem 3.4.4.

This proves that the Gram–Schmidt process provides another means of calculating a QR decomposition: Given a matrix $V \in \mathbb{R}^{n \times m}$ ($n \geq m$) of full rank, whose QR decomposition is required. Let $v_1, \ldots, v_m$ denote the columns of V. Carry out the Gram–Schmidt procedure on $v_1, \ldots, v_m$ to produce an orthonormal basis $q_1, \ldots, q_m$ and the coefficients r_{jk}. Then define Q and R by (3.4.13) to get the QR decomposition of V. Conversely, any method for calculating the QR decomposition can be used to orthonormalize vectors: Let $v_1, \ldots, v_m$ be a linearly independent set that is to be orthonormalized. Define $V = [v_1 \cdots v_m] \in \mathbb{R}^{n \times m}$ and use reflectors, rotators, or any other method to produce the unique QR decomposition of V guaranteed by Theorem 3.4.4. Let $q_1, \ldots, q_m$ denote the columns of the resulting isometry Q. Then by the uniqueness of the QR decomposition, $q_1, \ldots, q_m$ are exactly the vectors that would be produced by the Gram–Schmidt process. Thus we have the main result of this section:

> The Gram–Schmidt orthonormalization is the same as the QR decomposition.

Exercise 3.4.10   (a) Let $v_1 = [3 \ -3 \ 3 \ -3]^T$, $v_2 = [1 \ 2 \ 3 \ 4]^T$, and $\mathcal{S} = \langle v_1, v_2 \rangle \subseteq \mathbb{R}^4$. Apply the Gram–Schmidt process to v_1 and v_2 to obtain an orthonormal basis of $\mathcal{S}$. Save the coefficients r_{jk}. (b) Let

$$V = \begin{bmatrix} 3 & 1 \\ -3 & 2 \\ 3 & 3 \\ -3 & 4 \end{bmatrix} \in \mathbb{R}^{4 \times 2}$$

Use the result of part (a) to find an isometric $Q \in \mathbb{R}^{4 \times 2}$ and an upper-triangular $R \in \mathbb{R}^{2 \times 2}$ with positive main-diagonal entries, such that $V = QR$.   ☐

When one computes a QR decomposition in practice, one usually does not bother to force the main-diagonal elements of R to be positive. In this case the columns of Q may differ from the Gram–Schmidt vectors, but it is not hard to show that the difference is trivial.

Exercise 3.4.11   Let $v_1, \ldots, v_m \in \mathbb{R}^n$ be linearly independent vectors, and let $V = [v_1 \cdots v_m] \in \mathbb{R}^{n \times m}$. Suppose $V = QR$, where $Q \in \mathbb{R}^{n \times m}$ is an isometry and $R \in \mathbb{R}^{m \times m}$ is upper triangular but does not necessarily have positive main-diagonal entries. Let $q_1, \ldots, q_m$ denote the columns of Q, and let $\hat{q}_1, \ldots, \hat{q}_m$ denote the vectors obtained from $v_1, \ldots, v_m$ by the Gram–Schmidt process. Show that $q_i = \pm \hat{q}_i$, $i = 1, \ldots, m$.   ☐

Since the Gram–Schmidt process yields a QR decomposition, it can, in principle, be used to solve least-squares problems. Unfortunately the Gram–Schmidt process turns out to be numerically unstable; small roundoff errors can sometimes cause the computed vectors to be far from orthogonal. Perhaps surprisingly, a slight modification of the algorithm suffices to make it stable. In the classical Gram–Schmidt algorithm, q_1 is taken to be a multiple of v_1. Then the appropriate multiple of q_1 is subtracted from v_2 to obtain a vector that is orthogonal to q_1. The *modified Gram–Schmidt procedure* calculates q_1 just as before. It then subtracts multiples of q_1 not just from v_2, but from $v_3, \ldots, v_m$ as well, so that the resulting vectors $v_2^{(1)}, \ldots, v_m^{(1)}$ are all orthogonal to q_1. Thus the first step of the modified Gram–Schmidt algorithm is as follows:

$$r_{11}' = \|v_1\|_2$$

$$q_1' = \frac{1}{r_{11}'} v_1$$

$$\left.\begin{array}{l} r_{1j}' = (v_j, q_1') \\[2mm] v_j^{(1)} = v_j - r_{1j}' q_1' \end{array}\right\} \quad j = 2, \ldots, m$$

You can easily verify that $v_2^{(1)}, \ldots, v_m^{(1)}$ are all orthogonal to q_1'. Since $v_2^{(1)}$ is orthogonal to q_1', we can normalize it to get q_2'. Then the appropriate multiple of q_2' is subtracted from each of $v_3^{(1)}, \ldots, v_m^{(1)}$ to obtain vectors $v_3^{(2)}, \ldots, v_m^{(2)}$ that are orthogonal to q_2'. Since each of $v_3^{(2)}, \ldots, v_m^{(2)}$ is a linear combination of vectors orthogonal to q_1', these vectors are also orthogonal to q_1'. After $k-1$ steps of the modified Gram–Schmidt process we have orthonormal vectors $q_1', \ldots, q_{k-1}'$ and vectors $v_k^{(k-1)}, \ldots, v_m^{(k-1)}$ that are orthogonal to $q_1', \ldots, q_{k-1}'$. The kth step scales $v_k^{(k-1)}$ to get q_k', then subtracts the appropriate multiple of q_k from each of $v_{k+1}^{(k-1)}, \ldots, v_m^{(k-1)}$ to produce vectors $v_{k+1}^{(k)}, \ldots, v_m^{(k)}$ that are orthogonal to $q_1', \ldots, q_k'$. Specifically, the following computations are carried out:

$$r_{kk}' = \|v_k^{(k-1)}\|_2$$

$$q_k' = \frac{1}{r_{kk}'} v_k^{(k-1)} \tag{3.4.14}$$

$$\left.\begin{array}{l} r_{kj}' = (v_j^{(k-1)}, q_k'), \\[2mm] v_j^{(k)} = v_j^{(k-1)} - r_{kj}' q_k' \end{array}\right\} \quad j = k+1, \ldots, m$$

You can verify that $v_{k+1}^{(k)}, \ldots, v_m^{(k)}$ are orthogonal to q_k'. They are also orthogonal to $q_1', \ldots, q_{k-1}'$, since each $v_j^{(k)}$ is a linear combination of vectors orthogonal to $q_1', \ldots, q_{k-1}'$.

Performing (3.4.14) for $k = 1, \ldots, m$ (with $v_j^{(0)} = v_j$), we produce an orthonormal sequence $q_1', \ldots, q_m'$. In exercises that follow, you will show that (ignoring roundoff errors) these vectors are identical to those produced by the classical Gram–Schmidt process. Furthermore $r_{jk}' = r_{jk}$ for all j and k. The algorithm can be summarized as follows:

Modified Gram–Schmidt Algorithm

Given linearly independent $v_1, \ldots, v_m \in \mathbb{R}^n$, this algorithm produces an orthonormal set $q_1, \ldots, q_m$ such that $\langle q_1, \ldots, q_i \rangle = \langle v_1, \ldots, v_i \rangle$, $i = 1, \ldots, m$. $q_1, \ldots, q_m$ are stored over $v_1, \ldots, v_m$ for $k = 1, \ldots, m$.

$$
\begin{aligned}
&\text{for } k = 1, \ldots, m \\
&\left[\begin{aligned}
&r_{kk} = \| v_k \|_2 \\
&\text{if } (r_{kk} = 0) \text{ set flag } (v_1, \ldots, v_k \text{ are dependent}), \text{ exit} \\
&v_k \leftarrow \frac{1}{r_{kk}} v_k \\
&\text{for } j = k + 1, \ldots, m \text{ (skipped when } k = m) \\
&\quad \left[\begin{aligned}
&r_{kj} = (v_j, v_k) \\
&v_j \leftarrow v_j - r_{kj} v_k
\end{aligned}\right\} \qquad (v_k \text{ is really } q_k)
\end{aligned}\right. \\
&\text{exit}
\end{aligned}
$$

Exercise 3.4.12   (Modified Gram–Schmidt = Classical Gram–Schmidt)

(a)   Prove by induction on k that for $k = 2, \ldots, m$,

$$
v_j^{(k-1)} = v_j - \sum_{i=1}^{k-1} r_{ij}' q_i' \qquad j = k, \ldots, m
$$

(b)   The algorithm breaks down unless $r_{kk}' \neq 0$ for all k. Prove by induction on k that for $k = 1, \ldots, m$, $r_{kk}' \neq 0$ and $\langle q_1', \ldots, q_k' \rangle = \langle v_1, \ldots, v_k \rangle$. (We are assuming here that $v_1, \ldots, v_m$ are linearly independent.)

(c)   Use the result of part (a) to prove that

$$
v_k = \sum_{i=1}^{k} q_i' r_{ik}' \qquad k = 1, \ldots, m
$$

(d)   Let

$$
V = [v_1, \ldots, v_m] \qquad Q' = [q_1' \quad \cdots \quad q_m']
$$

$$
\text{and} \qquad R' = \begin{bmatrix} r_{11}' & \cdots & r_{1m}' \\ & \ddots & \vdots \\ 0 & & r_{mm}' \end{bmatrix}
$$

Then Q' is isometric and R' is upper triangular with positive entries on the main diagonal. Show that $V = Q'R'$. Conclude that the vectors $q_1', \ldots, q_m'$ and the coefficients r_{jk}' are identical to those produced by the classical Gram–Schmidt algorithm.   □

In this exercise you invoked the uniqueness of the QR decomposition to prove that the two forms of the Gram–Schmidt algorithm are equivalent in principle. This is the convenient approach; of course, one can also prove the result directly:

Exercise 3.4.13   Prove by induction on k that for $k = 1, \ldots, m$,

$$
r_{ik}' = r_{ik} \qquad i = 1, \ldots, k - 1
$$

$$
v_k^{(k-1)} = \tilde{q}_k
$$

$$
r_{kk}' = r_{kk}
$$

$$
q_k' = q_k
$$

All symbols are as defined in the developments of the two algorithms. ($v_1^{(0)}$ and $\tilde{q}_1$ are defined by $v_1^{(0)} = v_1 = \tilde{q}_1$.) You will probably find that the result of part (a) of Exercise 3.4.12 is useful.   □

Now is a good time to summarize and clarify the exact difference between the classical and modified Gram–Schmidt procedures. In the classical algorithm v_k is not touched until the kth step, at which point a multiple of each of the vectors $q_1, \ldots, q_{k-1}$ is subtracted from v_k to form the vector

$$\tilde{q}_k = v_k - \sum_{i=1}^{k-1} r_{ik} q_i$$

which is orthogonal to $q_1, \ldots, q_{k-1}$. Let us call $r_{ik} q_i$ the *component of v_k in the direction of q_i*. This is exactly the multiple of q_i that when subtracted from v_k, yields a vector orthogonal to q_i. Using this new terminology, we can say that at the kth step the component of v_k in the direction of q_i is subtracted from v_k, for $i = 1, \ldots, k - 1$. This takes place all at once.

In contrast the modified algorithm changes v_k little by little. As soon as q_1 has been calculated, the component of v_k in the direction of q_1 is subtracted from v_k to yield

$$v_k^{(1)} = v_k - r_{ik} q_1$$

As soon as q_2 is available, the component in the direction of q_2 is subtracted to yield

$$v_k^{(2)} = v_k^{(1)} - r_{2k} q_2$$

and so on. After $k - 1$ steps we have

$$v_k^{(k-1)} = v_k - \sum_{i=1}^{k-1} r_{ik} q_i = \tilde{q}_k$$

It is hard to see how the two versions of the algorithm could possibly differ in their numerical (stability) properties, since they seem to be performing exactly the same calculations, though not in the same sequence. The difference lies in the manner in which the coefficients r_{ik} are calculated. The classical procedure makes the calculation

$$r_{ik} = (v_k, q_i)$$

whereas the modified procedure calculates

$$r_{ik} = (v_k^{(i-1)}, q_i)$$

In other words, the modified version determines the components of v_k in the direction of q_i, not using v_k, but instead using the vector $v_k^{(i-1)}$, from which the components in the directions $q_1, \ldots, q_{i-1}$ have already been subtracted. This makes an enormous difference in the numerical behavior of the two algorithms.

Instead of trying to explain in general why the modified algorithm is superior to the classical algorithm, we will content ourselves with an example (Exercise 3.4.14) that illustrates the difference. For a proof that the modified Gram–Schmidt method provides a stable means of solving the least-squares problem, see the original paper of Björck (1967).

Exercise 3.4.14   The classical and modified Gram–Schmidt algorithms are identical as far as the calculation of q_1 and q_2 is concerned, so any example that illustrates their difference must have at least three vectors. Let

$$
v_1 = \begin{bmatrix} 1 \\ \epsilon \\ 0 \\ 0 \end{bmatrix} \qquad
v_2 = \begin{bmatrix} 1 \\ 0 \\ \epsilon \\ 0 \end{bmatrix} \qquad
v_3 = \begin{bmatrix} 1 \\ 0 \\ 0 \\ \epsilon \end{bmatrix}
$$

where $\epsilon \ll 1$. (Note that these vectors are nearly linearly dependent.) Suppose ϵ is so small that $\epsilon^2 < u$, where u is the unit roundoff on whatever computer is being used. Then $\mathrm{fl}(1 + \epsilon^2) = 1$.

    (a)   Use the classical Gram–Schmidt method to compute q_1, q_2, and q_3, making the approximation $1 + \epsilon^2 \cong 1$. Although a computer would make additional rounding errors, you may do the rest of the calculations exactly for simplicity. Note that the computed q_2 and q_3 satisfy $(q_2, q_3) = 1/2$. Thus they are far from orthogonal.

    (b)   Repeat part (a) using the modified Gram–Schmidt method, and note that the computed q_2 and q_3 satisfy $(q_2, q_3) = 0$. Since $(q_1, q_2) = -\epsilon/\sqrt{2}$ and $(q_1, q_3) = -\epsilon/\sqrt{6}$, we see that the modified Gram–Schmidt algorithm does a reasonable job of producing orthonormal vectors from the very ill-conditioned starting vectors.   □

Let us take a moment to analyze what happened in Exercise 3.4.14. The two algorithms compute, respectively,

$$
\tilde{q}_3 = v_3 - r_{13}q_1 - r_{23}q_2
$$
$$
v_3^{(2)} = v_3 - r_{13}q_1 - r'_{23}q_2
$$

The two computations are identical in theory and practice, except that $r'_{23} \neq r_{23}$. In principle the numbers $r_{23} = (v_3, q_2)$ and $r'_{23} = (v_3^{(1)}, q_2)$ should be equal because v_3 and $v_3^{(1)}$ differ by a multiple of q_1, and q_1 is orthogonal to q_2. In practice, however, q_1 and q_2 are not quite orthogonal, so there is a small discrepancy between r_{23} and r'_{23}. While this discrepancy is small relative to v_3, it is not small relative to $v_3^{(1)}$, which can be seen to be much smaller than v_3. ($v_3^{(1)}$ is small because cancellation occurs in the computation $v_3^{(1)} = v_3 - r_{13}q_1$, owing to the fact that v_3 and q_1 point in nearly the same direction.) Consequently the "small" discrepancy between $r_{23}q_2$ and $r'_{23}q_2$ actually leads to a large relative difference between $\tilde{q}_3$ and $v_3^{(2)}$. The vector produced by the modified procedure is much more nearly orthogonal to q_2 because the computation $r'_{23} = (v_3^{(1)}, q_2)$

uses $v_3^{(1)}$ and therefore takes into account and corrects for the error in the component in the q_2 direction that was introduced during the computation $v_3^{(1)} = v_3 - r_{13}q_1$. The classical procedure does not take this error into account and therefore fails.

Because of the good numerical properties of reflectors and rotators, the QR decomposition method using either of these types of transformations solves the least-squares problem (full-rank case) with as much accuracy as we could hope for. Björck (1967) showed that the modified Gram–Schmidt method is equally accurate. As the next exercise shows, the modified Gram–Schmidt method is slightly more expensive than the QR decomposition using reflectors.

Exercise 3.4.15 (a) Do a flop count for both the classical and modified Gram–Schmidt algorithms and show that both algorithms require nm^2 flops. (b) Compare this figure with the cost of performing a QR decomposition using reflectors in two cases: (i) $n \gg m$ and (ii) $n = m$. ☐

The Gram–Schmidt procedure was originally presented as an algorithm for orthonormalizing vectors. Only after the connection with the QR decomposition had been made, did we begin to view it as a method for solving the least-squares problem. We can also reverse our viewpoint and view the QR decomposition algorithm using reflectors as an algorithm for calculating orthonormal vectors. The stability of the reflectors implies that it does an excellent job at this task. The vectors produced generally satisfy $|(q_i, q_j)| \approx u$ (if $i \neq j$) where u is the unit roundoff. By contrast the vectors produced by the modified Gram–Schmidt algorithm deviate from orthogonality roughly in proportion to a condition number associated with the vectors $v_1, \ldots, v_m$. Specifically, let $V = [v_1 \;\cdots\; v_m] \in \mathbb{R}^{n \times m}$. The condition number of a nonsquare matrix with linearly independent columns can be defined by

$$\kappa(V) = \frac{\max\limits_{x \neq 0} \dfrac{\|Vx\|}{\|x\|}}{\min\limits_{x \neq 0} \dfrac{\|Vx\|}{\|x\|}} = \frac{\text{maxmag}(V)}{\text{minmag}(V)}$$

The vectors produced by the modified Gram–Schmidt method satisfy roughly $|(q_i, q_j)| \approx u\kappa_2(V)$ for $i \neq j$. See Björck (1967). Intuitively $\kappa_2(V)$ is large whenever $v_1, \ldots, v_m$ are nearly linearly dependent. Thus if the modified Gram–Schmidt process is applied to vectors that are nearly linearly dependent, the resulting vectors can deviate significantly from orthogonality. We conclude that the QR decomposition by reflectors is a superior method for computing orthonormal vectors.

Unfortunately there is an additional cost associated with the QR decomposition by reflectors. The matrix Q is obtained as a product of reflectors: $Q = Q_1 \cdots Q_m$. If we really want to see the orthonormal vectors (the columns of Q), we must assemble Q. The ith column of Q is $q_i = Qe_i = Q_1Q_2 \cdots Q_me_i$, so it can be obtained by applying the reflectors $Q_m, \ldots, Q_1$ to e_i. This operation is less expensive than it at first appears, because most of these reflectors are fundamentally of dimension less than n. Recall that, for example,

$$Q_2 = \begin{bmatrix} 1 & 0 & \cdots & 0 \\ \hline 0 & & & \\ \vdots & & \hat{Q}_2 & \\ 0 & & & \end{bmatrix}$$

where $\hat{Q}_2 \in R^{(n-1)\times(n-1)}$ is a reflector. In general

$$Q_j = \begin{bmatrix} I & 0 \\ \hline 0 & \hat{Q}_j \end{bmatrix}$$

where $\hat{Q}_j$ is a reflector in $R^{(n-j+1)\times(n-j+1)}$. It follows immediately that $Q_j e_i = e_i$ if $j > i$. Therefore $q_1 = Q_1 Q_2 \cdots Q_m e_1 = Q_1 e_1$, $q_2 = Q_1 Q_2 e_2$, and in general

$$q_i = Q_1 Q_2 \cdots Q_i e_i \qquad i = 1, \ldots, m$$

Thus the calculation of q_1 requires $2n$ flops, that of q_2 requires $2n + 2(n-1)$ flops, and so on. You can easily show that the total flop count for calculating $q_1, \ldots, q_m$ is about $nm^2 - m^3/3$ flops, which is as much as the QR decomposition cost originally. Thus the QR decomposition by reflectors is up to twice as expensive as the modified Gram–Schmidt method, considered as a method for calculating orthonormal vectors. Nevertheless the modified Gram–Schmidt method should not be used unless the starting vectors $v_1, \ldots, v_m$ are reasonable well conditioned.

Example 3.4.15   Use a QR decomposition by reflectors to orthonormalize the vectors

$$v_1 = \begin{bmatrix} 3 \\ 0 \\ 4 \end{bmatrix} \qquad \text{and} \qquad v_2 = \begin{bmatrix} -2 \\ 3 \\ 4 \end{bmatrix}$$

In Example 3.3.6 we calculated a QR decomposition of the matrix

$$V = \begin{bmatrix} 3 & -2 \\ 0 & 3 \\ 4 & 4 \end{bmatrix}$$

Q was determined as a product of reflectors, $Q_1 Q_2$, where

$$Q_1 = I - \gamma_1 u^{(1)} u^{(1)T} \qquad \gamma_1 = \frac{1}{40}, \qquad u^{(1)} = \begin{bmatrix} 8 \\ 0 \\ 4 \end{bmatrix}$$

$$Q_2 = \begin{bmatrix} 1 & 0 & 0 \\ \hline 0 & & \\ 0 & & \hat{Q}_2 \end{bmatrix}$$

$$\hat{Q}_2 = I - \gamma_2 u^{(2)} u^{(2)T} \qquad \gamma_2 = \frac{1}{40}, \qquad u^{(2)} = \begin{bmatrix} 8 \\ 4 \end{bmatrix}$$

Thus

$$q_1 = Q_1 e_1 = (I - \gamma_1 u^{(1)} u^{(1)T}) e_1 = e_1 - \gamma_1 (u^{(1)T} e_1) u^{(1)}$$

$$= \begin{bmatrix} 1 \\ 0 \\ 0 \end{bmatrix} - \frac{1}{40}(8) \begin{bmatrix} 8 \\ 0 \\ 4 \end{bmatrix} = \begin{bmatrix} -3/5 \\ 0 \\ -4/5 \end{bmatrix}$$

$$q_2 = Q_1 Q_2 e_2 \quad \text{and} \quad Q_2 e_2 = \begin{bmatrix} 1 & 0 & 0 \\ 0 & & \\ 0 & \hat{Q}_2 & \end{bmatrix} \begin{bmatrix} 0 \\ e_1 \end{bmatrix} = \begin{bmatrix} 0 \\ \hat{Q}_2 e_1 \end{bmatrix}$$

$$\hat{Q}_2 e_1 = (I - \gamma_2 u^{(2)} u^{(2)T}) e_1 = e_1 - \gamma_2 (u^{(2)T} e_1) u^{(2)} = \begin{bmatrix} -3/5 \\ -4/5 \end{bmatrix}$$

so

$$q_2 = Q_1 \begin{bmatrix} 0 \\ -3/5 \\ -4/5 \end{bmatrix} = \begin{bmatrix} 0 \\ -3/5 \\ -4/5 \end{bmatrix} - \frac{1}{40}\left(-\frac{16}{5}\right) \begin{bmatrix} 8 \\ 0 \\ 4 \end{bmatrix} = \begin{bmatrix} 16/25 \\ -15/25 \\ -12/25 \end{bmatrix}$$

Thus

$$q_1 = \begin{bmatrix} -3/5 \\ 0 \\ -4/5 \end{bmatrix} \quad \text{and} \quad q_2 = \begin{bmatrix} 16/25 \\ -15/25 \\ -12/25 \end{bmatrix}$$

Exercise 3.4.16 In Chapter 1 we saw that the various ways of calculating the LU decomposition can be derived by partitioning the equation $A = LU$ in different ways. It is natural to try to do the same thing with the QR decomposition. Consider a decomposition $V = QR$, where $V = [v_1, \ldots, v_k] \in \mathbb{R}^{n \times k}$, $n \geq k$, and V has full rank; $Q = [q_1, \ldots, q_k] \in \mathbb{R}^{n \times k}$ is an isometry; and $R \in \mathbb{R}^{k \times k}$ is upper triangular with positive entries on the main diagonal. Partition the matrices as follows:

$$V = [\tilde{V} \mid v_k] \qquad Q = [\tilde{Q} \mid q_k] \qquad \text{and} \qquad R = \begin{bmatrix} \tilde{R} & s \\ 0 & r_{kk} \end{bmatrix}$$

Derive an algorithm to calculate q_k, given $q_1, \ldots, q_{k-1}$. You may take the following steps:

(a) Use the equation $V = QR$ in partitioned form to derive a formula for $q_k r_{kk}$ in terms of known quantities and s.

(b) The condition that q_k is orthogonal to $q_1, \ldots, q_{k-1}$ can be expressed as $\tilde{Q}^T q_k = 0$. Use this equation to derive a formula for s.

(c) Show that q_k and r_{kk} are uniquely determined by $q_k r_{kk}$ and the conditions $\|q_k\|_2 = 1$ and $r_{kk} > 0$.

(d) Parts (a), (b), and (c) can be combined to yield an algorithm to calculate q_k, given $q_1, \ldots, q_{k-1}$. Show that this algorithm is exactly the classical Gram–Schmidt algorithm.   □

Exercise 3.4.17 Let V, Q, and R be as in the previous exercise, and consider the partition

$$V = [v_1 \mid \hat{V}] \qquad Q = [q_1 \mid \hat{Q}] \qquad \text{and} \qquad R = \left[\begin{array}{c|c} r_{11} & r^T \\ \hline 0 & \hat{R} \end{array} \right]$$

(a) Use this partition to derive an algorithm for calculating Q and R.

(b) Show that this algorithm is exactly the modified Gram–Schmidt procedure.   □

Exercise 3.4.18 State and prove an analog of Theorem 3.4.2 for complex matrices.   □

Exercise 3.4.19 Let $v_1, \ldots, v_m \in \mathbb{R}^n$ and $w_1, \ldots, w_m \in \mathbb{R}^n$ be two linearly independent sets of vectors, and let $V = [v_1 \cdots v_m] \in \mathbb{R}^{n \times m}$ and $W = [w_1 \cdots w_m] \in \mathbb{R}^{n \times m}$. Show that $\langle v_1, \ldots, v_i \rangle = \langle w_1, \ldots, w_i \rangle$ for $i = 1, \ldots, m$ if and only if there exists a nonsingular, upper-triangular matrix $R \in \mathbb{R}^{m \times m}$ such that $V = WR$.   □

3.5
GEOMETRIC APPROACH TO THE LEAST-SQUARES PROBLEM

In this section we introduce a few basic concepts and prove some fundamental theorems that yield a clear geometric picture of the least-squares problem. The tools developed here will also be used in later chapters. It is traditional and possibly more logical to place this material at the beginning of the chapter, but that arrangement would have caused an unnecessary delay in the introduction of the algorithms.

Let $\mathscr{S}$ be any subset of $\mathbb{R}^n$. The *orthogonal complement* of $\mathscr{S}$, denoted $\mathscr{S}^\perp$ (pronounced $\mathscr{S}$ *perp*), is defined to be the set of vectors in $\mathbb{R}^n$ that are orthogonal to $\mathscr{S}$. That is,

$$\mathscr{S}^\perp = \{ x \in \mathbb{R}^n \mid (x, y) = 0 \text{ for all } y \in \mathscr{S} \}$$

The set $\mathscr{S}^\perp$ is nonempty since it always contains the vector 0.

Exercise 3.5.1 (a) Show that the sum of two members of $\mathscr{S}^\perp$ is also in $\mathscr{S}^\perp$. (b) Show that every scalar multiple of a member of $\mathscr{S}^\perp$ is also a member of $\mathscr{S}^\perp$. Thus $\mathscr{S}^\perp$ is a subspace of $\mathbb{R}^n$.   □

Exercise 3.5.2 Let $q_1, \ldots, q_n$ be an orthonormal basis of $\mathbb{R}^n$ and $\mathscr{S} = \langle q_1, \ldots, q_k \rangle$, where $1 \leq k \leq n - 1$. Show that $\mathscr{S}^\perp = \langle q_{k+1}, \ldots, q_n \rangle$.   □

THEOREM 3.5.1 Let $\mathcal{S}$ be a subspace of $\mathbb{R}^n$. Then for every $x \in \mathbb{R}^n$, there exist unique elements $s \in \mathcal{S}$ and $s^\perp \in \mathcal{S}^\perp$ for which

$$x = s + s^\perp$$

Proof Let $v_1, \ldots, v_k$ be a basis for $\mathcal{S}$. From elementary linear algebra we know that there exist vectors $v_{k+1}, \ldots, v_n$ such that $v_1, \ldots, v_k, v_{k+1}, \ldots, v_n$ is a basis of $\mathbb{R}^n$. Let $q_1, \ldots, q_n$ be the orthonormal basis of $\mathbb{R}^n$ obtained by applying the Gram–Schmidt procedure to $v_1, \ldots, v_n$. Then the nesting property (3.4.5) of the Gram–Schmidt process guarantees that $\langle q_1, \ldots, q_k \rangle = \langle v_1, \ldots, v_k \rangle$. Thus, by Exercise 3.5.2,

$$\mathcal{S} = \langle q_1, \ldots, q_k \rangle \qquad \text{and} \qquad \mathcal{S}^\perp = \langle q_{k+1}, \ldots, q_n \rangle$$

Let $x \in \mathbb{R}^n$. We must find $s \in \mathcal{S}$ and $s^\perp \in \mathcal{S}^\perp$ such that $x = s + s^\perp$. With the help of the basis $q_1, \ldots, q_n$, this is easy. x can be expressed uniquely as a linear combination of $q_1, \ldots, q_n$: $x = c_1 q_1 + \cdots + c_n q_n$. Let $s = c_1 q_1 + \cdots + c_k q_k$ and $s^\perp = c_{k+1} q_{k+1} + \cdots + c_n q_n$. Then $s \in \mathcal{S}$, $s^\perp \in \mathcal{S}^\perp$, and $x = s + s^\perp$. To see that this decomposition is unique, suppose $x = \hat{s} + \hat{s}^\perp$, where $\hat{s} \in \mathcal{S}$ and $\hat{s}^\perp \in \mathcal{S}^\perp$. We will show that $s = \hat{s}$ and $s^\perp = \hat{s}^\perp$. Let $r = (s - \hat{s}) \in \mathcal{S}$. Then the equation $s + s^\perp = \hat{s} + \hat{s}^\perp$ implies that $r = (\hat{s}^\perp - s^\perp) \in \mathcal{S}^\perp$. Thus $r \in S \cap \mathcal{S}^\perp$. As you can easily verify, $\mathcal{S} \cap \mathcal{S}^\perp = \{0\}$. Therefore $r = 0$, which implies $s = \hat{s}$ and $s^\perp = \hat{s}^\perp$. □

The unique elements s and $s^\perp$ whose existence is guaranteed by Theorem 3.5.1 are called the *orthogonal projections* of x into $\mathcal{S}$ and $\mathcal{S}^\perp$, respectively.

Given two subspaces $\mathcal{S}$ and $\mathcal{T}$ of $\mathbb{R}^n$ the *sum* $\mathcal{S} + \mathcal{T}$ is defined by

$$\mathcal{S} + \mathcal{T} = \{s + t \mid s \in \mathcal{S}, t \in \mathcal{T}\}$$

You can easily verify that $\mathcal{S} + \mathcal{T}$ is a subspace of $\mathbb{R}^n$. If $\mathcal{U} = \mathcal{S} + \mathcal{T}$, then every $u \in \mathcal{U}$ can be expressed as a sum $u = s + t$, where $s \in \mathcal{S}$ and $t \in \mathcal{T}$. The sum is said to be a *direct sum* if for every $u \in \mathcal{U}$ the decomposition $u = s + t$ is unique. Direct sums are denoted $\mathcal{S} \oplus \mathcal{T}$. Theorem 3.5.1 states that $\mathbb{R}^n$ is the direct sum of $\mathcal{S}$ and $\mathcal{S}^\perp$:

$$\mathbb{R}^n = \mathcal{S} \oplus \mathcal{S}^\perp$$

Exercise 3.5.3 Let $\mathcal{S}$ and $\mathcal{T}$ be subspaces of $\mathbb{R}^n$.

  (a) Prove that $\mathcal{S} + \mathcal{T}$ is a subspace of $\mathbb{R}^n$.
  (b) Prove that $\mathcal{S} + \mathcal{T}$ is a direct sum if and only if $\mathcal{S} \cap \mathcal{T} = \{0\}$. □

Exercise 3.5.4 Let $v_1 = [1 \ 0 \ 0]^T$, $v_2 = [0 \ 1 \ 1]^T$, and $\mathcal{S} = \langle v_1, v_2 \rangle \subseteq \mathbb{R}^3$.

  (a) Find (a basis for) $\mathcal{S}^\perp$.
  (b) Find (a basis for) a subspace $\mathcal{T}$ of $\mathbb{R}^3$ such that $\mathcal{T} \neq \mathcal{S}^\perp$ but $\mathcal{S} \oplus \mathcal{T} = \mathbb{R}^3$. □

Let $A \in \mathbb{R}^{n \times m}$. As you are probably aware, A can be viewed as a linear transformation mapping $\mathbb{R}^m$ into $\mathbb{R}^n$: The vector $x \in \mathbb{R}^m$ is mapped to $Ax \in \mathbb{R}^n$. Two fundamental subspaces associated with a linear transformation are its null space (or kernel) and its range. The *null space*, denoted $\mathcal{N}(A)$, is a subspace of $\mathbb{R}^m$ defined by

$$\mathcal{N}(A) = \{x \in \mathbb{R}^m \mid Ax = 0\}$$

The *range* denoted $\mathcal{R}(A)$, is a subspace of $\mathbb{R}^n$ defined by

$$\mathcal{R}(A) = \{Ax \mid x \in \mathbb{R}^m\}$$

Exercise 3.5.5   (a) Show that $\mathcal{N}(A)$ is a subspace of $\mathbb{R}^m$.
(b) Show that $\mathcal{R}(A)$ is a subspace of $\mathbb{R}^n$.   □

LEMMA 3.5.2   Let $A \in \mathbb{R}^{n \times m}$. Then for all $x \in \mathbb{R}^m$ and $y \in \mathbb{R}^n$

$$(Ax, y) = (x, A^T y)$$

Proof $(Ax, y) = y^T (Ax) = (y^T A)x = (A^T y)^T x = (x, A^T y)$.   □

Note that the inner product on the left is the inner product on $\mathbb{R}^n$, while that on the right is the inner product on $\mathbb{R}^m$.

The matrix $A^T \in \mathbb{R}^{m \times n}$ can be viewed as a linear operator mapping $\mathbb{R}^n$ into $\mathbb{R}^m$. Thus it has a nullspace $\mathcal{N}(A^T) \subseteq \mathbb{R}^n$ and a range $\mathcal{R}(A^T) \subseteq \mathbb{R}^m$. There is an interesting relationship between these spaces and the corresponding spaces associated with A.

THEOREM 3.5.3 $\mathcal{R}(A)^{\perp} = \mathcal{N}(A^T)$.

Proof   If $y \in \mathcal{R}(A)^{\perp}$, then $(Ax, y) = 0$ for all $x \in \mathbb{R}^m$. By Lemma 3.5.2, $(Ax, y) = (x, A^T y)$, so $(x, A^T y) = 0$ for all $x \in \mathbb{R}^m$. Thus in particular we can take $x = A^T y$ and get $(A^T y, A^T y) = 0$. Thus $\| A^T y \|_2^2 = 0$, which implies $A^T y = 0$; that is, $y \in \mathcal{N}(A^T)$. Therefore $\mathcal{R}(A)^{\perp} \subseteq \mathcal{N}(A^T)$. The converse is left as an exercise for you.   □

Exercise 3.5.6   Prove that $\mathcal{N}(A^T) \subseteq \mathcal{R}(A)^{\perp}$.   □

You can easily verify that for any subspace $\mathcal{S} \subseteq \mathbb{R}^n$, $\mathcal{S}^{\perp\perp} = \mathcal{S}$. Therefore an alternate statement of Theorem 3.5.3 is $\mathcal{N}(A^T)^{\perp} = \mathcal{R}(A)$. Since $(A^T)^T = A$, the roles of A and A^T can be reversed in Theorem 3.5.3 to yield the following corollary.

COROLLARY 3.5.4 $\mathcal{N}(A)^{\perp} = \mathcal{R}(A^T)$.

An equivalent statement is $\mathcal{R}(A^T)^{\perp} = \mathcal{N}(A)$. Finally

$$\mathbb{R}^n = \mathcal{R}(A) \oplus \mathcal{N}(A^T)$$
$$\mathbb{R}^m = \mathcal{N}(A) \oplus \mathcal{R}(A^T)$$

In the important special case in which $A \in \mathbb{R}^{n \times n}$ is symmetric, we have $\mathcal{N}(A)^\perp = \mathcal{R}(A)$ and $\mathbb{R}^n = \mathcal{N}(A) \oplus \mathcal{R}(A)$.

Exercise 3.5.7   (a) The *column space* of A is defined to be the subspace of $\mathbb{R}^n$ spanned by the columns of A. Prove that the column space of A is just $\mathcal{R}(A)$.
(b) The *row space* of A is the subspace of $\mathbb{R}^m$ spanned by the rows of A. Prove that the row space of A is just $\mathcal{N}(A)^\perp$.
(c) Obviously the column space of A is the same as the row space of A^T. Use this observation to produce a second proof of Theorem 3.5.3 and Corollary 3.5.4. $\square$

The Discrete Least-Squares Problem

Now let $A \in \mathbb{R}^{n \times m}$ and $b \in \mathbb{R}^n$, $n \geq m$,[§] and consider the least-squares problem for the overdetermined system $Ax = b$. The problem is to find $x \in \mathbb{R}^m$ such that

$$\| b - Ax \|_2 = \min_{w \in \mathbb{R}^m} \| b - Aw \|_2$$

The set of all Aw such that $w \in \mathbb{R}^m$ is $R(A)$, so this problem is obviously closely related to that of finding $y \in R(A)$ such that

$$\| b - y \|_2 = \min_{s \in \mathcal{R}(A)} \| b - s \|_2$$

The next theorem shows that this problem has a unique solution, and it gives a simple characterization of the solution. First we have to prove a lemma.

LEMMA 3.5.5 *(Pythagorean Theorem)*   Let u and v be orthogonal vectors in $\mathbb{R}^n$. Then

$$\| u + v \|^2 = \| u \|^2 + \| v \|^2$$

Proof $\| u + v \|^2 = (u + v, u + v) = (u, u) + (v, u) + (u, v) + (v, v) = (u, u) + (v, v)$, because $(u, v) = (v, u) = 0$. Thus $\| u + v \|^2 = \| u \|^2 + \| v \|^2$.
$\square$

THEOREM 3.5.6   Let $\mathcal{S}$ be a subspace of $\mathbb{R}^n$, and let $b \in \mathbb{R}^n$. Then there exists a unique $y \in \mathcal{S}$ such that

$$\| b - y \|_2 = \min_{s \in \mathcal{S}} \| b - s \|_2 \tag{3.5.7}$$

y is the unique element in $\mathcal{S}$ such that $b - y \in \mathcal{S}^\perp$. In other words, y is the orthogonal projection of b into $\mathcal{S}$.

Proof By Theorem 3.5.1 there exist unique elements $y \in \mathcal{S}$ and $z \in \mathcal{S}^\perp$ such that $b = y + z$. y is the orthogonal projection of b into $\mathcal{S}$. Notice that $b - y = z \in \mathcal{S}^\perp$. There can be no other $w \in \mathcal{S}$ such that $b - w \in \mathcal{S}^\perp$ because then

---

[§] Although $n \gg m$ is typical for least-squares problems, the results that follow are also valid for $n < m$.

the decomposition $b = w + (b - w)$ $(w \in \mathcal{S}, b - w \in \mathcal{S}^\perp)$ would violate the uniqueness part of Theorem 3.5.1. To see that y satisfies (3.5.7), let $s \in \mathcal{S}$ and consider $\| b - s \|_2^2$. Since $b - s = (b - y) + (y - s)$, where $b - y \in \mathcal{S}^\perp$ and $y - s \in \mathcal{S}$, we have by Lemma 3.5.5

$$\| b - s \|_2^2 = \| b - y \|_2^2 + \| y - s \|_2^2$$

As s runs through $\mathcal{S}$, the term $\| b - y \|_2^2$ remains constant, while the term $\| y - s \|_2^2$ remains strictly positive, except that it equals zero when $y = s$. Thus $\| b - s \|_2^2$, and hence also $\| b - s \|_2$, is minimized when and only when $s = y$. $\quad\square$

Exercise 3.5.8 Draw a picture that illustrates Theorem 3.5.6 in the following cases: (a) $\dim(\mathcal{S}) = 1$, (b) $\dim(\mathcal{S}) = 2$. (In the first case the theorem states that the shortest distance from a point, the tip of b, to a straight line $\mathcal{S}$ is along the perpendicular $b - y$. In the second case the line is replaced by a plane.) $\quad\square$

Now let us see what Theorem 3.5.6 tells us about the least-squares problem. Taking $\mathcal{S} = \mathcal{R}(A)$, we find that there is a unique $y \in \mathcal{R}(A)$ such that $\| b - y \|_2 = \min_{s \in \mathcal{R}(A)} \| b - s \|_2 = \min_{w \in \mathbb{R}^m} \| b - Aw \|_2$. Any $x \in \mathbb{R}^m$ for which $Ax = y$ will be a solution to the least-squares problem for the system $Ax = b$. Since $y \in \mathcal{R}(A)$, there must be at least one such x. This proves that the least-squares problem always has at least one solution—a fact that we already proved in Section 3.3 by different means.

Exercise 3.5.9

(a) Suppose $x \in \mathbb{R}^m$ satisfies $Ax = y$. Show that $A\hat{x} = y$ if and only if $x - \hat{x} \in \mathcal{N}(A)$.

(b) Show that the least-squares problem has a unique solution if and only if $\mathcal{N}(A) = \{0\}$. $\quad\square$

We now have two necessary and sufficient conditions for the uniqueness of a solution of the least-squares problem. In Section 3.3 we saw that the solution is unique if and only if A has full rank, and now we see that the solution is unique if and only if $\mathcal{N}(A) = \{0\}$. These two conditions must therefore be equivalent. In the following exercise you are asked to prove directly that they are.

Exercise 3.5.10 Let $A \in \mathbb{R}^{n \times m}$. Prove that $\operatorname{rank}(A) = m$ if and only if $\mathcal{N}(A) = \{0\}$. $\quad\square$

The part of Theorem 3.5.6 that characterizes the minimizing vector yields the following corollary.

COROLLARY 3.5.8 Let $x \in \mathbb{R}^m$. Then $\| b - Ax \|_2 = \min_{w \in \mathbb{R}^m} \| b - Aw \|_2$ if and only if $b - Ax \in \mathcal{R}(A)^\perp$.

Combining this corollary with Theorem 3.5.3, we see that x solves the least-squares problem if and only if $b - Ax \in \mathcal{N}(A^T)$; that is, $A^T(b - Ax) = 0$. Rewriting this last equation, we obtain the following result.

THEOREM 3.5.9   Let $x \in \mathbb{R}^m$. Then x solves the least-squares problem for the system $Ax = b$ if and only if

$$A^T Ax = A^T b \qquad\qquad (3.5.10)$$

The matrix $A^T A$ is in $\mathbb{R}^{m \times m}$, and $A^T b \in \mathbb{R}^m$, so (3.5.10) is a system of m linear equations in m unknowns, known as the *normal equations*.

The coefficient matrix of the normal equations is *positive semidefinite*; that is, it is symmetric, and $x^T(A^T A)x \geq 0$ for all $x \in \mathbb{R}^m$. If rank$(A) = m$, then $A^T A$ is positive definite.

Exercise 3.5.11   (a) Prove that $A^T A$ is positive semidefinite.
(b) Suppose $A \in \mathbb{R}^{n \times m}$, with $n < m$. Show that $A^T A$ is not positive definite.
(c) Suppose $A \in \mathbb{R}^{n \times m}$, with $n \geq m$. Show that $A^T A$ is positive definite if and only if A has full rank. (We already covered the case $n = m$ in Theorem 1.5.4.)
□

In the full-rank case (with $n \geq m$), the unique solution of the least-squares problem can be found by solving the positive definite system (3.5.10) by Cholesky's method. Indeed, up until about 1970 this was the standard technique for solving least-squares problems. It has the advantage of being inexpensive.

Exercise 3.5.12   Count how many flops are required to solve the least-squares problem using the normal equations and Cholesky's method. Don't forget to count the cost of assembling $A^T A$ and $A^T b$. Setting up $A^T A$ is the expensive part of the algorithm, but the symmetry of $A^T A$ halves the cost. Show that when $n \gg m$, this method costs about half what it costs to solve the problem using a QR decomposition by reflectors. Note that in this case (3.5.10) is a small system and the cost of solving it is negligible. How do the costs compare when $n = m$? □

The disadvantage of the normal-equation approach is that it is sometimes less accurate than the QR approach. In fact, critical information can be lost when $A^T A$ is formed.

Example 3.5.11   Let

$$A = \begin{bmatrix} 1 & 1 \\ \epsilon & 0 \\ 0 & \epsilon \end{bmatrix}$$

where $\epsilon > 0$ is small. Clearly A has full rank, and

$$A^T A = \begin{bmatrix} 1 + \epsilon^2 & 1 \\ 1 & 1 + \epsilon^2 \end{bmatrix}$$

which is positive definite. However if ϵ is small enough that $\epsilon^2 < u$, then the computed $A^T A$ will be $\begin{bmatrix} 1 & 1 \\ 1 & 1 \end{bmatrix}$, which is singular.

One way to take care of the problem illustrated by this example is to accumulate and solve the normal equations in double-precision arithmetic. On most computers the resulting increase in computer time will more than offset the flop count advantage.

In spite of its inferior numerical properties, the normal-equation approach is still frequently used to solve least-squares problems. It can be used safely whenever A is reasonably well conditioned. This issue will be covered in more detail in the next section, in which we discuss the sensitivity of the least-squares problem.

Exercise 3.5.13 (a) Let $B \in \mathbb{R}^{n \times m}$ be any matrix such that $\mathcal{R}(B) = \mathcal{R}(A)$. Show that x is a solution of the least-squares problem for the system $Ax = b$ if and only if $B^T Ax = B^T b$.
(b) Show that $\mathcal{R}(A) = \mathcal{R}(B)$ if there exists a nonsingular $C \in \mathbb{R}^{m \times m}$ such that $A = BC$. $\qquad\square$

The most obvious instance of the situation in Exercise 3.5.13 is the case $B = A$, for which the system $B^T Ax = B^T b$ is nothing other than the normal equations. Another interesting instance stems from the decomposition $A = Q\hat{R}$, where $Q \in \mathbb{R}^{n \times m}$ is an isometry and $\hat{R} \in \mathbb{R}^{m \times m}$ is upper triangular. If A has full rank, then $\hat{R}$ is nonsingular, and by part (b) of Exercise 3.5.13, $\mathcal{R}(A) = \mathcal{R}(Q)$. Therefore, by part (a), the unique solution of the least-squares problem satisfies

$$Q^T Ax = Q^T b$$

It is exactly this system that is solved, in the guise $\hat{R}x = \hat{c}$, when the least-squares problem is solved by the QR decomposition method.

Exercise 3.5.14 (Derivation of the normal equations using calculus). The function $f(x_1, x_2, \ldots, x_m) = f(x) = \| Ax - b \|_2^2$ is a differentiable function of m variables. It has a minimum only when $\nabla f = (\partial f / \partial x_1, \ldots, \partial f / \partial x_m)^T = 0$. Calculate ∇f and note that the equations $\nabla f = 0$ are just the normal equations. $\qquad\square$

The Continuous Least-Squares Problem

We introduced the discrete least-squares problem by considering the problem of approximating a discrete point set $\{(t_i, y_i) \mid i = 1, \ldots, n\}$ by a simple curve such as a straight line. In the continuous least-squares problem the discrete point set is replaced by continuous data $\{(t, f(t)) \mid t \in [a, b]\}$. Thus, given a function f defined on some bounded interval $[a, b]$, we seek a simple function ϕ (e.g., a linear polynomial defined on $[a, b]$) such that ϕ approximates f well. The goodness of the approximation is measured, not by calculating a *sum* of squares, but by calculating the *integral* of the square of the error:

$$\int_a^b |f(x) - \phi(x)|^2 dx \qquad (3.5.12)$$

The continuous least-squares problem is solved by minimizing this integral over whichever set of functions ϕ we are allowing as approximations of f. For example, if the approximating function is to be a first-degree polynomial, we minimize (3.5.12) over the set of functions $\{\phi \mid \phi(t) = a_0 + a_1 t; a_0, a_1 \in \mathbb{R}\}$. More generally, if the approximating function is to be a polynomial of degree less than m, we minimize (3.5.12) over the set of functions

$$P_{m-1} = \left\{ \phi \middle| \phi(t) = \sum_{k=0}^{m-1} a_k t^k; a_k \in \mathbb{R}, k = 0, \ldots, m-1 \right\}$$

The set P_{m-1} is an m-dimensional vector space of functions. Still more generally we can let $\mathcal{S}$ be any m-dimensional vector space of functions defined on $[a, b]$ and minimize (3.5.12) over $\mathcal{S}$.

We can solve the continuous least-squares problem by introducing an inner product and a norm for functions and utilizing the geometric ideas introduced in this section. The *inner product* of two functions f and g on $[a, b]$ is defined by

$$(f, g) = \int_a^b f(x) g(x) \, dx$$

This inner product enjoys the same algebraic properties as the inner product on $\mathbb{R}^n$. For example, $(f, g) = (g, f)$, and $(c_1 f_1 + c_2 f_2, g) = c_1(f_1, g) + c_2(f_2, g)$ for any c_1 and $c_2 \in \mathbb{R}$. Two functions f and g are said to be *orthogonal* if $(f, g) = 0$. The *norm* of f is defined by

$$\| f \| = \left(\int_a^b |f(x)|^2 \, dx \right)^{1/2}$$

Notice that the norm and inner product are related by $\| f \| = (f, f)^{1/2}$. Furthermore (3.5.12) can be expressed in terms of the norm as $\| f - \phi \|^2$. Thus the continuous least-squares problem is to find $\phi \in \mathcal{S}$ such that

$$\| f - \phi \| = \min_{\psi \in \mathcal{S}} \| f - \psi \| \tag{3.5.13}$$

We proved Theorem 3.5.6 only for $\mathcal{S} \subseteq \mathbb{R}^n$, but it is valid for spaces of functions as well. See for example Kreyszig (1978), Chapter 3. By Theorem 3.5.6, (3.5.13) has a unique solution, which is characterized by $f - \phi \in \mathcal{S}^\perp$; that is,

$$(f - \phi, \psi) = 0 \qquad \text{for all } \psi \in \mathcal{S} \tag{3.5.14}$$

Let $\phi_1, \ldots, \phi_m$ be a basis for $\mathcal{S}$. Then $\phi = \sum_{j=1}^m x_j \phi_j$ for some unknown coefficients $x_1, \ldots, x_m$. From (3.5.14) it follows that

$$\left(f - \sum_{j=1}^m x_j \phi_j, \phi_i \right) = 0 \qquad i = 1, \ldots, m$$

This is a system of m equations in m unknowns, which can be rewritten as

$$\sum_{j=1}^{m} (\phi_j, \phi_i) x_j = (f, \phi_i) \qquad i = 1, \ldots, m \qquad (3.5.15a)$$

or

$$C x = d \qquad (3.5.15b)$$

where

$$C = \begin{bmatrix} (\phi_1, \phi_1) & \cdots & (\phi_m, \phi_1) \\ \vdots & & \vdots \\ (\phi_1, \phi_m) & \cdots & (\phi_m, \phi_m) \end{bmatrix} \qquad x = \begin{bmatrix} x_1 \\ \vdots \\ x_m \end{bmatrix} \quad \text{and} \quad d = \begin{bmatrix} (f, \phi_1) \\ \vdots \\ (f, \phi_m) \end{bmatrix}$$

The matrix C is clearly symmetric. In fact, it is positive definite, so (3.5.15) can be solved by Cholesky's method to yield the solution of the continuous least-squares problem.

Exercise 3.5.15   With each nonzero $y = [y_1 \; \cdots \; y_m]^T \in \mathbb{R}^m$ associate the nonzero function $\psi = \sum_{j=1}^{m} y_j \phi_j$. Prove that $y^T C y = (\psi, \psi)$. Combining this with the fact that $(\psi, \psi) > 0$, we conclude that C is positive definite.   ☐

The next exercise shows that the equations (3.5.15) are analogous to the normal equations of the discrete least-squares problem.

Exercise 3.5.16   Find $v_1, \ldots, v_m \in \mathbb{R}^n$ for which the normal equations (3.5.10) have the form

$$\sum_{j=1}^{m} (v_j, v_i) x_j = (b, v_i) \qquad i = 1, \ldots, m \qquad ☐$$

Thus the normal equations have the same general form as (3.5.15a).

Exercise 3.5.17   Find the polynomial $\phi(t) = x_1 + x_2 t$ of degree 1 that best approximates $f(t) = t^2$ in the least-squares sense on $[0, 1]$. Check your answer by verifying that $f - \phi$ is orthogonal to $S = \{a_0 + a_1 t \mid a_0, a_1 \in \mathbb{R}\}$.   ☐

Exercise 3.5.18   Let $[a, b] = [0, 1]$, $S = P_{m-1}$, and $\phi_1, \ldots, \phi_m$ be the basis of P_{m-1} defined by $\phi_1(t) = 1, \phi_2(t) = t, \phi_3(t) = t^2, \ldots, \phi_m(t) = t^{m-1}$. Calculate C and note that C is just the m-by-m member of the family of Hilbert matrices introduced in Section 2.6. This is the context in which the Hilbert matrices first arose.   ☐

In Section 2.6 the Hilbert matrices were used as an example of a family of ill-conditioned matrices $H_m \in \mathbb{R}^{m \times m}$, $m = 1, 2, \ldots$, whose conditions get steadily worse with increasing m. Now we can observe, at least intuitively, that the ill conditioning originates with the basis $\phi_1(t) = 1, \phi_2(t) = t, \phi_3(t) = t^2, \ldots$. If you plot the functions t^k on $[0, 1]$, you will see that with increasing k they look more and more alike. That is, they come closer and closer to being linearly dependent. Thus the basis $\phi_1, \ldots, \phi_m$ is (in some sense) ill conditioned and becomes increasingly ill conditioned as m increases. This ill conditioning is inherited by the Hilbert matrices.

3.6
SENSITIVITY OF THE
LEAST-SQUARES PROBLEM

Consider the full-rank least-squares problem. Given $A \in I\!\!R^{n \times m}$ and $b \in I\!\!R^n$, with $n \geq m$ and rank$(A) = m$, there is a unique $x \in I\!\!R^m$ such that

$$\| b - Ax \|_2 = \min_{w \in I\!\!R^m} \| b - Aw \|_2$$

If we now perturb A and b slightly, the solution x will also be altered. In this section we will consider the question of how sensitive x is to perturbations in A and b. We will also consider the sensitivity of the residual $r = b - Ax$. The sensitivity analysis can be combined with backward error analyses to provide an assessment of the accuracy of various methods for solving the least-squares problem. Before you reach the end of this section, you will see that we have not yet developed all the tools we need; we will have to borrow some results from Chapter 7. Nevertheless, now is a good time to discuss this question, while you still have the least-squares problem firmly in mind. As you will see, we can say quite a bit without invoking any results from Chapter 7.

In the previous section it became clear that we can think of the solution of the least-squares problem as a two-stage process. First we find a $y \in \mathcal{R}(A)$ whose distance from b is minimal:

$$\| b - y \|_2 = \min_{s \in \mathcal{R}(A)} \| b - s \|_2$$

Then the least-squares solution $x \in I\!\!R^m$ is found by solving the equation $Ax = y$ exactly. Because A has full rank, the solution is unique. Even though $A \in I\!\!R^{n \times m}$ and $y \in I\!\!R^n$, the equation $Ax = y$ is essentially a system of m equations in m unknowns, for y lies in the m-dimensional space $\mathcal{R}(A)$, and $A : I\!\!R^m \to \mathcal{R}(A)$ can be viewed as a mapping of one m-dimensional space onto another.

Exercise 3.6.1 The m-dimensional nature of the system can be revealed by a judicious change of coordinates. Show that the QR decomposition does just this. How is the vector y determined? How is the system $Ax = y$ solved? $\qquad \square$

Let $r = b - y = b - Ax$. This is the residual of the least-squares problem. By Theorem 3.5.6, r is orthogonal to $\mathcal{R}(A)$. In particular, b, y, and r form a right

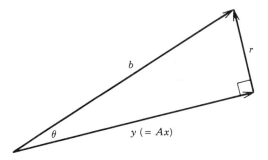

Figure 3.6

triangle, as illustrated in Figure 3.6. The angle θ between b and y is the smallest angle between b and an element of $\mathcal{R}(A)$. It is clear that

$$\| r \|_2 = \| b \|_2 \sin \theta \tag{3.6.1}$$

$$\| y \|_2 = \| b \|_2 \cos \theta \tag{3.6.2}$$

These equations and Figure 3.6 will prove useful in our sensitivity analysis.

Exercise 3.6.2 Let $s \in \mathcal{R}(A)$, and let α be the angle between b and s. Show that $\theta \leq \alpha$. ☐

The Effect of Perturbations in b

At first we will examine the effects of perturbations in b only. Given a perturbation δb, let $y + \delta y$ denote the element of $\mathcal{R}(A)$ that is closest to $b + \delta b$, and let $\hat{x} = x + \delta x$ be the exact solution of $A\hat{x} = y + \delta y$. Then $\hat{x}$ is the minimizer of $\| (b + \delta b) - Aw \|_2$. We would like to be able to say that if $\| \delta b \|_2 / \| b \|_2$ is small, then $\| \delta x \|_2 / \| x \|_2$ is also small. As we shall see, there are two reasons why this might fail to be the case. The first is that $\| \delta y \|_2 / \| y \|_2$ may fail to be small. To see this, refer to Figure 3.6 and consider what happens when b is orthogonal or nearly orthogonal to $\mathcal{R}(A)$. Then y is either zero or very small. It is clear that a small perturbation of b in a direction parallel to $\mathcal{R}(A)$ will cause a perturbation δy that is large relative to $\| y \|_2$. This is illustrated in Figure 3.7. The second problem is that even if $\| \delta y \|_2 / \| y \|_2$ is small, $\| \delta x \|_2 / \| x \|_2$ can be large if the linear system $Ax = y$ is ill conditioned. Any error bound for $\| \delta x \|_2 / \| x \|_2$ must reflect both of these factors. In the next exercise you will derive such a bound.

The condition of the system $Ax = y$ can be measured using the condition number for full-rank nonsquare matrices introduced in Section 3.4:

$$\kappa_2(A) = \frac{\text{maxmag}(A)}{\text{minmag}(A)}$$

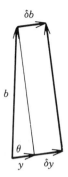

Figure 3.7

where

$$\text{maxmag}(A) = \max_{x \neq 0} \frac{\|Ax\|_2}{\|x\|_2} \qquad \text{and} \qquad \text{minmag}(A) = \min_{x \neq 0} \frac{\|Ax\|_2}{\|x\|_2}$$

We will also use the notation

$$\|A\|_2 = \max_{x \neq 0} \frac{\|Ax\|_2}{\|x\|_2}$$

Exercise 3.6.3

(a) Defining $\delta r = \delta b - \delta y$, we have $\delta b = \delta y + \delta r$. Show that δr is orthogonal to δy. Conclude that

$$\|\delta y\|_2 \leq \|\delta b\|_2$$

(b) Show that provided $\cos\theta \neq 0$,

$$\frac{\|\delta y\|_2}{\|y\|_2} \leq \frac{1}{\cos\theta} \frac{\|\delta b\|_2}{\|b\|_2} \qquad (3.6.3)$$

(c) Show that $\|y\|_2 \leq \text{maxmag}(A)\|x\|_2$, $\|\delta y\|_2 \geq \text{minmag}(A)\|\delta x\|_2$, and

$$\frac{\|\delta x\|_2}{\|x\|_2} \leq \kappa_2(A)\frac{\|\delta y\|_2}{\|y\|_2} \qquad (3.6.4)$$

(d) Conclude that

$$\frac{\|\delta x\|_2}{\|x\|_2} \leq \frac{\kappa_2(A)}{\cos\theta} \frac{\|\delta b\|_2}{\|b\|_2} \qquad (3.6.5)$$

$\square$

Inequality (3.6.3) shows that if θ is not close to $\pi/2$, that is, if b is not nearly orthogonal to the range of A, a small perturbation in b results in a small perturbation in y. Of course, if θ is close to $\pi/2$, we can have a disaster. Such disasters almost never occur in real problems. The angle θ, like $\|r\|_2$, is a measure of how well the least-squares solution fits the data. Indeed, by (3.6.1), $\sin\theta$ equals $\|r\|_2/\|b\|_2$, which is the size of the residual relative to the size of the data b. The angle θ will not be close to $\pi/2$ unless all vectors in $\mathcal{R}(A)$ approximate b very poorly. Thus, as long as the problem is formulated so that the least-squares solution fits the data reasonably well, (3.6.3) will give a useful bound on $\|\delta y\|_2/\|y\|_2$.

Inequality (3.6.4) shows the effect of $\kappa_2(A)$. This inequality is really no different from inequality (2.3.3'), which we derived during the discussion of the sensitivity of square, nonsingular systems.

Inequality (3.6.5) combines the two effects. We see that if $\kappa_2(A)$ is not large and $\cos\theta$ is not close to zero, then a small perturbation in b results in a small perturbation in x.

Inequality (3.6.5) can also be derived using the normal equations (3.5.10). We will work through this second derivation as preparation for the discussion of the effects of perturbations in A. Here we will make our first use of results from Chapter 7. In the derivation we will use the notation $\sigma_1 = \text{maxmag}(A) = \|A\|_2$ and $\sigma_m = \text{minmag}(A)$. Thus $\kappa_2(A) = \sigma_1/\sigma_m$. The significance of this notation is explained in Chapter 7.

From Theorem 3.5.9 we know that the solution of the least-squares problem satisfies the normal equations

$$A^T A x = A^T b \tag{3.6.6}$$

The solution of the perturbed problem satisfies the perturbed normal equations

$$A^T A(x + \delta x) = A^T(b + \delta b) \tag{3.6.7}$$

Subtracting (3.6.6) from (3.6.7), we find that $(A^T A)\delta x = A^T \delta b$, or $\delta x = (A^T A)^{-1} A^T \delta b$. Thus $\|\delta x\|_2 \leq \|(A^T A)^{-1} A^T\|_2 \|\delta b\|_2$. By Exercise 7.3.7, $\|(A^T A)^{-1} A^T\|_2 = \sigma_m^{-1}$, so

$$\|\delta x\|_2 \leq \sigma_m^{-1}\|\delta b\|_2 \tag{3.6.8}$$

On the other hand we see from (3.6.2) that $\|b\|_2 \cos\theta = \|y\|_2 = \|Ax\|_2 \leq \|A\|_2 \|x\|_2 = \sigma_1\|x\|_2$. Thus

$$\frac{1}{\|x\|_2} \leq \frac{\sigma_1}{\cos\theta}\frac{1}{\|b\|_2} \tag{3.6.9}$$

Multiplying (3.6.8) by (3.6.9), we get (3.6.5).

The Effect of Perturbations in A

Unfortunately, perturbations in A have a more severe effect than perturbations in b. As the following theorem shows, the sensitivity of x to perturbations in A depends, to some extent, on the square of the condition number of A. Our approach will be to state the theorem, discuss its implications, then prove it.

THEOREM 3.6.10 Let $A \in \mathbb{R}^{n \times m}$, $b \in \mathbb{R}^n$, $n \geq m$, and $\text{rank}(A) = m$. Let $x \in \mathbb{R}^m$ be the unique solution of

$$\| b - Ax \|_2 = \min_{w \in \mathbb{R}^m} \| b - Aw \|_2$$

Let θ denote the angle between b and Ax, and assume $\theta \neq \pi/2$. Let $\delta A \in \mathbb{R}^{n \times m}$, $\delta b \in \mathbb{R}^n$, $\epsilon_A = \| \delta A \|_2 / \| A \|_2$, $\epsilon_b = \| \delta b \|_2 / \| b \|_2$, and $\epsilon = \max\{\epsilon_a, \epsilon_b\}$. Assume $\epsilon \ll 1$ and, in particular, $\epsilon_A + \frac{1}{2}\epsilon_A^2 \leq 1/(4\kappa_2(A)^2)$. Let $\hat{x} = x + \delta x \in \mathbb{R}^m$ be the unique solution of

$$\| (b + \delta b) - (A + \delta A)\hat{x} \|_2 = \min_{w \in \mathbb{R}^m} \| (b + \delta b) - (A + \delta A)w \|_2$$

Then

$$\frac{\| \delta x \|_2}{\| x \|_2} \leq \frac{2\kappa_2(A)}{\cos \theta}\epsilon_b + 2[\kappa_2(A)^2 \tan \theta + \kappa_2(A)]\epsilon_A + O(\epsilon^2) \qquad (3.6.11)$$

Let $r = b - Ax$ and $\hat{r} = r + \delta r = (b + \delta b) - (A + \delta A)\hat{x}$. Then

$$\frac{\| \delta r \|_2}{\| b \|_2} \leq 2\epsilon_b + 3\kappa_2(A)\epsilon_A + O(\epsilon^2) \qquad (3.6.12)$$

Discussion: In both of the bounds the term $O(\epsilon^2)$ stands for terms that contain factors ϵ_A^2 or $\epsilon_A\epsilon_b$ and are negligible. If we take $\epsilon_A = 0$ in (3.6.11), we get (3.6.5), except for a factor 2. A careful examination of the proof reveals that the factor 2 is an overestimate of a factor that approaches 1 as $\epsilon_A \to 0$. Since ϵ_A is small, the condition $\epsilon_A + \frac{1}{2}\epsilon_A^2 \leq 1/(4\kappa_2(A)^2)$ is about the same as $\epsilon_A < 1/(4\kappa_2(A)^2)$.

The most striking feature of (3.6.11) is that it depends on the square of $\kappa_2(A)$. This is not surprising if one considers that a perturbation in A causes a perturbation in $\mathcal{R}(A)$ that turns out to be (at worst) proportional to $\kappa_2(A)$. Thus a perturbation ϵ_A can cause a change of magnitude $\kappa_2(A)\epsilon_A$ in y [$= Ax \in R(A)$], which can in turn cause a change of magnitude $\kappa_2(A)^2\epsilon$ in x.

The presence of $\kappa_2(A)^2$ in (3.6.11) means that even if A is only mildly ill conditioned, a small perturbation in A can cause a large change in x. An exception is the class of problems for which the least-squares solution fits the data exactly; that is, $r = 0$. Then $\tan \theta = 0$ as well, and the offending term disappears. While a perfect fit is unusual, a good fit is not. If the fit is good, $\tan \theta$ will be small and will tend to cancel out $\kappa_2(A)^2$ to some extent.

Since we measure the size of residuals relative to b, it is not unreasonable to measure the change in the residual relative to $\| b \|_2$, as is done in (3.6.12). Notice that (3.6.12) depends on $\kappa_2(A)$, not $\kappa_2(A)^2$. This means that the residual, and the goodness of fit, is generally much less sensitive to perturbations than the least-squares solution is.

Keeping the Condition Number under Control   From Theorem 3.6.10 it is clear that it is important to avoid having to solve least-squares problems for which the coefficient matrix is ill conditioned. This is something over which you have some control. For the purpose of illustration, consider the problem of fitting a set of data points by a polynomial of low degree. Then the system to be solved has the form (3.1.3); the coefficient matrix is determined by the point abscissas $t_1, t_2, \ldots, t_n$ and the basis functions $\phi_1, \phi_2, \ldots, \phi_m$. In principle any basis for the space of approximating polynomials can be used. However, the choice of basis affects the condition number of the system.

Example 3.6.13   Suppose we need to find the least-squares first-degree polynomial for a set of seven points whose abscissae are

t_1	t_2	t_3	t_4	t_5	t_6	t_7
1.01	1.02	1.03	1.04	1.05	1.06	1.07

The obvious choice of basis for the linear polynomials is $\phi_1(t) = 1$ and $\phi_2(t) = t$. Let us contrast the behavior of this basis with that of the more carefully constructed basis $\tilde{\phi}_1(t) = 1$ and $\tilde{\phi}_2(t) = 30(t - 1.04)$. The two bases give rise to the coefficient matrices

$$A = \begin{bmatrix} 1 & 1.01 \\ 1 & 1.02 \\ 1 & 1.03 \\ 1 & 1.04 \\ 1 & 1.05 \\ 1 & 1.06 \\ 1 & 1.07 \end{bmatrix} \qquad \tilde{A} = \begin{bmatrix} 1 & -.9 \\ 1 & -.6 \\ 1 & -.3 \\ 1 & 0 \\ 1 & .3 \\ 1 & .6 \\ 1 & .9 \end{bmatrix}$$

respectively. The condition numbers are $\kappa_2(A) \approx 104$ and $\kappa_2(\tilde{A}) \approx 1.67$. It is not hard to see why $\tilde{A}$ is so much better than A. The abscissas are concentrated in the interval $[1.01, 1.07]$, on which the function ϕ_2 varies very little; it looks a lot like ϕ_1. Consequently the two columns of A are nearly equal, and A is (mildly) ill conditioned. By contrast the basis $\tilde{\phi}_1$, $\tilde{\phi}_2$ was chosen with the interval $[1.01, 1.07]$ in mind. $\tilde{\phi}_2$ is centered on the interval and varies from -0.9 to $+0.9$ while $\tilde{\phi}_1$ remains constant. As a consequence the columns of $\tilde{A}$ are not nearly dependent. (In fact they are orthogonal.) The factor 30 in the definition of $\tilde{\phi}_2$ guarantees that the second column of $\tilde{A}$ has roughly the same magnitude as the first column, so the columns are not out of scale. Thus $\tilde{A}$ is well conditioned.

In general one wants to choose functions that are not close to being dependent on the interval of interest. Then the columns of the resulting A will not be close to being dependent. As a consequence A will be well conditioned.

Accuracy of Techniques for Solving the Least-Squares Problem   A standard backward error analysis [SLS] shows that the QR decomposition method using reflectors or rotators is stable. Thus the computed solution is the exact solution of a perturbed problem $\min \|(b + \delta b) - (A + \delta A)w\|_2$, where the perturbations are of the same order of magnitude as the unit roundoff u. This means that this method works as well as we could hope. The computed solution and residual satisfy the bounds (3.6.11) and (3.6.12) with $\epsilon_A \approx \epsilon_b \approx u$.

As Björck (1967) has shown, the modified Gram–Schmidt method is about as accurate as the QR decomposition by reflectors.

The method of normal equations has somewhat different characteristics. Suppose $A^T A$ and $A^T b$ have been calculated exactly, and the system $A^T A x = A^T b$ has been solved by Cholesky's method. From Section 2.5 we know that the computed solution $\hat{x} = x + \delta x$ is the exact solution of a perturbed equation $(A^T A + E)\hat{x} = A^T b$, where $\|E\|_2 / \|A^T A\|_2$ is of about the same magnitude as u. It follows from Theorem 2.3.17 that $\|\delta x\|_2 / \|x\|_2$ is roughly $\kappa_2(A^T A)u$. By Exercise 7.3.4, $\kappa_2(A^T A) = \kappa_2(A)^2$. Given that the factor $\kappa_2(A)^2$ also appears in (3.6.11), it looks as if the normal-equations method is about as accurate as the other methods. However, in problems with a small residual, the $\kappa_2(A)^2 \tan\theta$ term in (3.6.11) is diminished. In these cases the other methods will be more accurate.

This analysis has ignored the effect of errors in computing $A^T A$ and $A^T b$. As Example 3.5.11 demonstrated, significant errors can occur in the computation of $A^T A$. Furthermore, if $\cos\theta$ is near zero, cancellations will occur in the calculation of $A^T b$ and result in large relative errors.

Proof of Theorem 3.6.10 First of all, the perturbed problem has a unique solution, since $A + \delta A$ has full rank. This is a consequence of Corollary 7.3.7.

Rather than proving the theorem in full generality, we will examine the case in which only A is perturbed; that is, $\delta b = 0$. This will simplify the manipulations somewhat without altering the spirit of the proof. The complete proof is left as an exercise for you.

The solution $\hat{x} = x + \delta x$ of the perturbed least-squares problem must satisfy the perturbed normal equations

$$(A + \delta A)^T (A + \delta A)(x + \delta x) = (A + \delta A)^T b$$

Subtracting the unperturbed normal equations $A^T A x = A^T b$ and making some routine manipulations, we find that

$$A^T A \delta x = \delta A^T (b - Ax) - A^T \delta A x - (A^T \delta A + \delta A^T A + \delta A^T \delta A)\delta x - \delta A^T \delta A x$$

or

$$\delta x = (A^T A)^{-1} \delta A^T r - (A^T A)^{-1} A^T \delta A x - (A^T A)^{-1}(A^T \delta A + \delta A^T A + \delta A^T \delta A)\delta x$$
$$- (A^T A)^{-1} \delta A^T \delta A x \tag{3.6.14}$$

Taking norms of both sides and using the fact that $\|B^T\|_2 = \|B\|_2$ for any B (Corollary 7.3.2), we obtain the inequality

$$\|\delta x\|_2 \leq \|(A^T A)^{-1}\|_2 \|\delta A\|_2 \|r\|_2 + \|(A^T A)^{-1} A^T\|_2 \|\delta A\|_2 \|x\|_2$$
$$+ \|(A^T A)^{-1}\|_2 (2 \|A\|_2 \|\delta A\|_2 + \|\delta A\|_2^2) \|\delta x\|_2$$
$$+ \|(A^T A)^{-1}\|_2 \|\delta A\|_2^2 \|x\|_2$$

Using the notation $\sigma_1 = \text{maxmag}(A)$, $\sigma_m = \text{minmag}(A)$, and $\kappa_2(A) = \sigma_1/\sigma_m$ introduced earlier, we have $\|(A^T A)^{-1}\|_2 = \sigma_m^{-2}$ and $\|(A^T A)^{-1} A^T\|_2 = \sigma_m^{-1}$, by Exercise 7.3.7, and also $\|\delta A\|_2 = \sigma_1 \epsilon_A$. Furthermore, by (3.6.1) and (3.6.2), $\|r\|_2 = \|b\|_2 \sin\theta = \|Ax\|_2 \tan\theta \le \sigma_1 \|x\|_2 \tan\theta$. Applying these results and moving all terms involving $\|\delta x\|_2$ to the left-hand side of the inequality, we have

$$[1 - 2\kappa_2(A)^2(\epsilon_A + \tfrac{1}{2}\epsilon_A^2)]\,\|\delta x\|_2 \le [\kappa_2(A)^2 \tan\theta + \kappa_2(A)]\epsilon_A \|x\|_2 + \kappa_2(A)^2 \epsilon_A^2 \|x\|_2$$

The condition $\epsilon_A + \tfrac{1}{2}\epsilon_A^2 \le 1/(4\kappa_2(A)^2)$ guarantees that the coefficient of $\|\delta x\|_2$ is at least 1/2. Therefore

$$\frac{\|\delta x\|_2}{\|x\|_2} \le 2[\kappa_2(A)^2 \tan\theta + \kappa_2(A)]\epsilon_A + O(\epsilon_A^2)$$

This is (3.6.11) in the case $\epsilon_b = 0$.

To obtain the bound (3.6.12) on the residual, we note first that since $r + \delta r = b - (A + \delta A)(x + \delta x)$ and $r = b - Ax$,

$$\delta r = -\delta A\, x - A\, \delta x - \delta A\, \delta x$$

Using (3.6.14) and noting that by (3.6.11) $\|\delta A\|_2 \|\delta x\|_2 = O(\epsilon^2)$, we obtain

$$\|\delta r\|_2 \le \|\delta A\|_2 \|x\|_2 + \|A(A^T A)^{-1}\|_2 \|\delta A\|_2 \|r\|_2$$
$$+\|A(A^T A)^{-1} A^T\|_2 \|\delta A\|_2 \|x\|_2 + O(\epsilon^2)$$

Since $\|Ax\|_2 \ge \sigma_m \|x\|_2$, we have $\|x\|_2 \le \sigma_m^{-1}\|Ax\|_2 = \sigma_m^{-1}\cos\theta\|b\|_2$. Also, by Exercise 7.3.7, $\|A(A^T A)^{-1}\|_2 = \sigma_m^{-1}$ and $\|A(A^T A)^{-1} A^T\|_2 = 1$. Therefore

$$\frac{\|\delta r\|_2}{\|b\|_2} \le (2\cos\theta + \sin\theta)\kappa_2(A)\epsilon_A + O(\epsilon^2)$$

$$< 3\kappa_2(A)\epsilon_A + O(\epsilon^2)$$

This is (3.6.12) in the case $\epsilon_b = 0$.   □

Exercise 3.6.4 Work out a complete proof of Theorem 3.6.10 with $\epsilon_A > 0$ and $\epsilon_b > 0$.
                                                                                    □

4

Eigenvalues and Eigenvectors I

4.1
SYSTEMS OF DIFFERENTIAL EQUATIONS

Eigenvalues and eigenvectors turn up in stability theory, theory of vibrations, quantum mechanics, continuum mechanics, statistical analysis, and many other areas. In many of these applications eigenvalues and eigenvectors arise naturally from the consideration of systems of differential equations. Let us look at a system of n first-order, linear, ordinary differential equations in n unknown functions $x_1(t), \ldots, x_n(t)$:

$$\frac{dx_1}{dt} = a_{11}x_1 + a_{12}x_2 + \cdots + a_{1n}x_n$$

$$\frac{dx_2}{dt} = a_{21}x_1 + a_{22}x_2 + \cdots + a_{2n}x_n$$

$$\vdots \qquad \vdots \qquad \vdots \qquad \qquad \vdots$$

$$\frac{dx_n}{dt} = a_{n1}x_1 + a_{n2}x_2 + \cdots + a_{nn}x_n$$

The coefficients a_{ij} are constant. The system can be expressed more compactly in matrix notation as

$$\frac{dx}{dt} = Ax \tag{4.1.1}$$

199

where

$$A = \begin{bmatrix} a_{11} & a_{12} & \cdots & a_{1n} \\ a_{21} & a_{22} & \cdots & a_{2n} \\ \vdots & \vdots & & \vdots \\ a_{n1} & a_{n2} & \cdots & a_{nn} \end{bmatrix} \qquad x = \begin{bmatrix} x_1 \\ x_2 \\ \vdots \\ x_n \end{bmatrix} \qquad \text{and} \qquad \frac{dx}{dt} = \begin{bmatrix} \dfrac{dx_1}{dt} \\ \dfrac{dx_2}{dt} \\ \vdots \\ \dfrac{dx_n}{dt} \end{bmatrix}$$

A common approach to solving linear differential equations is to begin by look-ing for solutions of a particularly simple form. Therefore let us look for solutions of the form

$$x(t) = f(t)v \tag{4.1.2}$$

where $f(t)$ is a nonzero scalar (real or complex) function of t, and v is a nonzero constant vector. The time-varying nature of $x(t)$ is expressed by $f(t)$, while the vector nature of $x(t)$ is expressed by v. Substituting the form (4.1.2) into (4.1.1), we obtain the equation

$$f'(t)v = f(t)Av$$

or equivalently

$$\frac{f'(t)}{f(t)} v = Av \tag{4.1.3}$$

Since v and Av are constant vectors, (4.1.3) implies that $f'(t)/f(t)$ must be constant. That is, there exists a (real or complex) constant λ such that

$$\frac{f'(t)}{f(t)} = \lambda \tag{4.1.4}$$

In addition (4.1.3) implies

$$Av = \lambda v$$

A nonzero vector v for which there exists a λ such that $Av = \lambda v$ is called an *eigenvector* of A. The number λ is called the *eigenvalue* of A associated with v. So far we have shown that if $x(t)$ is a solution of (4.1.1) of the form (4.1.2), then v must be an eigenvector of A and $f(t)$ must satisfy the differential equation (4.1.4), where λ is the eigenvalue of A associated with v. The general solution of the scalar differential equation (4.1.4) is $f(t) = ce^{\lambda t}$, where c is an arbitrary constant. Conversely, if v is an eigenvector of A with eigenvalue λ, then

$$e^{\lambda t}v$$

is a solution of (4.1.1), as you can easily verify. Thus each eigenvector of A gives rise to a solution of (4.1.1). If A has enough eigenvectors, then every solution of (4.1.1) can be realized as a linear combination of these simple solutions. Specifically, suppose A has a set of n linearly independent eigenvectors $v_1, \ldots, v_n$ with associated eigenvalues $\lambda_1, \ldots, \lambda_n$. Then for any constants $c_1, \ldots, c_n$,

$$x(t) = c_1 e^{\lambda_1 t} v_1 + \cdots + c_n e^{\lambda_n t} v_n \qquad (4.1.5)$$

is a solution of (4.1.1). (This is so regardless of whether or not $v_1, \ldots, v_n$ are independent.) Since $v_1, \ldots, v_n$ are independent, (4.1.5) turns out to be the general solution of (4.1.1); that is, every solution of (4.1.1) has the form (4.1.5).

Exercise 4.1.1 Let $\hat{x} \in \mathbb{C}^n$ be an arbitrary vector. (a) Show that there exist (unique) $c_1, \ldots, c_n$ such that the function $x(t)$ of (4.1.5) has the property $x(0) = \hat{x}$. (b) It is a basic fact of the theory of differential equations that for a given $\hat{x} \in \mathbb{C}^n$, the *initial value problem* $dx/dt = Ax$, $x(0) = \hat{x}$, has exactly one solution (see any text on differential equations). Use this fact together with the result of part (a) to show that every solution of (4.1.1) has the form (4.1.5). $\qquad \square$

An n-by-n matrix that possesses a set of n linearly independent eigenvectors is said to be a *simple matrix*.[§] In the next section we will see that in some sense "most" matrices are simple. Thus for "most" systems of the form (4.1.1), (4.1.5) is the general solution.

It is easy to show (see Section 4.2) that λ is an eigenvalue of A if and only if λ is a solution of the *characteristic equation* $\det(\lambda I - A) = 0$. For each eigenvalue λ, a corresponding eigenvector (or eigenvectors) can be found by solving the equation $(\lambda I - A)v = 0$. For small enough (e.g., 2-by-2) systems of differential equations it is possible to solve the characteristic equation exactly and thereby solve the differential equation.

Example 4.1.6 Find the general solution of

$$\dot{x}_1 = 2x_1 + 3x_2$$
$$\dot{x}_2 = x_1 + 4x_2$$

The coefficient matrix is $A = \begin{bmatrix} 2 & 3 \\ 1 & 4 \end{bmatrix}$, whose characteristic equation is $0 = \det(\lambda I - A) = \lambda^2 - 6\lambda + 5 = (\lambda - 1)(\lambda - 5)$. Therefore the eigenvalues are $\lambda_1 = 1$ and $\lambda_2 = 5$. Solving the equation $(\lambda I - A)v = 0$ with $\lambda = \lambda_1$ and $\lambda = \lambda_2$, we find (nonunique) solutions

$$v_1 = \begin{bmatrix} 3 \\ -1 \end{bmatrix} \quad \text{and} \quad v_2 = \begin{bmatrix} 1 \\ 1 \end{bmatrix}$$

[§] A synonym for *simple* is *nondefective*. Matrices that are not simple are usually called *defective*, although the term *nonsimple* is also used.

Since these two $(=n)$ vectors are obviously linearly independent, we conclude that the general solution of the system is

$$x(t) = c_1 e^t \begin{bmatrix} 3 \\ -1 \end{bmatrix} + c_2 e^{5t} \begin{bmatrix} 1 \\ 1 \end{bmatrix}$$

Stability of Differential Equations

In stability theory the behavior of solutions of differential equations as $t \to \infty$ is studied. The characteristics of the system (4.1.1) can be determined by examining (4.1.5). Suppose for example that all λ_k are real and negative. Then every solution of (4.1.1) tends to zero as $t \to \infty$. On the other hand, if some λ_k is positive, then there will be solutions for which $\| x(t) \| \to \infty$ as $t \to \infty$. In the former case (4.1.1) is said to be *asymptotically stable*, in the latter case *unstable*. What happens if some λ_k is complex? Say $\lambda_k = \alpha_k + \beta_k i$. Then $e^{\lambda_k t} = e^{\alpha_k t} e^{i \beta_k t} = e^{\alpha_k t} (\cos \beta_k t + i \sin \beta_k t)$, which grows if $\alpha_k > 0$ and decays if $\alpha_k < 0$. Since $\alpha_k = \text{Re}(\lambda_k)$, we conclude that (4.1.1) is asymptotically stable if all eigenvalues of A satisfy $\text{Re}(\lambda_k) < 0$, and (4.1.1) is unstable if there is an eigenvalue for which $\text{Re}(\lambda_k) > 0$. We have come to this conclusion under the assumption that the coefficient matrix A is simple. Fortunately the conclusion turns out to be valid for all A.

We note in passing that complex eigenvalues signal oscillations in the solution. The rate of oscillation of a component is governed by $\beta_k = \text{Im}(\lambda_k)$.

Now consider a nonlinear system of n first-order differential equations

$$\begin{aligned}
\dot{x}_1 &= f_1 (x_1, x_2, \ldots, x_n) \\
\dot{x}_2 &= f_2 (x_1, x_2, \ldots, x_n) \\
&\ \vdots \qquad\qquad \vdots \\
\dot{x}_n &= f_n (x_1, x_2, \ldots, x_n)
\end{aligned}$$

or in vector notation

$$\dot{x} = f(x) \tag{4.1.7}$$

A vector $\hat{x}$ is called an *equilibrium point* of (4.1.7) if $f(\hat{x}) = 0$, since then the constant function $x(t) = \hat{x}$ is a solution. An equilibrium point $\hat{x}$ is called *asymptotically stable* if there is a neighborhood N of $\hat{x}$ such that for every solution of (4.1.7) for which $x(t) \in N$ for some t, it must happen that $\lim_{t \to \infty} x(t) = \hat{x}$. Physically this means that a system whose state is perturbed slightly from an equilibrium point will return to that equilibrium point. We can analyze the stability of each equilibrium point by linearizing (4.1.8) about that point.

Specifically, consider a multivariate Taylor expansion of f about $\hat{x}$:

$$f_i(x_1, \ldots, x_n) = f_i(\hat{x}) + \sum_{j=1}^{n} \frac{\partial f_i}{\partial x_j} (\hat{x})(x_j - \hat{x}_j)$$

$$+ \frac{1}{2} \sum_{j=1}^{n} \sum_{k=1}^{n} \frac{\partial^2 f_i}{\partial x_j \, \partial x_k} (\hat{x})(x_j - \hat{x}_j)(x_k - \hat{x}_k) + \cdots \qquad i = 1, \ldots, n$$

In vector notation we have

$$f(x) = f(\hat{x}) + \frac{\partial f}{\partial x}(\hat{x})(x - \hat{x}) + \text{higher-order terms}$$

where $\partial f / \partial x$ is the *Jacobian matrix*, whose i, j entry is $\partial f_i / \partial x_j$. For x near $\hat{x}$ the quadratic and higher terms are very small. Also $f(\hat{x}) = 0$, so $f(x) \approx \partial f / \partial x (\hat{x})(x - \hat{x})$. It is therefore reasonable to expect that for x near $\hat{x}$ the solutions of (4.1.7) are approximated well by solutions of

$$\frac{d}{dt}(x - \hat{x}) = \dot{x} = \frac{\partial f}{\partial x}(\hat{x})(x - \hat{x}) \qquad (4.1.8)$$

which is linear with coefficient matrix $\partial f / \partial x (\hat{x})$. From our study of linear systems we know that all solutions of (4.1.8) satisfy $x - \hat{x} \to 0$ as $t \to \infty$ if all eigenvalues of $\partial f / \partial x (\hat{x})$ satisfy $\text{Re}(\lambda_k) < 0$. We conclude that $\hat{x}$ is an asymptotically stable equilibrium point of (4.1.7) if the eigenvalues of the Jacobian $\partial f / \partial x (\hat{x})$ all have negative real parts.

Exercise 4.1.2 The differential equation of a damped pendulum (Figure 4.1) is

$$\ddot{\theta} + k_1 \dot{\theta} + k_2 \sin \theta = 0$$

where k_1 and k_2 are positive constants and θ is the angle of the pendulum from its vertical resting position. Introducing new variables $x_1 = \theta$ and $x_2 = \dot{\theta}$, rewrite the second-order differential equation as a system of two first-order differential equations in x_1 and x_2. Find all equilibrium points of the system and note that they correspond to $\dot{\theta} = 0$, $\theta = n\pi$, $n = 0, \pm 1, \pm 2, \dots$. Calculate the Jacobian of the system and show that the equilibrium points $\theta = n\pi$ are asymptotically stable for even n and unstable for odd n. Interpret your result physically. □

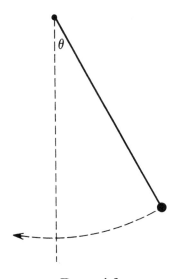

Figure 4.1

The small examples that we have considered in this section can be solved by forming and factoring the characteristic polynomial. For larger problems this approach is generally impractical because (1) $\det(\lambda I - A)$ is expensive to compute and (2) the problem of finding all solutions of $\det(\lambda I - A) = 0$ can be difficult. It will behoove us therefore to investigate some other approaches to solving the eigenvalue problem.

We have looked at two applications of eigenvalues and eigenvectors involving differential equations. To solve a linear system of differential equations, we require both eigenvalues and eigenvectors; to assess the stability of a stationary point of a nonlinear system, we require only a knowledge of the eigenvalues. This begins to give us an idea of what sorts of algorithms will be useful. There is a need for algorithms that calculate all eigenvalues and eigenvectors of a matrix, but algorithms that calculate only the eigenvalues are also useful. In addition there are applications for which only some of the eigenvalues (and possibly eigenvectors) are needed.

4.2
BASIC PROPERTIES OF EIGENVALUES
AND EIGENVECTORS

Most applications require the eigenvalues of a real matrix A. Nevertheless, as we shall soon see, the natural setting for the study of eigenvalues is the field of complex numbers. Thus let $A \in \mathbb{C}^{n \times n}$. As we already noted in the previous section, a vector $v \in \mathbb{C}^n$ is called an *eigenvector* of A if $v \neq 0$ and Av is a multiple of v; that is, there exists a $\lambda \in \mathbb{C}$ such that

$$Av = \lambda v \qquad (4.2.1)$$

The scalar λ is called the *eigenvalue* of A associated with v. Similarly v is called *an* eigenvector of A *associated with* the eigenvalue λ. Whereas each eigenvector has a unique eigenvalue associated with it, each eigenvalue is associated with many eigenvectors: if v is an eigenvector of A associated with λ, then every nonzero multiple of v is also an eigenvector of A associated with λ.

Frequently we will wish to work with eigenvectors with unit norm. This is always possible, for given any eigenvector v of A, the vector $\hat{v} = (1/\|v\|)v$ is also an eigenvector of A associated with the same eigenvalue, and $\|\hat{v}\| = 1$. This normalization process can be carried out with any vector norm.

Exercise 4.2.1 Let λ be an eigenvalue of A and consider the set $\mathcal{S}_\lambda = \{v \in \mathbb{C}^n \mid Av = \lambda v\}$, which consists of all eigenvectors associated with λ, together with the zero vector. (a) Show that if v_1 and $v_2 \in \mathcal{S}_\lambda$, then $v_1 + v_2 \in \mathcal{S}_\lambda$. (b) Show that if $v \in \mathcal{S}_\lambda$ and $c \in \mathbb{C}$, then $cv \in \mathcal{S}_\lambda$. Thus $\mathcal{S}_\lambda$ is a subspace of $\mathbb{C}^n$.   □

The space $\mathcal{S}_\lambda$ introduced in Exercise 4.2.1 is called the *eigenspace* of A associated with λ. Even if λ is not an eigenvalue, we can still define the space $\mathcal{S}_\lambda = \{v \in \mathbb{C}^n \mid Av = \lambda v\}$, but in that case $\mathcal{S}_\lambda = \{0\}$. In fact it is clear that λ

is an eigenvalue of A if and only if $\mathscr{S}_\lambda \neq \{0\}$. Recall from Chapter 3 that the null space of A is defined by $\mathcal{N}(A) = \{v \in \mathbb{C}^n \mid Av = 0\}$. Notice that $\mathcal{N}(A) = \mathscr{S}_0$, and 0 is an eigenvalue of A if and only if $\mathcal{N}(A) \neq \{0\}$. This is true in turn if and only if A is singular. More generally, note that Eq. 4.2.1 can be rewritten as

$$(\lambda I - A)v = 0$$

Thus a nonzero v is an eigenvector of A if and only if $(\lambda I - A)v = 0$; that is, $v \in \mathcal{N}(\lambda I - A)$. In other words $\mathscr{S}_\lambda = \mathcal{N}(\lambda I - A)$. It follows then that λ is an eigenvalue of A if and only if $\mathcal{N}(\lambda I - A) \neq \{0\}$. This holds in turn if and only if $\lambda I - A$ is singular; that is, $\det(\lambda I - A) = 0$. Thus we have the following two theorems.

THEOREM 4.2.2 λ is an eigenvalue of A if and only if $\mathcal{N}(\lambda I - A) \neq \{0\}$.

THEOREM 4.2.3 λ is an eigenvalue of A if and only if

$$\det(\lambda I - A) = 0 \tag{4.2.4}$$

Equation 4.2.4 is called the *characteristic equation* of A. It is not hard to see that $\det(\lambda I - A)$ is a polynomial in λ of degree n. It is called the *characteristic polynomial* of A. The characteristic equation therefore has n complex roots, so A has n eigenvalues, some of which may be repeated. If A is a real matrix, then the characteristic polynomial has real coefficients. But the zeros of a real polynomial are not necessarily real, so a real matrix can have complex eigenvalues. This is why it makes sense to work with complex numbers from the beginning. Notice, however, that if a real matrix has complex eigenvalues, they must occur in complex conjugate pairs; that is, if $\lambda = \alpha + \beta i$ (α and $\beta \in \mathbb{R}$) is an eigenvalue of $A \in \mathbb{R}^{n \times n}$, then $\bar{\lambda} = \alpha - \beta i$ is also an eigenvalue.

Exercise 4.2.2 Prove the above assertion two different ways: (a) Let $p(\lambda) = a_0 + a_1\lambda + a_2\lambda^2 + \cdots + a_n\lambda^n$ be a polynomial with real coefficients $a_0, \ldots, a_n$. Show that for any $z = \alpha + \beta i \in \mathbb{C}$, $p(\bar{z}) = \overline{p(z)}$. Conclude that $p(z) = 0$ if and only if $p(\bar{z}) = 0$.
(b) Let $A \in \mathbb{R}^{n \times n}$. Let λ be a complex eigenvalue of A with an associated eigenvector v. Let $\bar{v}$ be the vector obtained by taking complex conjugates of the components of v. Show that $\bar{v}$ is an eigenvector of A whose associated eigenvalue is $\bar{\lambda}$. ☐

Now is a good time to mention one other important point about real matrices. Let $A \in \mathbb{R}^{n \times n}$ and suppose it has a real eigenvalue λ. Then in the homogeneous equation $(\lambda I - A)v = 0$, the coefficient matrix $\lambda I - A$ is real. The fact that $\lambda I - A$ is singular implies that the equation $(\lambda I - A)v = 0$ has nontrivial *real* solutions. We conclude that every real eigenvalue of a real matrix has a real eigenvector associated with it.

Exercise 4.2.3 Let $A \in \mathbb{C}^{n \times n}$. (a) Use the characteristic equation to show that A and A^T have the same eigenvalues.
(b) Show that λ is an eigenvalue of A if and only if $\bar{\lambda}$ is an eigenvalue of A^*. (A^* denotes the complex conjugate transpose of A.) ☐

Exercise 4.2.4 There are some types of matrices for which the eigenvalues are obvious. For example, show that if T is a (lower- or upper-) triangular matrix, then its main-diagonal entries are its eigenvalues. Give two proofs: (a) using Theorem 4.2.2, (b) using Theorem 4.2.3. ☐

The next theorem provides a useful generalization of Exercise 4.2.4.

THEOREM 4.2.5 Let A be a block triangular matrix, say

$$A = \begin{bmatrix} A_{11} & A_{12} & \cdots & A_{1m} \\ & A_{22} & & A_{2m} \\ & & \ddots & \\ & & & A_{mm} \end{bmatrix}$$

Then the set of eigenvalues of A, counting multiplicity, equals the union of the sets of eigenvalues of $A_{11}, \ldots, A_{mm}$.

Proof Because $\lambda I - A$ is block triangular, $\det(\lambda I - A) = \det(\lambda I - A_{11}) \det(\lambda I - A_{22}) \ldots \det(\lambda I - A_{mm})$. Thus the set of roots of the characteristic polynomial of A equals the union of the roots of the characteristic polynomials of $A_{11}, \ldots, A_{mm}$. (Although A has been depicted as being block upper triangular, the result is obviously valid for block lower-triangular matrices as well.) ☐

If you did not like the determinant argument used in the proof of Theorem 4.2.5, you can work the next exercise, which gives a second proof based on eigenvectors. This second proof does not, however, yield the fact that the multiplicities are equal.

Exercise 4.2.5 Let $A \in \mathbb{C}^{n \times n}$ be a block triangular matrix:

$$A = \begin{bmatrix} A_{11} & A_{12} \\ 0 & A_{22} \end{bmatrix}$$

where $A_{11} \in \mathbb{C}^{j \times j}$ and $A_{22} \in \mathbb{C}^{k \times k}$, $j + k = n$.

(a) Let λ be an eigenvalue of A_{11} with eigenvector v. Show that there exists $w \in \mathbb{C}^k$ such that $\begin{bmatrix} v \\ w \end{bmatrix}$ is an eigenvector of A with associated eigenvalue λ.

(b) Let λ be an eigenvalue of A_{22} with associated eigenvector w. Suppose λ is not also an eigenvector of A_{11}. Show that there is a unique $v \in \mathbb{C}^j$ such that $\begin{bmatrix} v \\ w \end{bmatrix}$ is an eigenvector of A associated with the eigenvalue λ.

(c) Let λ be an eigenvalue of A with associated eigenvector $\begin{bmatrix} v \\ w \end{bmatrix}$. Show that either w is an eigenvector of A_{22} with associated eigenvalue λ or v is an eigenvector of A_{11} with associated eigenvalue λ.

(d) Combining parts (a), (b), and (c), show that λ is an eigenvalue of A if and only if it is an eigenvalue of A_{11} or A_{22}. $\square$

THEOREM 4.2.6   Let $v_1, \ldots, v_k$ be eigenvectors of $A \in \mathbb{C}^{n \times n}$ associated with distinct eigenvalues $\lambda_1, \ldots, \lambda_k$. Then $v_1, \ldots, v_k$ are linearly independent.

Proof The proof is by induction on k. The result is trivial in the case of a single vector. Now let us assume the result for sets of $k - 1$ vectors and show that it follows for sets of k vectors. Suppose

$$c_1 v_1 + c_2 v_2 + \cdots + c_k v_k = 0 \qquad (4.2.7)$$

We must show that $c_1 = c_2 = \cdots = c_k = 0$. Applying A to (4.2.7) and using the equations $A v_i = \lambda_i v_i$, $i = 1, \ldots, k$, we find that

$$\lambda_1 c_1 v_1 + \lambda_2 c_2 v_2 + \cdots + \lambda_k c_k v_k = 0 \qquad (4.2.8)$$

Multiplying (4.2.7) by λ_k, we find that

$$\lambda_k c_1 v_1 + \lambda_k c_2 v_2 + \cdots + \lambda_k c_k v_k = 0 \qquad (4.2.9)$$

Subtracting (4.2.8) from (4.2.9), we cause the kth term to vanish:

$$(\lambda_k - \lambda_1) c_1 v_1 + (\lambda_k - \lambda_2) c_2 v_2 + \cdots + (\lambda_k - \lambda_{k-1}) c_{k-1} v_{k-1} = 0 \quad (4.2.10)$$

By the induction hypothesis $v_1, \ldots, v_{k-1}$ are linearly independent, so (4.2.10) implies that $(\lambda_k - \lambda_i) c_i = 0$, $i = 1, \ldots, k - 1$. Since $\lambda_k - \lambda_i \neq 0$, we get $c_i = 0$, $i = 1, \ldots, k - 1$. Now (4.2.7) reduces to $c_k v_k = 0$, from which it follows that $c_k = 0$. $\square$

Recall from the previous section that a matrix $A \in \mathbb{C}^{n \times n}$ is *simple* if there exists a set of n linearly independent eigenvectors of A.

COROLLARY 4.2.11   If A has n distinct eigenvalues, then A is simple.

If we pick a matrix "at random," its characteristic equation is almost certain to have distinct roots. After all, the occurrence of repeated roots is an exceptional event. Thus a matrix chosen "at random" is almost certain to be simple. This is the basis of the claim made in the previous section that "most" matrices are simple. In addition it is not hard to show that every nonsimple matrix has simple matrices arbitrarily close to it. Given a nonsimple matrix (which must then have repeated eigenvalues), a slight perturbation suffices to split the repeated eigenvalues and yield a simple matrix. (See Exercise 4.4.26.) In the language of topology, the set of simple matrices is *dense* in the set of all matrices.

A matrix that is not simple is called *defective*. Whereas a defective matrix must have repeated eigenvalues, the existence of repeated eigenvalues does not guarantee that a matrix is defective.

Exercise 4.2.6   Find all eigenvalues and eigenvectors of the identity matrix $I \in \mathbb{C}^{n \times n}$. Conclude that (if $n > 1$) the identity matrix has repeated eigenvalues but is not defective. ☐

So far we do not even know that defective matrices exist. The next exercise takes care of that.

Exercise 4.2.7   (a) The matrix $\begin{bmatrix} 0 & 1 \\ 0 & 0 \end{bmatrix}$ has only the eigenvalue zero. (Why?) Show that it does not have two linearly independent eigenvectors.
(b) Find a matrix in $\mathbb{C}^{2 \times 2}$ that has only the eigenvalue zero but that does have two linearly independent eigenvectors. (There is only one such matrix.)
(c) Show that the matrix

$$\begin{bmatrix} c & 1 & & & \\ & c & 1 & & \\ & & \ddots & \ddots & \\ & & & c & 1 \\ & & & & c \end{bmatrix}$$

is defective. A matrix of this form is called a *Jordan block*. ☐

We have seen that finding the eigenvalues of A is equivalent to finding the zeros of the polynomial $\det(\lambda I - A)$. Therefore, if we have a way of finding the zeros of an arbitrary polynomial, then in principle we can find the eigenvalues of an arbitrary matrix. The next exercise essentially proves that the converse is true: If we have a way of finding the eigenvalues of an arbitrary matrix, then in principle we can find the zeros of an arbitrary polynomial.

Exercise 4.2.8   A *monic* polynomial is one whose highest-degree term has coefficient 1. Thus $p(\lambda) = 1\lambda^n + a_{n-1}\lambda^{n-1} + a_{n-2}\lambda^{n-2} + \cdots + a_1\lambda + a_0$ is monic. With the monic polynomial p we associate a matrix $A \in \mathbb{C}^{n \times n}$, called the *companion matrix* of p, defined by

$$A = \begin{bmatrix} & 1 & & & \\ & & 1 & & \\ & & & \ddots & \\ & & & & 1 \\ -a_0 & -a_1 & -a_2 & \cdots & -a_{n-1} \end{bmatrix}$$

This matrix has ones on the superdiagonal, the opposites of the coefficients of p in the bottom row, and zeros elsewhere. Prove that $\det(\lambda I - A) = p(\lambda)$. Thus the zeros of p are the eigenvalues of A. ☐

Exercise 4.2.9 Let A be the companion matrix of the polynomial p, as in Exercise 4.2.8, and let λ be an eigenvalue of A.

(a) Verify that $v = [1 \ \lambda \ \lambda^2 \ \cdots \ \lambda^{n-1}]^T$ is an eigenvector of A associated with the eigenvalue λ.

(b) Show that every eigenvector of A associated with λ is a multiple of v. (Thus the eigenspaces of A are one dimensional.) □

Now let $q(\lambda) = b_n \lambda^n + b_{n-1} \lambda^{n-1} + \cdots + b_0$, $b_n \neq 0$, be an arbitrary polynomial. Let $p(\lambda)$ be the monic polynomial obtained by dividing the coefficients of q by b_n. Then p has the same zeros as q, so if we can find the eigenvalues of the companion matrix of p, we will have the zeros of q. Therefore, in principle, the eigenvalue problem is equivalent to that of finding the zeros of a polynomial.

We have already suggested, at the end of the previous section, that (most of) the algorithms that we will consider do not make direct use of this equivalence. However, it does have an important theoretical implication. The problem of finding the zeros of a polynomial is an old one that has attracted the interest of many great mathematicians. In particular, early in the nineteenth century, Niels Henrik Abel proved that there is no general formula[§] for the zeros of a polynomial of degree n if $n > 4$. See Hadlock (1978). It follows that there is no general formula for the eigenvalues of an n-by-n matrix if $n > 4$.

Numerical methods can be divided into two broad categories—direct methods and iterative methods. A *direct method* is one that produces the result in a prescribed, finite number of steps. All the algorithms that we have considered up to now are direct methods. By contrast an *iterative method* is one that produces a sequence of approximants that (hopefully) converges to the true solution of the problem. Abel's theorem implies that there are no direct methods for solving the general eigenvalue problem, for the existence of a finite, specified procedure would imply the existence of a (perhaps very complicated) formula for the zeros of an arbitrary polynomial. Thus all eigenvalue algorithms are iterative. In the next section we will examine some of the simplest of these iterative methods.

Convergence of Sequences of Vectors

In the discussion of iterative methods, the question of convergence arises. What do we mean by convergence of a sequence of vectors? Let (x_j) be a sequence of vectors in $\mathbb{C}^n$, and let x_{ij} denote the ith component of x_j. Thus $x_j = [x_{1j} \ x_{2j} \ \cdots \ x_{nj}]^T$. We say that (x_j) *converges* to $x = [x^{(1)} \ x^{(2)} \ \cdots \ x^{(n)}]^T \in \mathbb{C}^n$ (and write $x_j \to x$ or $\lim_{j \to \infty} x_j = x$) provided that $\lim_{j \to \infty} |x_{ij} - x^{(i)}| = 0$ for $i = 1, \ldots, n$. In other words, $x_j \to x$ if and only if each component of x_j converges to the corresponding component of x. It is obvious that $x_j \to x$ if and only if $\lim_{j \to \infty} \|x_j - x\|_\infty = 0$ or $\lim_{j \to \infty} \|x_j - x\|_1 = 0$ or $\lim_{j \to \infty} \|x_j - x\|_2 = 0$.[∥]

[§] We allow formulas that involve addition, subtraction, multiplication, division, and the extraction of roots.

[∥] Our discussion of vector norms in Section 2.2 was restricted to norms on $\mathbb{R}^n$. However that entire discussion carries over to the complex setting.

In fact it is not too hard to show [see Kreyszig (1978), page 75] that for any two norms $\|\cdot\|_a$ and $\|\cdot\|_b$ on $\mathbb{C}^n$, there exist positive constants C_1 and C_2 such that for all $z \in \mathbb{C}^n$

$$C_1\|z\|_a \le \|z\|_b \le C_2\|z\|_a$$

It follows that $\|x_j - x\|_a \to 0$ if and only if $\|x_j - x\|_b \to 0$. Thus for any norm $\|\cdot\|$ on $\mathbb{C}^n$ it holds that $x_j \to x$ if and only if $\lim_{j\to\infty}\|x_j - x\| = 0$.

Although the question of whether or not a sequence of iterates converges is important, the *rate* of convergence is even more important. The knowledge that a given sequence converges to some desired vector is of no practical use if the rate of convergence is so slow that millions of iterations are required before a reasonable approximation is obtained. Convergence rates will be discussed in the next section in conjunction with the numerical methods introduced there.

4.3
THE POWER METHOD
AND SOME SIMPLE EXTENSIONS

Let $A \in \mathbb{C}^{n \times n}$. To avoid complicating the analysis, we assume that A is simple. Let $v_1, \ldots, v_n$ be a basis of $\mathbb{C}^n$, consisting of eigenvectors of A, and let $\lambda_1, \ldots, \lambda_n$ be the associated eigenvalues. We assume that $v_1, \ldots, v_n$ are ordered so that $|\lambda_1| \ge |\lambda_2| \ge \cdots \ge |\lambda_n|$. If $|\lambda_1| > |\lambda_2|$, λ_1 is called the *dominant eigenvalue* and v_1 is called a *dominant eigenvector* of A.

If A has a dominant eigenvalue, then we can find it and an associated dominant eigenvector by the *power method*. The basic idea is to pick a vector q and form the sequence

$$q, Aq, A^2q, A^3q, \ldots$$

To calculate this sequence, it is not necessary to form the powers of A explicitly since $A^{j+1}q = A(A^jq)$. It is easy to show that this sequence converges, in some sense, to a dominant eigenvector, for almost all choices of q. Since $v_1, \ldots, v_n$ form a basis for $\mathbb{C}^n$, there exist constants $c_1, \ldots, c_n$ such that

$$q = c_1v_1 + c_2v_2 + \cdots + c_nv_n$$

We do not know the values of $c_1, \ldots, c_n$ because we do not know $v_1, \ldots, v_n$, but it is clear that for practically any choice of q, c_1 will be nonzero. The argument that follows is valid for every q for which c_1 is not zero. Multiplying q by A, we have

$$Aq = c_1Av_1 + c_2Av_2 + \cdots + c_nAv_n$$
$$= c_1\lambda_1v_1 + c_2\lambda_2v_2 + \cdots + c_n\lambda_nv_n$$

Similarly

$$A^2 q = c_1 \lambda_1^2 v_1 + c_2 \lambda_2^2 v_2 + \cdots + c_n \lambda_n^2 v_n$$

and in general

$$A^j q = c_1 \lambda_1^j v_1 + c_2 \lambda_2^j v_2 + \cdots + c_n \lambda_n^j v_n \qquad (4.3.1)$$

Since λ_1 dominates the other eigenvalues, the component of $A^j q$ in the direction of v_1 becomes steadily greater relative to the other components as j increases. This can be made clearer by rewriting (4.3.1) in the form

$$A^j q = \lambda_1^j \, [c_1 v_1 + c_2 (\lambda_2/\lambda_1)^j v_2 + \cdots + c_n (\lambda_n/\lambda_1)^j v_n] \qquad (4.3.2)$$

The fact that every multiple of an eigenvector is also an eigenvector means that the magnitude of an eigenvector is unimportant, only the direction matters. Therefore the factor λ_1^j in (4.3.2) is, in principle, unimportant. Thus, instead of the sequence $(A^j q)$, let us consider the rescaled sequence (q_j), where $q_j = A^j q / \lambda_1^j$. From (4.3.2) it is clear that $q_j \to c_1 v_1$. Indeed, for any vector norm

$$\begin{aligned}
\| q_j - c_1 v_1 \| &= \| c_2 (\lambda_2/\lambda_1)^j v_2 + \cdots + c_n (\lambda_n/\lambda_1)^j v_n \| \\
&\leq |c_2| \, |\lambda_2/\lambda_1|^j \, \| v_2 \| + \cdots + |c_n| \, |\lambda_n/\lambda_1|^j \, \| v_n \| \\
&\leq \Big(|c_2| \, \| v_2 \| + \cdots + |c_n| \, \| v_n \| \Big) |\lambda_2/\lambda_1|^j
\end{aligned}$$

Here we have used the fact that $|\lambda_i| \leq |\lambda_2|$ for $i = 3, \ldots, n$. Letting $C = |c_2| \, \| v_2 \| + \cdots + |c_n| \, \| v_n \|$, we have

$$\| q_j - c_1 v_1 \| \leq C |\lambda_2/\lambda_1|^j \qquad j = 1, 2, 3, \ldots \qquad (4.3.3)$$

From the fact that $|\lambda_1| > |\lambda_2|$, it follows that $|\lambda_2/\lambda_1|^j \to 0$ as $j \to \infty$. Thus $\| q_j - c_1 v_1 \| \to 0$. This means that for sufficiently large j we can view q_j as an approximation of the dominant eigenvector $c_1 v_1$. The number $\| q_j - c_1 v_1 \|$ gives a measure of the error in this approximation. From (4.3.3) we see that, roughly speaking, the magnitude of the error is decreased by a factor $|\lambda_2/\lambda_1|$ with each iteration. Therefore the ratio $|\lambda_2/\lambda_1|$ is an important indicator of the rate of convergence.

The convergence behavior exhibited by the power method is called *linear convergence*. In general a sequence (x_j) that converges to x is said to converge *linearly* if there is a number r satisfying $0 < r < 1$, such that

$$\lim_{j \to \infty} \frac{\| x_{j+1} - x \|}{\| x_j - x \|} = r \qquad (4.3.4)$$

This means that $\| x_{j+1} - x \| \approx r \| x_j - x \|$ for sufficiently large j.[§] The number r is called the *convergence ratio* or *contraction number* of the sequence.

[§] Many authors also allow $r = 1$, but it seems preferable to think of that case as representing *slower than linear* convergence. A sequence that satisfies (4.3.4) with $r = 1$ converges very slowly indeed.

Exercise 4.3.1 Show that the sequence (q_j) generated by the power method converges linearly to $c_1 v_1$ with convergence ratio $r = |\lambda_2/\lambda_1|$ provided that $|\lambda_2| > |\lambda_3|$ and $c_2 \neq 0$. ☐

Exercise 4.3.1 shows that the power method typically converges linearly.[§]

In practice the sequence $q_j = A^j q/(\lambda_1)^j$ is inaccessible because we do not know λ_1 in advance. On the other hand, it is impractical to work with $A^j q$ because $\|A^j q\| \to \infty$ if $|\lambda_1| > 1$ and $\|A^j q\| \to 0$ if $|\lambda_1| < 1$. In order to avoid the danger of overflow or underflow and to recognize when the iterates are converging, we must employ some kind of scaling strategy. Thus we let $q_0 = q$ and define

$$q_{j+1} = (Aq_j)/\sigma_{j+1}$$

where σ_{j+1} is some convenient scaling factor. The exact choice of scale factor is unimportant, since we are interested in the directions of vectors, not the lengths. A convenient and simple choice is to take σ_{j+1} to be that entry of Aq_j whose absolute value is largest. The effect of this choice is that the largest component of each q_j is 1, and the sequence converges to a dominant eigenvector whose largest component is 1.

Example 4.3.5 We will use the power method to calculate a dominant eigenvector of

$$A = \begin{bmatrix} 9 & 1 \\ 1 & 2 \end{bmatrix}$$

starting with the vector $q_0 = [1, 1]^T$. On the first step we have $Aq_0 = [10, 3]^T$. Dividing by the scale factor $\sigma_1 = 10$, we get $q_1 = [1, 0.3]^T$. Then $Aq_1 = [9.3, 1.6]^T$, $\sigma_2 = 9.3$, and $q_2 = [1, 0.172043]^T$. Subsequent iterates are listed in Table 4.1. Only the second component of each q_j is listed because the first component is always 1. We see that after 10 iterations the sequence (q_j) has converged to six decimal places. Thus (to six decimal places)

$$v_1 = \begin{bmatrix} 1.0 \\ 0.140055 \end{bmatrix}$$

The sequence (σ_j) converges to the dominant eigenvalue $\lambda_1 = 9.140055$.

In Example 4.3.5 and also in Exercise 4.3.3 the iterates converge reasonably rapidly. This is so because in both cases the ratio $|\lambda_2/\lambda_1|$ is fairly small. For most larger matrices this is not the case, so convergence is usually slow. Not uncommonly $|\lambda_2/\lambda_1| \approx 0.9$ or 0.99.

Exercise 4.3.2 Calculate the characteristic polynomial of the matrix of Example 4.3.5, and use the quadratic formula to find the eigenvalues. Calculate the dominant eigenvector as well and verify that the results of Example 4.3.5 are correct. Calculate the

---

[§] With ingenuity one can construct certain special examples (satisfying $|\lambda_2| = |\lambda_3|$) that violate (4.3.4). These examples do, however, satisfy (4.3.3).

TABLE 4.1
Power Method

j	σ_j	q_j (second component)
3	9.172043	0.146541
4	9.146541	0.141374
5	9.141374	0.140323
6	9.140323	0.140110
7	9.140110	0.140066
8	9.140066	0.140057
9	9.140057	0.140055
10	9.140055	0.140055

ratios of errors $\|q_{j+1} - v_1\|/\|q_j - v_1\|$, $j = 0, 1, 2, \ldots$, and note that they agree very well with the theoretical convergence ratio $|\lambda_2/\lambda_1|$. Why is the agreement so good? Would you expect the agreement to be as good if A were 3 by 3 or larger? □

Exercise 4.3.3 Let $A = \begin{bmatrix} 8 & 1 \\ -2 & 1 \end{bmatrix}$.

(a) Use the power method with $q_0 = [1, 1]^T$ to calculate the dominant eigenvalue and eigenvector of A. Tabulate q_i and σ_i at each step. Iterate until the q_i and σ_i have converged to at least six decimal places.

(b) Now that you have calculated the eigenvector v, calculate the ratios $\|q_{i+1} - v\|/\|q_i - v\|$, $i = 0, 1, 2, \ldots$, to find the observed rate of convergence. Using the characteristic equation, solve for the eigenvalues and calculate the ratio $|\lambda_2/\lambda_1|$, which gives the theoretical rate of convergence. How well does theory agree with practice in this case? □

Exercise 4.3.4 Suppose the power method is applied to $A \in \mathbb{C}^{n \times n}$ using some scaling strategy that produces a sequence (q_j) that converges to an eigenvector. Prove that the sequence of scale factors (σ_j) converges to the dominant eigenvalue. □

Exercise 4.3.5 Let $A = \begin{bmatrix} 0 & 1 \\ 1 & 0 \end{bmatrix}$. Carry out the power method with starting vector $q_0 = [a, b]^T$, where $a \neq b$. Explain why the sequence fails to converge. □

Exercise 4.3.6 Let $A = \begin{bmatrix} 0.9 & 0 \\ 0 & 1 \end{bmatrix}$.

(a) Find the eigenvalues of A and associated eigenvectors.

(b) Carry out power iterations starting with $q_0 = [1, 1]^T$. Derive a general expression for q_j.

(c) How many iterations are required in order to obtain

$$\|q_j - v_1\|_\infty/\|v_1\|_\infty < 10^{-6}$$ □

Exercise 4.3.7 It was claimed that if a vector q is chosen at random, then in the expansion

$$q = c_1 v_1 + c_2 v_2 + \cdots + c_n v_n$$

the coefficient c_1 is almost certain to be nonzero. If you have trouble believing that claim, then this exercise is for you.

    (a) Picture the situation in $\mathbb{R}^2$. Let v_1 and v_2 be any two linearly independent vectors in $\mathbb{R}^2$. The set of all vectors $q = c_1 v_1 + c_2 v_2$ such that $c_1 = 0$ is just the subspace $\langle v_2 \rangle$. Sketch this subspace of $\mathbb{R}^2$ and note that it is a very small subset of $\mathbb{R}^2$. A vector chosen "at random" is almost certain not to lie in it.

    (b) Repeat part (a) in $\mathbb{R}^3$.   □

    In general any proper subspace is a very small subset of the space in which it lies.

Exercise 4.3.8

    (a) What happens if the power method is applied with a starting vector $q = c_1 v_1 + c_2 v_2 + \cdots + c_n v_n$ for which $c_1 = 0$?

    (b) If the calculations are done on a calculator or computer, roundoff errors will be made at each step. How do you think roundoff errors would affect your conclusions for part (a)?   □

    Iterations of the power method are relatively inexpensive. The cost of multiplying the n-by-n matrix A by q_j is n^2 flops. The normalizing operations require only $O(n)$ work, as you can easily verify, so the total cost of a power iteration is about n^2 flops. Thus m iterations will cost $n^2 m$ flops. This count assumes that A is not a sparse matrix. If A is sparse, the cost of calculating $A q_j$ will be considerably less than n^2 flops.

    Our development of the power method assumes that the matrix is simple. This turns out not to be crucial; essentially the same conclusions can be drawn for defective matrices.

The Inverse Power Method

We continue to assume that $A \in \mathbb{C}^{n \times n}$ is simple with linearly independent eigenvectors $v_1, \ldots, v_n$ and associated eigenvalues $\lambda_1, \ldots, \lambda_n$, arranged in order of descending magnitude. If A is nonsingular, we can apply the power method to A^{-1}. This is the *inverse power method* or *inverse iteration*. (By contrast, the power method applied to A is sometimes called *direct iteration*.)

Exercise 4.3.9 Suppose $A \in \mathbb{C}^{n \times n}$ is nonsingular. Then all eigenvalues of A are nonzero. Show that if v is an eigenvector of A with associated eigenvalue λ, then v is also an eigenvector of A^{-1} with associated eigenvalue λ^{-1}.   □

From Exercise 4.3.9 we see that A^{-1} has linearly independent eigenvectors $v_n, v_{n-1}, \ldots, v_1$ with eigenvalues $\lambda_n^{-1}, \lambda_{n-1}^{-1}, \ldots, \lambda_1^{-1}$. If $|\lambda_n^{-1}| > |\lambda_{n-1}^{-1}|$ (i.e., $|\lambda_{n-1}| > |\lambda_n|$) and we start with a vector $q = c_n v_n + \cdots + c_1 v_1$ for which $c_n \neq 0$, then the inverse power iterates will converge to a multiple of v_n, an eigenvector associated with the smallest eigenvalue of A. The convergence ratio is $|\lambda_{n-1}^{-1}/\lambda_n^{-1}| = |\lambda_n/\lambda_{n-1}|$, so convergence will be fast when $|\lambda_{n-1}| \gg |\lambda_n|$. This suggests that we *shift* the eigenvalues so that the smallest eigenvalue is very close to zero. To understand how this works, we need only the following simple result.

Exercise 4.3.10   Let $A \in \mathbb{C}^{n \times n}$ and $\rho \in \mathbb{C}$. Show that if v is an eigenvector of A with eigenvalue λ, then v is also an eigenvector of $A - \rho I$ with eigenvalue $\lambda - \rho$.   ☐

From Exercise 4.3.10 we see that if A has eigenvalues $\lambda_1, \ldots, \lambda_n$, then $A - \rho I$ has eigenvalues $\lambda_1 - \rho, \lambda_2 - \rho, \ldots, \lambda_n - \rho$. The scalar ρ is called a *shift*. If the shift is chosen so that it is a good approximation to λ_n, then $|\lambda_{n-1} - \rho| \gg |\lambda_n - \rho|$, and inverse iteration applied to $A - \rho I$ will converge rapidly to a multiple of v_n. Actually there is nothing special about λ_n; the shift ρ can be chosen to approximate any one of the eigenvalues of A. If ρ is a good enough approximation of λ_i that $\lambda_i - \rho$ is much smaller than any other eigenvalue of $A - \rho I$, then (for almost every starting vector q) inverse iteration applied to $A - \rho I$ will converge to a multiple of the eigenvector v_i. The convergence ratio is $|(\lambda_i - \rho)/(\lambda_k - \rho)|$, where $\lambda_k - \rho$ is the second smallest eigenvalue of $A - \rho I$. The closer ρ is to λ_i, the swifter the convergence will be.

Shifts can be used in conjunction with direct iteration as well, but they are not nearly so effective. The combination of shifting and inverse iteration works well because there exist shifts for which one eigenvalue of the shifted matrix is much smaller than all other eigenvalues. In contrast there (usually) do not exist shifts for which one eigenvalue is much larger than all other eigenvalues.

Exercise 4.3.11   Diagonal matrices such as

$$A = \begin{bmatrix} 2.99 & 0 & 0 \\ 0 & 1.99 & 0 \\ 0 & 0 & 1.00 \end{bmatrix}$$

have particularly simple eigensystems.

(a)   Find the eigenvalues and eigenvectors of A.

(b)   Find the eigenvalues of $A - \rho I$ and $(A - \rho I)^{-1}$, where $\rho = 0.99$. Perform both direct and inverse iteration on $A - \rho I$, starting with $q_0 = [1, 1, 1]^T$. To which eigenvector does each sequence converge? Which converges faster?

(c)   Perform inverse iteration with the values (i) $\rho = 2.00$ and (ii) $\rho = 3.00$. Use the same starting vector as in part (b). To which eigenvectors do these sequences converge?   ☐

Let us consider some of the practical aspects of inverse iteration. The iterates will satisfy

$$q_{j+1} = (A - \rho I)^{-1} q_j / \sigma_{j+1}$$

but it is not necessary to calculate $(A - \rho I)^{-1}$ explicitly. Instead one can solve the linear system $(A - \rho I)\hat{q}_{j+1} = q_j$ and then set $q_{j+1} = \hat{q}_{j+1}/\sigma_{j+1}$, where σ_{j+1} equals the component of $\hat{q}_{j+1}$ that is largest in magnitude. If the system is to be solved by Gaussian elimination, then the LU decomposition of $A - \rho I$ has to be done only once. Then each iteration consists of forward substitution, back substitution, and normalization. For a full n-by-n matrix the cost is $n^3/3$ flops for the LU decomposition plus n^2 flops per iteration.

Example 4.3.6   Let us apply inverse iteration with a shift to the matrix

$$A = \begin{bmatrix} 9 & 1 \\ 1 & 2 \end{bmatrix}$$

We know from Example 4.3.5 that 9 is a good approximation to an eigenvalue of A. Even if we did not know this, we might expect as much from the dominant 9 in the $(1, 1)$ position of the matrix. Thus it is reasonable to use the shift $\rho = 9$. Starting with $q_0 = [1, 1]^T$, we solve the system $(A - 9I)\hat{q}_1 = q_0$ to get $\hat{q}_1 = [8.0, 1.0]^T$. We then rescale by taking $\sigma_1 = 8$ to get $q_1 = [1.0, 0.125]^T$. Solving $(A - 9I)\hat{q}_2 = q_1$, we get $\hat{q}_2 = [7.125, 1.0]^T$. Subsequent iterates are listed in Table 4.2. As in Table 4.1 we have listed only the second component of q_j because the first component is always 1. After five iterations we have the eigenvector $v_1 = [1.0, 0.140055]$ correct to six decimal places. The good choice of shift gave much faster convergence than in Example 4.3.5. The scale factors converge to an eigenvalue of $(A - 9I)^{-1}$: $(\lambda_1 - 9)^{-1} = 7.140055$. Solving for λ_1, we have $\lambda_1 = 9.140055$.

Exercise 4.3.12   Let $A = \begin{bmatrix} 8 & 1 \\ -2 & 1 \end{bmatrix}$.

(a)   Use inverse iteration with $\rho = 8$ and $q_0 = [1, 1]^T$ to calculate an eigenvalue and eigenvector of A. (On this small problem it may be easiest simply to calculate $B = (A - 8I)^{-1}$ and perform direct iterations on B.)

TABLE 4.2
Inverse Power Method

j	σ_j	q_j (second component)
3	7.140351	0.140049
4	7.140049	0.140055
5	7.140055	0.140055

(b)   Now that you have calculated an eigenvector v, calculate the ratios $\| q_{j+1} - v \| / \| q_j - v \|$ for $j = 0, 1, 2, \ldots$ to find the observed rate of convergence. Solve for the eigenvalues using the characteristic equation, and calculate the theoretical convergence rate $|(\lambda_1 - 8)/(\lambda_2 - 8)|$. How well does theory agree with practice?   □

The introduction of shifts allows us to find any eigenvector, not just those associated with the largest and smallest eigenvalues. However, in order to find a given eigenvector, we must have a good approximation to the associated eigenvalue. Therefore one of the most important applications of inverse iteration is to find an eigenvector associated with an eigenvalue that has already been found by other means. Using the computed eigenvalue as a shift, we obtain an excellent eigenvector in just one or two iterations. A detailed discussion of this use of inverse iteration is given in contribution II/18 of [HAC].

The previous paragraph seems to contain a contradiction. The use of an eigenvalue as a shift in conjunction with inverse iteration appears to be impossible: If ρ is an eigenvalue of A, then $A - \rho I$ is singular; so $(A - \rho I)^{-1}$ does not exist. However, in practice ρ is (almost) never exactly an eigenvalue; there is (almost) always some error. Thus $A - \rho I$ is (usually) nonsingular but ill conditioned (see Exercise 4.3.13). Recall from Chapter 2 that in numerical practice it is impossible to distinguish between singular and nonsingular matrices. Therefore, even if $A - \rho I$ is singular, we probably will not be able to detect the fact. Thus the method does not break down in practice. Still, one might think that the ill condition of $A - \rho I$ would spoil the computation, since at each step a system of the form $(A - \rho I)\, \hat{q}_{j+1} = q_j$ must be solved.

Because of the ill conditioning, a small perturbation in q_j can lead to a large perturbation in $\hat{q}_{j+1}$ and hence also in q_{j+1}. It is not hard to see that this does not cause any difficulties. Suppose for example that $\rho \approx \lambda_1$, and we are going to use inverse iteration with shift ρ to try to find (a multiple of) v_1. If we start with the vector $q_0 = c_1 v_1 + c_2 v_2 + \cdots + c_n v_n$, then

$$\hat{q}_1 = (A - \rho I)^{-1} q_0 = (\lambda_1 - \rho)^{-1} c_1 v_1 + (\lambda_2 - \rho)^{-1} c_2 v_2 + \cdots + (\lambda_n - \rho)^{-1} c_n v_n$$

Because $\rho \approx \lambda_1$, $(\lambda_1 - \rho)^{-1}$ is enormous in comparison with $(\lambda_2 - \rho)^{-1}, \ldots, (\lambda_n - \rho)^{-1}$. If our choice of q_0 was not unlucky, then the magnitude of $c_1 v_1$ will be roughly the same as that of the other components of q_0. In this case $(\lambda_1 - \rho)^{-1} c_1 v_1$ is huge compared with the other components of q_1. Thus q_1 is almost exactly a multiple of v_1, which is just what we want.

Now suppose that instead of starting with q_0, we start with the slightly perturbed vector $q_0' = c_1' v_1 + c_2' v_2 + \cdots + c_n' v_n$. Since $q_0' \approx q_0$, it follows that $c_i' \approx c_i$, $i = 1, \ldots, n$, but this doesn't really matter. Instead of getting $\hat{q}_1$, we get

$$\hat{q}_1' = (A - I)^{-1} q_0' = (\lambda_1 - \rho)^{-1} c_1' v_1 + (\lambda_2 - \rho)^{-1} c_2' v_2 + \cdots + (\lambda_n - \rho)^{-1} c_n' v_n$$

By the same argument as before, the first component of $\hat{q}_1'$ is much larger than all other components, so $\hat{q}_1'$ is almost exactly a multiple of v_1. Thus $\hat{q}_1'$ and $\hat{q}_1$ are both good estimates of an eigenvector. The fact that there might be a big difference between $\hat{q}_1$ and $\hat{q}_1'$ is unimportant; the difference is almost exactly a multiple of v_1.

This argument is incomplete because it accounts only for perturbations in q_j and ignores the effect of roundoff errors in solving $(A - \rho I)\hat{q}_{j+1} = q_j$. Nevertheless it gives a reasonable picture of what happens during inverse iteration. A more complete argument must await our discussion of the sensitivity of eigenvalues and eigenvectors in Section 5.5.

Exercise 4.3.13   Let $A \in \mathbb{C}^{n \times n}$ be a nonsingular matrix with eigenvalues λ and μ such that $\lambda \neq \mu$.

   (a)   Show that $\|A\| \geq |\lambda|$, where $\|\cdot\|$ denotes an induced matrix norm.

   (b)   Show that $\kappa(A) \geq |\lambda|/|\mu|$, where κ denotes the condition number with respect to any induced matrix norm.

   (c)   Let (ρ_j) be a sequence of shifts such that $\rho_j \to \mu$. Prove that $\kappa(A - \rho_j I) \to \infty$ as $j \to \infty$. $\square$

The Rayleigh Quotient

There is no reason why, in using inverse iteration to calculate an eigenvector of A, one could not use a different shift at each step. This can be useful in situations in which the eigenvalue is not known in advance. It seems reasonable that if the iterate q_j is close enough to an eigenvector, it should somehow be possible to use q_j to obtain an estimate of the associated eigenvalue. This estimate could then be used as a shift for the next iteration. In this way we would get an improved shift, and hence an improved convergence ratio, for each iteration.

Thus we shall consider the following problem: Given a $q \in \mathbb{C}^n$ that approximates an eigenvector of $A \in \mathbb{C}^{n \times n}$, use q to obtain an estimate of the associated eigenvalue. We can imagine numerous approaches to this problem. Our approach will be to minimize a certain residual. If q is exactly an eigenvector, then there exists exactly one number ρ for which

$$Aq = \rho q \qquad (4.3.7)$$

This number is the eigenvalue. If q is not an eigenvector, then there is no value of ρ for which (4.3.7) is satisfied. Equation 4.3.7 is a vector equation; that is, it is a system of n scalar equations. Thus we can view (4.3.7) as an overdetermined system of n equations in the single unknown ρ. Letting r denote the *residual*, $r = Aq - \rho q$, we can find the value of ρ for which the 2-norm $\|r\|_2$ assumes a minimum. In the case when q is an eigenvector, the minimizing ρ is exactly the associated eigenvalue. It therefore seems reasonable that if q approximates an eigenvector, then the minimizing ρ should yield a good estimate of the associated eigenvalue.

The choice of the 2-norm means that the minimization problem is a least-squares problem, so we can solve it by one of the techniques developed in Chapter 3. Thus we can use a QR decomposition or the normal equations, for example. It is probably simplest to use the normal equations. In Chapter 3 we restricted our attention to real matrices, but all of the developments of that chapter can be carried over to the complex setting. The only modification that must be made is that whenever a transpose, say B^T, occurs, it must be replaced by the conjugate transpose B^*. This is the matrix obtained by transposing B and then taking complex conjugates of all entries. Making this modification to Theorem 3.5.9, we find that the normal equations for the complex overdetermined system $Cz = b$ are $C^*Cz = C^*b$. Rewriting (4.3.7) as

$$q\rho = Aq \qquad (4.3.8)$$

we see that the role of C is played by q, that of z is played by ρ, and that of b is played by Aq. Thus the normal equations for (4.3.8) are

$$(q^*q)\rho = q^*Aq$$

Actually there is only one normal equation in this case because there is only one unknown. Its solution is

$$\rho = \frac{q^*Aq}{q^*q}$$

This number is called the *Rayleigh quotient* associated with q and A. We have proved the following theorem.

THEOREM 4.3.9 Let $A \in \mathbb{C}^{n \times n}$ and $q \in \mathbb{C}^n$. The unique complex number that minimizes $\|Aq - \rho q\|_2$ is the Rayleigh quotient $\rho = q^*Aq/q^*q$.

In particular, if q is an eigenvector of A, then the Rayleigh quotient equals the associated eigenvalue.

Exercise 4.3.14 (QR approach) Use a complex QR decomposition of $q \in \mathbb{C}^{n \times 1}$ to prove that the Rayleigh quotient minimizes $\|Aq - \rho q\|_2$. (The Q factor in the QR decomposition satisfies $Q^*Q = I$.) □

Exercise 4.3.15 (Calculus approach) Letting $\rho = \alpha + \beta i$, the function $f(\alpha, \beta) = \|Aq - \rho q\|_2^2 = (Aq - \rho q, Aq - \rho q)$ is a smooth function in the two real variables α and β. Use differential calculus to show that f is minimized if and only if $\rho = \alpha + \beta i$ is the Rayleigh quotient. □

Rayleigh Quotient Iteration

Rayleigh quotient iteration is that variant of inverse iteration in which the Rayleigh quotient for each q_j is calculated and used as the shift for the next iteration. Thus a step of Rayleigh iteration is as follows:

$$\rho_{j+1} = \frac{q_j^*Aq_j}{q_j^*q_j} \qquad (A - \rho_{j+1}I)\hat{q}_{j+1} = q_j \qquad q_{j+1} = \frac{\hat{q}_{j+1}}{\sigma_{j+1}}$$

where σ_{j+1} is any convenient scaling factor. The computation can be simplified by choosing σ_{j+1} so that $|\sigma_{j+1}| = \|\hat{q}_{j+1}\|_2$, since then $q_{j+1}^*q_{j+1} = \|q_{j+1}\|^2 = 1$. If we also choose q_0 so that $\|q_0\|_2 = 1$, then at each step we have $q_j^*q_j = 1$; so the Rayleigh quotient simplifies to $\rho_{j+1} = q_j^*Aq_j$.

TABLE 4.3
Rayleigh Quotient Iteration

j	q_j (second component)	ρ_j
0	1.0	6.5
1	-0.272727273	8.007692311
2	0.221558346	9.094844261
3	0.139551306	9.140053168
4	0.140054945	9.140054945
5	0.140054945	9.140054945

Because a different shift is used at each step, it is difficult to analyze the global convergence properties of Rayleigh quotient iteration. The algorithm is not guaranteed to converge to an eigenvector, but experience suggests that it will converge for almost any choice of starting vector.[§] When it does converge, it generally converges rapidly. The next example illustrates the swift convergence of Rayleigh quotient iteration.

Example 4.3.10 Consider once again the matrix $A = \begin{bmatrix} 9 & 1 \\ 1 & 2 \end{bmatrix}$. We will use Rayleigh quotient iteration starting with $q_0 = [1 \ 1]^T$ to calculate an eigenvector of A. For easy comparison with Examples 4.3.5 and 4.3.6, we will normalize the iterates so that the first component is 1. (Thus they will not satisfy $\| q_j \|_2 = 1$.) The results are listed in Table 4.3. You can easily verify that q_4 is an eigenvector and ρ_4 is an eigenvalue, correct to nine digits after the decimal point. Notice the rapid convergence: q_2 has no correct digits, q_3 has (essentially) three correct digits, and q_4 has nine correct digits.

Exercise 4.3.16 Calculate the ratios $\| q_{j+1} - v_1 \| / \| q_j - v_1 \|$, $j = 0, 1, 2, 3$, from Table 4.3, where $v_1 = q_5 =$ the eigenvector. Notice that the ratios decrease dramatically with increasing j. □

Exercise 4.3.16 shows that the rate of convergence observed in Example 4.3.10 is better than linear. This is to be expected since the convergence ratio depends on a ratio of shifted eigenvalues that improves from one step to the next. Let us take a closer look at the convergence rate of Rayleigh quotient iteration. We will begin with a theorem that shows that, as expected, the Rayleigh quotient of an approximate eigenvector approximates the associated eigenvalue. The vectors in the theorem are normalized so that they have Euclidean norm 1. This simplifies both the statement and the proof of the theorem, but it is a matter of convenience, not necessity.

[§] Batterson and Smillie (1990) discovered an open set of matrices and corresponding open sets of starting vectors for which Rayleigh quotient iteration fails to converge—the shifts wander chaotically. It would appear that the sets on which this behavior is observed are quite small.

THEOREM 4.3.11   Let $A \in \mathcal{C}^{n \times n}$ and let v be an eigenvector of A with associated eigenvalue λ. Assume $\| v \|_2 = 1$. Let $q \in \mathcal{C}^n$ with $\| q \|_2 = 1$, and let $\rho = q^*Aq$ be the associated Rayleigh quotient. Then

$$|\lambda - \rho| \leq 2 \| A \|_2 \| v - q \|_2$$

Proof   Since v is an eigenvector with $v^*v = 1$, $\lambda = v^*Av$. Thus $\lambda - \rho = v^*Av - q^*Aq = v^*Av - v^*Aq + v^*Aq - q^*Aq = v^*A(v - q) + (v - q)^*Aq$. Therefore $|\lambda - \rho| \leq |v^*A(v - q)| + |(v - q)^*Aq|$. By the Schwarz inequality (3.2.1), $|v^*A(v - q)| \leq \| v \|_2 \| A(v - q) \|_2 = \| A(v - q) \|_2$. By Theorem 2.2.16, $\| A(v - q) \|_2 \leq \| A \|_2 \| v - q \|_2$. Thus $|v^*A(v - q)| \leq \| A \|_2 \| v - q \|_2$. By similar reasoning $|(v - q)^*Aq| \leq \| A \|_2 \| v - q \|_2$. The assertion of the theorem follows. $\square$

Thus if $\| v - q \|_2 < \epsilon$, then $|\lambda - \rho| < 2 \| A \|_2 \epsilon$. This fact is sometimes expressed briefly by saying that if $\| v - q \|_2 = O(\epsilon)$, then $|\lambda - \rho| = O(\epsilon)$.

We are now ready to study the convergence rate of Rayleigh quotient iteration. However, what follows should not be mistaken for a rigorous analysis. Let (q_j) be a sequence generated by Rayleigh quotient iteration, normalized so that $\| q_j \|_2 = 1$ for all j, and suppose $q_j \to v_i$ as $j \to \infty$. Then also $\| v_i \|_2 = 1$. Note that this normalization implies that the relative error $\| v_i - q_j \|_2 / \| v_i \|_2$ equals the absolute error $\| v_i - q_j \|_2$, so it suffices to study the absolute error. Assume further that λ_i is not a multiple eigenvalue, and let λ_k denote the closest eigenvalue to λ_i with $k \neq i$. Since the jth step of Rayleigh quotient iteration is just a power iteration with matrix $(A - \rho_j I)^{-1}$, we know that

$$\| v_i - q_{j+1} \|_2 \approx r_j \| v_i - q_j \|_2 \tag{4.3.12}$$

where r_j is the ratio of the two eigenvalues of $(A - \rho_j I)^{-1}$ of largest absolute value. By Theorem 4.3.11 the Rayleigh quotients ρ_j converge to λ_i. Once ρ_j is close enough to λ_i, the two largest eigenvalues of $(A - \rho_j I)^{-1}$ will be $(\lambda_i - \rho_j)^{-1}$ and $(\lambda_k - \rho_j)^{-1}$. Thus

$$r_j = |(\lambda_k - \rho_j)^{-1}/(\lambda_i - \rho_j)^{-1}| = |(\lambda_i - \rho_j)/(\lambda_k - \rho_j)|$$

By Theorem 4.3.11, $|\lambda_i - \rho_j| \leq 2 \| A \|_2 \| v_i - q_j \|_2$. Also, since $\rho_j \approx \lambda_i$, we can make the approximation $|\lambda_k - \rho_j| \approx |\lambda_k - \lambda_i|$. Thus

$$r_j \approx \frac{2 \| A \|_2}{|\lambda_k - \lambda_i|} \| v_i - q_j \|_2 = C \| v_i - q_j \|_2$$

where $C = 2 \| A \|_2 / |\lambda_k - \lambda_i|$. Substituting this estimate of r_j into (4.3.12), we obtain the estimate

$$\| v_i - q_{j+1} \|_2 \approx C \| v_i - q_j \|_2^2 \tag{4.3.13}$$

Thus the error after $j + 1$ iterations is roughly proportional to the square of the error after j iterations. Another way to express this is to say that if $\|v_i - q_j\|_2 = O(\epsilon)$, then $\|v_i - q_{j+1}\|_2 = O(\epsilon^2)$. This means that if the error after j iterations is somewhat small, then the error after $j + 1$ iterations will be much smaller. A sequence whose convergence rate satisfies (4.3.13) is said to converge *quadratically*. Thus we have shown that Rayleigh quotient iteration typically converges quadratically when it does converge.

A rule of thumb that one sometimes hears is this: quadratic convergence means that the number of correct digits doubles with each iteration. This is true if $C \approx 1$ in (4.3.13), for if q_j agrees with v_i to s_j decimal places, then $\|v_i - q_j\|_2 \approx 10^{-s_j}$. Thus by (4.3.12), $\|v_i - q_{j+1}\|_2 \approx 10^{-2s_j}$; that is, q_{j+1} agrees with v_i to about $2s_j$ decimal places. Even if $C \neq 1$, the rule of thumb is valid in the limit: If $C \approx 10^t$, then $\|v_i - q_{j+1}\|_2 \approx 10^{t-2s_j}$. Thus q_{j+1} agrees with v_i to about $2s_j - t$ decimal places. As s_j grows, t becomes increasingly insignificant. Once s_j is large enough, t can be ignored.

In Example 4.3.10 the rate of convergence appears to be better than quadratic, since the number of correct digits roughly triples at each iteration. This is not an accident, rather it is a consequence of the special form of the matrix. Note that in this example A is a (real) symmetric matrix. The real symmetric matrices belong to a class of matrices for which the Rayleigh quotient approximates the eigenvalue better than Theorem 4.3.11 would indicate. In the next section we will see (Exercise 4.4.13) that for symmetric matrices, $|\lambda - \rho| = O\left(\|v - q\|_2^2\right)$. (The notation is that of Theorem 4.3.11.) Thus if $\|v - q\|_2 = O(\epsilon)$, then $|\lambda - \rho| = O(\epsilon^2)$. Using this estimate instead of Theorem 4.3.11 to estimate r_j, we find that

$$\|v_i - q_{j+1}\|_2 \approx C\|v_i - q_j\|_2^3$$

This type of convergence is called *cubic* convergence. It implies that the number of correct digits roughly triples with each iteration.

Quite a lot can be said about Rayleigh quotient iteration in the symmetric case that cannot be said in general. Not only is the convergence cubic when it occurs, it is also known that Rayleigh quotient iteration converges for almost all choices of starting vector. Of course the eigenvector that is reached depends on the choice of starting vector. Unfortunately there is no simple characterization of this dependence. See [SEP], Chapter 4, for details.

Rayleigh quotient iteration can be expensive. Since a different shift is used at each iteration, a new *LU* decomposition will be needed at each step, assuming that Gaussian elimination is used. This costs $n^3/3$ flops per iteration for a full matrix and thus makes the method too expensive. However, there are some classes of matrices for which Rayleigh quotient iteration is economical. For example, in Exercise 1.9.4 you showed that for a matrix in upper Hessenberg form an *LU* decomposition costs only about $n^2/2$ flops. For such a matrix Rayleigh quotient iteration is economical.

Hessenberg matrices, which up to now have surfaced only in Exercise 1.9.4, will play an important role in this chapter. In Section 4.5 we will see that the eigenvalue problem for an arbitrary matrix can be reduced to that of finding the eigenvalues (and eigenvectors if desired) of a related upper Hessenberg matrix. We could use Rayleigh quotient iteration to attack this Hessenberg matrix, but instead we will use a more powerful algorithm, the *QR* algorithm, which is based on the

QR decomposition. Rayleigh quotient iteration takes place implicitly within certain versions of the QR algorithm.

Exercise 4.3.17 Let $A = \begin{bmatrix} 8 & 1 \\ -2 & 1 \end{bmatrix}$. Write a short computer or calculator program to perform Rayleigh quotient iteration on A.

(a) Use $q_0 = [1\ 1]^T$ as a starting vector. Iterate until the limits of the machine precision are reached. Notice that the iterates converge to a different eigenvector than in Exercises 4.3.3 and 4.3.12. Notice also that once the iterates get close to the limit, the number of correct digits roughly doubles with each iteration. This is true of both the iteration vector and the Rayleigh quotient. Thus both converge quadratically.

(b) Repeat part (a) with the starting vector $q_0 = [1\ 0]^T$. This time the iterates converge to the same eigenvector as in Exercises 4.3.3 and 4.3.12. ☐

Exercise 4.3.18 Consider the real, symmetric matrix $A = \begin{bmatrix} 0 & 1 \\ 1 & 0 \end{bmatrix}$.

(a) Apply Rayleigh quotient iteration to A with starting vector $q_0 = [1\ 0]^T$. (You will not need a computer program for this one.) This is an exceptional case in which Rayleigh quotient iteration fails to converge.

(b) Calculate the eigenvalues and eigenvectors of A by some other means and show that (i) the iterates $q_0, q_1, q_2, \ldots$ exactly bisect the angle between the two linearly independent eigenvectors and (ii) the Rayleigh quotients $\rho_0, \rho_1, \rho_2, \ldots$ lie exactly half way between the two eigenvalues. Thus the sequence "can't decide" which eigenvector to approach. ☐

Exercise 4.3.19 Let $A = \begin{bmatrix} a & b \\ -b & a \end{bmatrix}$, where $a, b \in \mathbb{R}$. Perform Rayleigh quotient iteration with starting vector $q_0 = [c\ d]^T$, where $c, d \in \mathbb{R}$ and $c^2 + d^2 = 1$. Analyze the problem. ☐

Exercise 4.3.20 Each of the following sequences of real numbers converges to zero. For each sequence (a_j), determine whether (a_j) converges (i) slower than linearly (i.e., $\lim_{j\to\infty} a_{j+1}/a_j = 1$), (ii) linearly (i.e., $0 < \lim_{j\to\infty} a_{j+1}/a_j = r < 1$ for some r), (iii) quadratically (i.e., $\lim_{j\to\infty} a_{j+1}/a_j^2 = C \neq 0$), or (iv) cubically (i.e., $\lim_{j\to\infty} a_{j+1}/a_j^3 = C \neq 0$).

(a) $10^{-1}, 10^{-2}, 10^{-3}, 10^{-4}, 10^{-5}, \ldots$

(b) $10^{-1}, 10^{-2}, 10^{-4}, 10^{-8}, 10^{-16}, \ldots$

(c) $1, 1/2, 1/3, 1/4, 1/5, \ldots$

(d) $10^{-3}, 10^{-6}, 10^{-9}, 10^{-12}, 10^{-15}, \ldots$

(e) $0.9, 0.81, 0.729, (0.9)^4, (0.9)^5, \ldots$

(f) $10^{-1}, 10^{-3}, 10^{-9}, 10^{-27}, 10^{-81}, \ldots$ ☐

Notice that quadratic and cubic convergence are qualitatively better than linear convergence. However, linear convergence can be quite satisfactory if r is small enough. Convergence that is slower than linear is too slow.

Exercise 4.3.21 The intent of this exercise is to show that quadratic convergence is nearly as good as cubic convergence. Suppose algorithm X produces a sequence of vectors (q_j) such that $\|q_j - v_i\| = a_j \to 0$ quadratically. Define a new algorithm Y for which one step consists of two steps of algorithm X. Then algorithm Y produces a sequence $(\hat{q}_j)$ for which $\|\hat{q}_j - v_i\| = b_j = a_{2j}$. Prove that $b_j \to 0$ *quartically*; that is, $\lim_{j \to \infty} b_{j+1}/b_j^4 = M \neq 0$. Thus the convergence of method Y is faster than cubic. $\qquad\square$

Exercise 4.3.22 Suppose $q_j \to v$ linearly with convergence ratio r. Show that on each iteration the number of correct decimal digits increases by about $m = -\log_{10} r$. (A rigorous argument is not required.) Thus the rate of increase in the number of correct digits is constant. $\qquad\square$

4.4
SIMILARITY TRANSFORMATIONS AND RELATED TOPICS

Two matrices $A, B \in \mathbb{C}^{n \times n}$ are said to be *similar* if there exists a nonsingular $P \in \mathbb{C}^{n \times n}$ such that

$$B = P^{-1}AP \qquad (4.4.1)$$

Equation 4.4.1 is called a *similarity transformation*, and P is called the *transforming matrix*.

As we shall soon see, similar matrices have the same eigenvalues, and their eigenvectors are related in a simple way. Some of the most important eigenvalue algorithms employ a sequence of similarity transformations to reduce a matrix to a simpler form. That is, they replace the original matrix by a similar one whose eigenvalues and eigenvectors are more easily determined. This section prepares us for these algorithms by setting out some of the basic facts about similarity transformations. In the process we will cover some important material on special classes of matrices such as symmetric and orthogonal matrices and their complex counterparts.

THEOREM 4.4.2 Similar matrices have the same eigenvalues.

Proof Suppose A and B are similar. Then there exists a nonsingular P such that $B = P^{-1}AP$. To show that A and B have the same eigenvalues, it suffices to show that they have the same characteristic polynomial. Now $\lambda I - B = P^{-1}\lambda I P - P^{-1}AP = P^{-1}(\lambda I - A)P$, so $\det(\lambda I - B) = \det(P^{-1})\det(\lambda I - A)\det(P) = \det(\lambda I - A)$. In the last equality we are using the facts that complex numbers commute and $\det(P^{-1})\det(P) = 1$. Thus A and B have the same characteristic polynomial. $\qquad\square$

This proof shows that the similarity transformation preserves the *algebraic multiplicity* of the eigenvalue. That is, if μ is a root of order k of the equation $\det(\lambda I - A) = 0$, then it is also a root of $\det(\lambda I - B) = 0$ of order k.

THEOREM 4.4.3 Suppose $B = P^{-1}AP$. Then v is an eigenvector of A with associated eigenvalue λ if and only if $P^{-1}v$ is an eigenvector of B with associated eigenvalue λ.

Proof Suppose $Av = \lambda v$. Then $B(P^{-1}v) = P^{-1}APP^{-1}v = P^{-1}Av = P^{-1}\lambda v = \lambda(P^{-1}v)$. The converse is proved by interchanging the roles of A and B. $\square$

Exercise 4.4.1 The *geometric multiplicity* of an eigenvalue is defined to be the dimension of the associated eigenspace $\{v \in \mathbb{C}^n \mid Av = \lambda v\}$. Show that a similarity transformation preserves the geometric multiplicity of each eigenvalue. $\square$

Exercise 4.4.2 Let A be a simple matrix. How are the geometric and algebraic multiplicities of the eigenvalues of A related? $\square$

Exercise 4.4.3 Suppose A is similar to B. Show that if A is nonsingular, then B is nonsingular and A^{-1} is similar to B^{-1}. $\square$

Exercise 4.4.4 Let D be any diagonal matrix. Find a set of n linearly independent eigenvectors of D thereby demonstrating that D is simple. $\square$

The next theorem shows that a matrix is simple if and only if it is similar to a diagonal matrix. Recall from Theorem 1.3.3 that a matrix is nonsingular if and only if its columns are linearly independent.

THEOREM 4.4.4 Let $A \in \mathbb{C}^{n \times n}$ be a simple matrix with linearly independent eigenvectors $v_1, v_2, \ldots, v_n$ and associated eigenvalues $\lambda_1, \lambda_2, \ldots, \lambda_n$. Define a diagonal matrix D and a nonsingular matrix V by

$$D = \begin{bmatrix} \lambda_1 & & & \\ & \lambda_2 & & \\ & & \ddots & \\ & & & \lambda_n \end{bmatrix} \quad \text{and} \quad V = [v_1, v_2, \ldots, v_n]$$

Then $V^{-1}AV = D$. Conversely, suppose A satisfies $V^{-1}AV = D$, where D is diagonal and V is nonsingular. Then the columns of V are n linearly independent eigenvectors of A, and the main-diagonal entries of D are the associated eigenvalues. In particular, A is simple.

Proof For $i = 1, \ldots, n$, let d_i denote the ith column of D. You can easily check that $Vd_i = \lambda_i v_i$. Since also $Av_i = \lambda_i v_i$, we have

$$Av_i = Vd_i \qquad i = 1, \ldots, n$$

Collecting these n vector equations into a single matrix equation, we get

$$A[v_1 v_2 \cdots v_n] = V[d_1 d_2 \cdots d_n]$$

or $AV = VD$. Multiplying this equation by V^{-1} on the left, we find that $V^{-1}AV = D$. Thus A is similar to the diagonal matrix D. The converse is proved by reversing the argument. The details are left as an exercise for you. □

Exercise 4.4.5   Complete the proof of Theorem 4.4.4. □

Theorem 4.4.4 tells us that we can solve the eigenvalue problem completely if we can find a similarity transformation that transforms A to diagonal form. Unfortunately the proof is not constructive; that is, it does not show us how to construct V and D without knowing the eigenvectors and eigenvalues in advance. The theorem does, however, give us some idea of a simple form toward which we might work. For example, we might try to construct a sequence of similar matrices $A = A_0, A_1, A_2, \ldots$ that converges to diagonal form.

Theorem 4.4.4 is about simple matrices. With considerable additional effort, we could obtain an extension of Theorem 4.4.4, valid for all matrices. See the discussion of the *Jordan canonical form* in Lancaster (1985) or Gantmacher (1959).

Geometric Viewpoint

Recall from elementary linear algebra that every matrix $A \in \mathbb{C}^{n \times n}$ can be viewed as a linear transformation whose action is to map $v \in \mathbb{C}^n$ to $Av \in \mathbb{C}^n$. If A is simple, then its action on $\mathbb{C}^n$ is easily pictured. A has a set of n linearly independent eigenvectors $v_1, \ldots, v_n$, which form a basis for $\mathbb{C}^n$. Every $v \in \mathbb{C}^n$ can be expressed as a linear combination $v = c_1 v_1 + \cdots + c_n v_n$. The action of A on each $c_i v_i$ is simply to multiply it by the scalar λ_i, and the action of A on v is just a sum of such simple actions: $Av = \lambda_1 c_1 v_1 + \lambda_2 c_2 v_2 + \cdots + \lambda_n c_n v_n$. Theorem 4.4.4 is actually just a restatement of this fact in matrix form. We will attempt to clarify this without going into too much detail.

Recall again from elementary linear algebra that a linear transformation $A : \mathbb{C}^n \to \mathbb{C}^n$ can be represented in numerous ways. Given any basis of $\mathbb{C}^n$, there is a unique matrix that represents A with respect to that basis. Two matrices represent the same linear transformation (with respect to different bases) if and only if they are similar. Thus a similarity transformation amounts just to a change of basis, that is, a change of coordinate system. Theorem 4.4.4 says that if A is simple, then there exists a coordinate system in which it is represented by a diagonal matrix. This coordinate system has eigenvectors of A as its coordinate axes.

Unitary Similarity Transformations

In Chapter 3 we introduced orthogonal matrices and noted that they have numerous desirable properties. The complex analog of the orthogonal matrix is the unitary matrix, which was also introduced and (briefly) discussed in Chapter 3. Recall that a matrix $U \in \mathbb{C}^{n \times n}$ is said to be *unitary* if $U^*U = I$; that is, $U^* = U^{-1}$. The class of unitary matrices contains the (real) orthogonal matrices. In a series of exercises at the end of Section 3.2 you showed that: (1) The product of unitary matrices is unitary. (2) The inverse of a unitary matrix is unitary. (3) The complex inner product $(x, y) = y^*x$ and Euclidean norm are preserved by unitary matrices; that is, $(Ux, Uy) = (x, y)$ and $\|Ux\|_2 = \|x\|_2$ for all $x, y \in \mathbb{C}^n$. (4) Rotators and reflectors have complex analogs. (5) Every $A \in \mathbb{C}^{n \times n}$ can be expressed as a product $A = QR$, where Q is unitary and R is upper triangular. You also showed (Exercise 3.4.18) that a matrix $U \in \mathbb{C}^{n \times n}$ is unitary if and only if its columns or its rows are orthonormal. Of course the orthonormality is with respect to the complex inner product.

Two matrices $A, B \in \mathbb{C}^{n \times n}$ are said to be *unitarily similar* if there exists a unitary $U \in \mathbb{C}^{n \times n}$ such that $B = U^{-1}AU$. Since $U^{-1} = U^*$, the unitary similarity can also be expressed as $B = U^*AU$. If A, B, and U are all real, then U is orthogonal and A and B are said to be *orthogonally similar*. Unitary similarity transformations have some nice properties not possessed by similarity transformations in general.

Exercise 4.4.6

(a)   Show that if U is unitary, then $\|U\|_2 = 1$ and $\kappa_2(U) = 1$.

(b)   Show that if A and B are unitarily similar, then $\|B\|_2 = \|A\|_2$ and $\kappa_2(B) = \kappa_2(A)$.

(c)   Suppose $B = U^*AU$, where U is unitary. Show that if A is perturbed, then the resulting perturbation in B is of the same magnitude. Specifically, show that if $B + \delta B = U^*(A + \delta A)U$, then $\|\delta B\|_2 = \|\delta A\|_2$. $\square$

This exercise shows that any errors that a matrix may contain will not be amplified by subsequent unitary similarity transformations.[§] The same cannot be said of arbitrary similarity transformations, as the following exercises show.

Exercise 4.4.7   Let $P = \begin{bmatrix} 1 & \alpha \\ 0 & 1 \end{bmatrix}$, where α is a real parameter.

(a)   Let $A = \begin{bmatrix} 2 & 0 \\ 0 & 1 \end{bmatrix}$ and $B = P^{-1}AP$. Calculate B and conclude that $\|B\|_\infty$ can be made arbitrarily large by taking α large. (The same is true of $\|B\|_2$. We calculate the ∞-norm here because we have not yet developed the tools to calculate the matrix 2-norm.)

(b)   Let $A = \begin{bmatrix} 1 & 0 \\ 0 & 1 \end{bmatrix}$ and $\delta A = \dfrac{\epsilon}{2}\begin{bmatrix} 1 & 1 \\ 1 & 1 \end{bmatrix}$. Notice that $\|\delta A\|_\infty / \|A\|_\infty = \epsilon$.

[§] Exercise 4.4.6 does not say anything about the magnitude of roundoff errors that may be introduced during the similarity transformation. This question was already covered in the previous chapter, at least for rotators and reflectors. In Section 3.2 we noted that the roundoff errors associated with multiplication by a rotator or a reflector are extremely small. Thus the same is true of unitary similarity transformations by (real or complex) rotators or reflectors.

Let $B = P^{-1}AP$ and $B + \delta B = P^{-1}(A + \delta A)P$. Calculate B and δB and show that $\|\delta B\|_\infty / \|B\|_\infty$ can be made arbitrarily large by taking α large. Notice that this example is even more extreme than that of part (a), since $\|\delta B\|_\infty$ is roughly proportional to α^2, not α. Optional exercise: How could this example be modified to make it even more spectacular? □

Exercise 4.4.8 Suppose $B = P^{-1}AP$ and $B + \delta B = P^{-1}(A + \delta A)P$. Show that

(a) $\dfrac{1}{\kappa(P)} \|A\| \leq \|B\| \leq \kappa(P)\|A\|$

(b) $\dfrac{1}{\kappa(P)^2} \dfrac{\|\delta A\|}{\|A\|} \leq \dfrac{\|\delta B\|}{\|B\|} \leq \kappa(P)^2 \dfrac{\|\delta A\|}{\|A\|}$

Here $\|\cdot\|$ denotes any matrix norm, and κ denotes the condition number associated with that norm. □

Thus we can control the growth of error terms by controlling the condition number of the transforming matrix. Note that the results of parts (b) and (c) for Exercise 4.4.6 follow from the results of Exercise 4.4.8.

In addition to their favorable error propagation properties, unitary similarity transformations also preserve certain desirable matrix properties. For example, consider the following theorem. (For other examples see Exercises 4.4.15 to 4.4.21.)

THEOREM 4.4.5 If $A = A^*$ and A is unitarily similar to B, then $B = B^*$.

Proof $B = U^*AU$ for some unitary U. Thus $B^* = (U^*AU)^* = U^*A^*U^{**} = U^*AU = B$. □

Exercise 4.4.9 Show by example that the conclusion of Theorem 4.4.5 does not hold for arbitrary similarity transformations. □

Recall that a matrix that satisfies $A = A^*$ is called *Hermitian*. Thus Theorem 4.4.5 states that the Hermitian property is preserved under unitary similarity transformations. The set of Hermitian matrices contains the (real) symmetric matrices, so every theorem about Hermitian matrices contains a theorem about symmetric matrices. In particular, Theorem 4.4.5 implies that if A is symmetric and B is orthogonally similar to A, then B is also symmetric. Eigenvalue problems for which the coefficient matrix is symmetric occur frequently in applications. Since symmetric and Hermitian matrices have special properties that make them easier to handle than general matrices, it is useful to have at hand classes of similarity transformations that preserve these properties.

The next result, Schur's theorem, is the most important result of this section. It states that every (square) matrix is unitarily similar to a triangular matrix.

THEOREM 4.4.6 *(Schur's Theorem)* Let $A \in \mathbb{C}^{n \times n}$. Then there exists a unitary matrix $U \in \mathbb{C}^{n \times n}$ and an upper-triangular matrix $T \in \mathbb{C}^{n \times n}$ such that $U^*AU = T$.

Proof   The proof is by induction on n. The result is trivial for $n = 1$. Now let us show that it holds for $n = k$, given that it holds for $n = k - 1$. Let $A \in \mathbb{C}^{k \times k}$. Let λ be an eigenvalue of A and v an associated eigenvector, chosen so that $\|v\|_2 = 1$. Let U_1 be any unitary matrix that has v as its first column. There are many such matrices: just take any orthonormal basis of $\mathbb{C}^k$ whose first member is v and let U_1 be the matrix whose columns are the members of the basis. Let $W \in \mathbb{C}^{k \times (k-1)}$ denote the submatrix of U_1 consisting of columns 2 through k, so that $U_1 = [v \; W]$. Since the columns of W are orthogonal to v, $W^*v = 0$. Let $A_1 = U_1^* A U_1$. Then

$$A_1 = \begin{bmatrix} v^* \\ W^* \end{bmatrix} A[v \; W] = \begin{bmatrix} v^*Av & v^*AW \\ W^*Av & W^*AW \end{bmatrix}$$

Since $Av = \lambda v$, it follows that $v^*Av = \lambda$ and $W^*Av = \lambda W^*v = 0$. Let $\hat{A} = W^*AW$. Then A_1 has the form

$$A_1 = \left[\begin{array}{c|ccc} \lambda & * & \cdots & * \\ \hline 0 & & & \\ \vdots & & \hat{A} & \\ 0 & & & \end{array}\right]$$

$\hat{A} \in \mathbb{C}^{(k-1) \times (k-1)}$, so by the induction hypothesis there exists a unitary matrix $\hat{U}_2$ and an upper-triangular matrix $\hat{T}$ such that $\hat{T} = \hat{U}_2^* \hat{A} \hat{U}_2$. Define $U_2 \in \mathbb{C}^{k \times k}$ by

$$U_2 = \left[\begin{array}{c|ccc} 1 & 0 & \cdots & 0 \\ \hline 0 & & & \\ \vdots & & \hat{U}_2 & \\ 0 & & & \end{array}\right]$$

Then U_2 is unitary, and

$$U_2^* A_1 U_2 = \left[\begin{array}{c|ccc} \lambda & * & \cdots & * \\ \hline 0 & & & \\ \vdots & & \hat{U}_2^* \hat{A} \hat{U}_2 & \\ 0 & & & \end{array}\right] = \left[\begin{array}{c|ccc} \lambda & * & \cdots & * \\ \hline 0 & & & \\ \vdots & & \hat{T} & \\ 0 & & & \end{array}\right]$$

which is upper triangular. Let us call this matrix T, and let $U = U_1 U_2$. Then $T = U_2^* A_1 U_2 = U_2^* U_1^* A U_1 U_2 = U^* A U$. $\square$

The main-diagonal entries of T are the eigenvalues of A. Schur's theorem suggests that it is reasonable to attempt to transform A to upper-triangular form by unitary similarity transformations in order to find its eigenvalues. However, the proof does not give us any recipe for doing so without knowing the eigenvalues and eigenvectors in advance. Nevertheless it gives us a reason to believe that we might be able to create an algorithm that would produce a sequence of unitarily similar matrices $A = A_0, A_1, A_2, \ldots$ that would converge to upper-triangular form. The QR algorithm, which we will discuss starting in Section 4.6, does essentially that.

It follows easily from the equation $T = U^*AU$ that the first column of U is an eigenvector of A. Indeed, this equation can be rewritten as $AU = UT$. The first column of AU is Au_1, where u_1 is the first column of U. The first column of AU must equal the first column of UT, which is $u_1 t_{11}$ because T is upper triangular. Thus $Au_1 = u_1 t_{11}$; that is, u_1 is an eigenvector of A associated with the eigenvalue t_{11}. In general the other columns of U are not eigenvectors of A. The extraction of the other eigenvectors requires a bit more work. (See Exercise 4.4.11 and Section 4.8.)

Exercise 4.4.10 Show that in Schur's theorem:

(a) T can be chosen so that the eigenvalues of A appear in any desired order on the main diagonal of T.

(b) U can be chosen so that its first column equals any desired eigenvector of A for which $\|v\|_2 = 1$. (*Hint:* Show that the first column of U equals the first column of U_1.) $\square$

Exercise 4.4.11 Let $T \in \mathbb{C}^{n \times n}$ be an upper-triangular matrix with distinct eigenvalues. Sketch an algorithm that calculates a set of n linearly independent eigenvectors of T. How many flops does the whole operation require? What difficulties can arise when the eigenvalues are not distinct? This exercise will be worked in Section 4.8. $\square$

Exercise 4.4.12 Let $v \in \mathbb{C}^n$ be any vector such that $\|v\|_2 = 1$. Show how to build a (complex) reflector whose first column is a multiple of v. Such a reflector could play the role of U_1 in the proof of Schur's theorem. $\square$

Schur's theorem is comparable in spirit to Theorem 4.4.4, which states, in part, that every simple matrix is similar to a diagonal matrix. Schur's theorem is more modest in the sense that the triangular form is not as simple and elegant as the diagonal form of Theorem 4.4.4. On the other hand, Schur's theorem is valid for all matrices, not just the simple ones. Furthermore, the unitary similarity transformations of Schur's theorem are well behaved from a numerical standpoint.

A class of matrices for which Schur's theorem and Theorem 4.4.4 overlap is the Hermitian matrices. If A is Hermitian, then the matrix $T = U^*AU$ is not only upper triangular but Hermitian as well (Theorem 4.4.5). This obviously implies that T is diagonal. This result is known as the *spectral theorem* for Hermitian matrices.

THEOREM 4.4.7 *(Spectral Theorem)* Suppose $A \in \mathbb{C}^{n \times n}$ is Hermitian. Then there exists a unitary matrix $U \in \mathbb{C}^{n \times n}$ and a diagonal matrix $D \in \mathbb{R}^{n \times n}$ such that $D = U^*AU$. The columns of U are eigenvectors of A, and the main-diagonal entries of D are the eigenvalues of A.

The fact that the columns of U are eigenvectors of A follows from the last part of Theorem 4.4.4. The diagonal matrix is real because $D = D^*$. This proves that the eigenvalues of a Hermitian matrix are real.

COROLLARY 4.4.8 (a) The eigenvalues of a Hermitian matrix are real.
(b) The eigenvalues of a real symmetric matrix are real.

COROLLARY 4.4.9 Every Hermitian matrix in $\mathbb{C}^{n \times n}$ has a set of n orthonormal eigenvectors. In particular every Hermitian matrix is simple.

In the previous section it was claimed that for real symmetric matrices, the Rayleigh quotient of an approximate eigenvector gives a particularly good approximation to the associated eigenvalue. In the following exercise you will verify that claim for Hermitian matrices.

Exercise 4.4.13 Let $A \in \mathbb{C}^{n \times n}$ be Hermitian, and let $q \in \mathbb{C}^n$ be a vector satisfying $\| q \|_2 = 1$ that approximates an eigenvector v of A. Assume that $\| v \|_2 = 1$ as well. By Corollary 4.4.9, A has n orthonormal eigenvectors $v_1, \ldots, v_n$ with associated eigenvalues $\lambda_1, \ldots, \lambda_n$. We can choose $v_1, \ldots, v_n$ in such a way that $v_1 = v$. The approximate eigenvector q can be expressed as a linear combination of $v_1, \ldots, v_n$: $q = c_1 v_1 + c_2 v_2 + \cdots + c_n v_n$.

    (a) Show that $\sum_{k=1}^{n} |c_k|^2 = \| q \|_2^2 = 1$.

    (b) Show that $\sum_{k=2}^{n} |c_k|^2 \leq \| v_1 - q \|_2^2$.

    (c) Derive an expression for the Rayleigh quotient $\rho = q^* A q$ in terms of the coefficients c_i and the eigenvalues λ_i.

    (d) Show that $|\lambda_1 - \rho| \leq C \| v_1 - q \|_2^2$, where $C = \max_{2 \leq i \leq n} |\lambda_1 - \lambda_i|$.

(*Hint:* $\lambda_1 = \sum_{k=1}^{n} \lambda_1 |c_k|^2$.) Thus if $\| v_1 - q \|_2 = O(\epsilon)$, then $|\lambda_1 - \rho| = O(\epsilon^2)$.
$\square$

Exercise 4.4.14 A second proof that the eigenvalues of a Hermitian matrix are real can be obtained by considering Rayleigh quotients. Let $A \in \mathbb{C}^{n \times n}$ be Hermitian.

    (a) Show that for all $x \in \mathbb{C}^n$, the number $x^* A x$ is real (cf., Exercise 1.5.29).

    (b) Show that for all nonzero $x \in \mathbb{C}^n$, the Rayleigh quotient $\rho = x^* A x / x^* x$ is real.

    (c) Show that every eigenvalue of A is real.
$\square$

Exercise 4.4.15 Recall that a Hermitian matrix $A \in \mathbb{C}^{n \times n}$ is called *positive definite* if for all nonzero $x \in \mathbb{C}^n$, $x^* A x > 0$. Prove that if A is positive definite and B is unitarily similar to A, then B is positive definite.
$\square$

Exercise 4.4.16 Use Rayleigh quotients to prove that the eigenvalues of a positive definite matrix are positive. The next exercise indicates a second way to prove this.
$\square$

Exercise 4.4.17 Let $A \in \mathbb{C}^{n \times n}$ be a Hermitian matrix. Use Theorem 4.4.7 and the result of Exercise 4.4.15 to prove that A is positive definite if and only if its eigenvalues are positive. □

Exercise 4.4.18 A Hermitian matrix $A \in \mathbb{C}^{n \times n}$ is said to be *positive semidefinite* if $x^*Ax \geq 0$ for all $x \in \mathbb{C}^n$. Formulate and prove results analogous to those of Exercises 4.4.15 to 4.4.17 for positive semidefinite matrices. □

Exercise 4.4.19 A matrix $A \in \mathbb{C}^{n \times n}$ is said to be *skew Hermitian* if $A^* = -A$.

(a) Prove that if A is skew Hermitian and B is unitarily similar to A, then B is skew Hermitian.

(b) What special form does Schur's theorem take when A is skew Hermitian?

(c) Prove that the eigenvalues of a skew Hermitian matrix are purely imaginary; that is, they satisfy $\bar{\lambda} = -\lambda$. Give two proofs, one based on Schur's theorem and one based on the Rayleigh quotient. □

Exercise 4.4.20

(a) Prove that if A is unitary and A is unitarily similar to B, then B is unitary.

(b) Prove that a matrix $T \in \mathbb{C}^{n \times n}$ that is both upper triangular and unitary must be a diagonal matrix. You will prove a more general result in Exercise 4.4.21.

(c) What special form does Schur's theorem take when A is unitary?

(d) Prove that the eigenvalues of a unitary matrix satisfy $\bar{\lambda} = \lambda^{-1}$. Equivalently $\lambda\bar{\lambda} = 1$ or $|\lambda|^2 = 1$; that is, the eigenvalues lie on the unit circle in the complex plane. Give two proofs. □

Exercise 4.4.21 A matrix $A \in \mathbb{C}^{n \times n}$ is said to be *normal* if $AA^* = A^*A$.

(a) Prove that all Hermitian, skew Hermitian, and unitary matrices are normal.

(b) Prove that if A is normal and B is unitarily similar to A, then B is normal.

(c) Prove that if the matrix $T \in \mathbb{C}^{n \times n}$ is both upper triangular and normal, then T is a diagonal matrix. (*Hint:* Partition the equation $TT^* = T^*T$ and use induction on n.)

(d) Prove that every diagonal matrix is normal.

(e) Prove that A is normal if and only if A is unitarily similar to a diagonal matrix. □

Part (e) of Exercise 4.4.21 is so important that we restate it as a theorem.

THEOREM 4.4.10 (*Spectral Theorem for Normal Matrices*) Let $A \in \mathbb{C}^{n \times n}$. Then A is normal if and only if there exists a unitary matrix U and a diagonal matrix D such that $D = U^*AU$.

Corollary **4.4.11**

(a) Let $A \in \mathbb{C}^{n \times n}$ be normal. Then A has a set of n orthonormal eigenvectors.

(b) Conversely, if $A \in \mathbb{C}^{n \times n}$ has a set of n orthonormal eigenvectors, then A is normal.

Exercise 4.4.22 Prove assertion (b) of Corollary 4.4.11. ☐

These results demonstrate the great importance of the class of normal matrices. This class contains all Hermitian matrices, positive semidefinite matrices, positive definite matrices,[§] skew Hermitian matrices, and unitary matrices, so every property that holds for normal matrices holds for these classes of matrices as well. For example, part (a) of Corollary 4.4.11 implies that every unitary matrix has a set of n orthonormal eigenvectors.

Exercise 4.4.23 Let $D \in \mathbb{C}^{n \times n}$ be diagonal. Show that:

(a) D is Hermitian if and only if its eigenvalues are real.

(b) D is positive semidefinite if and only if its eigenvalues are nonnegative.

(c) D is positive definite if and only if its eigenvalues are positive.

(d) D is skew Hermitian if and only if its eigenvalues lie on the imaginary axis of the complex plane.

(e) D is unitary if and only if its eigenvalues lie on the unit circle of the complex plane. ☐

Exercise 4.4.24 Let $A \in \mathbb{C}^{n \times n}$ be normal. Show that:

(a) A is Hermitian if and only if its eigenvalues lie on the real axis.

(b) A is positive semidefinite if and only if its eigenvalues are nonnegative.

(c) A is positive definite if and only if its eigenvalues are positive.

(d) A is skew Hermitian if and only if its eigenvalues lie on the imaginary axis.

(e) A is unitary if and only if its eigenvalues lie on the unit circle. ☐

Exercise 4.4.25 Verify that the results of Exercise 4.4.13 carry over verbatim to normal matrices. Thus the good approximation properties of the Rayleigh quotient hold for normal matrices in general, not just Hermitian matrices. ☐

Exercise 4.4.26 Let $A \in \mathbb{C}^{n \times n}$ be a defective matrix. Use Schur's theorem to show that for every $\epsilon > 0$, there exists a simple matrix $A_\epsilon \in \mathbb{C}^{n \times n}$ such that $\|A - A_\epsilon\|_2 < \epsilon$. Thus the set of simple matrices is dense in $\mathbb{C}^{n \times n}$. ☐

[§] *Reminder:* Some books, notably [MC], allow "positive definite" matrices that are not Hermitian.

Real Matrices

If we wish to find the eigenvalues and eigenvectors of a real matrix, we have to be prepared to work with complex numbers. Nevertheless it is a good idea to try to minimize our contact with the complex number system, because complex arithmetic is much more expensive than real arithmetic. Therefore we might well ask what sort of reduction we can obtain, using real, orthogonal similarity transformations. It turns out that we can get very close to upper-triangular form.

THEOREM 4.4.12 *(Real Schur Theorem)* Let $A \in \mathbb{R}^{n \times n}$. Then there exists an orthogonal $U \in \mathbb{R}^{n \times n}$ and a block triangular $T \in \mathbb{R}^{n \times n}$ such that $T = U^T A U$. The precise form of T is

$$
T = \begin{bmatrix}
T_{11} & T_{12} & \cdots & T_{1m} \\
0 & T_{22} & & \cdot \\
\vdots & \ddots & \ddots & \cdot \\
0 & \cdots & 0 & T_{mm}
\end{bmatrix}
$$

where each main diagonal block T_{ii} is in either $\mathbb{R}^{1 \times 1}$ or $\mathbb{R}^{2 \times 2}$. Each 1-by-1 block is a real eigenvalue of A. Each 2-by-2 block has as its eigenvalues a complex conjugate pair of eigenvalues of A.

Thus the real Schur theorem is just like the complex Schur theorem, except that 2-by-2 blocks are allowed on the main diagonal in order to accommodate pairs of complex conjugate eigenvalues.

With a little effort we could prove the real Schur theorem right now, but it is more sensible to delay the proof until after we have discussed invariant subspaces. Thus, see Section 5.1 for the proof.

The eigenvalues of a 2-by-2 matrix can be found very easily by applying the quadratic formula to the characteristic equation. Therefore the block triangular form in Theorem 4.4.12 is just as useful as triangular form for exposing the eigenvalues of a matrix. (See Theorem 4.2.5.)

There are important classes of real matrices for which the form of Theorem 4.4.12 simplifies considerably. The most important class is that of symmetric matrices. Let $A \in \mathbb{R}^{n \times n}$ be symmetric. Then all eigenvalues of A are real, so all of the main-diagonal blocks in the associated matrix T are 1 by 1. This means that T is actually upper triangular. Since T is also symmetric, it must in fact be diagonal. Thus we have the following spectral theorem for real symmetric matrices.

THEOREM 4.4.13 Let $A \in \mathbb{R}^{n \times n}$ be symmetric. Then there exists an orthogonal $U \in \mathbb{R}^{n \times n}$ and a diagonal $D \in \mathbb{R}^{n \times n}$ such that $D = U^T A U$.

COROLLARY 4.4.14 Let $A \in \mathbb{R}^{n \times n}$ be symmetric. Then A has a set of n real orthonormal eigenvectors.

These results show that we can completely solve the eigenvalue/eigenvector problem for a real symmetric matrix without going outside the real number system.

Although we have chosen to postpone the proof of Theorem 4.4.12 and have presented Theorem 4.4.13 as a corollary of Theorem 4.4.12, the good reasons for postponing the proof do not apply in this special case. A proof of Theorem 4.4.13 can be obtained directly by examining the proof of the complex Schur theorem and observing that in the case of a real symmetric matrix, the construction can be carried out entirely within the real number system. Let us quickly sketch the proof.

Assume that A is real and symmetric, and let λ be any eigenvalue of A. Since λ is real, it has a real eigenvector v associated with it, which may be chosen so that $\| v \|_2 = 1$. Let U_1 be a real orthogonal matrix whose first column is v, and let $A_1 = U_1^T A U_1$. Then A_1 is real and symmetric, and (as in the proof of Schur's theorem)

$$
A = \begin{bmatrix} \lambda & 0 & \cdots & 0 \\ 0 & & & \\ \vdots & & \hat{A} & \\ 0 & & & \end{bmatrix}
$$

Since $\hat{A} \in \mathbb{R}^{(n-1)\times(n-1)}$ is symmetric, we can assume inductively that there is an orthogonal matrix $\hat{U}_2$ and a diagonal matrix $\hat{D}$ such that $\hat{D} = \hat{U}_2^T \hat{A} \hat{U}_2$. Let

$$
U_2 = \begin{bmatrix} 1 & 0 & \cdots & 0 \\ 0 & & & \\ \vdots & & \hat{U}_2 & \\ 0 & & & \end{bmatrix} \qquad D = \begin{bmatrix} \lambda & 0 & \cdots & 0 \\ 0 & & & \\ \vdots & & \hat{D} & \\ 0 & & & \end{bmatrix}
$$

and $U = U_1 U_2$. Then U is orthogonal and $D = U^T A U$. The details are left as an exercise for you.

Exercise 4.4.27   Write out a detailed proof of Theorem 4.4.13.   ☐

A matrix $A \in \mathbb{R}^{n \times n}$ is said to be *skew symmetric* if $A^T = -A$. Thus a skew symmetric matrix is just a skew Hermitian matrix that is real. In particular it follows from Exercise 4.4.19 that the eigenvalues of a skew symmetric matrix lie on the imaginary axis of the complex plane.

Exercise 4.4.28

(a)   Let $A \in \mathbb{R}^{n \times n}$ be skew symmetric. Show that A is singular if n is odd. (Examine the eigenvalues.)

(b)   What does a 2-by-2 skew symmetric matrix look like?

(c)   What special form does the real Schur theorem take when A is skew symmetric? Be as specific as you can.   ☐

A real matrix is normal if it satisfies $AA^T = A^T A$. The real Schur theorem simplifies nicely when A is normal. We can carry out this simplification with the help of the trace function. The *trace* of a matrix $A \in \mathbb{C}^{n \times n}$ is defined to be the sum of the main-diagonal entries. Thus $\operatorname{tr}(A) = \sum_{i=1}^{n} a_{ii}$.

Exercise 4.4.29

    (a)   Show that for $C, D \in \mathbb{C}^{n \times n}$, $\text{tr}(C + D) = \text{tr}(C) + \text{tr}(D)$.

    (b)   Show that $\text{tr}(CD) = \text{tr}(DC)$.

    (c)   Recall that the Frobenius norm of $B \in \mathbb{C}^{n \times m}$ is defined by $\| B \|_F = \left(\sum_{i=1}^{n} \sum_{j=1}^{m} |b_{ij}|^2 \right)^{1/2}$. Show that $\| B \|_F^2 = \text{tr}(B^*B) = \text{tr}(BB^*)$.   □

Exercise 4.4.30

    (a)   Suppose $T \in \mathbb{C}^{n \times n}$ is normal and has the block triangular form

$$
T = \begin{matrix} & j & k & \\ & \begin{bmatrix} T_{11} & T_{12} \\ 0 & T_{22} \end{bmatrix} & \begin{matrix} j \\ k \end{matrix} \end{matrix} \qquad j + k = n
$$

Write the equation $TT^* = T^*T$ in partitioned form and apply the trace function to the (1, 1) block of this equation. Use the properties established in Exercise 4.4.29 to infer that $T_{12} = 0$. Show that T_{11} and T_{22} are normal.

    (b)   Suppose $T \in \mathbb{C}^{n \times n}$ is normal and has the block triangular form

$$
T = \begin{bmatrix} T_{11} & T_{12} & \cdots & T_{1m} \\ 0 & T_{22} & & \vdots \\ & \ddots & \ddots & \vdots \\ 0 & \cdots & 0 & T_{mm} \end{bmatrix}
$$

Use induction on m and the result of part (a) to prove that T is block diagonal and the main-diagonal blocks are normal.   □

Exercise 4.4.31

    (a)   Let $A = \begin{bmatrix} a & b \\ c & d \end{bmatrix} \in \mathbb{R}^2$. Show that A is normal if and only if either A is symmetric or A has the form $A = \begin{bmatrix} a & b \\ -b & a \end{bmatrix}$.

    (b)   In the symmetric case A has real eigenvalues. Find the eigenvalues of $A = \begin{bmatrix} a & b \\ -b & a \end{bmatrix} \in \mathbb{R}^{2 \times 2}$.

    (c)   What form does the real Schur theorem take when $A \in \mathbb{R}^{n \times n}$ is normal?   □

       Every orthogonal matrix is normal, so the result of Exercise 4.4.31 applies to all orthogonal matrices. (What else can be said of T in this special case?) Every skew symmetric matrix is normal. In this special case the result of Exercise 4.4.31 reduces to that of Exercise 4.4.28.

Exercise 4.4.32

(a)   Show that if A and B are similar matrices, then $\mathrm{tr}(A) = \mathrm{tr}(B)$. [*Hint:* $B = S^{-1}AS$. Apply part (b) of Exercise 4.4.29 with $C = S^{-1}$ and $D = AS$.]

(b)   Show that the trace of a matrix equals the sum of its eigenvalues. $\square$

Exercise 4.4.33   Prove that the determinant of a matrix equals the product of its eigenvalues. $\square$

4.5
REDUCTION TO HESSENBERG
AND TRIDIAGONAL FORMS

The results of the previous section encourage us to seek algorithms that reduce a matrix to triangular form by similarity transformations, as a means of finding the eigenvalues of the matrix. On theoretical grounds we must rule out the possibility of an algorithm that does so in a finite number of steps; such an algorithm would violate Abel's classical theorem, cited in Section 4.2. It turns out, however, that there are finite algorithms, that is, direct methods, that bring a matrix very close to upper-triangular form.

Recall that a matrix $A = (a_{ij})$ is called an *upper Hessenberg* matrix if $a_{ij} = 0$ whenever $i > j + 1$. Thus an upper Hessenberg matrix has the form

$$\begin{bmatrix} * & * & * & \cdots & & * & * \\ * & * & * & \cdots & & \cdot & \cdot \\ 0 & * & * & \cdots & & \cdot & \cdot \\ 0 & 0 & * & & & \cdot & \cdot \\ \vdots & \vdots & & \ddots & & & \\ 0 & 0 & 0 & \cdots & 0 & * & * \end{bmatrix}$$

where the asterisks denote entries that may be nonzero.

In this section we will examine two algorithms that use similarity transformations to reduce a matrix to upper Hessenberg form in a finite number of flops. Such algorithms do not of themselves solve the eigenvalue problem, but they are extremely important because they reduce the problem to a form that can be manipulated inexpensively. For example, you showed in Exercise 1.9.4 that the LU decomposition of an upper Hessenberg matrix can be calculated in about $n^2/2$ flops. (If you have not worked that problem yet, you should work it right now.) Thus Rayleigh quotient iteration can be performed at a cost of about n^2 flops per iteration. The QR decomposition of an upper Hessenberg matrix can also be calculated inexpensively.

Exercise 4.5.1   Let $A \in \mathbb{R}^{n \times n}$ be in upper Hessenberg form. Show that the QR decomposition of A can be performed using rotators at a cost of about $2n^2$ multiplications and n^2 additions. (Assume that Q is not formed explicitly.) $\square$

Of the two reductions that we shall consider, one uses nonunitary similarity transformations and the other uses unitary similarity transformations. The unitary reduction is especially useful when the matrix to be reduced is Hermitian. The unitary transformations preserve the Hermitian property, so the resulting reduced matrix is not merely upper Hessenberg, it is *tridiagonal*; that is, it has the form

$$
\begin{bmatrix}
* & * & & & & & \\
* & * & * & & & \text{\Large 0} & \\
  & * & * & & & & \\
  & & & \ddots & & & \\
  & \text{\Large 0} & & & * & * \\
  & & & & * & *
\end{bmatrix}
$$

Tridiagonal matrices can be manipulated very inexpensively. Furthermore, the symmetry of the matrix can be exploited to reduce the cost of the reduction.

Exercise 4.5.2   Let $A \in \mathbb{C}^{n \times n}$ be tridiagonal.

(a)   Show that the cost of an *LU* decomposition of A is $O(n)$ flops.

(b)   Show that the cost of a *QR* decomposition of A is $O(n)$ flops.   □

Nonunitary Reduction

We will begin with a reduction based on Gaussian elimination that is relatively inexpensive and generally works well. Let us suppose we wish to reduce the matrix $A \in \mathbb{C}^{n \times n}$ to upper Hessenberg form. The first step of the reduction creates the desired zeros in the first column. We begin by checking the elements $a_{21}, a_{31}, \ldots, a_{n1}$ to determine which one has the greatest absolute value. If all are zero, we can skip step 1 because the desired zeros are already in place. Otherwise we interchange rows 2 and m, where a_{m1} is the element of greatest absolute value. Performing this interchange is the same as multiplying A on the left by the matrix

$$
S_1 = \quad \overset{\begin{matrix} & & & & m \\ & & & & \downarrow \end{matrix}}{
\underset{m \,\rightarrow}{
\begin{bmatrix}
1 & & & & & & & & \\
  & 0 & & & & 1 & & & \\
  & & 1 & & & & & & \\
  & & & \ddots & & & & & \\
  & & & & 1 & & & & \\
  & 1 & & & & 0 & & & \\
  & & & & & & 1 & & \\
  & & & & & & & \ddots & \\
  & & & & & & & & 1
\end{bmatrix}}}
$$

obtained by interchanging rows 2 and m of the identity matrix. Of course the point of this pivoting operation is to make the algorithm as stable as possible.

For $i = 3, \ldots, n$ we subtract the appropriate multiple of row 2 from row i to create a zero in the $(i, 1)$ position. Notice that we are not touching the first row.

Subtracting m_{i1} times the second row from the ith row is equivalent to multiplying on the left by the matrix

$$
L_{i1} = \begin{array}{c} \\ \\ \\ i \to \\ \\ \\ \end{array}
\overset{\overset{\displaystyle 2}{\downarrow}}{
\begin{bmatrix}
1 & & & & & & \\
& 1 & & & & & \\
& & \ddots & & & & \\
& -m_{i1} & & 1 & & & \\
& & & & \ddots & & \\
& & & & & & 1
\end{bmatrix}}
$$

obtained by performing the same row operation on the identity matrix. So far we have transformed A to the form

$$
A_{1/2} = L_{n1} \cdots L_{41} L_{31} S_1 A = \begin{bmatrix}
* & * & \cdots & * \\
* & * & \cdots & * \\
0 & * & \cdots & * \\
\vdots & \vdots & & \vdots \\
0 & * & \cdots & *
\end{bmatrix}
$$

Since we wish to perform a similarity transformation, we must now multiply on the right by the inverse of these elementary matrices to obtain

$$
A_1 = A_{1/2} S_1^{-1} L_{31}^{-1} L_{41}^{-1} \cdots L_{n1}^{-1}
$$

which is similar to A. The following exercises, which duplicate in part exercises from Chapter 1, show how to carry out these operations.

Exercise 4.5.3   Let S be the matrix obtained by interchanging rows i and j of the identity matrix.

(a)   Show that $S^{-1} = S$.

(b)   Suppose $BS = C$. Show that C is obtained from B by interchanging *columns* i and j of B.   □

Exercise 4.5.4   Let L be the matrix obtained by subtracting m times the jth row of the identity matrix from the ith row:

$$
L = \begin{array}{c} \\ \\ \\ j \to \\ \\ i \to \\ \\ \\ \end{array}
\begin{array}{c}
\overset{\displaystyle j}{\downarrow} \qquad \overset{\displaystyle i}{\downarrow}
\end{array}
\begin{bmatrix}
1 & & & & & & \\
& \ddots & & & & & \\
& & 1 & & & & \\
& & & \ddots & & & \\
& & -m & & 1 & & \\
& & & & & \ddots & \\
& & & & & & 1
\end{bmatrix}
$$

(a)   Show that

$$
L^{-1} =
\begin{array}{c}
\\
\\
j \rightarrow \\
\\
i \rightarrow \\
\\
\\
\end{array}
\begin{bmatrix}
1 & & & & & & & \\
 & \ddots & & & & & & \\
 & & 1 & & & & & \\
 & & & \ddots & & & & \\
 & & +m & & 1 & & & \\
 & & & & & \ddots & & \\
 & & & & & & 1 \\
\end{bmatrix}
$$

with column labels j and i above.

(b)   Suppose $BL^{-1} = C$. Show that C is obtained from B by adding m times the ith column to the jth column.   □

Thus we obtain A_1 from $A_{1/2}$ as follows: First we interchange columns 2 and m (S_1^{-1}). Then for $i = 3, \ldots, m$ (in any order) we add m_{i1} times column i to column 2. Thus in the second half of step 1 we perform column operations (somewhat) analogous to the row operations performed in the first half of the step. The fact that the row operations do not touch the first row implies that the corresponding column operations do not touch the first column. This means that the zeros that were created in the first half of the step are not destroyed during the second half.

Step 2 creates the desired zeros in the second column. It looks a lot like step 1: A maximal pivot is moved into the $(3, 2)$ position, and the third row is used as the pivotal row to create zeros in positions $(4, 2), \ldots, (n, 2)$. Rows 1 and 2 are not touched, so the zeros in column 1 are undisturbed. This gives

$$
A_{3/2} = L_{n2} \cdots L_{42} S_2 A_1 =
\begin{bmatrix}
* & * & * & \cdots & * \\
* & * & * & & * \\
0 & * & * & \cdots & * \\
\vdots & \vdots & \vdots & & \vdots \\
0 & 0 & * & \cdots & *
\end{bmatrix}
$$

Then corresponding column operations are performed to complete the similarity transformation:

$$
A_2 = A_{3/2} S_2^{-1} L_{42}^{-1} \cdots L_{n2}^{-1}
$$

These column operations do not touch columns 1 and 2, so the zeros are not destroyed.

Step 3 performs similar operations to create zeros in column 3, and so on. After $n - 2$ steps the reduction is complete. The result is an upper Hessenberg matrix $B = S^{-1}AS$, where

$$
S = S_1^{-1} L_{31}^{-1} \cdots L_{n1}^{-1} S_2^{-1} L_{42}^{-1} \cdots L_{n2}^{-1} S_3^{-1} \cdots L_{n,n-2}^{-1} \tag{4.5.1}
$$

Example 4.5.2   Let

$$
A =
\begin{bmatrix}
1 & 3 & 1 & 4 \\
2 & 1 & 2 & 3 \\
4 & 2 & 3 & 2 \\
2 & 2 & 4 & 1
\end{bmatrix}
$$

We will find an upper Hessenberg matrix similar to A. We begin by interchanging rows 2 and 3 to obtain a maximal pivot in position $(2, 1)$. We then subtract 1/2 times row 2 from each of rows 3 and 4 to get zeros in positions $(3, 1)$ and $(4, 1)$. Storing the multipliers 1/2 and 1/2 in these positions instead of zeros, we have

$$\begin{bmatrix} 1 & 3 & 1 & 4 \\ 4 & 2 & 3 & 2 \\ 1/2 & 0 & 1/2 & 2 \\ 1/2 & 1 & 5/2 & 0 \end{bmatrix}$$

We complete step 1 by interchanging columns 2 and 3 and adding 1/2 times each of columns 3 and 4 to column 2. The result is

$$\begin{bmatrix} 1 & 9/2 & 3 & 4 \\ 4 & 5 & 2 & 2 \\ 1/2 & 3/2 & 0 & 2 \\ 1/2 & 3 & 1 & 0 \end{bmatrix}$$

We begin step 2 by interchanging rows 3 and 4 in order to get a maximal pivot in the $(3, 2)$ position. We then subtract 1/2 times the third row from the fourth row to create a zero in the $(4, 2)$ position:

$$\begin{bmatrix} 1 & 9/2 & 3 & 4 \\ 4 & 5 & 2 & 2 \\ 1/2 & 3 & 1 & 0 \\ 1/2 & 1/2 & -1/2 & 2 \end{bmatrix}$$

Finally we interchange columns 3 and 4 and add 1/2 column 4 to column 3 to obtain

$$\begin{bmatrix} 1 & 9/2 & 11/2 & 3 \\ 4 & 5 & 3 & 2 \\ 1/2 & 3 & 1/2 & 1 \\ 1/2 & 1/2 & 7/4 & -1/2 \end{bmatrix}$$

Thus the matrix

$$B = \begin{bmatrix} 1 & 9/2 & 11/2 & 3 \\ 4 & 5 & 3 & 2 \\ 0 & 3 & 1/2 & 1 \\ 0 & 0 & 7/4 & -1/2 \end{bmatrix}$$

is an upper Hessenberg matrix that is similar to A.

Exercise 4.5.5   Construct an upper Hessenberg matrix similar to

$$\begin{bmatrix} 1 & 4 & 7 \\ 2 & 5 & 8 \\ 3 & 6 & 9 \end{bmatrix}$$ $\square$

The reduction we have just developed is based on Gaussian elimination with partial pivoting, so it possesses the same stability properties. That is, the procedure is virtually always stable, but stability is not unconditionally guaranteed. Instability is signaled by severe growth of the entries of the matrix during the reduction.

A computer program to perform the reduction can be organized just as a Gaussian elimination program is. Thus the multipliers for the row (and column) operations can be stored in the positions where zeros are created (as we did in Example 4.5.2), and a one-dimensional integer array can be used to keep track of the row (and column) interchanges. The multipliers and the record of interchanges together constitute a complete record of the similarity transformation. Such a record is needed if we wish to calculate the eigenvectors of A.

In Chapter 1 we noted that there are numerous variants of Gaussian elimination—inner-product form, outer-product form, etc. An equal number of variants of the reduction to Hessenberg form exists. The outer-product form, column oriented, looks as follows:

Reduction to Upper Hessenberg Form by Gaussian Elimination (column oriented)

for $k = 1, \ldots, n - 2$ (step k creates zeros in column k)

    find (smallest) $m \geq k + 1$ such that $|a_{mk}| = \max\{|a_{ik}| \mid k + 1 \leq i \leq n\}$

    $intch(k) \leftarrow m$ (at step k rows (and columns) $k + 1$ and m will be interchanged)

    if $a_{mk} \neq 0$ then   (if $a_{mk} = 0$, step k is complete)

        if $(k + 1 \neq m)$ then

            for $j = k, \ldots, n$ (row interchange)

                interchange $a_{k+1,j}$ and a_{mj}

        for $i = k + 2, \ldots, n$ (multipliers stored where

            $a_{ik} \leftarrow a_{ik}/a_{k+1,k}$ zeros would appear)   (4.5.3)

        for $j = k + 1, \ldots, n$ (row operations

            for $i = k + 2, \ldots, n$ performed by columns)

                $a_{ij} \leftarrow a_{ij} - a_{ik}a_{k+1,j}$

        if $(k + 1 \neq m)$ then

            for $i = 1, \ldots, n$ (column interchange)

                interchange $a_{i,k+1}$ and a_{im}

        for $j = k + 2, \ldots, n$

            for $i = 1, \ldots, n$ (column operations)

                $a_{i,k+1} \leftarrow a_{i,k+1} + a_{ij}a_{jk}$

exit

Examining the limits of the loop indices in (4.5.3), we find that the total number of flops devoted to row operations is

$$\sum_{k=1}^{n-2} \sum_{j=k+1}^{n} \sum_{i=k+2}^{n} 1 = \sum_{k=1}^{n-2}(n-k)(n-k-1) \approx \sum_{k=1}^{n-1}(n-k)^2 = \sum_{l=1}^{n-1} l^2 \approx \frac{n^3}{3}$$

The total number of flops devoted to column operations is

$$\sum_{k=1}^{n-2} \sum_{j=k+2}^{n} \sum_{i=1}^{n} 1 = n \sum_{k=1}^{n-2}(n-k-1) = n \sum_{l=1}^{n-2} l \approx n \frac{n^2}{2} = \frac{n^3}{2}$$

You can easily check that the costs of interchanges and the calculations of multipliers are comparatively insignificant, so the total flop count for the algorithm is about $n^3/3 + n^3/2 = 5n^3/6$.

Exercise 4.5.6 Why do the column operations cost more than the row operations? ☐

Exercise 4.5.7 Write a Fortran program that implements algorithm (4.5.3). Test it on the matrix of Example 4.5.2 and other test problems of your own devising. ☐

Suppose we have transformed A to upper Hessenberg form $B = S^{-1}AS$, using algorithm (4.5.3). Suppose further that we have somehow found some eigenvectors of B and we would now like to calculate the corresponding eigenvectors of A. For each eigenvector v of B, Sv is an eigenvector of A. From (4.5.1) we see that

$$Sv = S_1^{-1}L_{31}^{-1}\cdots L_{n1}^{-1}S_2^{-1}L_{42}^{-1}\cdots L_{n2}^{-1}S_3^{-1}\cdots L_{n,n-2}^{-1}v \qquad (4.5.4)$$

so we can calculate Sv by performing a sequence of elementary operations on v. The following algorithm performs these operations.

Algorithm to Transform v to Sv [uses multipliers and interchange record generated by (4.5.3)]

$$
\begin{array}{l}
\text{for } k = n - 2, \ldots, 1 \\
\quad \left[
\begin{array}{l}
\text{for } i = k + 2, \ldots, n \\
\quad \left[\, v_i \leftarrow v_i + a_{ik}v_{k+1} \right. \\
m \leftarrow intch(k) \\
\text{if } (m \neq k + 1) \text{ then} \\
\quad \left[\, \text{interchange } v_{k+1} \text{ and } v_m \right.
\end{array}
\right.
\end{array}
\qquad (4.5.5)
$$

The i loop should run from $n, \ldots, k + 2$ rather than $k + 2, \ldots, n$, if it is to agree exactly with (4.5.4). However the order in which this loop is executed is irrelevant. The cost of executing (4.5.5) is about $n^2/2$ flops per vector.

Exercise 4.5.8 Prove the following interpretation of (4.5.3).

   (a) Given $A \in \mathbb{C}^{n \times n}$, there exists a matrix $\hat{A}$ that is similar to A and obtained by permuting rows and columns, a unit-lower-triangular matrix L, and an upper Hessenberg matrix B such that $B = L^{-1}\hat{A}L$.

   (b) What additional properties does L have?

   (c) How is (4.5.5) to be interpreted? ☐

Unitary Reduction

Now let us look at a unitary reduction based on reflectors. As we shall see, this reduction costs twice as much as the reduction based on Gaussian elimination, but the use of reflectors guarantees stability in all cases. The general plan of the reduction is the same as before. The first step creates the desired zeros in the first column, but this time we use a reflector instead of Gaussian elimination. To see how to do this, partition A as

$$
A = \begin{bmatrix} a_{11} & c^T \\ b & \hat{A} \end{bmatrix}
$$

Let $\hat{Q}_1 \in \mathbb{C}^{(n-1)\times(n-1)}$ be a reflector such that $\hat{Q}_1 b = [-\sigma_1 \ 0 \ \cdots \ 0]^T$ ($|\sigma_1| = \|b\|_2$), and let

$$
Q_1 = \left[\begin{array}{c|ccc}
1 & 0 & \cdots & 0 \\
\hline
0 & & & \\
\vdots & & \hat{Q}_1 & \\
0 & & &
\end{array}\right]
$$

Then

$$
A_{1/2} = Q_1 A = \left[\begin{array}{c|c}
a_{11} & c^T \\
\hline
-\sigma_1 & \\
0 & \\
\vdots & \hat{Q}_1 \hat{A} \\
0 &
\end{array}\right]
$$

which has the desired zeros in the first column. We complete the similarity transformation by multiplying on the right by $Q_1^{-1} = Q_1^* = Q_1$.

$$
A_1 = A_{1/2} Q_1 = \left[\begin{array}{c|c}
a_{11} & c^T \hat{Q}_1 \\
\hline
-\sigma_1 & \\
0 & \\
\vdots & \hat{Q}_1 \hat{A} \hat{Q}_1 \\
0 &
\end{array}\right] = \left[\begin{array}{c|ccc}
a_{11} & * & \cdots & * \\
\hline
-\sigma_1 & & & \\
0 & & & \\
\vdots & & \hat{A}_1 & \\
0 & & &
\end{array}\right]
$$

Because of the form of Q_1, this operation does not destroy the zeros in the first column.

The second step creates zeros in the second column of A_1, that is, in the first column of $\hat{A}_1$. Thus we pick a reflector $\hat{Q}_2 \in \mathbb{C}^{(n-2)\times(n-2)}$ in just the same way as in the first step, except that A is replaced by $\hat{A}_1$. Let

$$
Q_2 = \left[\begin{array}{cc|c}
1 & 0 & \\
0 & 1 & 0 \\
\hline
 & 0 & \hat{Q}_2
\end{array}\right]
$$

Then

$$
A_{3/2} = Q_2 A_1 = \left[\begin{array}{c|c|ccc}
a_{11} & * & * & \cdots & * \\
\hline
-\sigma_1 & * & * & \cdots & * \\
\hline
0 & -\sigma_2 & & & \\
. & 0 & & & \\
. & \vdots & & \hat{Q}_2 \hat{A}_2 & \\
0 & 0 & & &
\end{array}\right]
$$

We complete the similarity transformation by multiplying on the right by $Q_2^{-1} = Q_2^* = Q_2$. Because the first two columns of Q_2 are equal to the first two columns of the identity matrix, this operation does not alter the first two columns of $A_{3/2}$. Thus

$$
A_2 = A_{3/2}Q_2 =
\begin{bmatrix}
* & * & * & \cdots & * \\
-\sigma_1 & * & * & \cdots & * \\
0 & -\sigma_2 & & & \\
0 & 0 & & & \\
\vdots & \vdots & & \hat{Q}_2\hat{A}_2\hat{Q}_2 & \\
0 & 0 & & &
\end{bmatrix}
$$

The third step creates zeros in the third column, and so on. After $n-2$ steps the reduction is complete. The result is an upper Hessenberg matrix B that is unitarily similar to A: $B = Q^*AQ$, where

$$
Q = Q_1Q_2\cdots Q_{n-2} \quad \text{and} \quad Q^* = Q_{n-2}Q_{n-3}\cdots Q_1
$$

If A is real, then all operations are real, Q is real and orthogonal, and B is orthogonally similar to A.

Example 4.5.6   Let

$$
A = \begin{bmatrix} 5 & 1 & 3 \\ 1 & 2 & 1 \\ 2 & 4 & 3 \end{bmatrix}
$$

We will find an upper Hessenberg matrix B that is orthogonally similar to A. We begin the first and only step by finding a reflector $\hat{Q}_1 \in \mathbb{R}^{2\times2}$ such that $\hat{Q}_1 \begin{bmatrix} 1 \\ 2 \end{bmatrix} = \begin{bmatrix} -\sqrt{5} \\ 0 \end{bmatrix}$. By (3.2.16) we find that $\hat{Q}_1 = I - \gamma uu^T$, where $\gamma = 1/(5 + \sqrt{5})$ and $u = \begin{bmatrix} 1 + \sqrt{5} \\ 2 \end{bmatrix}$. Normally we would work with $\hat{Q}_1$ in this form, but for this small example we will calculate $\hat{Q}_1$ explicitly. You can check that

$$
\hat{Q}_1 = \frac{1}{\sqrt{5}} \begin{bmatrix} -1 & -2 \\ -2 & 1 \end{bmatrix}
$$

Let

$$
Q_1 =
\begin{bmatrix}
1 & 0 & 0 \\
0 & & \\
0 & & \hat{Q}_1
\end{bmatrix}
$$

The first half of the step is to form $A_{1/2} = Q_1 A$. Since

$$\hat{Q}_1 \begin{bmatrix} 1 \\ 2 \end{bmatrix} = \begin{bmatrix} -\sqrt{5} \\ 0 \end{bmatrix} \qquad \hat{Q}_1 \begin{bmatrix} 2 \\ 4 \end{bmatrix} = \begin{bmatrix} -10/\sqrt{5} \\ 0 \end{bmatrix} \qquad \hat{Q}_1 \begin{bmatrix} 1 \\ 3 \end{bmatrix} = \begin{bmatrix} -7/\sqrt{5} \\ 1/\sqrt{5} \end{bmatrix}$$

it follows that

$$A_{1/2} = \begin{bmatrix} 5 & 1 & 3 \\ -\sqrt{5} & -10/\sqrt{5} & -7/\sqrt{5} \\ 0 & 0 & 1/\sqrt{5} \end{bmatrix}$$

Finally $B = A_1 = A_{1/2} Q_1$. Since

$$[1 \ 3] \hat{Q}_1 = \begin{bmatrix} -7/\sqrt{5} & 1/\sqrt{5} \end{bmatrix} \qquad \begin{bmatrix} -10/\sqrt{5} & -7/\sqrt{5} \end{bmatrix} \hat{Q}_1 = [24/5 \quad 13/5]$$

and $\qquad \begin{bmatrix} 0 & 1/\sqrt{5} \end{bmatrix} \hat{Q}_1 = [-2/5 \quad 1/5]$

we have

$$B = \begin{bmatrix} 5 & -7/\sqrt{5} & 1/\sqrt{5} \\ -\sqrt{5} & 24/5 & 13/5 \\ 0 & -2/5 & 1/5 \end{bmatrix}$$

Exercise 4.5.9 Find an upper Hessenberg matrix that is orthogonally similar to

$$\begin{bmatrix} 1 & 3 & 2 \\ 2 & 2 & 0 \\ 3 & 1 & 1 \end{bmatrix} \qquad \square$$

A computer program to perform the reduction can be organized in just the same way as a QR decomposition by reflectors is. For simplicity we will restrict the discussion to the real case. Most of the details do not require discussion because they were already covered in Section 3.2. The one way in which the present algorithm is significantly different from the QR decomposition is that it involves multiplication by reflectors on the right as well as the left. Let us see how this can be done in practice.

Suppose we are given a matrix C, and we wish to calculate $C\hat{Q}$, where $\hat{Q}$ is a reflector. Letting $x_1^T, \ldots, x_n^T$ denote the rows of C, we have

$$C = \begin{bmatrix} x_1^T \\ x_2^T \\ \vdots \\ x_n^T \end{bmatrix} \qquad \text{and} \qquad C\hat{Q} = \begin{bmatrix} x_1^T \hat{Q} \\ x_2^T \hat{Q} \\ \vdots \\ x_n^T \hat{Q} \end{bmatrix}$$

Now $(x_i^T \hat{Q})^T = \hat{Q} x_i$, so the computation of each row of $C\hat{Q}$ is equivalent to multiplication of $\hat{Q}$ by a column vector. Therefore algorithm (3.2.17) can be used. The rows of C can be modified one at a time to yield $C\hat{Q}$. If this process is coded in a straightforward manner, the resulting program accesses the matrix by rows. If a column-oriented programming language such as Fortran is used, it is preferable to access the matrix by columns. This can be achieved by interchanging the loop indices carefully.

A column-oriented algorithm that uses reflectors to reduce a real matrix to upper Hessenberg form is given below. Loop interchanges were made only in the right multiplication portion of the code, since the left multiplication portion is already oriented by columns. The price of interchanging loops is that an extra one-dimensional array τ is needed, in which n inner products are accumulated simultaneously.

Real Householder Reduction to Upper Hessenberg Form (column-oriented version)

$$
\begin{aligned}
&\text{for } k = 1, \ldots, n - 2 \\
&\quad \left[\begin{array}{l}
\left.\begin{array}{l}
m \leftarrow \max\{|a_{ik}| \,\big|\, i = k + 1, \ldots, n\} \\
\gamma_k \leftarrow 0 \\
\text{if } (m \neq 0) \text{ then} \\
\quad \left[\begin{array}{l}
\text{for } i = k + 1, \ldots, n \\
\quad \left[\, a_{ik} \leftarrow a_{ik}/m \right. \\
\sigma_k \leftarrow \sqrt{a_{k+1,k}^2 + \cdots + a_{nk}^2} \\
\text{if } (a_{k+1,k} < 0) \ \sigma_k \leftarrow -\sigma_k \\
a_{k+1,k} \leftarrow a_{k+1,k} + \sigma_k \\
\gamma_k \leftarrow 1/(\sigma_k a_{k+1,k}) \\
\sigma_k \leftarrow \sigma_k m
\end{array}\right.
\end{array}\right\} \text{(set up the reflector } \hat{Q}_k) \\[6pt]
\left.\begin{array}{l}
\text{for } j = k + 1, \ldots, n \\
\quad \left[\begin{array}{l}
\tau_1 \leftarrow 0 \\
\text{for } i = k + 1, \ldots, n \\
\quad \left[\, \tau_1 \leftarrow \tau_1 + a_{ik} a_{ij} \right. \\
\tau_1 \leftarrow -\tau_1 \gamma_k \\
\text{for } i = k + 1, \ldots, n \\
\quad \left[\, a_{ij} \leftarrow a_{ij} + \tau_1 a_{ik} \right.
\end{array}\right.
\end{array}\right\} \text{(multiply on the left by } \hat{Q}_k) \quad (4.5.7) \\[6pt]
\left.\begin{array}{l}
\text{for } i = 1, \ldots, n \\
\quad \left[\, \tau_i \leftarrow 0 \right. \\
\text{for } j = k + 1, \ldots, n \\
\quad \left[\begin{array}{l}
\text{for } i = 1, \ldots, n \\
\quad \left[\, \tau_i \leftarrow \tau_i + a_{ij} a_{jk} \right.
\end{array}\right. \\
\text{for } i = 1, \ldots, n \\
\quad \left[\, \tau_i \leftarrow -\tau_i \gamma_k \right. \\
\text{for } j = k + 1, \ldots, n \\
\quad \left[\begin{array}{l}
\text{for } i = 1, \ldots, n \\
\quad \left[\, a_{ij} \leftarrow a_{ij} + \tau_i a_{jk} \right.
\end{array}\right.
\end{array}\right\} \text{(multiply on the right by } \hat{Q}_k)
\end{array}\right.
\end{aligned}
$$

$\sigma_{n-1} \leftarrow -a_{n,n-1}$
exit

This algorithm takes as input an array that originally contains $A \in \mathbb{R}^{n \times n}$. It returns an upper Hessenberg matrix $B = Q^T A Q$ in the following form. The main diagonal and upper triangle of B are stored in the corresponding locations of the array that contained A originally. The subdiagonal of B is stored in the array σ.

Specifically $b_{i+1,i} = -\sigma_i$, $i = 1, \ldots, n - 1$. In other words, B is stored over A, except that the subdiagonal of B is stored elsewhere and the portion of B that consists entirely of zeros is not stored at all. The portion of the array below the main diagonal is used to store information about the reflectors. The part of the kth column that lies below the main diagonal contains the vector u_k such that $\hat{Q}_k = I - \gamma_k u_k u_k^T$, where $\hat{Q}_k$ is the $(n - k)$-by-$(n - k)$ reflector used at step k. The scalar γ_k ($k = 1, \ldots, n - 2$) is stored in a separate array γ. Thus the information needed to construct the orthogonal transforming matrix Q is available.

Exercise 4.5.10

 (a)   Count the flops in (4.5.7) and show that the algorithm requires about $5n^3/3$ flops, twice as many as the reduction based on Gaussian elimination.

 (b)   Note that again right multiplication is more expensive than left multiplication. Why?   ☐

Exercise 4.5.11   Write a Fortran subroutine that implements (4.5.7).   ☐

Suppose we have transformed $A \in \mathbb{R}^{n \times n}$ to upper Hessenberg form $B = Q^T A Q$ using (4.5.7). Suppose further that we have somehow found some eigenvectors of B and we would now like to calculate the corresponding eigenvectors of A. For each eigenvector v of B, Qv is an eigenvector of A. Since

$$Qv = Q_1 Q_2 \cdots Q_{n-2} v$$

we can easily calculate Qv by applying $n - 2$ reflectors in succession. The following algorithm, which is based on (3.2.17), does exactly that.

Algorithm that transforms v to Qv [uses reflectors generated by (4.5.7) or (4.5.9)]

$$
\begin{aligned}
&\text{for } k = n - 2, n - 3, \ldots, 1 \\
&\quad \left[
\begin{array}{l}
\tau \leftarrow 0 \\
\text{for } i = k + 1, \ldots, n \\
\quad \left[\tau \leftarrow \tau + v_i a_{ik} \right. \\
\tau \leftarrow -\tau \gamma_k \\
\text{for } i = k + 1, \ldots, n \\
\quad \left[v_i \leftarrow v_i + \tau a_{ik} \right.
\end{array}
\right. \\
&\text{exit}
\end{aligned}
\qquad (4.5.8)
$$

The cost of using this algorithm is about n^2 flops per eigenvector.

The Symmetric Case

If A is Hermitian, then the matrix B produced by the unitary reduction is not merely a Hessenberg matrix; it is tridiagonal. Furthermore it is possible to exploit the symmetry of A to reduce the cost of the reduction to $2n^3/3$ flops. This is less than half the cost of the unitary reduction in the non-Hermitian case. In the interest of simplicity, we will restrict our attention to the real symmetric case. We begin with the matrix

$$A = \begin{bmatrix} a_{11} & b^T \\ b & \hat{A} \end{bmatrix}$$

In the first step of the reduction, we transform A to $A_1 = Q_1 A Q_1$, where

$$\begin{bmatrix} 1 & 0 & \cdots & 0 \\ \hline 0 & & & \\ \vdots & & \hat{Q}_1 & \\ 0 & & & \end{bmatrix}$$

and $\hat{Q}_1$ is a reflector chosen so that $\hat{Q}_1 b = [-\sigma_1 \; 0 \; \cdots \; 0]^T$. Thus

$$A_1 = \begin{bmatrix} a_{11} & b^T \hat{Q}_1 \\ \hline \hat{Q}_1 b & \hat{Q}_1 \hat{A} \hat{Q}_1 \end{bmatrix} = \begin{bmatrix} a_{11} & -\sigma_1 & 0 & \cdots & 0 \\ \hline -\sigma_1 & & & & \\ 0 & & & & \\ \vdots & & & \hat{A}_1 & \\ 0 & & & & \end{bmatrix}$$

We save a little bit here by not performing the computation $b^T \hat{Q}_1$, which duplicates the computation $\hat{Q}_1 b$.

The bulk of the computational effort in this step is expended in the computation of the symmetric submatrix $\hat{A}_1 = \hat{Q}_1 \hat{A} \hat{Q}_1$. We must perform this computation efficiently if we wish to realize significant savings. If symmetry is not exploited, it costs about $2n^2$ flops to calculate $\hat{Q}_1 \hat{A}$ and another $2n^2$ flops to calculate $(\hat{Q}_1 \hat{A})\hat{Q}_1$, as you can easily verify. Thus the entire computation of $\hat{A}_1$ costs about $4n^2$ flops. It turns out that we can cut this figure in half by very carefully exploiting symmetry.

$\hat{Q}_1$ is a reflector given in the form $\hat{Q}_1 = I - \gamma u u^T$. Thus

$$\begin{aligned} \hat{A}_1 &= (I - \gamma u u^T)\hat{A}(I - \gamma u u^T) \\ &= \hat{A} - \gamma \hat{A} u u^T - \gamma u u^T \hat{A} + \gamma^2 u u^T \hat{A} u u^T \end{aligned}$$

The terms in this expression admit considerable simplification if we introduce the auxiliary vector $v = -\gamma \hat{A} u$. Thus $-\gamma \hat{A} u u^T = v u^T$, $-\gamma u u^T \hat{A} = u v^T$, and $\gamma^2 u u^T \hat{A} u u^T = -\gamma u u^T v u^T$. Introducing the scalar $\delta = -\frac{1}{2}\gamma u^T v$, we can rewrite this last term as $2\delta u u^T$. Thus

$$\hat{A}_1 = \hat{A} + v u^T + u v^T + 2\delta u u^T$$

The final manipulation is to split the last term into two pieces in order to combine one piece with the term $v u^T$ and the other piece with the term $u v^T$. Specifically, let $w = v + \delta u$. Then

$$\hat{A}_1 = \hat{A} + w u^T + u w^T$$

This equation translates into the code segment

$$\text{for } j = 2, \ldots, n$$
$$\quad \text{for } i = j, \ldots, n$$
$$\quad\quad \left[a_{ij} \leftarrow a_{ij} + w_i u_j + u_i w_j \right.$$

which costs two flops per updated array entry. By symmetry we need only update the main diagonal and lower triangle. Thus the total number of flops in this segment is $2(n-1)n/2 \approx n^2$. This does not include the cost of calculating w. First of all, the computation $v = -\gamma \hat{A} u$ costs about n^2 flops. The computation $\delta = -\frac{1}{2}\gamma u^T v$ costs about n flops, and finally the computation $w = v + \delta u$ costs about n flops. Thus the total flop count for the first step is roughly $2n^2$, as claimed.

The second step of the reduction has no effect on the first column of A_1 just as in the nonsymmetric case. By symmetry it has no effect on the first row of A_1 either. This was not so in the nonsymmetric case; it is this difference that makes the symmetric reduction *less than* half as expensive as the nonsymmetric reduction. The second step is identical to the first step, except that it acts on the submatrix $\hat{A}_1$. Thus the flop count for this step is roughly $2(n-1)^2$. After $n-2$ steps the reduction is complete. The total flop count is approximately $2[n^2 + (n-1)^2 + \cdots] \approx 2n^3/3$.

Reduction of a Real Symmetric Matrix to Tridiagonal Form by an Orthogonal Similarity Transformation (column-oriented code)

$$\text{for } k = 1, \ldots, n-2$$
$$\quad m \leftarrow \max\{|a_{ik}| \mid i = k+1, \ldots, n\}$$
$$\quad \gamma_k \leftarrow 0$$
$$\quad \text{if } (m \neq 0) \text{ then}$$
$$\quad\quad \text{set up the reflector } \hat{Q}_k \text{ exactly as in (4.5.7)}$$
$$\quad\quad \text{for } i = k+1, \ldots, n$$
$$\quad\quad\quad \left[w_i \leftarrow 0 \right.$$
$$\quad\quad \text{for } j = k+1, \ldots, n$$
$$\quad\quad\quad \text{for } i = j, \ldots, n$$
$$\quad\quad\quad\quad \left[w_i \leftarrow w_i + a_{ij} a_{jk} \right.$$
$$\quad\quad \text{for } i = k+1, \ldots, n$$
$$\quad\quad\quad \text{for } j = i+1, \ldots, n$$
$$\quad\quad\quad\quad \left[w_i \leftarrow w_i + a_{ji} a_{jk} \right.$$

(calculate $\hat{A}u$, accessing only the lower part of $\hat{A}$, $u_j = a_{jk}$)

$$\quad\quad \delta \leftarrow 0$$
$$\quad\quad \text{for } i = k+1, \cdots, n$$
$$\quad\quad\quad \left[\delta \leftarrow \delta + w_i a_{ik} \right.$$
$$\quad\quad \delta \leftarrow -\gamma_k \delta / 2$$
$$\quad\quad \text{for } i = k+1, \ldots, n$$
$$\quad\quad\quad \left[w_i \leftarrow w_i + \delta a_{ik} \right.$$

(calculate w)

$$\quad\quad \text{for } j = k+1, \ldots, n$$
$$\quad\quad\quad \text{for } i = j, \ldots, n$$
$$\quad\quad\quad\quad \left[a_{ij} \leftarrow a_{ij} + w_i a_{jk} + a_{ik} w_j \right.$$

(update A)

(4.5.9)

$$\sigma_{n-1} \leftarrow -a_{n,n-1}$$
$$\text{for } i = 1, \ldots, n$$
$$\quad \left[d_i \leftarrow a_{ii} \right.$$
$$\text{for } i = 1, \ldots, n-1$$
$$\quad \left[s_i \leftarrow -\sigma_i \right.$$
$$\text{exit}$$

This algorithm accesses only the main diagonal and lower triangle of A. It stores the main-diagonal entries of the tridiagonal matrix B in a one-dimensional array d ($d_i = b_{ii}$, $i = 1, \ldots, n$), and it stores the off-diagonal entries in a one-dimensional array s ($s_i = b_{i+1,i} = b_{i,i+1}$, $i = 1, \ldots, n - 1$). The information about the reflectors used in the similarity transformation is stored exactly as in (4.5.7). Once some eigenvectors of B have been found, (4.5.8) can be used to calculate the corresponding eigenvectors of A.

Exercise 4.5.12 Confirm that the execution of (4.5.9) costs about $2n^3/3$ flops. □

Exercise 4.5.13 Write a Fortran subroutine that implements (4.5.9). □

4.6
THE *QR* ALGORITHM

In recent years the most widely used algorithm for calculating the complete set of eigenvalues of a matrix has been the *QR* algorithm. The present section is devoted to a description of the algorithm, and in the section that follows we will examine several implementations. The explanation of why the algorithm works is largely postponed to Sections 5.1 to 5.3. You can read those sections first if you want to.

Consider a matrix $A \in \mathbb{C}^{n \times n}$ whose eigenvalues we would like to compute. For now let us assume that A is nonsingular, a restriction that we will remove later. The basic *QR* algorithm is very easy to describe. It starts with $A_0 = A$ and generates a sequence of matrices (A_j) by the following prescription:

$$A_{m-1} = Q_m R_m \qquad R_m Q_m = A_m \qquad (4.6.1)$$

That is, A_{m-1} is decomposed into factors Q_m and R_m, where Q_m is unitary and R_m is upper triangular with positive entries on the main diagonal. These factors are uniquely determined (Theorem 3.2.24). The factors are then multiplied back together in the reverse order to produce A_m. You can easily verify that $A_m = Q_m^* A_{m-1} Q_m$.

Exercise 4.6.1

(a) Show that $A_m = Q_m^* A_{m-1} Q_m$.

(b) Show that $A_m = R_m A_{m-1} R_m^{-1}$. [Note that part (a) is valid for singular and nonsingular A, whereas part (b) requires that A be nonsingular. (Why?)] □

Thus all matrices in the sequence (A_j) are unitarily similar and therefore have the same eigenvalues. In Section 5.2 we will see that the *QR* algorithm is just a clever implementation of a procedure known as simultaneous iteration, which is itself a natural, easily understood extension of the power method. As a consequence, the sequence (A_j) converges, under suitable conditions, to upper-triangular form

$$
\begin{bmatrix}
\lambda_1 & & & * \\
& \lambda_2 & & \\
0 & & \ddots & \\
& & & \lambda_n
\end{bmatrix}
$$

where the eigenvalues appear in order of decreasing magnitude on the main diagonal. (As we shall see, what actually happens is often more complicated than this, but there is no point in discussing the details now.)

From now on it will be important to make a careful distinction between the terms *QR decomposition* and *QR algorithm*. The *QR algorithm* is an iterative procedure for finding eigenvalues. It is based on the *QR decomposition*, which is a direct procedure related to the Gram–Schmidt process. A single iteration of the *QR* algorithm will be called a *QR step* or *QR iteration*.

Each *QR* step performs a unitary similarity transformation. We noted in Section 4.4 that numerous matrix properties are preserved under such transformations, the most important of which is the Hermitian property. Thus if A is Hermitian, then all iterates A_j will be Hermitian and the sequence (A_j) will converge to diagonal form.

If A_{m-1} is real, then Q_m, R_m, and A_m are also real. Thus, if A is real, then the basic *QR* algorithm (4.6.1) remains within the real number system.

Example 4.6.2   Let us apply the basic *QR* algorithm to the real, symmetric matrix

$$
A = \begin{bmatrix} 8 & 2 \\ 2 & 5 \end{bmatrix}
$$

whose eigenvalues are easily seen to be $\lambda_1 = 9$ and $\lambda_2 = 4$. Letting $A_0 = A$, we have $A_0 = Q_1 R_1$, where

$$
Q_1 = \frac{1}{\sqrt{68}} \begin{bmatrix} 8 & -2 \\ 2 & 8 \end{bmatrix} \quad \text{and} \quad R_1 = \frac{1}{\sqrt{68}} \begin{bmatrix} 68 & 26 \\ 0 & 36 \end{bmatrix}
$$

Thus

$$
A_1 = R_1 Q_1 = \frac{1}{68} \begin{bmatrix} 596 & 72 \\ 72 & 288 \end{bmatrix} \approx \begin{bmatrix} 8.7647 & 1.0588 \\ 1.0588 & 4.2353 \end{bmatrix}
$$

Notice that A_1 is real and symmetric, just as A_0 is. Notice also that A_1 is closer to diagonal form than A_0 is, in the sense that its off-diagonal entries are closer to zero. Furthermore, the main-diagonal entries of A_1 are closer to the eigenvalues. On subsequent iterations the main-diagonal entries of A_j give progressively better estimates of the eigenvalues, until after 10 iterations they agree with the true eigenvalues to seven decimal places.

Exercise 4.6.2   Let $A_0 = A = \begin{bmatrix} 8 & 1 \\ -2 & 1 \end{bmatrix}$. Find an orthogonal Q_1 (a rotator) and an upper-triangular R_1 such that $A_0 = Q_1 R_1$. Calculate $A_1 = R_1 Q_1$. Notice that A_1

is closer to upper-triangular form than A_0 is, and the main-diagonal entries of A_1 approximate the eigenvalues of A better than those of A_0 do.   ☐

The assumption that A is nonsingular guarantees that every matrix in the sequence (A_j) is nonsingular. This fact allows us to specify the decomposition $A_{m-1} = Q_m R_m$ uniquely by requiring that the main-diagonal entries of R_m be positive. Thus the *QR* algorithm, as described above, is well defined. However, it is not always convenient in practice to arrange the computations so that each R_m has positive main-diagonal entries. It is therefore reasonable to ask how the sequence (A_j) is affected if this requirement is dropped. The next two exercises show that nothing bad happens.

Exercise 4.6.3 Let $B = (b_{ij})$ and $C = (c_{ij})$ be matrices in $\mathbb{C}^{n \times n}$, and suppose $C = D^{-1}BD = D^*BD$, where D is diagonal and unitary.

    (a)   Show that $|c_{ij}| = |b_{ij}|$ for all i and j.

    (b)   Show that $c_{ii} = b_{ii}$ for all i.   ☐

Exercise 4.6.4 A matrix that is both unitary and upper triangular must be a diagonal matrix (Exercise 4.4.20). You can use this fact to prove the following result by induction on m: Suppose A is nonsingular. Let $A_0 = A$ and $\tilde{A}_0 = A$, and let (A_j) and $(\tilde{A}_j)$ be sequences that satisfy

$$A_{m-1} = Q_m R_m \qquad R_m Q_m = A_m$$
$$\tilde{A}_{m-1} = \tilde{Q}_m \tilde{R}_m \qquad \tilde{R}_m \tilde{Q}_m = \hat{A}_m$$

where Q_m and $\tilde{Q}_m$ are unitary and R_m and $\tilde{R}_m$ are upper triangular. Since it is not required that R_m and $\tilde{R}_m$ have positive main-diagonal entries, the decompositions are not uniquely determined. Thus it can happen that $\tilde{A}_m \neq A_m$. Show that there exists a sequence of unitary diagonal matrices (D_j) such that $\tilde{A}_m = D_m^* A_m D_m$ for all m.   ☐

Exercise 4.6.4 shows that $\tilde{A}_m$ and A_m are related by a trivial similarity transformation of the type discussed in Exercise 4.6.3. It follows from that exercise that (A_j) converges to upper-triangular form if and only if $(\tilde{A}_j)$ does, and the convergence rates of the two sequences are the same. Therefore, in actual implementations of the *QR* algorithm we will not require that the main-diagonal entries of each R_m be positive.

There are two reasons why the basic *QR* algorithm (4.6.1) is too inefficient for general use. First, the cost of each *QR* step is high. Each *QR* decomposition costs $2n^3/3$ flops, and the matrix multiplication that follows also costs $O(n^3)$ flops. This is a very high price to pay, considering that we expect to have to perform quite a few steps. The second problem is that convergence is generally quite slow; a very large number of iterations is needed before A_m is sufficiently close to triangular form that we are willing to accept its main-diagonal entries as eigenvalues of A. Thus we need to make *QR* steps less expensive, and we need to accelerate the convergence somehow, so that fewer *QR* steps are needed.

QR Algorithm with Hessenberg Matrices

The problem of the high cost of QR iterations can be solved by first reducing the matrix to upper Hessenberg form. This works because the upper Hessenberg form is preserved by the QR algorithm, as the following theorem shows.

THEOREM 4.6.3   Let A_{m-1} be a nonsingular upper Hessenberg matrix, and suppose A_m is obtained from A_{m-1} by one QR step. Then A_m is also in upper Hessenberg form.

Proof   The equation $A_{m-1} = Q_m R_m$ can be rewritten as $Q_m = A_{m-1} R_m^{-1}$. By Exercise 1.7.15, R_m^{-1} is upper triangular. You can easily show (Exercise 4.6.5) that the product of an upper Hessenberg matrix with an upper-triangular matrix, in either order, is upper Hessenberg. Therefore Q_m is an upper Hessenberg matrix. But then $A_m = R_m Q_m$ must also be upper Hessenberg.   □

Exercise 4.6.5   Suppose $H \in \mathbb{C}^{n \times n}$ is upper Hessenberg and $R \in \mathbb{C}^{n \times n}$ is upper triangular. Prove that HR and RH are both upper Hessenberg.   □

It would be nice to be able to drop our requirement that A be nonsingular. Of course there is no reason why we cannot carry out (4.6.1) starting with a singular matrix; the only problem is that the QR decompositions are not uniquely determined. In Exercise 4.6.4 we found out that in the nonsingular case, nonuniqueness does not cause any problems. However, in the singular case there is greater freedom in the choice of Q and R, and the result of Exercise 4.6.4 is no longer valid. This is demonstrated by the following example, which also shows that the Hessenberg form is not necessarily preserved by the QR algorithm in the singular case. This is certainly a cause for concern.

Example 4.6.4   Let

$$A_0 = \begin{bmatrix} 0 & 0 & 1 \\ 0 & 0 & 0 \\ 0 & 0 & 0 \end{bmatrix}$$

which is obviously upper Hessenberg and singular. You can easily check that $A_0 = Q_1 R_1$, where

$$Q_1 = \begin{bmatrix} 0 & 0 & 1 \\ 0 & 1 & 0 \\ 1 & 0 & 0 \end{bmatrix} \quad \text{and} \quad R_1 = \begin{bmatrix} 0 & 0 & 0 \\ 0 & 0 & 0 \\ 0 & 0 & 1 \end{bmatrix}$$

Clearly Q_1 is unitary and R_1 is upper triangular. Then

$$A_1 = R_1 Q_1 = \begin{bmatrix} 0 & 0 & 0 \\ 0 & 0 & 0 \\ 1 & 0 & 0 \end{bmatrix}$$

which is not an upper Hessenberg matrix. Thus Theorem 4.6.3 is not valid for singular matrices. Of course the problem here is that we made a strange choice of Q_1 and R_1. A more obvious choice is $Q_1 = I$ and $R_1 = A_0$, which gives $A_1 = A_0$, which is upper Hessenberg.

Exercise 4.6.6   Show where the proof of Theorem 4.6.5 breaks down in the singular case.   □

It turns out that it is always possible, in fact easy, to avoid the phenomenon demonstrated by Example 4.6.4. If one carries out the QR decomposition in the most straightforward and obvious manner, the upper Hessenberg form is preserved. Indeed, suppose A is an upper Hessenberg matrix for which we wish to perform a QR step. We can transform A to upper-triangular form by using $n - 1$ rotators to transform the $n - 1$ subdiagonal entries to zero. (For what follows you might find it useful to refer back to the material on rotators in Section 3.2. If you find it helpful to think only in terms of the real case, then do so by all means.)

We begin by finding a (complex) rotator Q_{21}, acting in the 2, 1 plane, such that $Q_{21}^* A$ has a zero in the $(2, 1)$ position (Exercise 3.2.29). This rotator alters only the first and second rows of the matrix.

Next we find a rotator Q_{32}, acting in the 3, 2 plane, such that $Q_{32}^* Q_{21}^* A$ has a zero in the $(3, 2)$ position. Q_{32}^* alters only the second and third rows of $Q_{21}^* A$. Since the intersection of these rows with the first column consists of zeros, these zeros will not be destroyed. (They will be recombined to create new zeros.) In particular, the zero in the $(2, 1)$ position, which was created by the previous rotator, is preserved. This is all quite easy to see if you draw a schematic diagram of A and note which elements are altered, where zeros are created, and so forth.

Next we find Q_{43}, acting in the 4, 3 plane, such that $Q_{43}^* Q_{32}^* Q_{21}^* A$ has a zero in position $(4, 3)$. You can easily check that this rotator does not destroy the zeros that were created previously or any other zeros below the main diagonal.

Continuing in this manner, we transform A to an upper-triangular matrix R given by

$$R = Q_{n,n-1}^* \cdots Q_{32}^* Q_{21}^* A$$

Letting

$$Q = Q_{21} Q_{32} \cdots Q_{n,n-1} \tag{4.6.5}$$

we have $R = Q^* A$ or $A = QR$, where Q is unitary and R is upper triangular.

Now we must calculate $A_1 = RQ$ to complete a step of the QR algorithm. By (4.6.5)

$$A_1 = R Q_{21} Q_{32} \cdots Q_{n,n-1}$$

so all we need to do is multiply R on the right by the rotators $Q_{21}, Q_{32}, \ldots, Q_{n,n-1}$, successively. Since Q_{21} acts in the 2, 1 plane, it recombines the first and second columns of the matrix. Since both of these columns consist entirely of zeros after the first two positions, these zeros will not be destroyed by Q_{21}. The only zero that can (and probably will) be destroyed is the one in the (2, 1) position. Similarly Q_{32}, which recombines columns 2 and 3, can destroy only the zero in the (3, 2) position, and so on. Thus the transformation from R to A_1 creates new nonzero entries below the main diagonal only in positions (2, 1), (3, 2), ..., (n, n − 1). We conclude that A_1 is an upper Hessenberg matrix.

Exercise 4.6.7   Show that the matrix Q defined by (4.6.5) is upper Hessenberg. (*Hint:* Start with the identity matrix and build up Q by applying rotators to I one at a time.)   □

The construction that we have just examined shows that, in spite of Example 4.6.4, the QR algorithm does, as a practical matter, preserve upper Hessenberg form, regardless of whether A is singular or nonsingular. This observation will be strengthened by Exercise 4.6.13 below. By the way, the construction can also be carried out using reflectors in place of plane rotators. Whichever we use, the construction also yields the following result.

Theorem 4.6.6   A QR step applied to an upper Hessenberg matrix requires not more than $O(n^2)$ flops.

Proof   The construction outlined above transforms A_{m-1} to A_m by applying $n - 1$ rotators on the left, followed by $n - 1$ rotators on the right. The cost of applying each of these rotators (or reflectors) is $O(n)$ flops. Thus the total flop count is $O(n^2)$. The exact count depends on the details of the implementation.   □

We have now solved the problem of the cost of QR steps in a tolerably good fashion. To summarize, we begin by reducing the matrix to upper Hessenberg form. This costs not more than about $5n^3/3$ flops. The exact count depends on which of the algorithms from Section 4.5 is used. While this is expensive, it only has to be done once because upper Hessenberg form is preserved by the QR algorithm. By Theorem 4.6.6 the QR steps are relatively cheap, each one costing only $O(n^2)$ flops.

The situation is even better when the matrix is Hermitian, since in this case it can be reduced to tridiagonal form. The fact that the QR algorithm preserves both the Hermitian property and upper Hessenberg form implies that the tridiagonal Hermitian form is preserved by QR steps. Such QR steps are extremely inexpensive.

Exercise 4.6.8   Show that a QR step applied to a tridiagonal Hermitian matrix requires only $O(n)$ flops.   □

Thus once we have reduced the matrix to tridiagonal form at a cost of $2n^3/3$ flops, we can perform QR steps at negligible cost.

Accelerating the Convergence of the QR Algorithm

Now let us turn to the problem of accelerating the convergence of the QR algorithm. We will suppose that the iterates A_m are all in upper Hessenberg form, and we will let $a_{ij}^{(m)}$ denote the (i, j) entry of A_m. Thus

$$
A_m = \begin{bmatrix}
a_{11}^{(m)} & a_{12}^{(m)} & \cdots & a_{1,n-1}^{(m)} & a_{1n}^{(m)} \\
a_{21}^{(m)} & a_{22}^{(m)} & \cdots & a_{2,n-1}^{(m)} & a_{2n}^{(m)} \\
 & a_{32}^{(m)} & \cdots & a_{3,n-1}^{(m)} & a_{3n}^{(m)} \\
 & & \ddots & \vdots & \vdots \\
 & & & a_{n,n-1}^{(m)} & a_{nn}^{(m)}
\end{bmatrix}
$$

Let $\lambda_1, \lambda_2, \ldots, \lambda_n$ denote the eigenvalues of A, ordered so that $|\lambda_1| \geq |\lambda_2| \geq |\lambda_3| \geq \cdots \geq |\lambda_n|$. In Section 5.2 we will see that (most of) the subdiagonal entries $a_{i+1,i}^{(m)}$ converge to zero as $m \to \infty$. More precisely, if $|\lambda_i| > |\lambda_{i+1}|$, then $a_{i+1,i}^{(m)} \to 0$ linearly with convergence ratio $|\lambda_{i+1}/\lambda_i|$. We can improve the rate of convergence by decreasing some or all of the ratios $|\lambda_{i+1}/\lambda_i|$, $i = 1, \ldots, n-1$. An obvious way to do this is to shift the matrix.

The shifted matrix $A - \sigma I$ has eigenvalues $\lambda_1 - \sigma, \lambda_2 - \sigma, \ldots, \lambda_n - \sigma$. If we renumber the eigenvalues so that $|\lambda_1 - \sigma| \geq |\lambda_2 - \sigma| \geq \cdots \geq |\lambda_n - \sigma|$, then the ratios associated with $A - \sigma I$ are $|(\lambda_{i+1} - \sigma)/(\lambda_i - \sigma)|$, $i = 1, \ldots, n-1$. The one ratio that can be made really small is $|(\lambda_n - \sigma)/(\lambda_{n-1} - \sigma)|$, which we can make as close to zero as we please (provided that $\lambda_{n-1} \neq \lambda_n$) by choosing σ very close to λ_n.

Of course it is equally good to choose σ so that it approximates any one of the eigenvalues of A; after the eigenvalues have been renumbered, that eigenvalue which is closest to σ will be called λ_n. Thus, if we can find a σ that approximates an eigenvalue well, we should apply the QR algorithm to $A - \sigma I$ instead of A. The entry $a_{n,n-1}^{(m)}$ will converge to zero very quickly. Once it is sufficiently small, it can be considered to be zero for practical purposes, and adding the shift back on, we have

$$
A_m + \sigma I = \left[
\begin{array}{ccc|c}
 & & & * \\
 & \hat{A}_m & & \vdots \\
 & & & * \\
\hline
0 & \cdots & 0 & a_{nn}^{(m)}
\end{array}
\right]
$$

By Theorem 4.2.5 $a_{nn}^{(m)}$ is an eigenvalue of A; indeed $a_{nn}^{(m)} = \lambda_n$, the eigenvalue closest to σ. The remaining eigenvalues of A are eigenvalues of $\hat{A}_m$, so we might as well perform subsequent iterations on this smaller matrix. This process, through which the size of the problem is reduced when an eigenvalue is found, is called *deflation*. If we can find a $\hat{\sigma}$ that approximates an eigenvalue of $\hat{A}_m$ well, then we can extract that eigenvalue quickly by performing QR iterations on $\hat{A}_m - \hat{\sigma} I$. Once that eigenvalue has been found, we can deflate the problem again and go after the next eigenvalue. Continuing in this fashion we eventually find all eigenvalues of A.

The catch to this argument is that we need good approximations to the eigenvalues. Where can we obtain these approximations? Suppose we begin by performing several QR steps with no shift. After a number of iterations the matrices will begin to approach triangular form and the main-diagonal entries will begin to approach the eigenvalues. In particular, $a_{nn}^{(m)}$ will approximate λ_n, the eigenvalue of A of least modulus. It is therefore reasonable to take $\sigma = a_{nn}^{(m)}$ at some point and perform subsequent iterations on the shifted matrix $A_m - \sigma I$. In fact we can do better than that. With each step we get a better approximation to the eigenvalue λ_n. There is no reason why we should not update the shift frequently in order to improve the convergence rate. In fact we can choose a new shift at each step if we want to. This is exactly what is done in the *shifted QR algorithm*:

$$A_{m-1} - \sigma_{m-1}I = Q_m R_m \qquad R_m Q_m + \sigma_{m-1}I = A_m \qquad (4.6.7)$$

where at each step σ_{m-1} is chosen to approximate whichever eigenvalue is emerging at the lower right-hand corner of the matrix. [In practice the shift can be restored after each iteration, as shown in (4.6.7), or accumulated.] We have tentatively decided that the choice $\sigma_m = a_{nn}^{(m)}$ is good. This is called the *Rayleigh quotient shift* because $a_{nn}^{(m)}$ can be viewed as a Rayleigh quotient.

Exercise 4.6.9

    (a)   Show that $a_{nn}^{(m)} = e_n^* A_m e_n$, where $e_n = [0, \ldots, 0, 1]^T$, the nth standard basis vector. Thus $a_{nn}^{(m)}$ is a Rayleigh quotient of A_m.

    (b)   Give an informal argument that shows that e_n is approximately an eigenvector of A_m^T; the approximation improves as $a_{n,n-1}^{(m)} \to 0$. $\square$

Since the shifts ultimately converge to λ_n, the convergence ratios $|(\lambda_n - \sigma_m)/(\lambda_{n-1} - \sigma_m)|$ tend to zero (provided that $\lambda_{n-1} \neq \lambda_n$). Therefore the convergence is faster than linear. In Section 5.2 we will see that the QR algorithm with the Rayleigh quotient shift carries out Rayleigh quotient iteration implicitly as a part of its action. From this it will follow that the convergence is quadratic. If the matrix is Hermitian (or, more generally, normal), convergence is cubic.

How many unshifted QR steps do we need to take before $a_{nn}^{(m)}$ can be accepted as a good approximation to λ_n? A good deal of experience has shown that there is no need to wait until $a_{nn}^{(m)}$ approximates λ_n well; it is generally safe to start shifting right from the very first step. The only effect this has is that the initial shifts alter the order of the eigenvalues, so they do not necessarily emerge in order of increasing magnitude.

Example 4.6.8   Consider again the matrix

$$A = \begin{bmatrix} 8 & 2 \\ 2 & 5 \end{bmatrix}$$

of Example 4.6.2, whose eigenvalues are $\lambda_1 = 9$ and $\lambda_2 = 4$. When the unshifted QR algorithm is applied to A, the $(2, 1)$ entry converges to zero linearly with convergence ratio $|\lambda_2/\lambda_1| = 4/9$. Now let us try the shifted QR algorithm with the Rayleigh quotient shift. Thus we take $\sigma_0 = 5$ and perform a QR step with $A - \sigma_0 I$, whose eigenvalues are $\sigma_1 - 5 = 4$ and $\sigma_2 - 5 = -1$. Notice that the ratio 1/4 is less than 4/9. $A - \sigma_0 I = Q_1 R_1$, where

$$Q_1 = \frac{1}{\sqrt{13}} \begin{bmatrix} 3 & 2 \\ 2 & -3 \end{bmatrix} \qquad R_1 = \frac{1}{\sqrt{13}} \begin{bmatrix} 13 & 6 \\ 0 & 4 \end{bmatrix}$$

Thus

$$A_1 = R_1 Q_1 + \sigma_0 I = \frac{1}{13} \begin{bmatrix} 51 & 8 \\ 8 & -12 \end{bmatrix} + \begin{bmatrix} 5 & 0 \\ 0 & 5 \end{bmatrix}$$

$$\approx \begin{bmatrix} 8.9231 & 0.6154 \\ 0.6154 & 4.0769 \end{bmatrix}$$

Comparing this result with that of Example 4.6.2, we see that with one shifted *QR* step we have made more progress toward convergence than we did with an unshifted step. Not only is the off-diagonal element smaller, but the main-diagonal entries are now quite close to the eigenvalues. The Rayleigh quotient shift for the next step is $\sigma_1 = 4.0769$. The eigenvalues of $A_1 = \sigma_1 I$ are $\lambda_1 - \sigma_1 = 4.9231$ and $\lambda_2 - \sigma_1 = -0.0769$, whose ratio is 0.0156. We therefore expect that A_2 will be substantially better than A_1. In fact,

$$A_2 \approx \begin{bmatrix} 8.999981 & 0.009766 \\ 0.009766 & 4.000019 \end{bmatrix}$$

On the next step the shift is $\lambda_2 = 4.000019$. which gives an even better ratio, and

$$A_3 \approx \begin{bmatrix} 9.000000 & 3.7 \times 10^{-7} \\ 3.7 \times 10^{-7} & 4.000000 \end{bmatrix}$$

Exercise 4.6.10   Let $A = \begin{bmatrix} 8 & 1 \\ -2 & 1 \end{bmatrix}$, as in Exercise 4.6.2. Perform one step of the *QR* algorithm with the Rayleigh quotient shift. Compare your result with that of Exercise 4.6.2. Which is better?   □

Example 4.6.9   This example shows that the Rayleigh quotient shifting strategy does not always work. Consider the real, symmetric matrix

$$A = \begin{bmatrix} 2 & 1 \\ 1 & 2 \end{bmatrix}$$

whose eigenvalues can easily be seen to be $\lambda_1 = 3$ and $\lambda_2 = 1$. The Rayleigh quotient shift is $\sigma = 2$, which lies exactly half way between the eigenvalues. The shifted matrix $A = \sigma I$ has eigenvalues $+1$ and -1, which have the same magnitude. Since $A - \sigma I = \begin{bmatrix} 0 & 1 \\ 1 & 0 \end{bmatrix}$, which is unitary, the *QR* factors of $A - \sigma I$ are $Q_1 = A - \sigma I$ and $R_1 = I$. Thus $A_1 = R_1 Q_1 + \sigma I = A$. That is, the *QR* step leaves A fixed. The algorithm "cannot decide" which eigenvalue to approach. We have already observed this phenomenon in connection with Rayleigh quotient iteration (Exercise 4.3.18).

Because the Rayleigh quotient shift fails occasionally, and for one other reason to be given below, a different shift, the *Wilkinson shift*, is used more frequently. The Wilkinson shift is defined to be that eigenvalue of the trailing 2-by-2 submatrix

$$
\begin{bmatrix}
a_{n-1,n-1}^{(m)} & a_{n-1,n}^{(m)} \\
a_{n,n-1}^{(m)} & a_{n,n}^{(m)}
\end{bmatrix}
\tag{4.6.10}
$$

that is closer to $a_{nn}^{(m)}$. It is not difficult to calculate this shift, since the eigenvalues of a 2-by-2 matrix can be found by the quadratic formula. Because the Wilkinson shift uses a greater amount of information from A_m, it is not unreasonable to expect that it would give a better approximation of the eigenvalue. In the case of Hermitian matrices this expectation is confirmed by a theorem that states that the *QR* algorithm with the Wilkinson shift always converges. The rate of convergence is usually better than cubic. See [SEP] for details. This reference also discusses a number of other shifting strategies for Hermitian matrices. For general matrices there still remain some very special cases for which the Wilkinson shift fails. An example will be given below.

For the vast majority of matrices the Wilkinson shift strategy works very well. Experience has shown that typically only about five to nine *QR* steps are needed before the first eigenvalue emerges. While $a_{n,n-1}^{(m)}$ converges to zero rapidly, the other subdiagonal entries of (A_m) also move slowly toward zero. As a consequence, by the time the first eigenvalue has emerged, some progress toward convergence of the other eigenvalues will already have been made. Therefore, on the average, fewer iterations will be needed for subsequent eigenvalues. It is common for many of the later eigenvalues to emerge after two or fewer steps. The average is in the range of three to five iterations per eigenvalue. For Hermitian matrices the situation is even better; about two to three iterations are needed per eigenvalue.

It sometimes happens in the course of *QR* iterations that one of the subdiagonal entries other than the bottom one becomes (practically) zero. Whenever this happens, the problem can be *reduced*; that is, it can be broken into two smaller problems. Suppose, for example, $a_{i+1,i}^{(m)} = 0$. Then A_m has the form

$$
A_m = \begin{bmatrix} B_{11} & B_{12} \\ 0 & B_{22} \end{bmatrix}
$$

where $B_{11} \in \mathbb{C}^{i \times i}$ and $B_{22} \in \mathbb{C}^{j \times j}$, $i + j = n$. By Theorem 4.2.5 the set of eigenvalues of A_m is just the union of the eigenvalues of B_{11} and B_{22}. Thus the eigenvalue problem of A_m can be solved by finding the eigenvalues of B_{11} and B_{22} separately. This saves arithmetic because it is no longer necessary to perform operations on B_{12}.

An upper Hessenberg matrix whose subdiagonal entries are all nonzero is called an *unreduced* or *properly* upper Hessenberg matrix. I prefer the latter term because the former is illogical. Because every upper Hessenberg matrix that has zeros on the subdiagonal can be broken into submatrices, it is never necessary to perform a *QR* step on a matrix that is not properly upper Hessenberg. This fact is crucial to the development of the implicit *QR* algorithm, which will be discussed in the next section.

Exercise 4.6.11   Let A be a nonsingular block triangular matrix:

$$A = \begin{bmatrix} A_{11} & A_{12} \\ 0 & A_{22} \end{bmatrix}$$

where the blocks A_{11} and A_{22} are square. Let us see what happens when we perform a *QR* step with A. Let Q be unitary and R upper triangular with positive main-diagonal entries such that $A = QR$. Partitioning Q and R conformably with A, we have

$$Q = \begin{bmatrix} Q_{11} & Q_{12} \\ Q_{21} & Q_{22} \end{bmatrix} \quad \text{and} \quad R = \begin{bmatrix} R_{11} & R_{12} \\ 0 & R_{22} \end{bmatrix}$$

Write the equation $A = QR$ in partitioned form and draw the following conclusions:

(a) $Q_{21} = 0$.

(b)   From the equation $Q_{21} = 0$ and the fact that Q is unitary, $Q_{12} = 0$.

(c) Q_{11} and Q_{22} are unitary.

(d) $A_{11} = Q_{11}R_{11}$ and $A_{22} = Q_{22}R_{22}$.

Let $A_1 = RQ$. Partition A_1 conformably with A, Q, and R:

$$A_1 = \begin{bmatrix} A_{11}^{(1)} & A_{12}^{(1)} \\ A_{21}^{(1)} & A_{22}^{(1)} \end{bmatrix}$$

Write the equation $A_1 = RQ$ in partitioned form and conclude that:

(e) $A_{21}^{(1)} = 0$.

(f) $A_{11}^{(1)} = R_{11}Q_{11}$ and $A_{22}^{(1)} = R_{22}Q_{22}$. Thus *QR* steps are performed on A_{11} and A_{22} separately.

(g)   Show that $A_{12} = Q_{11}R_{12}$ and $A_{12}^{(1)} = R_{12}Q_{22}$. However, these operations need not be performed.

If A is singular, it cannot be concluded that $Q_{21} = 0$ (Example 4.6.4), so none of the other conclusions hold either.

(h)   Show that if A is singular, Q and R can be chosen in such a way that $Q_{21} = 0$. Then everything else follows. $\square$

The Singular Case

In our development of the *QR* algorithm, singular matrices have required special attention. Since matrices that are exactly singular hardly ever arise in practice, it is tempting to save some effort by ignoring them completely. However, to do so would be to risk leaving you with a misconception about the algorithm. If every result were prefaced by the words, "Assume A is nonsingular," you might get the impression that the singular case is a nasty one that we hope to avoid at all costs. In fact, just the opposite is true. The goal of the shifting strategies is to find shifts that are as close to eigenvalues as possible. The closer σ is to an eigenvalue, the closer the shifted matrix $A - \sigma I$ is to having a zero eigenvalue, that is, to being singular. If a shift that is exactly an eigenvalue should be chosen, $A - \sigma I$ would be singular. From this viewpoint the singular case appears to be very desirable and worthy of our special attention. This is confirmed by the next theorem, which shows that if the QR algorithm is applied to a properly upper Hessenberg matrix that is singular, then, in principle, just one QR step suffices to extract the zero eigenvalue. In preparation for the theorem, you should work the following easy exercise.

Exercise 4.6.12 Prove the following two statements.

(a) If $A \in \mathbb{C}^{n \times n}$ is properly upper Hessenberg, then the first $n - 1$ columns of A are linearly independent.

(b) If $R \in \mathbb{C}^{n \times n}$ is an upper-triangular matrix whose first $n - 1$ columns are linearly independent, then the first $n - 1$ main-diagonal entries $r_{11}, r_{22}, \ldots, r_{n-1,n-1}$ must all be nonzero. $\square$

Theorem 4.6.11 Let $A \in \mathbb{C}^{n \times n}$ be a singular, properly upper Hessenberg matrix, and let B be the result of one QR step, starting with A. Then the entire last row of B consists of zeros. In particular, $b_{nn} = 0$ is an eigenvalue and can be removed by deflation.

Proof $B = RQ$, where $A = QR$, Q is unitary, and R is upper triangular. Since A is singular and Q is nonsingular, R must be singular. This implies that at least one of the main-diagonal entries r_{ii} is zero. (Given this and Exercise 4.6.12 as a start, you should now be able to complete the proof yourself.)

From Exercise 4.6.12, part (a), we know that the first $n - 1$ columns of A are linearly independent. Let $a_1 \ldots, a_{n-1}$ denote these columns, and let $r_1, \ldots, r_{n-1}$ denote the first $n - 1$ columns of R. From the equation $R = Q^*A$, we see that $r_1 = Q^*a_1$ and $r_2 = Q^*a_2, \ldots, r_{n-1} = Q^*A_{n-1}$. Since Q^* is nonsingular and $a_1, \ldots, a_{n-1}$ are independent, $r_1, \ldots, r_{n-1}$ must also be linearly independent. Therefore, by Exercise 4.6.12, part (b), the first $n - 1$ main-diagonal entries of R are nonzero. We have already observed that at least one of the main-diagonal entries of R must be zero, so $r_{nn} = 0$. Since R is upper triangular, this means that the entire last row of R consists of zeros. Since $B = RQ$, the entire last row of B must also consist of zeros. $\square$

Corollary 4.6.12 Let λ be an eigenvalue of the properly upper Hessenberg matrix $A \in \mathbb{C}^{n \times n}$. Let B be the result of one step of the shifted QR algorithm with shift λ. Then the last row of B is $[0 \cdots 0 \ \lambda]$. Thus the eigenvalue λ can be removed immediately by deflation.

Theorem 4.6.11 and Corollary 4.6.12 are theoretical results. In practice, round-off errors will cause $b_{n,n-1}$ to be not quite zero. In fact, in most cases $b_{n,n-1}$ will be far enough from zero to prevent deflation. When this happens, one additional QR step is usually enough to allow deflation.

Exercise 4.6.13 Let $A \in \mathbb{C}^{n \times n}$ be a properly upper Hessenberg matrix with QR decomposition $A = QR$.

(a) Let $\hat{A}$, $\hat{Q} \in \mathbb{C}^{n \times (n-1)}$ consist of the first $n - 1$ columns of A and Q, respectively. Show that there is a nonsingular, upper-triangular $\hat{R} \in \mathbb{C}^{(n-1) \times (n-1)}$, such that $\hat{Q} = \hat{A}\hat{R}^{-1}$.

(b) Use the result of part (a) to show that Q is an upper Hessenberg matrix.

(c) Conclude that B is upper Hessenberg, where $B = RQ$. Thus the QR algorithm applied to a properly upper Hessenberg matrix always preserves upper Hessenberg form, regardless of whether or not the matrix is singular. $\square$

Complex Eigenvalues of Real Matrices

Most eigenvalue problems that arise in practice involve real matrices. The fact that real matrices can have complex eigenvalues exposes another weakness of the Rayleigh quotient shift. The Rayleigh quotient shift associated with a real matrix is real. A real shift cannot approximate a complex eigenvalue well. The Wilkinson shift, in contrast, can be complex, for it is an eigenvalue of a 2-by-2 matrix. Thus this shifting strategy allows the possibility of approximating complex eigenvalues well.

This brings us to an important point. When working with real matrices, we would prefer to work within the real number system as much as possible. The use of complex shifts would appear to force us into the complex number field. It turns out that we can avoid complex numbers by some clever manipulations. First of all, recall that complex eigenvalues of a real matrix occur in conjugate pairs. Thus, as soon as we know one complex eigenvalue λ, we immediately know another, namely $\bar{\lambda}$. In the interest of staying within the real numbers, we might try to design an algorithm that seeks these two eigenvalues simultaneously. Such an algorithm would not extract these eigenvalues one at a time, deflating twice, rather it would extract a real 2-by-2 block whose eigenvalues are λ and $\bar{\lambda}$. How might we build such an algorithm?

The Wilkinson shift strategy is to compute the eigenvalues of the trailing 2-by-2 submatrix (4.6.10). If one eigenvalue of (4.6.10) is complex, then so is the other; they are complex conjugates σ_m and $\bar{\sigma}_m$. Since $A_m - \sigma_m I$ is complex, a QR step with shift σ_m will result in a complex matrix A_{m+1}. However, it can (and will) be shown that if this step is followed immediately by a QR step with shift $\bar{\sigma}_m$, then the resulting matrix A_{m+2} is real again.

In the next section we will develop the *double-step QR algorithm*, which constructs A_{m+2} directly from A_m, bypassing A_{m+1}. The computation is carried out entirely in real arithmetic and costs about as much as two ordinary (real) QR steps.

The iterates satisfy $a^{(m)}_{n-1,n-2} \to 0$ instead of $a^{(m)}_{n,n-1} \to 0$. The rate of convergence is quadratic, so after a fairly small number of iterations $a^{(m)}_{n-1,n-2}$ is small enough that it can be considered to be zero for practical purposes. Then

$$
A_m = \left[\begin{array}{c|cc} & & * & * \\ \hat{A}_m & \vdots & \vdots \\ & & * & * \\ \hline 0 & \Lambda_m \end{array} \right]
$$

where $\Lambda_m \in \mathbb{R}^{2\times 2}$ has eigenvalues λ and $\bar{\lambda}$, which can be computed by the quadratic formula. The remaining eigenvalues of A_m are eigenvalues of $\hat{A}_m$, so subsequent iterations can operate on $\hat{A}_m$. Thus a double deflation takes place.

Exceptional Shifts

There are some very special situations in which the Wilkinson shift fails. Consider the following example.

Example 4.6.13   Let

$$
A = \begin{bmatrix} 0 & 0 & 0 & \cdots & 0 & 1 \\ 1 & 0 & 0 & \cdots & 0 & 0 \\ 0 & 1 & 0 & & & \\ \vdots & & \ddots & \ddots & \vdots & \vdots \\ & & & & 0 & 0 \\ 0 & & \cdots & & 1 & 0 \end{bmatrix}
$$

This is a properly upper Hessenberg matrix that is also unitary, since its columns form an orthonormal set. Therefore in the QR decomposition of A, $Q = A$ and $R = I$. Thus $A_1 = RQ = A$; the QR step goes nowhere. Clearly this will happen whenever A is unitary. This does not contradict the convergence results cited earlier; all eigenvalues of a unitary matrix lie on the unit circle of the complex plane, so the ratios $|\lambda_{i+1}/\lambda_i|$ are all 1. This symmetric situation will be broken up by any nonzero shift. Notice, however, that for the matrix under consideration here, both the Rayleigh quotient shift and the Wilkinson shift are 0. Thus both of these shifting strategies fail.

Because of the existence of matrices like the one in Example 4.6.13, the standard QR programs for the nonsymmetric eigenvalue problem include an *exceptional-shift* feature. Whenever it appears that the algorithm may not be converging, one step with an exceptional shift is taken. The point of the exceptional shift is to break up any unfortunate symmetries that might be impeding convergence. The exact value of the exceptional shift is unimportant, although it should be of the same magnitude as elements of the matrix. In the double-step QR program given in contribution II/14 of [HAC], a double step is taken with shifts equal to the roots of the polynomial $x^2 - rx + s = 0$, where $r = 1.5(|a_{n-1,n-2}| + |a_{n,n-1}|)$ and $s = (|a_{n-1,n-2}| + |a_{n,n-1}|)^2$.

The exact strategy used by that program is as follows. The standard shift is the Wilkinson shift. If 10 double *QR* steps have been taken and no eigenvalue has been found, one double step with exceptional shifts is taken. After that up to 9 more double steps with the standard shift are taken. If an eigenvalue still has not been found, another exceptional double step is taken. Then up to 9 more double steps are taken. If convergence still has not occurred, the program gives up.

While there is no theorem that guarantees this strategy will always work, it has been found to work very well in practice. Exceptional shifts are needed only rarely.

Exercise 4.6.14 Let $A \in \mathbb{C}^{n \times n}$ be the matrix of Example 4.6.13.

(a) Show that the characteristic polynomial of A is $\lambda^n - 1$. Thus the eigenvalues of A are the nth roots of unity. (The *nth roots of unity* are the complex numbers ω that satisfy $\omega^n = 1$. There are n of them, and they are spaced evenly around the unit circle.)

(b) Let ω be an nth root of unity. Show that $v = [\omega^{n-1} \ \omega^{n-2} \ \cdots \ \omega \ 1]^T$ is an eigenvector of A with eigenvalue ω. □

4.7
IMPLEMENTATION OF THE *QR* ALGORITHM

In this section we introduce the implicit *QR* algorithm and discuss some practical implementation details.

Implicit *QR* Algorithm

A shifted *QR* step has the form

$$A - \sigma I = QR \qquad RQ + \sigma I = A_1 \qquad (4.7.1)$$

The resulting matrix A_1 is similar to A via the unitary similarity transformation

$$A_1 = Q^* A Q \qquad (4.7.2)$$

We shall assume that A and A_1 are in upper Hessenberg form and A is properly upper Hessenberg. If the step (4.7.1) is programmed in a straightforward manner, with Q formed as a product of rotators or reflectors, a very satisfactory algorithm results. It is called the *explicit QR* algorithm. It has been found, however, that on rare occasions (see [HAC], page 230) the operation of subtracting σI from A causes vital information to be lost by cancellation. The *implicit QR* algorithm is based on a different way of carrying out the *QR* step. Instead of performing (4.7.1), the implicit *QR* algorithm carries out the similarity transformation (4.7.2) directly by applying a sequence of rotators or reflectors to A. Since the shift is never actually subtracted from A, this algorithm is more stable. The cost of an implicit QR step turns out to be almost exactly the same as that of an explicit step, so the implicit algorithm is generally preferred. Furthermore, the promised double-step QR algorithm is based on the same ideas.

The implicit QR step is based on a theorem whose careful statement and proof we will defer to Section 5.3. The theorem says roughly that Q is uniquely determined by its first column: Let $\tilde{Q}$ be any unitary matrix such that the first column is the same as that of Q, and $\tilde{Q}^*A\tilde{Q}$ is in upper Hessenberg form. Then $\tilde{Q}$ is (essentially) the same as Q, and the matrix $\tilde{A}_1 = \tilde{Q}^*A\tilde{Q}$ is (essentially) the same as A_1. The exact relationship is that $\tilde{Q} = QD$, where D is a unitary diagonal matrix. Thus A_1 and $\tilde{A}_1$ differ by a trivial similarity transformation of the type discussed in Exercise 4.6.3. (See Section 5.3 for the exact conditions under which this holds.) We will construct a transforming matrix $\tilde{Q}$ satisfying the specified conditions. $\tilde{Q}$ will be a product of rotators (reflectors could be used just as well), but we will not assemble $\tilde{Q}$ explicitly. The important thing is to make the transformation from A to $\tilde{A}_1$.

We begin with a rotator whose first column is the same as that of Q. This crucial column can be determined by inspecting the first column of the equation $A - \sigma I = QR$ from (4.7.1). Let v denote the first column of $A - \sigma I$. Then, since R is upper triangular, $v = q_1 r_{11}$. That is, q_1 is proportional to

$$v = \begin{bmatrix} a_{11} - \sigma \\ a_{21} \\ 0 \\ \vdots \\ 0 \end{bmatrix}$$

Since $r_{11} > 0$, $\|v\|_2 = \|q_1\|_2 r_{11} = r_{11}$. Thus $q_1 = v/r_{11} = v/\|v\|_2$. Let $c = (a_{11} - \sigma)/\|v\|_2$ and $s = a_{21}/\|v\|_2$. Then the rotator

$$V = \left[\begin{array}{cc|c} c & -\overline{s} & \\ & & 0 \\ s & \overline{c} & \\ \hline & 0 & I \end{array} \right]$$

has the same first column as Q. Notice that $a_{21} \neq 0$ because A is properly upper Hessenberg. Therefore $s \neq 0$; that is, V is a nontrivial rotator.

We now use V to carry out a similarity transformation on A. Multiplication of A on the left by V^* alters only the first two rows of A. Consequently the resulting matrix V^*A is still in upper Hessenberg form. Multiplication of V^*A by V on the right recombines the first two columns. The resulting matrix has the form

$$V^*AV = \left[\begin{array}{cc|ccccc} * & * & * & \cdot & \cdot & \cdot & * \\ * & * & * & \cdot & \cdot & \cdot & * \\ \hline * & * & a_{33} & \cdot & \cdot & \cdot & a_{3,n} \\ 0 & 0 & a_{43} & & & & \\ \vdots & \vdots & \vdots & & \ddots & & \vdots \\ 0 & 0 & 0 & & \cdots & a_{n,n-1} & a_{n,n} \end{array} \right]$$

where the asterisks denote entries that have been altered. This matrix fails to be in upper Hessenberg form because the (3, 1) entry is nonzero. This entry is called the *bulge*. The bulge is certainly nonzero because $a_{32} \neq 0$ and $a_{31} = 0$ (originally), and V is a nontrivial rotator. (You can easily check that the value of the bulge is $a_{32}s$.)

The remainder of the implicit *QR* step consists of reducing V^*AV to upper Hessenberg form. The procedure is called *chasing the bulge*, for reasons that will become obvious. The first step in this phase of the algorithm is to choose a rotator U_{32}, acting in the 3, 2 plane, such that $U_{32}^*(V^*AV)$ is in upper Hessenberg form. That is, U_{32}^* annihilates the bulge. Because the bulge is nonzero, U_{32} is a nontrivial rotator. U_{32}^* recombines the second and third rows. In particular, it combines the (3, 1) bulge with the (2, 1) entry to produce a new (2, 1) entry. This new entry must be nonzero (Exercise 4.7.3). It will remain unchanged for the remainder of the *QR* step. We complete the similarity transformation by multiplying on the right by U_{32}. This causes the second and third columns to be recombined, and the resulting matrix has the form

$$
U_{32}^*V^*AVU_{32} = \left[\begin{array}{ccc|cccccc}
* & * & * & * & \cdot & \cdot & \cdot & * \\
* & * & * & * & \cdot & \cdot & \cdot & * \\
0 & * & * & * & \cdot & \cdot & \cdot & * \\
\hline
0 & * & * & a_{44} & \cdot & \cdot & \cdot & a_{4n} \\
0 & 0 & 0 & a_{54} & & & & \\
\vdots & \vdots & \vdots & \vdots & & \ddots & & \vdots \\
0 & 0 & 0 & 0 & & \cdots & a_{n,n-1} & a_{nn}
\end{array}\right]
$$

This matrix has a new bulge in the (4, 2) position. The same reasoning as was used to show that the (3, 1) bulge was nonzero can be employed to show that this new bulge is also certainly nonzero.

The next step is to set up a rotator U_{43}, acting in the 4, 3 plane, such that left multiplication by U_{43}^* annihilates the bulge. This nontrivial rotator recombines rows 3 and 4 in such a way that the new (3, 2) entry is certainly nonzero, just as in the previous step. This entry will not be disturbed by subsequent rotators. Right multiplication by U_{43} recombines columns 3 and 4, creating a new bulge (which is certainly nonzero) in the (5, 3) position. The next rotator U_{54} annihilates the (5, 3) bulge and creates a new bulge in the (6, 4) position, and so on.

In general, the rotator $U_{i+1,i}$ acts in the $(i + 1), i$ plane and chases the bulge from position $(i + 1, i - 1)$ to position $(i + 2, i)$. In the process it deposits in the $(i, i - 1)$ position a nonzero entry that is not altered by subsequent rotators. The last rotator is $U_{n,n-1}$, which chases the bulge off the bottom of the matrix and deposits a nonzero entry in position $(n - 1, n - 2)$. The resulting matrix is upper Hessenberg and has nonzero entries on the subdiagonal, except possibly in the $(n, n - 1)$ position. We will call this matrix $\tilde{A}_1$.

Let $\tilde{Q}$ denote the accumulated transforming matrix:

$$
\tilde{Q} = VU_{32}U_{43} \cdots U_{n,n-1}
$$

Thus

$$\tilde{A}_1 = \tilde{Q}^* A \tilde{Q}$$

It is not hard to see that the first column of $\tilde{Q}$ is the same as the first column of V: Since the rotator U_{32} acts in the 3–2 plane, VU_{32} differs from V only in columns 2 and 3. Column 1 is unchanged. Similarly $(VU_{32})U_{43}$ differs from VU_{32} only in columns 3 and 4. Again column 1 is unchanged. Clearly all subsequent rotators also leave column 1 unchanged. Therefore the first column of $\tilde{Q} = VU_{32}\cdots U_{n,n-1}$ is the same as the first column of V. Recalling that V was chosen so that its first column is the same as that of the matrix Q of (4.7.1) and (4.7.2), we conclude that $\tilde{Q}$ is (essentially) the same as Q, and $\tilde{A}_1$ is (essentially) the same as A_1. Thus we have performed a shifted QR step implicitly.

Exercise 4.7.1   Show that V^* is the same as Q_{21}^*, the first rotator applied in the reduction of $A - \sigma I$ to upper-triangular form.   □

Exercise 4.7.2   Prove that the (3, 1) bulge is nonzero.   □

Exercise 4.7.3

    (a)   Let $Q \in \mathbb{C}^{2\times 2}$ be a complex rotator such that for some nonzero vector $\begin{bmatrix} a \\ b \end{bmatrix} \in \mathbb{C}^2$, $Q\begin{bmatrix} a \\ b \end{bmatrix} = \begin{bmatrix} x \\ 0 \end{bmatrix}$. Prove that $x \neq 0$. For what larger class of matrices is this assertion valid?

    (b)   Prove that the (2, 1) entry of $U_{32}^*(V^*AV)$ is nonzero.   □

Exercise 4.7.4   Show that the explicit and implicit QR steps employ the same number of rotators (acting in the same planes). Therefore the two algorithms have about the same cost.   □

Exercise 4.7.5   Write an algorithm that performs an implicit QR step.   □

The Symmetric Case

The implicit QR step is simplified considerably when A is Hermitian. Since most Hermitian problems that occur in practice are in fact real, let us assume that A is real, symmetric, and tridiagonal. Then (if we have the good sense to choose a real shift) each of the rotators V, $U_{32}, \ldots, U_{n,n-1}$ is real, so $\tilde{Q}$ is real and orthogonal. Furthermore, $\tilde{A}_1 = \tilde{Q}^T A \tilde{Q}$ is real, symmetric, and tridiagonal. Each of the intermediate matrices $V^T A V$, $U_{32}^T V^T A V U_{32}, \ldots$ is also symmetric. Each fails to be tridiagonal only in that it has a bulge below the subdiagonal and a reflection of the bulge above the superdiagonal. A good implementation of the implicit QR step will take full advantage of the symmetry. For example, in the transformation from A to $V^T A V$, it is inefficient to calculate the unsymmetric intermediate result $V^T A$. It is better to make the transformation from A to $V^T A V$ directly, exploiting the symmetry of both matrices to minimize the computational effort.

In a computer program the matrices can be stored very compactly. A one-dimensional array of length n suffices to store the main diagonal, an array of length $n - 1$ can store the subdiagonal ($=$ superdiagonal), and a single additional storage location can hold the bulge. Each intermediate matrix can be stored over the previous one, so only one such set of arrays is required. The imposition of this data structure not only saves space, it has the additional virtue of forcing the programmer to exploit the symmetry and streamline the computations. The following exercises lead to the development of an implicit *QR* program for symmetric, tridiagonal matrices.

Exercise 4.7.6 Calculate the product

$$\begin{bmatrix} c & s \\ -s & c \end{bmatrix} \begin{bmatrix} d & g \\ g & e \end{bmatrix} \begin{bmatrix} c & -s \\ s & c \end{bmatrix} \qquad \square$$

Exercise 4.7.7 Consider the symmetric matrix

$$\hat{A} = \begin{bmatrix} d_1 & g_1 & & & & & & & \\ g_1 & d_2 & g_2 & & & & & & \\ & g_2 & d_3 & & & & & & \\ & & & \ddots & & & & & \\ & & & & d_{i-1} & g_{i-1} & b & & \\ & & & & g_{i-1} & d_i & g_i & & \\ & & & & b & g_i & d_{i+1} & & \\ & & & & & & & \ddots & g_{n-1} \\ & & & & & & & g_{n-1} & d_n \end{bmatrix}$$

which would be tridiagonal if it did not have the bulge b in the $(i + 1, i - 1)$ and $(i - 1, i + 1)$ positions. Determine $U = \begin{bmatrix} c & -s \\ s & c \end{bmatrix}$ such that $U^T \begin{bmatrix} g_{i-1} \\ b \end{bmatrix} = \begin{bmatrix} * \\ 0 \end{bmatrix}$.

Define $U_{i+1,i} \in \mathbb{R}^{n \times n}$ by

$$U_{i+1,i} = \begin{bmatrix} I & 0 & 0 \\ 0 & U & 0 \\ 0 & 0 & I \end{bmatrix} \begin{matrix} \leftarrow i \\ \leftarrow i+1 \end{matrix}$$
$$\phantom{U_{i+1,i} = \begin{bmatrix} I & 0 & 0 \end{bmatrix}} \begin{matrix} \uparrow \ \uparrow \\ i, \ i+1 \end{matrix}$$

Calculate $U_{i+1,i}^T \hat{A} U_{i+1,i}$ explicitly, and note that the resulting matrix has a bulge in positions $(i + 2, i)$ and $(i, i + 2)$; the old bulge is gone. Show that taking symmetry into account, the transformation from $\hat{A}$ to $U_{i+1,i}^T \hat{A} U_{i+1,i}$ requires the calculation of only six new entries. $\square$

Exercise 4.7.8 We continue to use the notation of Exercise 4.7.7.

(a) Suppose $c \neq 0$. Show that U can be chosen so that $c > 0$. Let $t = s/c$. Show that (if $c > 0$)

$$c^2 = \frac{1}{1 + t^2} \qquad s^2 = \frac{t^2}{1 + t^2} \qquad cs = \frac{t}{1 + t^2}$$

$$c = \frac{1}{\sqrt{1 + t^2}} \qquad \text{and} \qquad s = \frac{t}{\sqrt{1 + t^2}}$$

(Keep in mind that the letters s, c, and t stand for sine, cosine, and tangent, respectively. Thus you can use various trigonometric identities such as $c^2 + s^2 = 1$.) Rewrite the formulas that transform $\hat{A}$ to $U^T_{i+1,i} \hat{A} U_{i+1,i}$ entirely in terms of t, omitting all reference to c and s.

(b) Now suppose $s \neq 0$. Show that U can be chosen so that $s > 0$. Let $k = c/s$. (k stands for kotangent.) Show that (if $s > 0$)

$$s^2 = \frac{1}{1 + k^2} \qquad c^2 = \frac{k^2}{1 + k^2} \qquad cs = \frac{k}{1 + k^2}$$

$$s = \frac{1}{\sqrt{1 + k^2}} \qquad \text{and} \qquad c = \frac{k}{\sqrt{1 + k^2}}$$

Rewrite the formulas that transform $\hat{A}$ to $U^T_{i+1,i} \hat{A} U_{i+1,i}$ entirely in terms of k, eliminating all reference to c and s.

(c) Show that if $|b| \leq |g_{i-1}|$, then $|t| \leq 1$; in this case $t^2 \leq 1$ and $1 \leq 1 + t^2 \leq 2$. Show that if $|b| \geq |g_{i-1}|$, then $|k| \leq 1$; in this case $k^2 \leq 1$ and $1 \leq 1 + k^2 \leq 2$. Thus all danger of overflow can be avoided by using the appropriate formulation. $\square$

Exercise 4.7.9 Using formulas and ideas from Exercises 4.7.7 and 4.7.8, write an algorithm that implements an implicit QR step for a symmetric, tridiagonal matrix. Try to minimize the number of arithmetic operations. The initial rotator, which creates the first bulge, can be handled in about the same way as the other rotators. $\square$

Numerous ways to organize a symmetric implicit QR step have been proposed. A number of them are discussed in [SEP] in the context of the equivalent QL algorithm.

Exercise 4.7.10 (Evaluation of the Wilkinson Shift)

(a) Show that the eigenvalues of $\begin{bmatrix} d & g \\ g & e \end{bmatrix}$ are the (real) numbers $(d + e)/2 \pm \sqrt{[(d - e)/2]^2 + g^2}$. The Wilkinson shift is that eigenvalue which is closer to e. Thus we take the positive square root if $e \geq d$ and the negative square root if $e < d$.

(b) Typically g will be small, and the shift will be quite close to e. Thus, a stable way to calculate it is by a formula of the form $\sigma = e + \delta e$, where

δe is a small correction term. Show that $\sigma = e + \delta e$, where

$$\delta e = \begin{cases} p + r & \text{if } p \le 0 \\ p - r & \text{if } p > 0 \end{cases}$$

$$p = \frac{d - e}{2} \quad \text{and} \quad r = \sqrt{p^2 + g^2}$$

This formula works quite well. Notice, however, that the calculation of δe always involves the addition of two numbers of opposite sign. Thus there will be a loss of accuracy through cancellation. Fortunately, there is a more accurate way to calculate δe.

(c) Show that

$$p \pm r = \frac{p^2 - r^2}{p \mp r} = \frac{-g^2}{p \mp r}$$

Thus

$$\delta e = \begin{cases} \dfrac{g^2}{r - p} & \text{if } p \le 0 \\[2mm] \dfrac{-g^2}{r + p} & \text{if } p > 0 \end{cases}$$

where p and r are as defined before. In this formulation, two positive numbers were always added. (*Note:* There is some danger of exaggerating the importance of this computation. In fact a shift calculated by a naive application of the quadratic formula will work almost as well. However, the computation outlined here is no more expensive, so we might as well use it.) □

Exercise 4.7.11 Write a Fortran subroutine that calculates the eigenvalues of a real, symmetric, tridiagonal matrix by the implicit *QR* algorithm. In order to appreciate cubic convergence fully, do all of the computations in double precision.

Since the algorithm requires properly tridiagonal matrices, before each iteration you must check the subdiagonal entries to see if the problem can be deflated or reduced. In practice any entry $a_{k+1,k}$ that is very close to zero should be regarded as zero. You can use the following criterion: $a_{k+1,k}$ is essentially zero whenever $|a_{k+1,k}| < u (|a_{k,k}| + |a_{k+1,k+1}|)$, where u is the machine's unit roundoff error. (A machine-independent way to do this is to set BIG $= |a_{k,k}| + |a_{k+1,k+1}|$ and SMALL $= a_{k+1,k}$ and then check whether or not BIG + SMALL is equal to BIG.)

Once the matrix has been broken into two or more pieces, the easiest way to keep track of which portions of the matrix still need to be processed is to work from the bottom up. Thus you should check the subdiagonal elements starting from the bottom, and as soon as you find a zero, work on the bottom matrix immediately. If you always work on the bottom matrix first, it will be obvious when you are done.

Take full advantage of symmetry. The only storage area you should use is a pair of one-dimensional arrays for the main diagonal and subdiagonal, and a few additional single storage locations for temporary variables such as the bulge, tangent, and cotangent.

The usual choice of shift is the Wilkinson shift, since its use guarantees convergence in the symmetric case. However, since we wish to use this program as a learning tool, build in three shift options: (1) zero shift, (2) Rayleigh quotient shift, and (3) Wilkinson shift. Since we wish to observe the convergence of the algorithm, the subroutine should optionally print out the matrix after each iteration and also print out how many iterations are required for each deflation or reduction. Since we wish to observe the convergence of the eigenvalues, print out the *main-diagonal* entries to at least 16 decimal places (depending on the value of u for your machine). Preferably use the exponential format for maximum flexibility. We are interested only in the magnitude of the *subdiagonal* entries, so we do not need to see 16 digits of the mantissa; two digits are plenty. However, the exponential format is essential.

This subroutine, like all iterative algorithms, must have some limit on the number of iterations allowed. As a suggestion do not allow more than 100 iterations between reductions or deflations. (The limit could be set much lower, say 20 or 30, if only the Wilkinson shift were to be used.)

If the iteration limit is reached, the subroutine should set an error flag and return.

Be sure to write clear, structured code, document it with a reasonable number of comments, and document clearly all the variables that are passed to and from the subroutine.

Try out your subroutine on the following two matrices using all three shift options.

$$
\begin{bmatrix}
16 & 1 & & & \\
1 & 8 & 1 & & \\
 & 1 & 4 & 1 & \\
 & & 1 & 2 & 1 \\
 & & & 1 & 1
\end{bmatrix}
$$

Try the above matrix first, since it converges fairly rapidly, even without shifts. The eigenvalues are approximately 16.124, 8.126, 4.244, 2.208, and 0.297. Print out the matrix after each iteration and observe the spectacular cubic convergence in the shifted cases. Now try this 7-by-7 matrix

$$
\begin{bmatrix}
2 & -1 & & & & & \\
-1 & 2 & -1 & & & & \\
 & -1 & 2 & -1 & & & \\
 & & -1 & 2 & -1 & & \\
 & & & -1 & 2 & -1 & \\
 & & & & -1 & 2 & -1 \\
 & & & & & -1 & 2
\end{bmatrix}
$$

whose eigenvalues are $\lambda_k = 4 \sin^2(k\pi/16)$, $k = 1, \ldots, 7$. This matrix will cause problems for the Rayleigh quotient shift. (These problems can be solved by using an exceptional shift, but a better solution is to use the Wilkinson shift, for which exceptional shifts are never necessary in the symmetric case.) Since this matrix requires many iterations in the unshifted case, don't print out the matrix after each iteration. Just keep track of the number of iterations required per eigenvalue.

If you would like to experiment with larger versions of this matrix, the n-by-n version has eigenvalues $\lambda_k = 4 \sin^2[k\pi/2(n + 1)]$, $k = 1, \ldots, n$. {The eigenvectors are $v^{(k)}$, where $v_i^{(k)} = \sin[ik\pi/(n + 1)]$, $i, k = 1, \ldots, n$.}

You should also test your subroutine's reduction mechanism by concocting some larger matrices with some zeros on the subdiagonal. For example, you could string together three or four copies of the matrices given above. $\square$

Double-Step *QR* Algorithm

Let $A_0 \in \mathbb{R}^{n \times n}$ be a real, properly upper Hessenberg matrix, and consider a pair of *QR* steps with shifts σ_0, σ_1:

$$A_0 - \sigma_0 I = Q_1 R_1 \qquad R_1 Q_1 + \sigma_0 I = A_1$$
$$A_1 - \sigma_1 I = Q_2 R_2 \qquad R_2 Q_2 + \sigma_1 I = A_2$$
(4.7.3)

Since $A_1 = Q_1^* A_0 Q_1$ and $A_2 = Q_2^* A_1 Q_2$, we have $A_2 = Q^* A_0 Q$, where $Q = Q_1 Q_2$. If σ_0 and σ_1 are both real, then all the intermediate matrices in (4.7.3) will be real, as will A_2. If complex shifts are used, complex matrices will result. However, if σ_0 and σ_1 are chosen so that $\sigma_1 = \overline{\sigma}_0$, then A_2 turns out to be real. The following lemma is the crucial result.

LEMMA 4.7.4   Let $Q = Q_1 Q_2$ and $R = R_2 R_1$, where Q_1, Q_2, R_1, and R_2 are given by (4.7.3). Then

$$(A_0 - \sigma_1 I)(A_0 - \sigma_0 I) = QR$$

Proof $(A_0 - \sigma_1 I)(A_0 - \sigma_0 I) = (A_0 - \sigma_1 I) Q_1 R_1 = Q_1 Q_1^* (A_0 - \sigma_1 I) Q_1 R_1 = Q_1 (A_1 - \sigma_1 I) R_1 = Q_1 Q_2 R_2 R_1$. $\square$

LEMMA 4.7.5   Suppose $\sigma_1 = \overline{\sigma}_0$ in (4.7.3). Then:

(a) $(A - \sigma_1 I)(A - \sigma_0 I)$ is real.

(b)   If σ_0 and $\overline{\sigma}_0$ are not eigenvalues of A_0, then the matrices Q and R of Lemma 4.7.4 are real and the matrix A_2 generated by (4.7.3) is real.

Proof   (a) $(A_0 - \overline{\sigma}_0 I)(A_0 - \sigma_0 I) = A_0^2 - (\sigma_0 + \overline{\sigma}_0)A + \sigma_0 \overline{\sigma}_0 I$. The coefficients $\sigma_0 + \overline{\sigma}_0 = 2 \operatorname{Re}(\sigma_0)$ and $\sigma_0 \overline{\sigma}_0 = |\sigma_0|^2$ are both real, so $(A_0 - \overline{\sigma}_0 I)(A_0 - \sigma_0 I)$ is real.

(b) Since $A_0 - \sigma_0 I$ is nonsingular, the equation $A_0 - \sigma_0 I = Q_1 R_1$ determines Q_1 and R_1 uniquely such that Q_1 is unitary and R_1 is upper triangular with real, positive main-diagonal entries. Since $A_1 - \sigma_1 I$ is also nonsingular, the same remarks apply to Q_2 and R_2. Thus $Q = Q_1 Q_2$ is unitary and $R = R_2 R_1$ is upper triangular with positive, real main-diagonal entries. So by Lemma 4.7.4, Q and R are the unique QR factors of the nonsingular matrix $(A_0 - \sigma_1 I)(A_0 - \sigma_0 I)$. But this matrix is real, so its QR decomposition must be real. Therefore Q and R are real. Since $A_2 = Q^* A_0 Q = Q^T A_0 Q$, A_2 must also be real. $\qquad\square$

Once again the singular case is troublesome because Q and R are not uniquely determined. It turns out that the problems can be surmounted; Q can always be chosen so that it is real, and thus the resulting A_2 is also real. Indeed, the algorithm we are about to consider always produces a real matrix, regardless of whether or not σ_0 and $\overline{\sigma}_0$ are eigenvalues of A_0.

The matrices A_0 and A_2 are linked by the similarity transformation

$$A_2 = Q^* A_0 Q \qquad (4.7.6)$$

The double QR step carries out (4.7.3) implicitly by (essentially) performing the transformation (4.7.6) directly. By the uniqueness theorem cited earlier, it suffices to build a real, orthogonal matrix $\tilde{Q}$ whose first column is the same as that of Q, such that the matrix $\tilde{A}_2 = \tilde{Q}^T A_0 \tilde{Q}$ is in upper Hessenberg form. Then $\tilde{A}_2$ is essentially the same as A_2. (See Section 5.3 for the details.)

Let q_1 denote the first column of Q. By Lemma 4.7.4, $q_1 r_{11}$ equals the first column of $(A_0 - \sigma_1 I)(A_0 - \sigma_0 I)$. Since A_0 is an upper Hessenberg matrix, the first column of $(A_0 - \sigma_1 I)(A_0 - \sigma_0 I)$ has a reasonably simple form that is not hard to calculate. You can easily verify that it is $v = [v_1 \ v_2 \ v_3 \ 0 \ \cdots \ 0]^T$, where

$$v_1 = a_{11}^2 - (\sigma_0 + \sigma_1)a_{11} + \sigma_0 \sigma_1 + a_{12} a_{21}$$

$$= a_{21} \left[\frac{a_{11}^2 - (\sigma_0 + \sigma_1)a_{11} + \sigma_0 \sigma_1}{a_{21}} + a_{12} \right]$$

$$v_2 = a_{21} \left[(a_{11} + a_{22}) - (\sigma_0 + \sigma_1) \right]$$

$$v_3 = a_{21}(a_{32})$$

One usually pulls out the nonzero factor a_{21} and works with the vector $\hat{v} = (1/a_{21})v$. The vector q_1 is proportional to v and $\hat{v}$. In fact, since $r_{11} > 0$, $q_1 = v/\|v\|_2$, but for our purposes we do not really need q_1 exactly. If we can build a $\tilde{Q}$ whose first column is $\pm q_1$, that will be good enough.

The shifts σ_0 and σ_1 appear in the vector $\hat{v}$ only in the combinations $\sigma_0 + \sigma_1$ and $\sigma_0 \sigma_1$. These combinations are real when $\sigma_1 = \overline{\sigma}_0$. Of course they are also real when both σ_0 and σ_1 are real. Thus in either of these cases $\hat{v}$ is a real vector. The shifts σ_0 and σ_1 are normally taken to be the eigenvalues of the lower right-hand 2-by-2 submatrix

$$\begin{bmatrix} a_{n-1,n-1} & a_{n-1,n} \\ a_{n,n-1} & a_{n,n} \end{bmatrix} \qquad (4.7.7)$$

It is easy to verify that with this choice of shifts,

$$\sigma_0 + \sigma_1 = a_{n-1,n-1} + a_{n,n} \qquad (= \text{trace})$$

$$\sigma_0\sigma_1 = a_{n-1,n-1}a_{n,n} - a_{n,n-1}a_{n-1,n} \qquad (= \text{determinant})$$

(4.7.8)

Thus it is possible to calculate $\sigma_0 + \sigma_1$ and $\sigma_0\sigma_1$ directly, without first calculating σ_0 and σ_1.

Exercise 4.7.12 Verify that the equations (4.7.8) are valid if σ_0 and σ_1 are the eigenvalues of (4.7.7). $\square$

You can easily show that with $\sigma_0 + \sigma_1$ and $\sigma_0\sigma_1$ given by (4.7.8), the expression $a_{11}^2 - (\sigma_0 + \sigma_1)a_{11} + \sigma_0\sigma_1$ (which occurs in the computation of $\hat{v}_1$) is equal to $(a_{11} - a_{n-1,n-1})(a_{11} - a_{nn}) - a_{n,n-1}a_{n-1,n}$. It usually does not matter which of these expressions we use, but the latter has superior roundoff properties that make a difference occasionally. Thus we take

$$\hat{v}_1 = \frac{(a_{11} - a_{n-1,n-1})(a_{11} - a_{nn}) - a_{n,n-1}a_{n-1,n}}{a_{21}} + a_{12}$$

$$\hat{v}_2 = a_{11} + a_{22} - a_{n-1,n-1} - a_{nn}$$

$$\hat{v}_3 = a_{32}$$

The first stage of the double QR step is to construct an orthogonal matrix V whose first column is q_1 or $-q_1$. To see how such a V can be constructed, note that this V must satisfy $Ve_1 = \pm q_1 = c\hat{v}$, where $e_1 = [1\ 0\ \cdots\ 0]^T$ and $c = \pm 1/\|\hat{v}\|_2$. This will hold if and only if $V^T\hat{v} = V^{-1}\hat{v} = \sigma e_1 = [\sigma\ 0\ \cdots\ 0]^T$, where $\sigma = \pm\|\hat{v}\|_2$. Since only the first three components of $\hat{v}$ are nonzero, this can be accomplished by a reflector acting only on components 1, 2, and 3. (A pair of rotators could be used just as well.)

Once we have V, we use it to perform an orthogonal similarity transformation on A_0. Let $B_1 = V^T A_0 V = VA_0 V$. The transformation $A_0 \rightarrow VA_0$ alters only the first three rows of A_0, and the transformation $VA_0 \rightarrow VA_0 V = B_1$ alters only the first three columns. The result has the form

$$B_1 = \begin{bmatrix} * & * & * & * & * & * & \cdots \\ * & * & * & * & * & * & \\ * & * & * & * & * & * & \\ * & * & * & a_{44} & a_{45} & a_{46} & \\ 0 & 0 & 0 & a_{54} & a_{55} & a_{56} & \\ 0 & 0 & 0 & 0 & a_{65} & a_{66} & \\ \vdots & & & & & & \ddots \end{bmatrix}$$

This matrix has a bulge consisting of three entries in the (3, 1), (4, 1), and (4, 2) positions. In Exercise 4.7.13 you will show that the (4, 1) entry is definitely nonzero.

The rest of the step consists of reducing B_1 to upper Hessenberg form and is, in principle, identical to the algorithm (4.5.7). Thus a reflector U_1 is chosen to annihilate the unwanted nonzero elements in the first column. Since only the $(3, 1)$ and $(4, 1)$ entries have to be annihilated, the transformation $B_1 \rightarrow U_1 B_1$ alters only the second through fourth rows. The transformation $U_1 B_1 \rightarrow U_1 B_1 U_1$, which completes a similarity transformation, recombines the second through fourth columns. The result is

$$
B_2 = U_1 B_1 U_1 = \begin{bmatrix}
* & * & * & * & * & * & \cdots \\
* & * & * & * & * & * \\
0 & * & * & * & * & * \\
0 & * & * & * & * & * \\
0 & * & * & * & a_{55} & a_{56} \\
0 & 0 & 0 & 0 & a_{65} & a_{66} \\
\vdots & & & & & & \ddots
\end{bmatrix}
$$

The bulge has been moved over and down one position. You can show (Exercise 4.7.13) that the $(5, 2)$ entry at the tip of the bulge is certainly nonzero, as is the new $(2, 1)$ entry. The latter will remain unchanged under subsequent transformations.

   The next step is to create a reflector U_2, acting on rows 3 through 5, that annihilates the $(4, 2)$ and $(5, 2)$ entries. The similarity transformation $B_3 = U_2 B_2 U_2$ annihilates those entries and creates new nonzero entries in positions $(6, 3)$ and $(6, 4)$. That is, it moves the bulge one more step down and to the right. The $(6, 3)$ entry at the tip of the bulge is certainly nonzero. The new $(3, 2)$ entry is also nonzero, and it will remain unchanged under subsequent transformations.

   The pattern is now clear. We make a sequence of transformations $B_i \rightarrow U_i B_i U_i = B_{i+1}$, where each U_i is a reflector acting in three dimensions that pushes the bulge one position further down and to the right. After $n - 3$ steps we will have begun to push the bulge off of the bottom; the matrix B_{n-2} fails to be upper Hessenberg only in that the $(n, n - 2)$ entry is nonzero. The last step is to choose a reflector U_{n-2}, acting on rows $n - 1$ and n, that annihilates that entry. The matrix $B_{n-1} = U_{n-2} B_{n-2} U_{n-2}$ is upper Hessenberg. Define

$$
\tilde{A}_2 = B_{n-1} \quad \text{and} \quad \tilde{Q} = V U_1 U_2 \cdots U_{n-2}
$$

Then

$$
\tilde{A}_2 = \tilde{Q}^T A_0 \tilde{Q}
$$

$\tilde{A}_2$ is almost a properly upper Hessenberg matrix: it has entries that are definitely nonzero in positions $(2, 1), (3, 2), \ldots, (n - 2, n - 3)$. $\tilde{Q}$ has the same first column as V, which is $\pm q_1$. Thus $\tilde{A}_2$ is essentially the same as A_2; we have performed a double QR step implicitly.

Exercise 4.7.13

   (a)   Determine the reflector V in the form $I - \gamma u u^T$.

   (b)   Derive an explicit expression for the $(4, 1)$ entry of B_1. Note that this entry is certainly nonzero.

(c) Show that the $(5, 2)$ entry of B_2 is nonzero.

(d) Show that the $(2, 1)$ entry of B_2 is nonzero.

[From considerations of this sort it follows that A_2 has nonzero entries in positions $(2, 1), (3, 2), \ldots, (n - 2, n - 3)$.] □

Most of the operations in a double *QR* step involve reflectors acting in three dimensions. Each premultiplication by a reflector amounts to a sequence of transformations of the form $x \to Qx$ (one for each column), where $Q \in \mathbb{R}^{3 \times 3}$ is a reflector, and $x \in \mathbb{R}^3$. Each postmultiplication amounts to a sequence of transformations $x^T \to x^T Q$ (one for each row). Thus transformations of the form $x \to Qx$ and $x^T \to x^T Q$ account for almost all the computational effort in the double step. It is therefore important to make these operations as efficient as possible.

Exercise 4.7.14 Let $Q = I - \gamma u u^T$ be a reflector in $\mathbb{R}^{3 \times 3}$.

(a) (Review) Show that $Qx = x - (\gamma u^T x)u$.

(b) Show that the transformation $x \to Qx$, organized as in part (a), requires seven multiplications and five additions. (Count subtractions as additions.)

(c) How many operations would be required to calculate Qx if Q were given explicitly (i.e., not in the form $I - \gamma u u^T$)?

(d) Show that a double *QR* step using reflectors, organized as in part (a), requires about $7n^2$ multiplications and $5n^2$ additions. □

The next exercise shows that some of the multiplications can be eliminated if the reflectors are expressed in a slightly different form.

Exercise 4.7.15 Given a nonzero $y \in \mathbb{R}^3$, let $Q \in \mathbb{R}^{3 \times 3}$ be the reflector such that

$$Q \begin{bmatrix} y_1 \\ y_2 \\ y_3 \end{bmatrix} = \begin{bmatrix} -\sigma \\ 0 \\ 0 \end{bmatrix} \quad \text{and} \quad \sigma = \begin{cases} \|y\|_2 & \text{if } y_1 \geq 0 \\ -\|y\|_2 & \text{if } y_1 < 0 \end{cases}$$

(a) (Review) Show that $Q = I - \gamma u u^T$, where $u = [y_1 + \sigma, y_2, y_3]^T$ and $\gamma = 1/(\sigma u_1)$.

(b) Show that $Q = I - v w^T$, where $v = (1/\sigma)u$ and $w = (1/u_1)u$. (Note that both σ and u_1 are safely far from zero).

(c) Noting that $w_1 = 1$, show that the operation $x \to Qx = x - v(w^T x)$ requires only five multiplications and five additions; that is, five flops. (Since $Q = Q^T = I - w v^T$, the operations $x^T \to x^T Q$ can be performed in exactly the same way.) □

If the operations are carried out as indicated in this exercise, a double *QR* step costs only about $5n^2$ flops. The modified form requires the formation and temporary storage of two vectors v and w, instead of the single vector u. This difference is of no importance because the vectors are very short and do not need to be saved.

Exercise 4.7.16   Let $Q = I - \gamma u u^T$ be a reflector in $\mathbb{R}^{2 \times 2}$.

(a)   Show that the transformation $x \to Qx = x - (\gamma u^T x)u$ requires five multiplications and three additions.

(b)   Show that Q can be expressed in the form $Q = I - v w^T$, where $w_1 = 1$. Show that the transformation $x \to Qx = x - v(w^T x)$ requires only three flops.

(c)   How much arithmetic does the transformation $x \to Qx$ require if Q is given explicitly? $\square$

Exercise 4.7.17   Write a detailed algorithm to perform a double QR step using reflectors. Set up the reflectors as indicated in Exercises 4.7.15 and 4.7.16. $\square$

Exercise 4.7.18   Write a Fortran subroutine that calculates the eigenvalues of an upper Hessenberg matrix by the double-step QR algorithm using reflectors. In order to appreciate quadratic convergence fully, do all the computations in double precision. Since the algorithm requires properly Hessenberg matrices, before each iteration you must check the subdiagonal entries to see if the problem can be deflated or reduced. The entry $a_{k+1,k}$ can be regarded as zero if $|a_{k+1,k}| < u(|a_{kk}| + |a_{k+1,k+1}|)$, where u is the unit roundoff error. (A machine-independent way to do this is to set BIG $= |a_{kk}| + |a_{k+1,k+1}|$ and SMALL $= a_{k+1,k}$ and check whether or not BIG $+$ SMALL equals BIG.) Once the matrix has been broken into two or more pieces, the easiest way to keep track of which pieces of the matrix still need to be processed is to work from the bottom up. Thus, you should check the subdiagonal elements starting from the bottom, and as soon you find a zero, work on the bottom matrix immediately. An isolated 1-by-1 block is a real eigenvalue. An isolated 2-by-2 block contains a pair of complex or real eigenvalues that can be found by the quadratic formula. Once you get to the top of the matrix, you are done.

Since we wish to observe the quadratic convergence of the algorithm, the subroutine should optionally print out the subdiagonal of the matrix after each iteration. It should also print out how many iterations are required for each deflation or reduction.

The algorithm should have a limit on the number of steps allowed, and an exceptional shift capability should be built in to deal with stubborn cases.

Write clear, structured code, document it with a reasonable number of comments, and document clearly all of the variables that are passed to and from the subroutine.

Try out your subroutine on the matrix of Example 4.6.13, whose eigenvalues are $\cos(2\pi j/n) + i \sin(2\pi j/n)$, $j = 1, \ldots, n$. Try several values of n. Also try the symmetric test matrices of Exercise 4.7.11. Finally, you might like to try the matrix given in [HAC], page 370, which has some ill-conditioned eigenvalues. $\square$

4.8
USE OF THE *QR* ALGORITHM TO
CALCULATE EIGENVECTORS

In Section 4.3 we noted that inverse iteration can be used to find eigenvectors associated with known eigenvalues. This is a powerful and important technique. In this section we will see how the *QR* algorithm can be used to perform the same task. In Section 5.2 it will be shown that this use of the *QR* algorithm can be viewed as a form of inverse iteration. Let us begin by discussing the symmetric case, since that case is relatively uncomplicated.

Symmetric Matrices

Let $A \in \mathbb{R}^{n \times n}$ be symmetric and tridiagonal. The *QR* algorithm can be used to find the eigenvalues and eigenvectors of A simultaneously. After some finite number of shifted *QR* steps, A will have been reduced essentially to diagonal form

$$D = Q^T A Q \tag{4.8.1}$$

where the main-diagonal entries of D are the eigenvalues of A. If a total of m *QR* steps are taken, then $Q = Q_1 Q_2 \cdots Q_m$, where Q_i is the transformation matrix for the ith step. Each Q_i is in turn a product of $n - 1$ rotators. Thus Q is the product of a large number of rotators. The importance of Q is that its columns are the eigenvectors of A, by Spectral Theorem 4.4.7. It is easy to accumulate Q in the course of performing the *QR* algorithm, thereby obtaining the eigenvectors along with the eigenvalues. An additional array is needed for the accumulation of Q. Calling this array Q also, we set $Q = I$ initially. Then for each rotator Q_{ij} that is applied to A (that is, $A \rightarrow Q_{ij}^T A Q_{ij}$), we multiply Q by Q_{ij} on the right $(Q \rightarrow Q Q_{ij})$. The end result is clearly the transforming matrix Q of (4.8.1).

How much does this accumulation procedure cost? Each transformation $Q \rightarrow Q Q_{ij}$ alters two columns of Q. If Q_{ij} is implemented as a fast, scaled rotator, then the cost of the arithmetic is about $2n$ flops. Each *QR* step uses $n - 1$ rotators, so the cost of updating Q for each complete step is about $2n^2$ flops. Recalling that the basic cost of a symmetric *QR* step is $O(n)$ flops, we conclude that the accumulation of Q increases the cost by an order of magnitude.

As a way of dealing with the high cost of updating Q, you might consider the following procedure, called the *ultimate shift* strategy. First use *QR* without accumulating Q to calculate the eigenvalues only. This costs $O(n^2)$ flops in all, which is negligible. Then (having saved a copy of A) perform the *QR* algorithm over again, accumulating Q, using the computed eigenvalues as shifts. With these excellent shifts, each *QR* step should, in principle, deliver an eigenvalue (Corollary 4.6.12). Because of roundoff errors, two steps are needed for most eigenvalues. A good order in which to apply the eigenvalues as shifts is the order in which they emerged originally. The number of *QR* steps needed to reduce A to diagonal form on this second pass is somewhat less than were needed on the first pass, so some computer time is saved by accumulating Q on the second pass.

Exercise 4.8.1   Assuming two QR steps are needed for each eigenvalue and taking deflation into account, show that if Q is accumulated by the ultimate shift strategy, then it is the product of n^2 rotators. Thus the total operation count (using fast rotators) is about $2n^3$ flops.   □

Exercise 4.8.2   Outline the steps that would be taken to calculate a complete set of eigenvalues and eigenvectors of a symmetric (nontridiagonal) matrix using the QR algorithm. State the flop count for each step and determine the total flop count.□

How does the QR procedure compare with inverse iteration? In inverse iteration a starting vector v_0 is chosen, and the iterate v_1 is calculated by $\hat{v}_1 = (A - \lambda I)^{-1}v_0$, $v_1 = \hat{v}_1/\|\hat{v}_1\|$. If λ is a computed eigenvalue of A, then two or three such iterations will almost always be enough to produce an eigenvector to working precision. In practice $\hat{v}_1$ is computed by solving $(A - \lambda I)\hat{v}_1 = v_0$. An LU decomposition of the tridiagonal matrix $A - \lambda I$ costs only $O(n)$ flops, as do the forward- and back-substitution and rescaling steps. Therefore the total cost of calculating an eigenvector of A is $O(n)$ flops. The cost of producing n eigenvectors is thus only $O(n^2)$ flops. Notice also that the n eigenvectors are computed independently, so they can be processed simultaneously if a parallel computer is used. We conclude that inverse iteration is less expensive.

The sole disadvantage of inverse iteration is that if some of the eigenvalues lie in a tight cluster, the resulting eigenvectors will not be orthogonal as they should be. By contrast the QR algorithm always delivers eigenvectors that are orthonormal to working precision, for these vectors are the columns of an orthogonal matrix that was determined by the stable process of accumulating rotators. This defect of inverse iteration can be corrected to some extent by including a feature that orthonormalizes vectors associated with eigenvalue clusters. See for example [HAC], contribution II/18. While this feature adds to the cost of inverse iteration (and to the complexity of the program), it nevertheless remains true that inverse iteration is generally much less expensive than the QR algorithm.

Focusing on the tridiagonal case tends to exaggerate the difference in cost between inverse iteration and the QR algorithm. If the tridiagonal matrix $\tilde{A}$ was obtained from some other symmetric matrix $\tilde{A}$ by the reduction algorithm (4.5.9) for example, then the eigenvectors of $\tilde{A}$ must be obtained from those of A by the back transformation algorithm (4.5.8) at a cost of n^3 flops for n eigenvectors. This is true regardless of how the eigenvectors of A were found. In addition the reduction to tridiagonal form itself costs $2n^3/3$ flops. Viewed against these background costs, the extra cost of computing the eigenvectors by the QR algorithm no longer seems so great.

It is frequently the case that only the eigenvectors associated with a few selected eigenvalues are desired. Of course inverse iteration works well here too, but the QR algorithm can also be used. If eigenvectors associated with eigenvalues $\lambda_1, \ldots, \lambda_k$ are desired, then the QR algorithm can be carried part way to completion using $\lambda_1, \ldots, \lambda_k$ as shifts. Approximately $2k$ QR steps will be needed to get to the form

$$
\begin{bmatrix} \hat{A} & 0 \\ & \lambda_k & 0 \\ 0 & & \ddots \\ & 0 & \lambda_1 \end{bmatrix} = Q^T A Q \tag{4.8.2}
$$

Then the last k columns of Q are the desired eigenvectors. Again inverse iteration is much less expensive, but the *QR* algorithm delivers orthonormal vectors.

Exercise 4.8.3   Verify that the last k columns of Q in (4.8.2) are eigenvectors of A.
□

Exercise 4.8.4   Write a Fortran program that calculates the eigenvectors of a symmetric tridiagonal matrix by the *QR* algorithm.
□

Unsymmetric Matrices

Let $A \in \mathbb{C}^{n \times n}$ be an upper Hessenberg matrix. If the *QR* algorithm is used to calculate the eigenvalues of A, then after some finite number of steps we have essentially

$$T = Q^*AQ \tag{4.8.3}$$

where T is in upper-triangular form. This is the Schur form (Theorem 4.4.6). As in the symmetric case, we can accumulate the transforming matrix Q. Again we might consider calculating the eigenvalues first without accumulating Q, then perform *QR* again, accumulating Q, using the eigenvalues as shifts. However, the basic cost of a nonsymmetric *QR* step is $O(n^2)$, not $O(n)$, so there is a significant overhead cost associated with performing the task in two passes. Sometimes the savings realized by accumulating Q on the second pass will not be enough to offset the overhead. Nevertheless the ultimate shift strategy merits consideration, especially if only a few of the eigenvectors are wanted.

There is an important difference in the way the *QR* algorithm is handled, depending on whether just eigenvalues are computed or eigenvectors as well. If a zero appears on the subdiagonal for some iterate, then the matrix has the block triangular form

$$\begin{bmatrix} A_{11} & A_{12} \\ 0 & A_{22} \end{bmatrix}$$

If only eigenvalues are needed, then A_{11} and A_{22} can be treated separately, and A_{12} can be ignored. However, if eigenvectors are wanted, then A_{12} cannot be ignored. We must continue to update it because it eventually forms part of the matrix T, which is needed for the computation of the eigenvectors. Thus if rows i and j of A_{11} are altered by some rotator, then the same rotator must also be applied to rows i and j of A_{12}. Similarly, if columns i and j of A_{22} are altered by some rotator, then the corresponding columns of A_{12} must also be updated.

Once we have obtained the form (4.8.3), we are still not finished; only the first column of Q is an eigenvector of A. To find the other eigenvectors, it suffices to calculate the eigenvectors of T, since for each eigenvector v of T, Q^*v is an eigenvector of A. Let us therefore examine the problem of calculating the eigenvectors of an upper-triangular matrix. The eigenvalues of T are $t_{11}, \ldots, t_{nn}$, which we will assume to be distinct. To find an eigenvector associated with the eigenvalue t_{ii}, we must solve the homogeneous equation $(T - t_{ii}I)v = 0$. It is convenient to make the partition

$$T - t_{ii}I = \begin{bmatrix} S_{11} & S_{12} \\ 0 & S_{22} \end{bmatrix} \qquad v = \begin{bmatrix} v_1 \\ v_2 \end{bmatrix}$$

where $S_{11} \in \mathbb{C}^{i \times i}$ and $v_1 \in \mathbb{C}^i$. Then the equation $(T - t_{ii})v = 0$ becomes

$$S_{11}v_1 + S_{12}v_2 = 0$$

$$S_{22}v_2 = 0$$

S_{11} and S_{22} are upper triangular. S_{22} is nonsingular because its main-diagonal entries are $t_{jj} - t_{ii}$, $j = i+1, \ldots, n$, all of which are nonzero. Therefore v_2 must equal zero, and the equations reduce to $S_{11}v_1 = 0$. S_{11} is singular because its (i, i) entry is zero. Making another partition

$$S_{11} = \left[\begin{array}{c|c} \hat{S} & r \\ \hline 0 \cdots 0 & 0 \end{array} \right] \qquad v_1 = \begin{bmatrix} \hat{v} \\ w \end{bmatrix}$$

where $\hat{S} \in \mathbb{C}^{(i-1) \times (i-1)}$, $r \in \mathbb{C}^{i-1}$, and $\hat{v} \in \mathbb{C}^{i-1}$, the equation $S_{11}v_1 = 0$ becomes

$$\hat{S}\hat{v} + rw = 0$$

The matrix $\hat{S}$ is upper triangular and nonsingular. Taking w to be any nonzero number, we can solve $\hat{S}\hat{v} = -rw$ for $\hat{v}$ by back substitution to obtain an eigenvector. For example, if we take $w = 1$, we get the eigenvector

$$v = \begin{bmatrix} -\hat{S}^{-1}r \\ 1 \\ 0 \\ \vdots \\ 0 \end{bmatrix}$$

Exercise 4.8.5 Calculate three linearly independent eigenvectors of the matrix

$$T = \begin{bmatrix} 1 & 1 & 1 \\ 0 & 2 & 1 \\ 0 & 0 & 3 \end{bmatrix} \qquad \square$$

Exercise 4.8.6 How many flops are needed to calculate n linearly independent eigenvectors of the upper-triangular matrix $T \in \mathbb{C}^{n \times n}$ with distinct eigenvalues? $\square$

The case in which T has repeated eigenvalues is more complicated, since in that case T may fail to have n linearly independent eigenvectors: that is, T may be defective. The case of repeated eigenvalues is worked out in part in Exercise 4.8.7. Even when the eigenvalues are distinct, the algorithm that we have just outlined can sometimes give inaccurate results because the eigenvectors can be ill-conditioned. Section 5.5 contains an elementary discussion of conditioning of eigenvalues and eigenvectors. For a more detailed treatment see [AEP].

Exercise 4.8.7 Let T be an upper-triangular matrix with a double eigenvalue $\lambda = t_{ii} = t_{jj}$. Give necessary and sufficient conditions under which T has two linearly independent eigenvectors associated with λ, and briefly outline an algorithm to calculate the eigenvectors. ☐

If A is a real matrix, we prefer to work within the real number system as much as possible. Therefore we use the double QR algorithm to reduce A to the form

$$T = Q^T A Q$$

where $Q \in \mathbb{R}^{n \times n}$ is orthogonal and T is block upper triangular with a 2-by-2 block for each pair of complex conjugate eigenvalues. This is the real Schur form (Theorem 4.4.12).

Exercise 4.8.8 Let $T \in \mathbb{R}^{n \times n}$ be a block upper triangular matrix that would be upper triangular except for 2-by-2 blocks on the main diagonal corresponding to pairs of complex conjugate eigenvalues. Assuming that T has distinct eigenvalues, briefly outline an algorithm for calculating the eigenvectors. ☐

Suppose $A \in \mathbb{R}^{n \times n}$ has complex eigenvalues λ and $\bar{\lambda}$. If we wish to calculate the associated eigenvectors, we must eventually be prepared to deal with complex vectors. Nevertheless it turns out that we can avoid complex arithmetic right up to the very end. Suppose $A = QTQ^T$, where T is the real Schur matrix. Using the algorithm you developed in Exercise 4.8.8, we can find an eigenvector v of T associated with λ: $Tv = \lambda v$. Then $w = Qv$ is an eigenvector of A. Each of v and w can be broken into its real and imaginary parts: $v = v_1 + iv_2$ and $w = w_1 + iw_2$, where v_1, v_2, w_1, and $w_2 \in \mathbb{R}^n$. Since Q is real, it follows easily that $w_1 = Qv_1$ and $w_2 = Qv_2$. Thus the computation of w does not require complex arithmetic; w_1 and w_2 can be calculated individually, entirely in real arithmetic. Notice also that $\bar{w} = w_1 - iw_2$ is the eigenvector of A associated with $\bar{\lambda}$. Thus the two real vectors w_1 and w_2 yield two complex eigenvectors.

Exercise 4.8.9 Rework Exercise 4.8.8 so that all calculations are done in real arithmetic. ☐

Exercise 4.8.10 Let $A \in \mathbb{R}^{n \times n}$ be in upper Hessenberg form, and suppose we have calculated the eigenvalues of A by the QR algorithm or some other technique. How much does it cost to calculate the eigenvectors of A by inverse iteration? Is this less or more expensive than by the QR algorithm? ☐

If only a few selected eigenvectors are desired, they can be found by inverse iteration very cheaply. Nevertheless it is interesting to note that a variant of the

QR algorithm can also do the job. We shall begin by noting that if we try to proceed exactly as in the symmetric case, we run into a snag. Suppose we seek eigenvectors associated with the eigenvalues $\lambda_1, \ldots, \lambda_k$. For simplicity we will assume that these are all real and distinct. After about $2k$ QR steps with shifts $\lambda_1, \lambda_2, \ldots, \lambda_k$, A is transformed to

$$
\left[
\begin{array}{c|c}
\hat{A} & * \\
\hline
 & \begin{matrix} \lambda_k & & * \\ & \ddots & \\ 0 & & \lambda_1 \end{matrix} \\
0 &
\end{array}
\right] = Q^T A Q
\tag{4.8.4}
$$

In the symmetric case, the last k columns of Q are eigenvectors of A, but this is not so in general. In fact, not even the last column of Q is an eigenvector of A; it is an eigenvector of A^T.

Exercise 4.8.11 Prove that the last column of Q in (4.8.4) is an eigenvector of A^T associated with λ_1. (This topic will be explored further in Section 5.2.) With a little effort we can extract $k - 1$ other eigenvectors of A^T as linear combinations of the last k columns of Q. This is no help if what we really want are eigenvectors of A. □

To get eigenvectors of A, we must shelve the QR algorithm in favor of the related RQ algorithm, to be described below. Some $2k$ steps of the RQ algorithm with shifts $\lambda_1, \lambda_2, \ldots, \lambda_k$ suffice to transform A to the form

$$
\left[
\begin{array}{c|c}
\hat{T} & * \\
\hline
0 & \hat{A}
\end{array}
\right] = Q^T A Q
\tag{4.8.5}
$$

where

$$
\hat{T} =
\begin{bmatrix}
\lambda_1 & & * \\
 & \ddots & \\
0 & & \lambda_k
\end{bmatrix}
$$

It is a simple matter to calculate the eigenvectors of the small, upper-triangular matrix $\hat{T}$. The following exercise shows how these eigenvectors can be used in conjunction with the first k columns of Q to produce eigenvectors of A associated with $\lambda_1, \ldots, \lambda_k$.

Exercise 4.8.12 Let $q_1, \ldots, q_k$ be the first k columns of Q in (4.8.5), and let $\hat{Q} = [q_1 \cdots q_k] \in \mathbb{R}^{n \times k}$.

(a) Show that $A\hat{Q} = \hat{Q}\hat{T}$. (As a consequence the space $\langle q_1, \ldots, q_k \rangle$ is an *invariant subspace* of A. This important concept will be discussed in Section 5.1.)

(b)   Let $\hat{v} \in \mathbb{R}^k$ be an eigenvector of $\hat{T}$ associated with the eigenvalue λ_k, and let $v = \hat{Q}\hat{v} \in \mathbb{R}^n$. Show that v is an eigenvector of A associated with λ_i.   □

We close this section with a brief description of the *RQ* algorithm. Every matrix $A \in \mathbb{C}^{n \times n}$ can be expressed as a product $A = RQ$, where R is upper triangular and Q is unitary. This can be proved directly, by arguments analogous to those used to establish the existence of the *QR* decomposition, or it can be deduced from the *QR* decomposition theorem. (See Exercise 4.8.17.) Associated with the *RQ* decomposition is the basic *RQ* algorithm, defined as follows. Starting with $A = A_0$, define a sequence (A_i) by

$$A_{m-1} = R_m Q_m \qquad Q_m R_m = A_m$$

Notice that the *RQ* algorithm is just the *QR* algorithm run in reverse. Since the *QR* algorithm tends to bring out the eigenvalues, we might at first think that the *RQ* algorithm would tend to bury them. Fortunately this turns out not to be the case. The *RQ* algorithm typically converges to upper-triangular form just as the *QR* algorithm does, but the order of the eigenvalues is reversed. The refinements of the *QR* algorithm that we developed are also applicable to the *RQ* algorithm: upper Hessenberg form is preserved, shifts can be used to speed convergence, and implicit implementations can be developed. The truth of all of these claims is verified in the following exercises.

Exercise 4.8.13   Given $X \in \mathbb{C}^{n \times n}$, let X^R be the matrix obtained by reversing the rows of X. Similarly X^C will denote the matrix obtained by reversing the columns. Let I denote the identity matrix, and define $\hat{I} = I^R = I^C$.

   (a)   Show that $X^R = \hat{I}X$ and $X^C = X\hat{I}$.
   (b)   Show that if $A = XY$, then $A^R = X^R Y$ and $A^C = XY^C$.
   (c)   Show that if $A = XY$, then $A = X^C Y^R$.   □

Exercise 4.8.14   Given $X \in \mathbb{C}^{n \times n}$, define X^B by $X^B = X^{RC} = X^{CR}$. (*B* stands for "both.") Show that if $A = XY$, then $A^B = X^R Y^C = X^B Y^B$.   □

Exercise 4.8.15   Given $X \in \mathbb{C}^{n \times n}$, define X^U by $X^U = X^{BT} = X^{TB}$. Show that:

   (a) $X^{UU} = X$.
   (b) $X^U = \hat{I}X^T \hat{I}$. (Since $\hat{I} = \hat{I}^T = \hat{I}^{-1}$ this means that X^U is unitarily similar to X^T.)
   (c)   The (i, j) entry of X^U is $x_{n+1-j, n+1-i}$. (This means that X^U is the matrix obtained by reflecting X through the secondary diagonal consisting of entries $x_{n1}, x_{n-1,2}, \ldots, x_{1n}$. The superscript *U* stands for "upside down.")   □

Exercise 4.8.16   Show that (a) if A is upper Hessenberg, then so is A^U; (b) if R is upper triangular, then so is R^U; (c) if Q is unitary, then so is Q^U.   □

Exercise 4.8.17

    (a)   Show that if $A = XY$, then $A^U = Y^U X^U$.

    (b)   Use part (a) in conjunction with the QR decomposition theorem to show that every $A \in \mathbb{C}^{n \times n}$ has an RQ decomposition.   ☐

Exercise 4.8.18   Let (A_m) be the sequence obtained from the shifted RQ algorithm with shift sequence (σ_m):

$$A_{m-1} - \sigma_{m-1} I = R_m Q_m \qquad Q_m R_m + \sigma_{m-1} I = A_m$$

with starting matrix $A_0 = A$. Show that (A_m^U) is the sequence obtained from the QR algorithm with the same shift sequence, with starting matrix $A_0^U = A^U$. (Thus the RQ algorithm is just the QR algorithm turned upside down. All claims about the RQ algorithm follow immediately from this observation.)

5

Eigenvalues and Eigenvectors II

5.1
INVARIANT SUBSPACES

The main business of the first sections of this chapter is to explain why the *QR* algorithm works. For this purpose it is essential to introduce the notion of invariant subspaces, a generalization of eigenspaces.

Let $\mathbb{F}$ denote either the real or complex number field. Thus $\mathbb{F}^n$ and $\mathbb{F}^{n \times n}$ will denote either $\mathbb{R}^n$ and $\mathbb{R}^{n \times n}$ or $\mathbb{C}^n$ and $\mathbb{C}^{n \times n}$, depending on the choice of $\mathbb{F}$. Let $A \in \mathbb{F}^{n \times n}$. A subspace $\mathcal{S}$ of $\mathbb{F}^n$ is said to be *invariant under A* if $Ax \in \mathcal{S}$ whenever $x \in \mathcal{S}$; in other words $A\mathcal{S} \subseteq \mathcal{S}$. We will also say that $\mathcal{S}$ is an *invariant subspace* of A. The spaces $\{0\}$ and $\mathbb{F}^n$ are invariant subspaces of every $A \in \mathbb{F}^{n \times n}$. The following exercises give some nontrivial examples of invariant subspaces.

First we recall some notation from Chapter 3. Given $x_1, \ldots, x_k \in \mathbb{F}^n$, $\langle x_1, \ldots, x_k \rangle$ denotes the set of all linear combinations of $x_1, \ldots, x_k$. That is,

$$\langle x_1, \ldots, x_k \rangle = \left\{ \sum_{i=1}^{k} c_i x_i \;\middle|\; c_i \in \mathbb{F}, \; i = 1, \ldots, k \right\}$$

Notice that there is now some ambiguity in the meaning of $\langle x_1, \ldots, x_k \rangle$; it depends on the choice of $\mathbb{F}$. If $\mathbb{F} = \mathbb{R}$, the linear combinations have real coefficients; whereas if $\mathbb{F} = \mathbb{C}$, complex coefficients are allowed. There was no such ambiguity in Chapter 3, because there we always had $\mathbb{F} = \mathbb{R}$.

Exercise 5.1.1

    (a) Let $v \in \mathbb{F}^n$ be any eigenvector of A. Show that $\langle v \rangle$ is invariant under A.

    (b) Let $\mathcal{S}$ be any eigenspace of A (of dimension 1 or greater). Thus $\mathcal{S} = \mathcal{N}(\lambda I - A)$ for some λ. Show that $\mathcal{S}$ is an invariant subspace of A.

(c)   Let $v_1, \ldots, v_k \in F^n$ be any k eigenvectors of A associated with eigenvalues $\lambda_1, \ldots, \lambda_k$. Show that $\mathcal{S} = \langle v_1, \ldots, v_k \rangle$ is invariant under A. (It can be shown that if A is simple, then every invariant subspace of A has this form.)   □

Exercise 5.1.2   Let

$$A = \begin{bmatrix} 1 & 0 & 0 \\ 0 & 2 & 1 \\ 0 & 0 & 2 \end{bmatrix} \in \mathbb{C}^{3 \times 3}$$

(This matrix is not simple.) Find an invariant subspace of A that is not of the form $\langle v_1, \ldots, v_k \rangle$, where $v_1, \ldots, v_k$ are eigenvectors.   □

If we think of A as a linear operator mapping F^n into F^n, then an invariant subspace can be used to reduce the operator, in the following sense. Suppose $\mathcal{S}$ is an invariant subspace of A that is neither $\{0\}$ nor F^n, and let $\hat{A}$ denote the restricted operator that acts on the subspace $\mathcal{S}$, that is, $\hat{A} = A|_{\mathcal{S}}$. Then $\hat{A}$ maps $\mathcal{S}$ into $\mathcal{S}$. It is clear that every eigenvector of $\hat{A}$ is an eigenvector of A. Thus we can obtain information about the eigensystem of A by studying $\hat{A}$, an operator that is simpler than A in the sense that it acts on a lower-dimensional space. The following sequence of theorems and exercises shows how this reduction can be carried out in practice using similarity transformations. We begin by characterizing invariant subspaces in the language of matrices.

THEOREM 5.1.1   Let $\mathcal{S}$ be a subspace of F^n with basis $x_1, \ldots, x_k$. Thus $\mathcal{S} = \langle x_1, \ldots, x_k \rangle$. Let $\hat{X} = [x_1 \;\cdots\; x_k] \in F^{n \times k}$. Then $\mathcal{S}$ is invariant under $A \in F^{n \times n}$ if and only if there exists $\hat{B} \in F^{k \times k}$ such that $A\hat{X} = \hat{X}\hat{B}$.

Proof   Suppose $\hat{B}$ exists such that $A\hat{X} = \hat{X}\hat{B}$. Let b_{ij} denote the (i, j) entry of $\hat{B}$. Equating the jth column of $A\hat{X}$ with the jth column $\hat{X}\hat{B}$, we see that $Ax_j = \sum_{i=1}^{k} x_i b_{ij} \in \langle x_1, \ldots, x_k \rangle = \mathcal{S}$. Since $Ax_j \in \mathcal{S}$ for $j = 1, \ldots, k$, and $x_1, \ldots, x_k$ span $\mathcal{S}$, it follows that $Ax \in \mathcal{S}$ for all $x \in \mathcal{S}$. Thus $\mathcal{S}$ is invariant. Conversely, assume $\mathcal{S}$ is invariant. We must construct $\hat{B}$ such that $A\hat{X} = \hat{X}\hat{B}$. For $j = 1, \ldots, k$, $Ax_j \in \mathcal{S} = \langle x_1, \ldots, x_k \rangle$, so there exist constants $b_{1j}, \ldots, b_{kj} \in F$ such that $Ax_j = x_1 b_{1j} + x_2 b_{2j} + \cdots + x_k b_{kj}$. Define $\hat{B} \in F^{k \times k}$ to be the matrix whose (i, j) entry is b_{ij}. Then $A[x_1 \;\cdots\; x_k] = [x_1 \;\cdots\; x_k]\hat{B}$; that is, $A\hat{X} = \hat{X}\hat{B}$.   □

Exercise 5.1.3   Let $\mathcal{S}$ be an invariant subspace of A with basis $x_1, \ldots, x_k$, and let $\hat{X} = [x_1 \;\cdots\; x_k]$. By Theorem 5.1.1 there exists $\hat{B} \in F^{k \times k}$ such that $A\hat{X} = \hat{X}\hat{B}$.

(a)   Show that if $\hat{v} \in F^k$ is an eigenvector of $\hat{B}$ associated with the eigenvalue λ, then $v = \hat{X}\hat{v}$ is an eigenvector of A associated with the eigenvalue λ. In particular every eigenvalue of $\hat{B}$ is an eigenvalue of A. What role does the linear independence of $x_1, \ldots, x_k$ play?

(b)   Show that $v \in \mathcal{S}$. Thus v is even an eigenvector of $A|_{\mathcal{S}}$.   □

Exercise 5.1.4   With $A, \hat{B},$ and $\mathcal{S}$ as in the previous exercise, show that $\hat{B}$ is a matrix representation of the linear operator $\hat{A} = A|_{\mathcal{S}}$.   □

Exercise 5.1.5   Let $x_1, \ldots, x_k \in \mathbb{F}^n$, $\hat{X} = [x_1 \cdots x_k] \in \mathbb{F}^{n \times k}$, and $\mathcal{S} = \langle x_1, \ldots, x_k \rangle \subseteq \mathbb{F}^n$. Viewing $\hat{X}$ as a linear operator mapping $\mathbb{F}^k$ into $\mathbb{F}^n$, the *range* of $\hat{X}$ (cf., Section 3.5) is defined by

$$R(\hat{X}) = \left\{ \hat{X}\hat{x} \mid \hat{x} \in \mathbb{F}^k \right\}$$

(a)   Prove that $R(\hat{X}) = \mathcal{S}$.

(b)   How can this observation be used to simplify the proof of Theorem 5.1.1?   □

THEOREM 5.1.2   Let $\mathcal{S}$ be invariant under $A \in \mathbb{F}^{n \times n}$, and let $x_1, \ldots, x_k$ be a basis for $\mathcal{S}$. Let $x_{k+1}, \ldots, x_n$ be any $n - k$ vectors such that $x_1, \ldots, x_n$ is a basis for $\mathbb{F}^n$. Let $X_1 = [x_1 \cdots x_k]$, $X_2 = [x_{k+1} \cdots x_n]$, and $X = [X_1 X_2] = [x_1 \cdots x_n] \in \mathbb{F}^{n \times n}$. Define $B = X^{-1}AX$. Then B is block upper triangular:

$$B = \begin{bmatrix} B_{11} & B_{12} \\ 0 & B_{22} \end{bmatrix} \text{ where } B_{11} \in \mathbb{F}^{k \times k}. \text{ Furthermore } AX_1 = X_1 B_{11}.$$

Thus this theorem shows how to construct the matrix $\hat{B}$ ($= B_{11}$) that was introduced in the previous theorem and exercises.

Proof   The equation $B = X^{-1}AX$ is equivalent to $AX = XB$. The jth column of this equation is $Ax_j = \sum_{i=1}^{n} x_i b_{ij}$. Since $x_1, \ldots, x_n$ are linearly independent, this sum is the unique representation of Ax_j as a linear combination of $x_1, \ldots, x_n$. On the other hand, for $j = 1, \ldots, k$, x_j lies in the invariant subspace $\mathcal{S} = \langle x_1, \ldots, x_k \rangle$. Thus $Ax_j = c_1 x_1 + \cdots + c_k x_k$ for some $c_1, \ldots, c_k \in \mathbb{F}$. This also is a linear combination of $x_1, \ldots, x_n$. By uniqueness we have $c_i = b_{ij}$, $i = 1, \ldots, k$, and more importantly $b_{ij} = 0$, $i = k + 1, \ldots, n$. In the partition $B = \begin{bmatrix} B_{11} & B_{12} \\ B_{21} & B_{22} \end{bmatrix}$, B_{21} consists exactly of those b_{ij} for which $k + 1 \leq i \leq n$ and $1 \leq j \leq k$. Thus $B_{21} = 0$. This proves that B is block triangular. The equation $AX_1 = X_1 B_{11}$ now follows immediately from the obvious partition of the equation $AX = XB$.   □

Exercise 5.1.6   Prove the converse of Theorem 5.1.2: If $B_{21} = 0$, then $\mathcal{S}$ is an invariant subspace.   □

The eigenvalues of B_{11} are exactly the eigenvalues of $A|_{\mathcal{S}}$. Exercise 5.1.3 shows how to retrieve the associated eigenvectors of $A|_{\mathcal{S}}$ from the eigenvectors of B_{11}. The remaining eigenvalues of A are eigenvalues of B_{22}, by Theorem 4.2.5. Extraction of the remaining eigenvectors requires a bit more work.

Exercise 5.1.7   In the notation of Theorem 5.1.2, let $\mathcal{T} = \langle x_{k+1}, \ldots, x_n \rangle$.

(a)   Show that $\mathcal{T}$ is invariant if and only if $B_{12} = 0$.

(b)   Suppose $\mathcal{T}$ is invariant. Show that if $\hat{v}$ is an eigenvector of B_{22} associated with the eigenvalue λ, then $v = X_2 \hat{v}$ is an eigenvector of A associated with the eigenvalue λ.   □

Exercise 5.1.8   Suppose the hypotheses of Theorem 5.1.2 are satisfied. Define $Y \in \mathbb{F}^{n \times n}$ by $Y^T = X^{-1}$. Let $y_1, \ldots, y_n$ denote the columns of Y, and define $Y_1 = [y_1 \cdots y_k] \in \mathbb{F}^{n \times k}$, $Y_2 = [y_{k+1} \cdots y_n] \in \mathbb{F}^{n \times (n-k)}$, and $\mathcal{U} = \langle y_{k+1}, \ldots, y_n \rangle$.

> **(a)**   Show that $\mathcal{U}$ is an invariant subspace of A^T.
>
> **(b)**   Show that if $\hat{w}$ is an eigenvector of B_{22}^T associated with the eigenvalue λ, then $w = Y_2 \hat{w}$ is an eigenvector of A^T associated with λ. (A and A^T have the same eigenvalues but in general different eigenvectors. Here we see that B_{22} can be used to deliver eigenvectors of A^T, not of A.)   □

Exercise 5.1.9   We repeat Exercise 5.1.8 with one minor alteration. Take $\mathbb{F} = \mathbb{C}$ and define $Z \in \mathbb{C}^{n \times n}$ by $Z^* = X^{-1}$. Let $z_1, \ldots, z_n$ denote the columns of Z, and define $Z_2 = [z_{k+1} \cdots z_n]$ and $\mathcal{V} = \langle z_{k+1}, \ldots, z_n \rangle$.

> **(a)**   Show that $\mathcal{V}$ is an invariant subspace of A^*.
>
> **(b)**   Show that if $\hat{w}$ is an eigenvector of B_{22}^* associated with the eigenvalue $\bar{\lambda}$, then $w = Z_2 \hat{w}$ is an eigenvector of A^* associated with the eigenvalue $\bar{\lambda}$.   □

In Theorem 5.1.2 it is always possible to choose $x_1, \ldots, x_n$ so that $x_1, \ldots, x_k$ is an *orthonormal* basis of $\mathcal{S}$ and $x_1, \ldots, x_n$ is an *orthonormal* basis of $\mathbb{F}^n$. In this case X is unitary and B is unitarily similar to A. Furthermore, in this case the matrix Z of Exercise 5.1.9 is identical to X, so $Z_2 = X_2$ and $\mathcal{V} = \mathcal{S}^\perp$. This proves that if $\mathcal{S}$ is invariant under A, then $\mathcal{S}^\perp$ is invariant under A^*, a fact that can also be proved directly.

Exercise 5.1.10   Recall that the inner product in $\mathbb{C}^n$ is defined by $(x, y) = \sum_{i=1}^n x_i \bar{y}_i$. Let $A \in \mathbb{C}^{n \times n}$.

> **(a)**   Show that for $x, y \in \mathbb{C}^n$, $(Ax, y) = (x, A^* y)$. (This is the complex analog of Lemma 3.5.2.)
>
> **(b)**   Use the result of part (a) to prove directly that if $\mathcal{S}$ is invariant under A, then $\mathcal{S}^\perp$ is invariant under A^*.
>
> **(c)**   Conclude from part (b) that if A is Hermitian ($A = A^*$), then $\mathcal{S}$ is invariant under A if and only of $\mathcal{S}^\perp$ is. Relate this result to Exercise 5.1.7 and Theorem 5.1.2.   □

Recall Schur's theorem (Theorem 4.4.6), which states that every $A \in \mathbb{C}^{n \times n}$ is unitarily similar to an upper-triangular matrix. More specifically, there exists a unitary matrix $U \in \mathbb{C}^{n \times n}$ and upper-triangular matrix $T \in \mathbb{C}^{n \times n}$ such that $T = U^* A U$. The developments of this section show that Schur's theorem is a theorem about invariant subspaces. Indeed, let $u_1, \ldots, u_n$ denote the columns of U. Then for each $k \in \{1, \ldots, n-1\}$ the space $\langle u_1, \ldots, u_k \rangle$ is invariant under A, since the matrix $T = \begin{bmatrix} T_{11} & T_{12} \\ 0 & T_{22} \end{bmatrix}$ ($T_{11} \in \mathbb{C}^{k \times k}$) is block triangular. (The

conclusion follows not from Theorem 5.1.2 but from its converse, Exercise 5.1.6.)
At the same time the spaces $\langle u_{k+1}, \ldots, u_n \rangle$, $k = 1, \ldots, n-1$ are invariant under
A^*. If A is Hermitian, then both $\langle u_1, \ldots, u_k \rangle$ and $\langle u_{k+1}, \ldots, u_n \rangle$ are invariant
under A. This is reflected in the fact that T is diagonal in this case. Of course the
vectors $u_1, \ldots, u_n$ are then eigenvectors (cf., Spectral Theorem 4.4.7).

Exercise 5.1.11 Let $A, B, X \in \mathbb{F}^{n \times n}$ with X nonsingular and $B = X^{-1}AX$. Let
$x_1, \ldots, x_n$ denote the columns of X.

   (a) Prove that B is upper triangular if and only if $\langle x_1, \ldots, x_k \rangle$ is invariant
   under A for $k = 1, \ldots, n-1$.

   (b) Prove that B is lower triangular if and only if $\langle x_{k+1}, \ldots, x_n \rangle$ is invariant
   under A for $k = 1, \ldots, n-1$. □

Proof of the Real Schur Theorem

The remainder of this section is devoted to some unfinished business from Section
4.4, the proof of the real Schur theorem (Theorem 4.4.12). We begin with an
exercise that develops the appropriate invariant subspace.

Exercise 5.1.12 Suppose $A \in \mathbb{R}^{n \times n}$ has a complex (nonreal) eigenvalue λ with an
associated eigenvector v. Let $v_1, v_2 \in \mathbb{R}^n$ denote the real and imaginary parts of
v. Thus $v = v_1 + iv_2$.

   (a) Show that $\bar{v} = v_1 - iv_2$ is an eigenvector of A associated with the
   eigenvalue $\bar{\lambda}$.

   (b) Show that v_1 and v_2 are linearly independent. (*Hint:* v and $\bar{v}$ are linearly
   independent by Theorem 4.2.6.)

   (c) Show that the *real* space $\mathcal{S} = \langle v_1, v_2 \rangle \subseteq \mathbb{R}^n$ is invariant under A.

   (d) Let $q_1, q_2 \in \mathbb{R}^n$ be an orthonormal basis of $\mathcal{S}$, and let $Q_1 = [q_1 \; q_2] \in$
   $\mathbb{R}^{n \times 2}$. Show that there exists $w \in \mathbb{C}^2$ (complex!) such that $v = Q_1 w$.
   Show that $w = Q_1^T v$ as well.

   (e) Let $B_{11} = Q_1^T A Q_1 \in \mathbb{R}^{2 \times 2}$. Show that the eigenvalues of B_{11} are λ and
   $\bar{\lambda}$ with associated eigenvectors w and $\bar{w}$. □

Recall that the real Schur theorem states that for every $A \in \mathbb{R}^{n \times n}$, there exist
matrices $U, T \in \mathbb{R}^{n \times n}$ such that $T = U^T A U$, U is orthogonal, and T is block
upper triangular with 1-by-1 and 2-by-2 blocks on its main diagonal. Each 1-by-1
block is a real eigenvalue of A, and each 2-by-2 block has as its eigenvalues a pair
of complex conjugate eigenvalues of A.

Proof of Theorem 4.4.12   The proof is by induction of n. The theorem is trivially
true for $n = 1$ and also for $n = 2$ in case A has a pair of complex conjugate
eigenvalues. Now we will show that the result holds for $n = m$, assuming that it
holds for all $n < m$. We will break the argument into two overlapping cases.

Case I *(A has a real eigenvalue, $m \geq 2$)*   This looks just like the proof of the complex Schur theorem, except that real numbers are used. Let λ be a real eigenvalue of $A \in \mathbb{R}^{m \times m}$ with associated eigenvector $q_1 \in \mathbb{R}^m$, chosen so that $\|q_1\|_2 = 1$. Pick $q_2, \ldots, q_m \in \mathbb{R}^m$ so that $q_1, \ldots, q_m$ is an orthonormal basis of $\mathbb{R}^m$. Then $Q = [q_1 \;\cdots\; q_m]$ is an orthogonal matrix. Let $B = Q^T A Q$. Since $\langle q_1 \rangle$ is invariant, it follows from Theorem 5.1.2 that B has the block triangular form

$$B = \begin{bmatrix} b_{11} & B_{12} \\ 0 & B_{22} \end{bmatrix}$$

It is a simple matter to show that $b_{11} = \lambda$. $B_{22} \in \mathbb{R}^{(m-1) \times (m-1)}$, so by the induction hypothesis there exists an orthogonal $W_{22} \in \mathbb{R}^{(m-1) \times (m-1)}$ such that the matrix $T_{22} = W_{22}^T B_{22} W_{22}$ has the block upper-triangular form described in the statement of the theorem. Let

$$W = \begin{bmatrix} 1 & 0 & \cdots & 0 \\ \hline 0 & & & \\ \vdots & & W_{22} & \\ 0 & & & \end{bmatrix} \in \mathbb{R}^{m \times m}$$

Then

$$W^T B W = \begin{bmatrix} \lambda & B_{12} W_{22} \\ \hline 0 & \\ \vdots & W_{22}^T B_{22} W_{22} \\ 0 & \end{bmatrix} = \begin{bmatrix} \lambda & * & \cdots & * \\ \hline 0 & & & \\ \vdots & & T_{22} & \\ 0 & & & \end{bmatrix}$$

This last matrix is in the desired block triangular form. Let us call it T, and let $U = QW$. Then U is orthogonal, and $U^T A U = W^T Q^T A Q W = W^T B W = T$.

Case II *(A has a complex eigenvalue, $m \geq 3$)*   Let λ be a complex eigenvalue of $A \in \mathbb{R}^{m \times m}$ with associated complex eigenvector $v = v_1 + iv_2$, where $v_1, v_2 \in \mathbb{R}^m$. By Exercise 5.1.12, $\langle v_1, v_2 \rangle$ is a two-dimensional real invariant subspace of A. Let $q_1, q_2 \in \mathbb{R}^m$ be orthonormal vectors that span $\langle v_1, v_2 \rangle$, and choose $q_3, \ldots, q_m \in \mathbb{R}^m$ so that $q_1, q_2, \ldots, q_m$ is an orthonormal basis of $\mathbb{R}^m$. Then $Q = [q_1 \;\cdots\; q_m]$ is an orthogonal matrix. Let $B = Q^T A Q$. Since $\langle q_1, q_2 \rangle$ is invariant, it follows from Theorem 5.1.2 that B has the block triangular form

$$B = \begin{bmatrix} B_{11} & B_{12} \\ 0 & B_{22} \end{bmatrix}$$

where $B_{11} \in \mathbb{R}^{2 \times 2}$. Clearly $B_{11} = Q_1^T A Q_1$, where $Q_1 = [q_1 \; q_2] \in \mathbb{R}^{m \times 2}$, so by part (e) of Exercise 5.1.12, B_{11} has the complex eigenvalues λ and $\bar{\lambda}$. $B_{22} \in \mathbb{R}^{(m-2) \times (m-2)}$, so by the induction hypothesis there exists an orthogonal

$W_{22} \in \mathbb{R}^{(m-2)\times(m-2)}$ such that the matrix $T_{22} = W_{22}^T B_{22} W_{22}$ has the block upper-triangular form described in the statement of the theorem. Let

$$W = \left[\begin{array}{cc|c} 1 & 0 & \\ 0 & 1 & 0 \\ \hline & 0 & W_{22} \end{array} \right] \in \mathbb{R}^{m\times m}$$

Let $U = QW$ and $T = W^T BW = U^T AU$. Then, just as in Case I, U and T have the desired properties. $\square$

5.2
SUBSPACE ITERATION, SIMULTANEOUS ITERATION, AND THE *QR* ALGORITHM

The principal aims of this section are to derive the *QR* algorithm in a natural, logical manner and to explain its convergence properties. On the way we will encounter simultaneous iteration, an algorithm that is important in its own right as a method for calculating a few dominant eigenvalues and eigenvectors of a large, sparse matrix. We will also examine an important duality principle, which states that whenever direct (power) iteration takes place, inverse iteration automatically takes place at the same time. This principle will allow us to connect the *QR* algorithm with inverse iteration and Rayleigh quotient iteration.

Subspace Iteration

We begin by returning to the power method and recasting it in terms of subspaces. Given a matrix $A \in \mathbb{C}^{n\times n}$ that has a dominant eigenvalue, we can use the power method to calculate the dominant eigenvector v_1 or a multiple thereof. It does not matter which multiple we obtain, since each multiple is as good an eigenvector as any other. In fact each multiple of v_1 is just a representative of the eigenspace $\langle v_1 \rangle$, which is the real object of interest. Likewise, in the power sequence

$$q, Aq, A^2q, A^3q, \ldots$$

each of the iterates $A^m q$ can be viewed as a representative of the space $\langle A^m q \rangle$, which it spans. The operation of rescaling a vector amounts to replacing one representative by another representative of the same one-dimensional subspace. Thus the power method can be viewed as a process of iterating on subspaces: First a one-dimensional subspace $\mathscr{S} = \langle q \rangle$ is chosen. Then the iterates

$$\mathscr{S}, A\mathscr{S}, A^2\mathscr{S}, A^3\mathscr{S}, \ldots \tag{5.2.1}$$

are formed. This sequence converges linearly to the eigenspace $\mathscr{T} = \langle v_1 \rangle$ in the sense that the angle between $A^m\mathscr{S}$ and $\mathscr{T}$ converges to zero as $m \to \infty$.

It is quite natural to generalize this process to spaces of dimension greater than one. Thus we can choose a subspace $\mathscr{S}$ of any dimension and form the sequence (5.2.1). It is perhaps not surprising that this sequence will generally converge to an invariant subspace of A. Before proceeding, you should work the following exercise, which reviews a number of concepts and covers the most basic properties of subspace iteration.

Exercise 5.2.1   Let $A \in \mathbb{F}^{n \times n}$, where $\mathbb{F} = \mathbb{R}$ or $\mathbb{C}$, and let $\mathscr{S}$ be a subspace of $\mathbb{F}^n$. Then $A\mathscr{S}$ is defined by $A\mathscr{S} = \{Ax \mid x \in \mathscr{S}\}$.

(a)   Recall that a nonempty subset $\mathscr{U}$ of $\mathbb{F}^n$ is a subspace if and only if (i) $x_1, x_2 \in \mathscr{U} \Rightarrow x_1 + x_2 \in \mathscr{U}$ and (ii) $\alpha \in \mathbb{F}$ and $x \in \mathscr{U} \Rightarrow \alpha x \in \mathscr{U}$. Show that if $\mathscr{S}$ is a subspace of $\mathbb{F}^n$, then $A\mathscr{S}$ is a subspace of $\mathbb{F}^n$.

(b)   By definition, $A^m\mathscr{S} = \{A^m x \mid x \in \mathscr{S}\}$. Show that $A^m\mathscr{S} = A(A^{m-1}\mathscr{S})$.

(c)   Show that if $\mathscr{S} = \langle x_1, \ldots, x_k \rangle$, then $A\mathscr{S} = \langle Ax_1, \ldots, Ax_k \rangle$.

(d)   Recall that the *null space* of A is $\mathscr{N}(A) = \{x \in \mathbb{F}^n \mid Ax = 0\}$. Let $\mathscr{S}$ be a subspace of $\mathbb{F}^n$ for which $\mathscr{S} \cap \mathscr{N}(A) = \{0\}$. Show that if $x_1, \ldots, x_k$ is a basis of $\mathscr{S}$, then $Ax_1, \ldots, Ax_k$ is a basis of $A\mathscr{S}$. Consequently $\dim(A\mathscr{S}) = \dim(\mathscr{S})$.   □

In order to talk about convergence of sequences of subspaces, we need to say what we mean by the distance between two subspaces. We will cover the real and complex cases simultaneously, so let $\mathscr{S}_1$ and $\mathscr{S}_2$ be two subspaces of $\mathbb{F}^n$ of the same dimension, where $\mathbb{F} = \mathbb{R}$ or $\mathbb{C}$. Given $s_1 \in \mathscr{S}_1$, the distance from s_1 to $\mathscr{S}_2$ is defined by $d(s_1, \mathscr{S}_2) = \min\{\|s_1 - s_2\|_2 \mid s_2 \in \mathscr{S}_2\}$ (a least-squares problem!). Now define the distance between $\mathscr{S}_1$ and $\mathscr{S}_2$ by

$$d(\mathscr{S}_1, \mathscr{S}_2) = \max_{\substack{s_1 \in \mathscr{S}_1 \\ \|s_1\|_2 = 1}} d(s_1, \mathscr{S}_2)$$

It is easy to check that $d(\mathscr{S}_1, \mathscr{S}_1) = 0$ and $d(\mathscr{S}_1, \mathscr{S}_2) \neq 0$ if $\mathscr{S}_1 \neq \mathscr{S}_2$. It turns out that this distance function satisfies a number of other desirable properties; for example, $d(\mathscr{S}_1, \mathscr{S}_2) = d(\mathscr{S}_2, \mathscr{S}_1)$ and $d(\mathscr{S}_1, \mathscr{S}_3) \leq d(\mathscr{S}_1, \mathscr{S}_2) + d(\mathscr{S}_2, \mathscr{S}_3)$. We will be able to prove these two properties after we have discussed the singular value decomposition in Chapter 7. Given a sequence of subspaces $(\mathscr{S}_m)$ and a subspace $\mathscr{T}$, all of the same dimension, we will say that $(\mathscr{S}_m)$ *converges* to $\mathscr{T}$ (denoted symbolically by $\mathscr{S}_m \to \mathscr{T}$) provided that $d(\mathscr{S}_m, \mathscr{T}) \to 0$ as $m \to \infty$.

Perhaps a more natural way to describe the proximity of two subspaces is to speak in terms of angles rather than distances. The relative orientation of two k-dimensional subspaces is described by k *principal angles* between them. This is another topic for Chapter 7, in which it will be shown that $d(\mathscr{S}_1, \mathscr{S}_2)$ is just the sine of the largest principal angle between $\mathscr{S}_1$ and $\mathscr{S}_2$. Thus $\mathscr{S}_m \to \mathscr{T}$ if and only if all principal angles between $\mathscr{S}_m$ and $\mathscr{T}$ converge to zero as $m \to \infty$.

We are now ready to state the main theorem on the convergence of subspace iteration.

Theorem 5.2.2   Let $A \in \mathbb{F}^{n \times n}$ be simple with linearly independent eigenvectors $v_1, \ldots, v_n$ and associated eigenvalues $\lambda_1, \ldots, \lambda_n$, satisfying $|\lambda_1| \geq |\lambda_2| \geq$

$\cdots \geq |\lambda_n|$. Suppose $|\lambda_k| > |\lambda_{k+1}|$ for some k. Let $\mathcal{T} = \langle v_1, \ldots, v_k \rangle$ and $\mathcal{U} = \langle v_{k+1}, \ldots, v_n \rangle$. Let $\mathcal{S}$ be any k-dimensional subspace of $\mathbf{F}^n$ such that $\mathcal{S} \cap \mathcal{U} = \{0\}$. Then there exists a constant C such that

$$d(A^m \mathcal{S}, \mathcal{T}) \leq C |\lambda_{k+1}/\lambda_k|^m \qquad m = 0, 1, 2, \ldots$$

Thus $A^m \mathcal{S} \to \mathcal{T}$ linearly with convergence ratio $|\lambda_{k+1}/\lambda_k|$.

Proofs of Theorem 5.2.2 and generalizations to defective matrices are given by Parlett and Poole (1973) and Watkins and Elsner (1990).

Exercise 5.2.2

(a) Show that the condition $|\lambda_k| > |\lambda_{k+1}|$ implies that $\mathcal{N}(A) \subseteq \mathcal{U}$.

(b) More generally, show that $\mathcal{N}(A^m) \subseteq \mathcal{U}$ for $m = 1, 2, 3, \ldots$.

(c) Conclude that $A^m \mathcal{S}$ has dimension k for $m = 1, 2, 3, \ldots$. $\square$

It is easy to argue the plausibility of Theorem 5.2.2. Let q be any nonzero vector in $\mathcal{S}$. We can easily show that the iterates $A^m q$ lie (relatively) closer and closer to $\mathcal{T}$ as m increases. Indeed q may be expresses uniquely in the form

$$q = c_1 v_1 + c_2 v_2 + \cdots + c_k v_k \qquad \text{(component in } \mathcal{T})$$
$$+ c_{k+1} v_{k+1} + \cdots + c_n v_n \qquad \text{(component in } \mathcal{U})$$

in which q has been expressed as a sum of a component in $\mathcal{T}$ and a component in $\mathcal{U}$. Since $q \notin \mathcal{U}$, at least one of the coefficients $c_1, \ldots, c_k$ must be nonzero. Now

$$A^m q / (\lambda_k)^m = c_1 (\lambda_1/\lambda_k)^m v_1 + \cdots + c_{k-1} (\lambda_{k-1}/\lambda_k)^m v_{k-1} + c_k v_k$$
$$+ c_{k+1} (\lambda_{k+1}/\lambda_k)^m v_{k+1} + \cdots + c_n (\lambda_n/\lambda_k)^m v_n$$

Note that the nonzero coefficients of the component in $\mathcal{T}$ increase, or at least do not decrease, as m increases. In contrast, the coefficients of the component in $\mathcal{U}$ tend to zero linearly with ratio $|\lambda_{k+1}/\lambda_k|$ or better. Thus each sequence $(A^m q)$ converges to $\mathcal{T}$ at the stated rate, and consequently the limit of $(A^m \mathcal{S})$ lies in $\mathcal{T}$. The limit cannot be a proper subspace of $\mathcal{T}$ because it has dimension k.

The hypothesis $\mathcal{S} \cap \mathcal{U} = \{0\}$ merits some comment. First let us see what it amounts to in the case $k = 1$. In this case we have $\mathcal{S} = \langle q \rangle$ for some nonzero $q \in \mathbf{F}^n$, and $\mathcal{U} = \langle v_2, \ldots, v_n \rangle$. Clearly $\mathcal{S} \cap \mathcal{U} = \{0\}$ if and only if $q \notin \mathcal{U}$. This just means that $c_1 \neq 0$ in the unique expansion $q = c_1 v_1 + c_2 v_2 + \cdots + c_n v_n$. You will recall from Section 4.3 that this is the condition for convergence of the basic power method. It is satisfied by almost every $q \in \mathbf{F}^n$. In the language of subspaces, a one-dimensional subspace $\mathcal{S}$ and an $(n - 1)$-dimensional subspace $\mathcal{U}$ chosen at random are almost certain to satisfy $\mathcal{S} \cap \mathcal{U} = \{0\}$. Fortunately the same is true in general, as long as the sum of the dimensions of $\mathcal{S}$ and $\mathcal{U}$ does not exceed n. Note that if the sum of the dimensions does exceed n, then $\mathcal{S}$ and $\mathcal{U}$ must intersect nontrivially. This is a consequence of the fundamental relationship $\dim(\mathcal{S} \cap \mathcal{U}) + \dim(\mathcal{S} + \mathcal{U}) = \dim(\mathcal{S}) + \dim(\mathcal{U})$, since $\dim(\mathcal{S} + \mathcal{U}) \leq \dim(\mathbf{F}^n) = n$.

In our present situation $\dim(\mathcal{S}) = k$ and $\dim(\mathcal{U}) = n - k$. The dimensions sum to exactly n, so there is enough room in $\mathbb{F}^n$ that $\mathcal{S}$ and $\mathcal{U}$ do not have to intersect nontrivially. If two subspaces do not have to intersect nontrivially, then they almost certainly will not. You will probably be able to convince yourself that this is so by considering the situation in $\mathbb{R}^3$. There any two two-dimensional subspaces (planes through the origin) are required to intersect nontrivially because the sum of their dimensions exceeds three. In contrast, a plane and a line are not required to intersect nontrivially, and it is obvious that they almost certainly will not. You may also find the following exercise helpful.

Exercise 5.2.3   Let $\mathcal{S}$ and $\mathcal{U}$ be subspaces of $\mathbb{F}^n$ of dimension k and $n - k$, respectively. Let $s_1, \ldots, s_k$ be any basis for $\mathcal{S}$ and let $u_{k+1}, \ldots, u_n$ be any basis for $\mathcal{U}$. Define $B \in \mathbb{F}^{n \times n}$ by $B = [s_1\ s_2\ \cdots\ s_k\ u_{k+1}\ \cdots\ u_n]$. Show that $\mathcal{S} \cap \mathcal{U} = \{0\}$ if and only if B is nonsingular. (B fails to be nonsingular when and only when $\det(B) = 0$. This is a very special relationship. If one chooses n vectors at random from $\mathbb{F}^n$ and builds a matrix whose columns are these vectors, its determinant is almost certain to be nonzero.)   □

Simultaneous Iteration

In order to carry out subspace iteration in practice, we must choose a basis for $\mathcal{S}$ and iterate on the basis vectors simultaneously. Let $q_1^0, q_2^0, \ldots, q_k^0$ be a basis for $\mathcal{S}$. From Exercises 5.2.1 and 5.2.2 we know that if $\mathcal{S} \cap \mathcal{U} = \{0\}$, then $A^m q_1^0, A^m q_2^0, \ldots, A^m q_k^0$ form a basis for $A^m \mathcal{S}$. Thus, in theory at least, we can simply iterate on a basis of $\mathcal{S}$ to obtain bases for $A\mathcal{S}, A^2\mathcal{S}, A^3\mathcal{S}$, and so on. There are two reasons why it is not advisable to do this in practice: (1) The vectors will have to be rescaled regularly in order to avoid overflow or underflow. (2) Each of the sequences $q_i^0, Aq_i^0, A^2 q_i^0, \ldots$ independently converges to the dominant eigenspace $\langle v_1 \rangle$ (assuming $|\lambda_1| > |\lambda_2|$). It follows that for large m the vectors $A^m(q_1^0), \ldots, A^m(q_k^0)$ all point in nearly the same direction. Thus they form a very ill-conditioned basis of $A^m \mathcal{S}$.

Ill-conditioned bases can be avoided by replacing the basis gotten at each step by a well-conditioned basis for the same subspace. This replacement operation can also incorporate the necessary rescaling. There are numerous ways to do this, but the most reliable is to orthonormalize. Thus the following procedure is recommended.

1.   Let $q_1^0, \ldots, q_k^0$ be an orthonormal basis for $\mathcal{S}$.
2.   For $m = 0, 1, 2, \ldots$?

> a.   Calculate $A(q_1^m), \ldots, A(q_k^m)$ (a basis for $A^{m+1}\mathcal{S}$).   (5.2.3)
> b.   Orthonormalize by Gram–Schmidt to obtain
> $q_1^{m+1}, \ldots, q_k^{m+1}$ (an orthonormal basis for $A^{m+1}\mathcal{S}$).

Of course the modified Gram–Schmidt algorithm or a QR decomposition by reflectors should be used to perform the orthonormalization. (See Section 3.4.)

The simultaneous iteration procedure (5.2.3) has the agreeable property of iterating on lower-dimensional subspaces at no extra cost. For $i = 1, 2, \ldots, k$, let $\mathcal{S}_i$ denote the i-dimensional subspace spanned by $q_1^0, \ldots, q_i^0$. Then $A\mathcal{S}_i = \langle A(q_1^0), \ldots, A(q_i^0) \rangle = \langle q_1^1, \ldots, q_i^1 \rangle$, by the subspace preserving property (3.4.5) of the Gram–Schmidt procedure. In general $A^m \mathcal{S}_i = \langle q_1^m, \ldots, q_i^m \rangle$, so $\langle q_1^m, \ldots, q_i^m \rangle$ converges to the invariant subspace $\langle v_1, \ldots, v_i \rangle$ provided that appropriate hypotheses are satisfied. Thus simultaneous iteration seeks not only an invariant subspace of dimension k, but subspaces of dimensions $1, 2, \ldots, k - 1$ as well.

Our main reason for discussing simultaneous iteration is to use it as a vehicle for the introduction of the QR algorithm. However, let us pause briefly to consider its practical use as an eigenvalue algorithm in its own right.

Exercise 5.2.4 Let $A \in \mathbb{F}^{n \times n}$ have linear independent eigenvectors $v_1, \ldots, v_k$ (and possibly others) and associated eigenvalues $\lambda_1, \ldots, \lambda_k$. Let $q_1, \ldots, q_k \in \mathbb{F}^n$ be orthonormal vectors [perhaps calculated using (5.2.3)] such that for $i = 1, \ldots, k, \langle q_1, \ldots, q_i \rangle = \langle v_1, \ldots, v_i \rangle$. Let $Q = [q_1 \ \cdots \ q_k] \in \mathbb{F}^{n \times k}$ and $B = Q^* A Q \in \mathbb{F}^{k \times k}$.

- **(a)**   Show that B is upper triangular with entries $\lambda_1, \ldots, \lambda_k$ on the main diagonal.

- **(b)**   Given $q_1, \ldots, q_k$, how many flops are needed to calculate $\lambda_1, \ldots, \lambda_k$, (i) assuming that A is not sparse and (ii) assuming A is banded with band-width b?   □

For years algorithms based on simultaneous iteration have been used to find a few dominant eigenvalues of large, sparse matrices. Thus typically $k \ll n$. Simultaneous iteration is attractive because matrix–vector products Aq can be calculated relatively inexpensively when A is sparse. Quite sophisticated implementations, which incorporate acceleration techniques to speed convergence, have been developed. However, the more recently developed Lanczos algorithms are generally considered to be superior to simultaneous iteration. For more information on simultaneous iteration see, for example, [HAC] contribution II/9, Stewart (1976), Jennings (1977), and [SEP], Chapter 14.

Returning to our approach to the QR algorithm, let us now consider what happens when simultaneous iteration is applied to a complete set of orthonormal eigenvectors $q_1^0, \ldots, q_n^0 \in \mathbb{F}^n$. We continue to assume that A is simple, with linearly independent eigenvectors $v_1, \ldots, v_n$. For $k = 1, \ldots, n - 1$, let $\mathcal{S}_k = \langle q_1^0, \ldots, q_k^0 \rangle$, $\mathcal{T}_k = \langle v_1, \ldots, v_k \rangle$, and $\mathcal{U}_k = \langle v_{k+1}, \ldots, v_n \rangle$, and assume $\mathcal{S}_k \cap \mathcal{U}_k = \{0\}$ and $|\lambda_k| > |\lambda_{k+1}|$. Then (for $k = 1, \ldots, n - 1$) $A^m \mathcal{S}_k = \langle q_1^m, \ldots, q_k^m \rangle \to \mathcal{T}_k$ linearly with convergence ratio $|\lambda_{k+1}/\lambda_k|$ as $m \to \infty$. There are a number of ways to recognize convergence of the iterates. One way is to carry out a similarity transformation.

Let $\hat{Q}_m \in \mathbb{F}^{n \times n}$ denote the unitary matrix whose columns are $q_1^m, \ldots, q_n^m$, and let

$$A_m = \hat{Q}_m^* A \hat{Q}_m \tag{5.2.4}$$

For large m the space spanned by the first k columns of $\hat{Q}_m$ is very close to the invariant subspace $\mathcal{T}_k$. In Theorem 5.1.2 we learned that if these columns exactly span $\mathcal{T}_k$, then A_m has the block upper-triangular form

$$
\begin{array}{c} k \\ n-k \end{array}
\left[\begin{array}{cc} A_{11}^{(m)} & A_{12}^{(m)} \\ 0 & A_{22}^{(m)} \end{array} \right]
\begin{array}{cc} \\ k \quad\quad n-k \end{array}
\tag{5.2.5}
$$

and the eigenvalues of $A_{11}^{(m)}$ are $\lambda_1, \ldots, \lambda_k$. Since the columns do not exactly span $\mathcal{T}_k$, we do not get a block of zeros, but we have reason to hope that the entries in the block $A_{21}^{(m)}$ will be close to zero. It is not unreasonable to expect that $A_{21}^{(m)} \to 0$ at the same rate as $A^m \mathcal{S}_k \to \mathcal{T}_k$. This turns out to be true and not hard to prove. See, for example, Watkins and Elsner (1990). Thus the sequence (A_m) converges to the block upper-triangular form (5.2.5).

This happens not just for one choice of k, but for all k simultaneously, so the limiting form is upper triangular. The main-diagonal entries of A_m converge to the eigenvalues $\lambda_1, \lambda_2, \ldots, \lambda_n$ in order. Should some of the inequalities $|\lambda_k| > |\lambda_{k+1}|$ fail to hold, the limit will be block triangular. This happens most frequently when A is real and has some pairs of complex conjugate eigenvalues. For each such pair $\lambda_i = \overline{\lambda}_{i+1}$, a 2-by-2 block will emerge in rows and columns i and $i+1$. The eigenvalues of this block are λ_i and $\overline{\lambda}_i$. The QR algorithm is a variant of simultaneous iteration that produces the sequence (A_m) directly.

The QR Algorithm

It is easy to recast simultaneous iteration in matrix form. After m iterations we have the orthonormal vectors $q_1^m, \ldots, q_n^m$, which are the columns of $\hat{Q}_m$. Let B_{m+1} be the matrix whose columns are $Aq_1^m, \ldots, Aq_n^m$. Then $B_{m+1} = A\hat{Q}_m$. To complete the step, we must orthonormalize $Aq_1^m, \ldots, Aq_n^m$. From Chapter 3 we know that this can be accomplished by a QR decomposition of B_{m+1}. Thus a step of simultaneous iteration can be expressed in matrix form as

$$
A\hat{Q}_m = B_{m+1} = \hat{Q}_{m+1} R_{m+1}
\tag{5.2.6}
$$

Exercise 5.2.5   Write down the matrix form of simultaneous iteration for the case when only $k \, (< n)$ vectors are used. What are the dimensions of the matrices involved? □

Now suppose we begin subspace iteration with the standard basis vectors $q_1^0 = e_1, q_2^0 = e_2, \ldots, q_n^0 = e_n$; that is, $\hat{Q}_0 = I$. Then by (5.2.6), $AI = B_1 = \hat{Q}_1 R_1$. Letting $Q_1 = \hat{Q}_1$, we have

$$
A = Q_1 R_1
\tag{5.2.7}
$$

Suppose that we feel very optimistic about the rate of convergence, and after this one step we wish to see how much progress we have made. One way to do this is to carry out the unitary similarity transformation $A_1 = \hat{Q}_1^* A \hat{Q}_1 = Q_1^* A Q_1$, as suggested above, and see how close to upper-triangular form A_1 is. Since $Q_1^* A = R_1$ by (5.2.7), A_1 can be calculated by

$$R_1 Q_1 = A_1 \qquad (5.2.8)$$

Equations 5.2.7 and 5.2.8 together constitute one step of the *QR* algorithm.

Finding that A_1 is not upper triangular, we decide to take another step. But now we have a choice. We can continue simultaneous iteration with A, or we can work instead with the similar matrix A_1. Recall that similar matrices can be regarded as representations of the same linear transformation with respect to different coordinate systems. Thus the choice between A and A_1 is simply a choice between coordinate systems. If we stay in the original coordinate system, the next step has the form

$$A \hat{Q}_1 = B_2 = \hat{Q}_2 R_2 \qquad (5.2.9)$$

If we wish to assess our progress toward convergence, we can make the similarity transformation

$$A_2 = \hat{Q}_2^* A \hat{Q}_2 \qquad (5.2.10)$$

Let us see what these operations look like in the coordinate system of A_1. Because $A_1 = \hat{Q}_1^* A \hat{Q}_1$, a vector that is represented by the coordinate vector x in the original system will be represented by $\hat{Q}_1^* x$ in the new system. Therefore the vectors $q_1^1, \ldots, q_n^1$ (the columns of $\hat{Q}_1$) are transformed to $\hat{Q}_1^* q_1^1 = e_1$, $\hat{Q}_1^* q_2^1 = e_2, \ldots, \hat{Q}_1^* q_n^1 = e_n$ (the columns of I) in the new system. Thus the equation $A \hat{Q}_1 = B_2$ is equivalent to $A_1 I = A_1$, and the *QR* decomposition $B_2 = \hat{Q}_2 R_2$ is equivalent to a *QR* decomposition of A_1:

$$A_1 = Q_2 R_2 \qquad (5.2.11)$$

If we now wish to see how much progress we have made in this one step, we can make the similarity transformation

$$A_2 = Q_2^* A_1 Q_2 \qquad (5.2.12)$$

Since $R_2 = Q_2^* A_1$, this transformation can be accomplished via

$$R_2 Q_2 = A_2 \qquad (5.2.13)$$

Equations (5.2.11) and (5.2.13) represent a second *QR* step.

Exercise 5.2.6   Assume (for convenience) A is nonsingular. Therefore B_2 is also nonsingular, so its *QR* decomposition is unique (Theorem 3.2.24). One expression for the *QR* decomposition of B_2 is given by (5.2.9). Obtain a second expression by rewriting the equation $B_2 = A \hat{Q}_1$ in terms of A_1 and using (5.2.11). Conclude that the matrices R_2 given by (5.2.9) and (5.2.11) are the same, $\hat{Q} = Q_1 Q_2$, and the matrices A_2 given by (5.2.10) and (5.2.12) are the same. $\square$

If we continue the process, we have the choice of working in the coordinate system A, A_1, or A_2. If we decide that at each step we will work in the newest coordinate system, we generate the sequence (A_m) by the elegant transformation

$$A_{m-1} = Q_m R_m \qquad R_m Q_m = A_m \qquad (5.2.14)$$

This is the QR algorithm. As we have just seen, it is nothing but simultaneous iteration with a change of coordinates at each step. In order to solidify your understanding of this fact, let us review the meaning of the equations in (5.2.14).

The columns of A_{m-1} are $A_{m-1}e_1, \ldots, A_{m-1}e_n$; that is, they are the result of one step of simultaneous iteration by A_{m-1} on the standard basis. The decomposition $A_{m-1} = Q_m R_m$ orthonormalizes these vectors. The columns of Q_m are the orthonormal basis vectors for the next step of simultaneous iteration. The step $R_m Q_m = A_m$ simply carries out the transformation to a new coordinate system: $A_m = Q_m^* A_{m-1} Q_m$. In this coordinate transformation, the orthonormal basis vectors for the next iteration are transformed to the standard basis vectors $e_1, \ldots, e_n$.

The global correspondence between the QR algorithm and simultaneous iteration without coordinate transformations is easy to establish. The matrices A_m generated by (5.2.14) are the same as those generated by (5.2.4) (assuming the iterations were initiated with $\hat{Q}_0 = I$). The R_m of (5.2.14) are the same as those of (5.2.6), and the Q_m of (5.2.14) are related to the $\hat{Q}_m$ of (5.2.6) by

$$\hat{Q}_m = Q_1 Q_2 \cdots Q_m \qquad (5.2.15)$$

Q_m is the matrix of the change of coordinates at step m, and $\hat{Q}_m$ is the matrix of the accumulated coordinate change after m steps.

We have established the equivalence of simultaneous iteration and the QR algorithm by looking at the processes one step at a time. Another way is to examine the cumulative effect of m steps. In this approach Q_m, R_m, and A_m are defined by (5.2.14), with $A_0 = A$, and $\hat{Q}_m$ is defined by (5.2.15). If, in addition, $\hat{R}_m$ is defined by

$$\hat{R}_m = R_m R_{m-1} \cdots R_1 \qquad (5.2.16)$$

$$A = Q_1 R_1 = \hat{Q}_1 \hat{R}_1$$
$$A^2 = Q_1 R_1 Q_1 R_1 = Q_1 Q_2 R_2 R_1 = \hat{Q}_2 \hat{R}_2$$
$$A^3 = Q_1 R_1 Q_1 R_1 Q_1 R_1 = Q_1 Q_2 R_2 Q_2 R_2 R_1 = Q_1 Q_2 Q_3 R_3 R_2 R_1 = \hat{Q}_3 \hat{R}_3$$

Clearly we can prove by induction that

$$A^m = \hat{Q}_m \hat{R}_m \qquad m = 1, 2, 3, \ldots \qquad (5.2.17)$$

Exercise 5.2.7 Suppose A_m, Q_m, and R_m are defined by (5.2.14) with $A_0 = A$, $\hat{Q}_m$ is defined by (5.2.15), and $\hat{R}_m$ is defined by (5.2.16).

(a) Prove that $A_m = \hat{Q}_m^* A \hat{Q}_m$, and hence $A \hat{Q}_m = \hat{Q}_m A_m$, for all m.

(b) Use the result of part (a) to prove (5.2.17).   □

Clearly $\hat{Q}_m$ is unitary and $\hat{R}_m$ is upper triangular. Therefore (5.2.17) is just the *QR* decomposition of A^m. This means that for all k, the first k columns of $\hat{Q}_m$ form an orthonormal basis for the space spanned by the first k columns of A^m. But the columns of A^m are just $A^m e_1, \ldots, A^m e_n$. Thus

$$A^m \langle e_1, \ldots, e_k \rangle = \langle q_1^m, \ldots, q_k^m \rangle \qquad k = 1, 2, \ldots, n$$

That is, the columns $\hat{Q}_m$ are just the result of m steps of simultaneous iteration, starting from the standard basis vectors $e_1, \ldots, e_n$.

If you decided to read this section before reading Section 4.6, now is a good time to go back and read that material. If you have already read Section 4.6, you might find it useful to review it at this point.

In Section 4.6 we introduced the *QR* algorithm without motivation and discussed the modifications that are needed to make the algorithm efficient. The first of these modifications is to transform that matrix to upper Hessenberg form before beginning the *QR* iterations. This is a good idea because (1) Hessenberg form is preserved under *QR* iterations and (2) the cost of a *QR* iteration is much less for a Hessenberg matrix than for a full matrix. Thus each of our iterates has the form

$$A_m = \begin{bmatrix} a_{11}^{(m)} & \cdot & \cdot & \cdot & a_{1n}^{(m)} \\ a_{21}^{(m)} & & & & \cdot \\ 0 & \ddots & & & \cdot \\ \vdots & \ddots & \ddots & & \cdot \\ 0 & \cdots & 0 & a_{n,n-1}^{(m)} & a_{nn}^{(m)} \end{bmatrix} \qquad m = 0, 1, 2 \ldots$$

We may assume further that each of the subdiagonal entries $a_{k+1,k}^{(m)}$ is nonzero, since otherwise we could reduce the eigenvalue problem to two smaller problems. Recall that an upper Hessenberg matrix that satisfies this condition is called a *properly upper Hessenberg matrix*.

Since the matrices A_m produced by the *QR* algorithm are identical to those given by (5.2.4), we can conclude that if $|\lambda_k| > |\lambda_{k+1}|$ and the subspace condition $\mathscr{S} \cap \mathscr{U} = \{0\}$ holds, then A_m converges to the block triangular form (5.2.5). Since all A_m are in upper Hessenberg form, the $(n - k)$-by-k block that is converging to zero contains only one entry that is not already zero, namely $a_{k+1,k}^{(m)}$. This one entry therefore gives a rough indication of the distance of $\langle q_1^m, \ldots, q_k^m \rangle$ from the invariant subspace $\mathscr{T}_k = \langle v_1, \ldots, v_k \rangle$. Under the given conditions $a_{k+1,k}^{(m)} \to 0$ linearly with convergence ratio $|\lambda_{k+1}/\lambda_k|$ as $m \to \infty$.

A pleasant side effect of using properly upper Hessenberg matrices is that the subspace conditions $\mathscr{S}_k \cap \mathscr{U}_k = \{0\}, k = 1, \ldots, n - 1$, are always satisfied (Exercise 5.2.8). Recall that $\mathscr{U}_k = \langle v_{k+1}, \ldots, v_n \rangle$ is an invariant subspace (of the starting matrix $A = A_0$) of dimension $n - k$, and $\mathscr{S}_k = \langle q_1^0, \ldots, q_k^0 \rangle$, where $q_1^0, \ldots, q_n^0$ are the starting vectors for simultaneous iteration. The *QR* algorithm starts with $e_1, e_2, \ldots, e_n$, so $\mathscr{S}_k = \langle e_1, \ldots, e_k \rangle$.

Exercise 5.2.8 (a) Let $A \in \mathbb{F}^{n \times n}$ and $v \in \mathbb{F}^n$. Show that every invariant subspace of A that contains v also contains $A^m v$ for $m = 1, 2, 3, \ldots$.

(b) Let A be a properly upper Hessenberg matrix, and let $v \in \langle e_1, \ldots, e_k \rangle = \mathcal{S}_k, v \neq 0$, where $1 \le k \le n - 1$. (What does v look like? What can you say about Av? A^2v?) Show that the vectors $v, Av, A^2v, \ldots, A^{n-k}v$ are linearly independent.

(c) Conclude that v cannot lie in $\mathcal{U}_k$. Thus $\mathcal{S}_k \cap \mathcal{U}_k = \{0\}$.   □

   While this result has little practical importance, it is certainly very satisfying from a theoretical standpoint. As a consequence of the subspace conditions being satisfied, we can state for example that if A is a properly upper Hessenberg matrix for which $|\lambda_1| > |\lambda_2| > \cdots > |\lambda_n|$, then the sequence of QR iterates starting from A will certainly converge to upper-triangular form.

   A word of caution about the nature of the convergence is in order. The statement that (A_m) converges to upper-triangular form means simply that the subdiagonal entries $a_{k+1,k}^{(m)}$ converge to zero. It follows that the main-diagonal entries converge to the eigenvalues, but nothing has been said about the entries above the main diagonal. These entries may fail to converge. Thus we *cannot* state that there is some upper triangular $R \in \mathbb{F}^{n \times n}$ such that $\| A_m - R \| \to 0$.

Exercise 5.2.9 Let $A = \begin{bmatrix} 2 & 1 \\ 0 & -1 \end{bmatrix}$. (This matrix is not properly upper Hessenberg, nor is there any reason why one would want to apply the QR algorithm to it, except that it illustrates the point of the previous paragraph simply and clearly.) Calculate the QR iterates $A = A_0, A_1, A_2, \ldots$. In each QR decomposition choose R_m so that its main-diagonal entries are positive. Note that $a_{12}^{(m)}$ does not converge as $m \to \infty$.   □

   Along with the use of upper Hessenberg matrices, the other major modification of the QR algorithm is the use of shifts to accelerate convergence. Now that we have established the convergence ratios $|\lambda_{k+1}/\lambda_k|, k = 1, \ldots, n - 1$, it is clear how shifts can be used. This was discussed in Section 4.6, and there is no need to repeat the discussion here. Perhaps one point does bear repeating. While it is clear why shifts of origin are helpful, their use does complicate the convergence analysis. Nobody has been able to prove that the shifted QR algorithm (with some specific shift strategy) always converges. It could conceivably happen that in some circumstances the shifts wander aimlessly and never converge to an eigenvalue. See Batterson and Smillie (1990).

Exercise 5.2.10 Let (A_m) be generated by the shifted QR algorithm:

$$A_{m-1} - \sigma_{m-1}I = Q_m R_m \qquad R_m Q_m + \sigma_{m-1}I = A_m \qquad m = 1, 2, 3, \ldots$$

where $A_0 = A$. Let $\hat{Q}_m = Q_1 Q_2 \cdots Q_m$ and $\hat{R}_m = R_m \cdots R_2 R_1$.

(a) Show that $A_m = \hat{Q}_m^* A \hat{Q}_m, m = 1, 2, \ldots$.

(b) Show that if $A_i - \sigma_i I$ is nonsingular for $i = 0, \ldots, m - 1$, then $A_m = \hat{R}_m A \hat{R}_m^{-1}$.

(c) From part (a) conclude that $(A - \sigma_m I)\hat{Q}_m = \hat{Q}_m(A_m - \sigma_m I)$ $m = 1, 2, \ldots$.

(d) Use the result of part (c) to prove that

$$(A - \sigma_{m-1}I)\cdots(A - \sigma_0 I) = \hat{Q}_m\hat{R}_m \qquad m = 1, 2, 3, \ldots \qquad (5.2.18)$$

□

Equation (5.2.18) generalizes (5.2.17). Notice that it is also a generalization of Lemma 4.7.4, which is one of the crucial results in the development of the double-step *QR* algorithm. We used that result to show that a double *QR* step with complex conjugate shifts transforms a real matrix to a real matrix. In the next section we will use the same equation to clarify the convergence properties of the double *QR* algorithm.

Duality in Subspace Iteration

The following duality theorem provides the link between the *QR* algorithm and inverse iteration. It shows that whenever direct (subspace) iteration takes place, inverse (subspace) iteration also takes place automatically.

THEOREM 5.2.19   Let $A \in \mathbb{F}^{n \times n}$ be nonsingular, and let $B = (A^*)^{-1}$. Let $\mathcal{S}$ be any subspace of $\mathbb{F}^n$. Then the sequences of subspaces

$$\mathcal{S}, \quad A\mathcal{S}, \quad A^2\mathcal{S}, \ldots$$
$$\mathcal{S}^\perp, \quad B\mathcal{S}^\perp, \quad B^2\mathcal{S}^\perp, \ldots$$

are equivalent in the sense that they yield orthogonal complements. That is $(A^m\mathcal{S})^\perp = B^m(\mathcal{S}^\perp)$.

Proof   For every $x, y \in \mathbb{F}^n$, $(A^m x, B^m y) = y^*(B^*)^m A^m x = y^*(A^{-1})^m A^m x = y^*x = (x, y)$. Thus $A^m x$ and $B^m y$ are orthogonal if and only if x and y are orthogonal. The assertion of the theorem follows directly from this observation. Work out the details as an exercise.   □

The second subspace sequence in Theorem 5.2.19 is a sequence of inverse iterates, since the iteration matrix is the inverse of A^*.

Let us first see how Theorem 5.2.19 applies to the unshifted *QR* algorithm. The *QR* algorithm is essentially simultaneous iteration with starting subspaces $\langle e_1, \ldots, e_k \rangle, k = 1, \ldots, n$. After m steps we have

$$A^m\langle e_1, \ldots, e_k \rangle = \langle q_1^m, \ldots, q_k^m \rangle \qquad k = 1, \ldots, n - 1 \qquad (5.2.20)$$

where $q_1^m, \ldots, q_n^m$ are the columns of $\hat{Q}_m$. Since $\langle e_1, \ldots, e_k \rangle^\perp = \langle e_{k+1}, \ldots, e_n \rangle$ and $\langle q_1^m, \ldots, q_k^m \rangle^\perp = \langle q_{k+1}^m, \ldots, q_n^m \rangle$, Theorem 5.2.19 applied to (5.2.20) yields

$$(A^*)^{-m}\langle e_{k+1}, \ldots, e_n \rangle = \langle q_{k+1}^m, \ldots, q_n^m \rangle \qquad k = 1, \ldots, n - 1 \qquad (5.2.21)$$

Of particular interest is the case $k = n - 1$: $(A^*)^{-1}\langle e_n \rangle = \langle q_n^m \rangle$. Thus the last column of $\hat{Q}_m$ is seen to be the result of single-vector inverse iteration on A^*. It should therefore be possible to cause this column to converge rapidly by applying shifts to A^*. Before we investigate this possibility, let us take time to see how (5.2.21) can be derived directly from the basic equations of the QR algorithm. This will reveal some interesting aspects of duality.

The QR sequence (A_m) is generated by (5.2.14) with $A_0 = A$. Let $B_m = (A_m^*)^{-1}$ and $L_m = (R_m^*)^{-1}$. Note that L_m is lower triangular with positive entries on the main diagonal. Take conjugate transposes and inverses in (5.2.14) to obtain

$$B_{m-1} = Q_m L_m \qquad L_m Q_m = B_m \qquad m = 1, 2, 3, \ldots$$

with $B_0 = B = (A^*)^{-1}$. This shows that the QR algorithm applied to A is equivalent to a QL algorithm applied to $(A^*)^{-1}$. The theory and algorithms based on the QL decomposition are identical to those based on QR. The next three exercises investigate some basic aspects of the QL theory.

Exercise 5.2.11   Let $B \in \mathbb{F}^{n \times n}$ be nonsingular. Prove that there exist unique matrices $Q, L \in \mathbb{F}^{n \times n}$ such that Q is unitary, L is lower triangular with positive entries on the main diagonal, and $B = QL$. (You are welcome to prove this from scratch if you want to, but it is easier to deduce it from the corresponding theorem about QR decompositions.)   □

Exercise 5.2.12   Let $B, Q, L \in \mathbb{F}^{n \times n}$ with $B = QL$, $B = [b_1 \ \cdots \ b_n]$ nonsingular, $Q = [q_1 \ \cdots \ q_n]$ unitary, and L lower triangular. Show that $\langle b_n \rangle = \langle q_n \rangle$, $\langle b_{n-1}, b_n \rangle = \langle q_{n-1}, q_n \rangle$, and in general

$$\langle b_{k+1}, \ldots, b_n \rangle = \langle q_{k+1}, \ldots, q_n \rangle \qquad k = n - 1, n - 2, \ldots, 0 \qquad \square$$

Exercise 5.2.12 shows that the QL decomposition is a Gram–Schmidt procedure that orthonormalizes the columns of B from right to left.

Exercise 5.2.13   The conclusion of Exercise 5.2.12 suggests an easy way to work Exercise 5.2.11, using some ideas developed at the end of Section 4.8. The basic idea is simply to reverse the order of the columns. Work through Exercises 4.8.13 and 4.8.14, if you have not already done so.

(a)   In the notation of those exercises, show that $B = QL$ if and only if $B^C = Q^C L^B$.

(b)   Use the result of part (a) to work Exercise 5.2.11.   □

Recalling that $\hat{Q}_m = Q_1 Q_2 \cdots Q_m$ and $\hat{R}_m = R_m R_{m-1} \cdots R_1$, define $\hat{L}_m = (\hat{R}_m^*)^{-1} = L_m L_{m-1} \cdots L_1$. Notice that $\hat{L}_m$ is lower triangular with positive main-diagonal entries. Taking conjugate transposes and inverses in (5.2.17), we obtain

$$B^m = \hat{Q}_m \hat{L}_m \qquad m = 1, 2, 3, \ldots$$

Applying the conclusion of Exercises 5.2.12 to this QL decomposition, we find that

$$B^m \langle e_{k+1}, \ldots, e_n \rangle = \langle q_{k+1}^m, \ldots, q_n^m \rangle \qquad k = 1, \ldots, n-1$$

This is just (5.2.21) again.

Now let us apply duality to the study of shifts of origin. Specifically, we will demonstrate that the QR algorithm with Rayleigh quotient shifts is, in part, Rayleigh quotient iteration. We can accomplish this by examining the cumulative effect of m steps or by taking a step-at-a-time point of view. The latter is probably more revealing, so we will adopt that approach.

Consider a single QR step

$$A_{m-1} - \sigma_{m-1}I = Q_m R_m \qquad R_m Q_m + \sigma_{m-1}I = A_m$$

where $\sigma_{m-1} = a_{nn}^{(m-1)} = e_n^* A_{m-1} e_n$. This is the Rayleigh quotient shift defined in Section 4.6. Then $\bar{\sigma}_{m-1}$ is a Rayleigh quotient of A_{m-1}^*, since $\bar{\sigma}_{m-1} = e_n^* A_{m-1}^* e_n$. If $|a_{n,n-1}^{(m)}|$ is small, then $\bar{\sigma}_{m-1}$ is a good approximation to an eigenvalue of A_{m-1}^*. It is also easy to see (informally) that e_n is approximately an eigenvector of A_{m-1}^* (cf., Exercise 4.6.9). Let $p_1, \ldots, p_n$ denote the columns of Q_m. From the equation $A_{m-1} - \sigma_{m-1}I = Q_m R_m$, it follows that

$$(A_{m-1} - \sigma_{m-1}I)\langle e_1, \ldots, e_{n-1} \rangle = \langle p_1, \ldots, p_{n-1} \rangle$$

Applying Theorem 5.2.19 to this equation, we obtain

$$(A_{m-1}^* - \bar{\sigma}_{m-1}I)^{-1}\langle e_n \rangle = \langle p_n \rangle \tag{5.2.22}$$

Since $\bar{\sigma}_{m-1}$ is the Rayleigh quotient formed from A_{m-1}^* and e_n, the left-hand side of (5.2.22) represents a step of Rayleigh quotient iteration. The result is p_n, the last column of Q_m.

Now suppose we were to take another step of Rayleigh quotient iteration using A_{m-1}^*. The shift would be $\bar{\sigma} = p_n^* A_{m-1}^* p_n$, and we would calculate

$$(A_{m-1}^* - \bar{\sigma}I)^{-1}\langle p_n \rangle \tag{5.2.23}$$

Recalling that $A_m = Q_m^* A_{m-1} Q_m$, we see easily that $\sigma = a_{nn}^{(m)}$. If, on the other hand, we take another QR step, we will operate with the matrix A_m and choose the shift $\sigma_m = e_n^* A_m e_n = a_{nn}^{(m)}$, which is the same as σ. This is hardly a coincidence. The coordinate transformation that transforms A_{m-1} to A_m also transforms p_n to e_n, so the computations $\sigma = p_n^* A_{m-1} p_n$ and $\sigma_m = e_n^* A_m e_n$ represent the same computation in two different coordinate systems. By (5.2.22) with $m-1$ replaced by m, the QR step on A_m calculates (among other things)

$$(A_m^* - \bar{\sigma}_m I)^{-1} e_n$$

This computation is the same as (5.2.23), except that the coordinate system has been changed. This proves that the QR algorithm is partly Rayleigh quotient iteration.

Exercise 5.2.14   Verify that $Q_m^* p_n = e_n$ and $Q_m^*[(A_{m-1}^* - \bar{\sigma}_m I)^{-1} p_n] = (A_m^* - \bar{\sigma}_m I)^{-1} e_n$. ☐

The next exercise develops the cumulative viewpoint.

Exercise 5.2.15   If m steps of the QR algorithm with Rayleigh quotient shift are performed, then (5.2.18) holds, where $\sigma_i = a_{nn}^{(i)}$, $i = 1, \ldots, m-1$. Apply Theorem 5.2.19 to (5.2.18) (or take conjugate transpose inverses) to obtain information about $\langle q_n^m \rangle$. Show that $\langle q_n^m \rangle$ is exactly the space obtained after m steps of Rayleigh quotient iteration applied to A^* with starting vector e_n. In particular, be sure to check that each shift is the appropriate Rayleigh quotient. ☐

Exercise 5.2.15 implies that if $\langle q_n^m \rangle$ converges to an eigenvector of A^*, then it (generally) converges quadratically, for that is the rate of convergence of Rayleigh quotient iteration. As a consequence $a_{nn}^{(m)}$ converges to an eigenvalue quadratically. If A is Hermitian (or even normal), convergence is cubic.

5.3
UNIQUENESS THEOREMS FOR HESSENBERG FORM; THE LANCZOS ALGORITHM

In Section 4.5 we showed that every $A \in \mathbb{C}^{n \times n}$ is unitarily similar to an upper Hessenberg matrix B. In the present section we will investigate the extent to which B is uniquely determined. The uniqueness theorems that we shall prove will be used to justify the implicit single- and double-step QR algorithms developed in Section 4.7. As an additional bonus, the construction used to prove the uniqueness theorems turns out to be an important algorithm in its own right. It is the basis of the Lanczos method, which is now frequently used to find eigenvalues of large, sparse, symmetric matrices.

Again we let $\mathbb{F} = \mathbb{C}$ or $\mathbb{R}$.

THEOREM 5.3.1   Let $A, B, Q \in \mathbb{F}^{n \times n}$, with $Q = [q_1 \cdots q_n]$ unitary, B properly upper Hessenberg, and $B = Q^* A Q$. Then Q and B are essentially uniquely determined by A and q_1, in the following sense: Suppose $\tilde{Q}, \tilde{B} \in \mathbb{F}^{n \times n}$, $\tilde{Q} = [\tilde{q}_1 \cdots \tilde{q}_n]$ is unitary, $\tilde{B}$ is upper Hessenberg, $\tilde{B} = \tilde{Q}^* A \tilde{Q}$, and $\tilde{q}_1 = q_1 d_1$ for some $d_1 \in \mathbb{F}$. Then there exist $d_2, \ldots, d_n \in \mathbb{F}$ such that $\tilde{q}_i = q_i d_i$, $i = 2, \ldots, n$.

Interpretation: The theorem states that if the first column of $\tilde{Q}$ is a multiple of the first column of Q, then each column of $\tilde{Q}$ is a multiple of the corresponding column of Q. Since all columns of Q and $\tilde{Q}$ have unit Euclidean norm, each of the d_i must satisfy $|d_i| = 1$. It is easy to state the conclusion of the theorem in matrix form: $\tilde{Q} = QD$, where D is the unitary, diagonal matrix whose main diagonal entries are $d_1, \ldots, d_n$. It follows that B and $\tilde{B}$ are related by the trivial similarity transformation $\tilde{B} = D^* B D$ (cf., Exercise 4.6.3). Thus it is reasonable to say that

Q and $\tilde{Q}$ are essentially the same and B and $\tilde{B}$ are essentially the same. Thus Q and B are essentially uniquely determined by A and q_1.

Proof   The proof is by induction. Assume $\tilde{q}_i = q_i d_i$ for $i = 1, \ldots, j$. We will prove that $\tilde{q}_{j+1} = q_{j+1} d_{j+1}$ for some $d_{j+1} \in \mathbb{F}$. The equation $B = Q^*AQ$ is equivalent to $AQ = QB$, whose jth column is $Aq_j = Qb_j = \sum_{i=1}^{n} q_i b_{ij}$. Since B is in upper Hessenberg form, $b_{ij} = 0$ for $i > j + 1$. Thus $Aq_j = \sum_{i=1}^{j} q_i b_{ij} + q_{j+1} b_{j+1,j}$, and

$$q_{j+1} b_{j+1,j} = Aq_j - \sum_{i=1}^{j} q_i b_{ij} \qquad (5.3.2)$$

Applying the same sequence of steps to the equation $\tilde{B} = \tilde{Q}^* A \tilde{Q}$, we obtain the analogous equation

$$\tilde{q}_{j+1} \tilde{b}_{j+1,j} = A\tilde{q}_j - \sum_{i=1}^{j} \tilde{q}_i \tilde{b}_{ij} \qquad (5.3.3)$$

For $i = 1, \ldots, j$, $\tilde{q}_i = q_i d_i$. Furthermore $\tilde{b}_{ij} = \tilde{q}_i^* A \tilde{q}_j = \overline{d}_i d_j (q_i^* A q_j) = \overline{d}_i d_j b_{ij}$. Substituting these expressions into the right-hand side of (5.3.3), bearing in mind that $d_i \overline{d}_i = |d_i|^2 = 1$, and comparing the result with (5.3.2), we find that

$$\tilde{q}_{j+1} \tilde{b}_{j+1,j} = q_{j+1} b_{j+1,j} d_j \qquad (5.3.4)$$

Since $b_{j+1,j} \neq 0$, it must be that $\tilde{b}_{j+1,j} \neq 0$ as well. Thus $\tilde{q}_{j+1} = q_{j+1} d_{j+1}$, where

$$d_{j+1} = \frac{b_{j+1,j}}{\tilde{b}_{j+1,j}} d_j$$

This completes the proof.   ☐

The assumption $b_{j+1,j} \neq 0$ is crucial. If $b_{j+1,j} = 0$ for some j, then Eq. 5.3.4 implies that $\tilde{b}_{j+1,j} = 0$ also. Thus (5.3.4) collapses to $0 = 0$, and we cannot infer that $\tilde{q}_{j+1}$ is a multiple of q_j.

The next exercise shows that a slight modification of the hypotheses turns Theorem 5.3.1 into a true uniqueness theorem.

Exercise 5.3.1   Suppose the hypotheses of Theorem 4.11.1 are satisfied, and B and $\tilde{B}$ satisfy the additional condition that the subdiagonal entries $b_{j+1,j}$ and $\tilde{b}_{j+1,j}$ are all real and positive. Show that if $\tilde{q}_1 = q_1$, then $\tilde{Q} = Q$ and $\tilde{B} = B$.   ☐

Exercise 5.3.2   Suppose $A, Q, B \in \mathbb{F}^{n \times n}$, Q is unitary, B is properly upper Hessenberg, and $B = Q^*AQ$. Show that there exist $\tilde{Q}, \tilde{B} \in \mathbb{F}^{n \times n}$ such that $\tilde{Q}$ is unitary, $\tilde{q}_1 = q_1 d_1$ for some $d_1 \in \mathbb{F}$, $\tilde{B}$ is upper Hessenberg, $\tilde{b}_{j+1,j} > 0$ for $j =$

$1, \ldots, n - 1$, and $\tilde{B} = \tilde{Q}^* A \tilde{Q}$. (From Theorem 5.3.1 we know that such a $\tilde{Q}$ must satisfy $\tilde{Q} = QD$ for some unitary, diagonal $D \in \mathbb{F}^{n \times n}$. Now find a D that works.) $\square$

Exercise 5.3.3 Verify that in the Householder reduction $B = Q^* A Q$ of Section 4.5, the first column of Q is e_1. Thus, if B is properly upper Hessenberg, this construction delivers the essentially unique unitary transformation to upper Hessenberg form for which $q_1 = e_1$. $\square$

To justify the implicit versions of the QR algorithm, we need a slightly more general version of Theorem 5.3.1

THEOREM 5.3.5 Suppose $A, B, Q, \tilde{B}, \tilde{Q} \in \mathbb{F}^{n \times n}$, $B = Q^* A Q$, $\tilde{B} = \tilde{Q}^* A \tilde{Q}$, $Q = [q_1 \cdots q_n]$ and $\tilde{Q} = [\tilde{q}_1 \cdots \tilde{q}_n]$ are unitary, B and $\tilde{B}$ are upper Hessenberg, and $b_{j+1,j} \neq 0$ for $j = 1, \ldots, k$, where k is some positive integer less than n. Suppose also that $\tilde{q}_1 = q_1 d_1$ for some $d_1 \in \mathbb{F}$. Then there exist $d_2, \ldots, d_{k+1} \in \mathbb{F}$ such that $\tilde{q}_i = q_i d_i, i = 1, \ldots, k + 1$.

Proof In the case $k = n - 1$, this theorem is precisely Theorem 5.3.1. When $k < n - 1$, the hypotheses are weaker and so is the conclusion. The construction that proved Theorem 5.3.1 also works here. The conclusion is weaker because the construction breaks down as soon as a j is reached for which $b_{j+1,j} = 0$. $\square$

Theorem 5.3.5 is exactly what is needed to justify the implicit QR algorithm. Recall that a single QR step transforms a properly upper Hessenberg matrix A to another upper Hessenberg matrix $B = Q^* A Q$, where $Q = [q_1 \cdots q_n]$ is the unitary factor in the decomposition $A - \sigma I = QR$. The implicit QR algorithm developed in Section 4.7 carries out an apparently different similarity transformation $\tilde{B} = \tilde{Q}^* A \tilde{Q}$, in which $\tilde{Q} = [\tilde{q}_1 \cdots \tilde{q}_n]$ is a unitary matrix such that $\tilde{q}_1 = q_1 d_1$, with $d_1 = 1$. In the development of that algorithm, we were careful to note that the resulting upper Hessenberg matrix $\tilde{B}$ satisfies $\tilde{b}_{j+1,j} \neq 0$ for $j = 1, \ldots, n - 2$. Therefore, applying Theorem 5.3.5 with the roles of B and $\tilde{B}$ reversed, there exist $d_2, \ldots, d_{n-1}$ such that $\tilde{q}_i = q_i d_i, i = 1, \ldots, n - 1$. But note in addition that $\langle \tilde{q}_n \rangle = \langle \tilde{q}_1, \ldots, \tilde{q}_{n-1} \rangle^\perp = \langle q_1, \ldots, q_{n-1} \rangle^\perp = \langle q_n \rangle$, so $\tilde{q}_n = q_n d_n$ for some $d_n \in \mathbb{F}$. Thus Q and $\tilde{Q}$ are essentially the same, and so are B and $\tilde{B}$.

The justification of the double-step QR algorithm is only slightly more involved. A pair of QR steps with real shifts or complex conjugate shifts transforms the real, properly upper Hessenberg matrix to another upper Hessenberg matrix $B = Q^* A Q$, where Q is unitary. Usually Q and B are real (cf., Lemma 4.7.5). The implicit double QR step developed in Section 4.7 carries out a real, orthogonal similarity transformation $\tilde{B} = \tilde{Q}^T A \tilde{Q}$, for which $\tilde{q}_1 = q_1 d_1$ with $d_1 = \pm 1$. The resulting $\tilde{B}$ is upper Hessenberg and satisfies $\tilde{b}_{j+1,j} \neq 0$ for $j = 1, \ldots, n - 3$. Therefore by Theorem 5.3.5 with the roles of B and $\tilde{B}$ reversed, there exist $d_1, \ldots, d_{n-2}$ such that $\tilde{q}_i = q_i d_i$ for $i = 1, \ldots, n - 2$. It is not necessarily true that there exist d_{n-1}, d_n such that $\tilde{q}_i = q_i d_i$ for $i = n - 1, n$. However, a look at the convergence properties of the double QR algorithm will show that this is not important.

Exercise 5.3.4 Given that $\tilde{q}_i = q_i d_i$, $i = 1, \ldots, n - 2$, what can be said about the relationship between B and $\tilde{B}$?   □

Recall that Q is the unitary factor in the decomposition

$$(A - \sigma_2 I)(A - \sigma_1 I) = QR \qquad (5.3.6)$$

where either $\sigma_2 = \bar{\sigma}_1$ or σ_1 and σ_2 are real. Let $C = (A - \sigma_2 I)(A - \sigma_1 I)$. You can easily verify that if λ is an eigenvalue of A with associated eigenvector v, then $(\lambda - \sigma_2)(\lambda - \sigma_1)$ is an eigenvalue of C with associated eigenvector v. To keep the analysis uncomplicated, we will assume that A is simple. Let $\lambda_1, \ldots, \lambda_n$ be eigenvalues of A with associated linearly independent eigenvectors $v_1, \ldots, v_n$. Then C is also simple, having the same eigenvectors associated with eigenvalues $\mu_1, \ldots, \mu_n$ where $\mu_i = (\lambda_i - \sigma_2)(\lambda_i - \sigma_1)$. Assume they are ordered so that $|\mu_1| \geq |\mu_2| \geq \cdots \geq |\mu_n|$. The ideas developed in the previous section can be used to interpret the double QR step. From the QR decomposition (5.3.6) we see that

$$\langle C e_1, \ldots, C e_i \rangle = \langle q_1, \ldots, q_i \rangle \qquad i = 1, \ldots, n$$

That is, the columns of Q represent the result of one step of simultaneous iteration with iteration matrix C and starting vectors $e_1, \ldots, e_n$. The similarity transformation $B = Q^* A Q$ is just a change of coordinates that maps $q_1, \ldots, q_n$ back to the standard basis $e_1, \ldots, e_n$. The progress toward convergence during this step is measured by the ratios

$$\left| \mu_{i+1} / \mu_i \right| \qquad i = 1, \ldots, n - 1$$

Exercise 5.3.5 Suppose λ_{n-1} and λ_n are distinct from the other eigenvalues of A. Show that if σ_1 and σ_2 approximate λ_{n-1} and λ_n sufficiently well, then C has two eigenvalues that are much smaller than the others.   □

Suppose the conditions of Exercise 5.3.5 are satisfied. Then $|\mu_{n-1}/\mu_{n-2}|$ will be small, which (typically) means that $\langle q_1, \ldots, q_{n-2} \rangle$ will be much closer to an invariant subspace than $\langle e_1, \ldots, e_{n-2} \rangle$ is. As a consequence the entry $b_{n-1,n-2}$ will (typically) be much smaller than $a_{n-1,n-2}$. Repeated double QR steps with these shifts or nearby shifts will cause the $(n - 1, n - 2)$ entry to tend rapidly toward zero. Notice that as this entry gets smaller, the 2-by-2 submatrix in the lower right-hand corner is increasingly isolated from the rest of the matrix; that is, its eigenvalues approach eigenvalues of A. If, as is usually the case, the shifts are chosen to be the eigenvalues of that submatrix, the quality of the shifts will improve dramatically as we approach an invariant subspace. The consequence is quadratic convergence. At the same time (most of) the subdiagonal entries $a_{j+1,j}^{(m)}$, $j = 1, \ldots, n - 3$, will also tend toward zero but much more slowly.

These conclusions apply not only to the explicitly computed double QR step $B = Q^* A Q$, but to the implicitly computed step $\tilde{B} = \tilde{Q}^T A \tilde{Q}$ as well, because

$$\langle \tilde{q}_1, \ldots, \tilde{q}_i \rangle = \langle q_1, \ldots, q_i \rangle \qquad i = 1, \ldots, n - 2$$

Of course the most important of these equations is that for which $i = n - 2$. The next exercise shows that the fact that this subspace equation may fail to be satisfied for $i = n - 1$ is of no importance.

Exercise 5.3.6 Suppose $\sigma_1 = \sigma$ and $\sigma_2 = \bar{\sigma}$ are complex conjugate shifts that are excellent approximations to a pair of complex conjugate eigenvalues of A. Show that the two smallest eigenvalues of C have the same magnitude: $|\mu_{n-1}| = |\mu_n|$. (Thus we cannot expect that $\langle q_1, \ldots, q_{n-1} \rangle$ is closer to an invariant subspace than $\langle e_1, \ldots, e_{n-1} \rangle$ is, nor can we expect that $b_{n,n-1}$ is closer to zero than $a_{n,n-1}$ is. Therefore we do not care whether or not $\langle \tilde{q}_1, \ldots, \tilde{q}_{n-1} \rangle$ and $\tilde{b}_{n,n-1}$ are linked to these quantities.) □

Exercise 5.3.7 Consider a double QR step for which the shifts σ_1 and σ_2 $(\sigma_1 \neq \sigma_2)$ are exactly eigenvalues of A. Show that B and $\tilde{B}$ have the general forms

$$
B = \left[\begin{array}{c|cc} & \vline & \\ \hline & & \\ 0 & \sigma_2 & \\ & 0 & \sigma_1 \end{array} \right]
\quad \text{and} \quad
\tilde{B} = \left[\begin{array}{c|cc} & \vline & \\ \hline & & \\ 0 & b & c \\ & d & e \end{array} \right]
$$

where $\begin{bmatrix} b & c \\ d & e \end{bmatrix}$ has eigenvalues σ_1 and σ_2. The fact that $\begin{bmatrix} b & c \\ d & e \end{bmatrix}$ may fail to be upper triangular is no loss, since its eigenvalues can be computed easily. [Notice that if σ_1 and σ_2 are complex, B is not a real matrix. This does not contradict any earlier assertions (cf., Lemma 4.7.5 and the discussion that follows it). The matrix $\tilde{B}$ produced by the implicit double QR step is always real.] □

As a final word on the double-step QR algorithm, it is important to point out that we did not prove it always converges. Because of Example 4.6.13 we cannot hope for a global convergence proof. Exceptional shifts take care of the problem posed by Example 4.6.13, but even with exceptional shifts nobody has shown that the algorithm always works; our knowledge that it works is based on practical experience. What we have shown here is that *if* the shifts get close enough to eigenvalues, then the double QR steps will converge very rapidly to these eigenvalues and an associated invariant subspace. While our arguments were informal, they can be made rigorous. In particular it can be proved rigorously that under appropriate conditions the asymptotic convergence rate is quadratic. See, for example, Watkins and Elsner (1990).

The Arnoldi Algorithm

The proof of Theorem 5.3.1 is based on the equation

$$
q_{j+1} b_{j+1,j} = A q_j - \sum_{i=1}^{j} q_i b_{ij} \tag{5.3.2}
$$

and the essentially identical equation (5.3.3). Equation 5.3.2 is is a consequence of the unitary similarity transformation $B = Q^*AQ$, where B is a properly upper Hessenberg matrix. Let us now reverse our viewpoint and consider (5.3.2) as an algorithm for constructing a unitary Q such that Q^*AQ is upper Hessenberg. First of all, if we wish to end up with a B for which $B = Q^*AQ$, we must define $b_{ij} = q_i^*Aq_j$ for $i = 1, \ldots, n$. This shows that the right-hand side of (5.3.2) is completely determined by A and $q_1, \ldots, q_j$. Consequently q_{j+1} is almost completely determined by A and $q_1, \ldots, q_j$. There is some freedom in the choice of the scalar $b_{j+1,j}$, but this freedom is severly limited by the requirement that $\|q_{j+1}\|_2 = 1$. Thus the following algorithm can be used to compute q_{j+1}, given A and the orthonormal vectors $q_1, \ldots, q_j$.

$$
\begin{aligned}
&b_{ij} \leftarrow q_i^*Aq_j && i = 1, \ldots, j \\
&\hat{q}_{j+1} \leftarrow Aq_j - \sum_{i=1}^{j} q_i b_{ij} \\
&b_{j+1,j} \leftarrow \|\hat{q}_{j+1}\|_2 \\
&q_{j+1} \leftarrow \hat{q}_{j+1}/b_{j+1,j} && \text{provided } b_{j+1,j} \neq 0
\end{aligned}
\tag{5.3.7}
$$

This algorithm yields a $b_{j+1,j}$ that is real and positive. This additional condition on $b_{j+1,j}$ causes q_{j+1} to be uniquely determined, a conclusion that is consistent with that of Exercise 5.3.1.

Exercise 5.3.8   Show that the vector $\hat{q}_{j+1}$ defined by (5.3.7) is orthogonal to $q_1, \ldots, q_j$. Thus $q_1, \ldots, q_j, q_{j+1}$ is an orthonormal set.   □

Exercise 5.3.9   Show that the $b_{j+1,j}$ defined by (5.3.7) satisfies $b_{j+1,j} = q_{j+1}^*\hat{q}_{j+1} = q_{j+1}^*Aq_j$. Thus the definition of $b_{j+1,j}$ in (5.3.7) is consistent with the requirement that $b_{ij} = q_i^*Aq_j$ for all i and j.   □

Starting from a single vector q_1 satisfying $\|q_1\|_2 = 1$, we can apply (5.3.7) with $j = 1, 2, \ldots, n - 1$ successively to produce an orthonormal basis $q_1, \ldots, q_n$ provided that $b_{j+1,j} \neq 0$ for $j = 1, \ldots, n - 1$. We can then define $Q = [q_1 \cdots q_n]$ and $B = Q^*AQ$. Since it may not be immediately evident that B is an upper Hessenberg matrix, the issues involved are clarified in Exercise 5.3.10. Of course the entries of B are exactly the b_{ij} computed in (5.3.7). This procedure for constructing a unitary similarity transformation to upper Hessenberg form is called the *Arnoldi algorithm*.

Exercise 5.3.10   Let $A, B, Q \in \mathbb{F}^{n \times n}$ with $Q = [q_1 \cdots q_n]$ unitary and $AQ = QB$.

(a)   Show that B is an upper Hessenberg matrix if and only if $Aq_j \in \langle q_1, \ldots, q_{j+1} \rangle$ for $j = 1, \ldots, n - 1$.

(b)   Show that if B is upper Hessenberg, then B is properly upper Hessenberg if and only if $Aq_j \notin \langle q_1, \ldots, q_j \rangle$ for $j = 1, \ldots, n - 1$.

(c)   Use the results of parts (a) and (b) to show that the matrix B produced by the Arnoldi algorithm is a properly upper Hessenberg matrix.   □

The Arnoldi algorithm produces not only an upper Hessenberg matrix but also an orthonormal basis $q_1, \ldots, q_n$. This might cause you to wonder whether the Arnoldi algorithm is related to the Gram–Schmidt process. This is indeed the case.

Exercise 5.3.11   Show that the Arnoldi algorithm with starting vector q_1 is exactly the classical Gram–Schmidt algorithm applied to the sequence $q_1, Aq_1, Aq_2, \ldots, Aq_{n-1}$. □

For the purpose of transforming a matrix to upper Hessenberg form, the Arnoldi algorithm is not as stable as the Householder reduction presented in Section 4.5, so the latter is much more frequently used. The problem with the Arnoldi algorithm is that the vectors $q_1, Aq_1, \ldots, Aq_j$ will frequently be ill conditioned; that is, they will be nearly linearly dependent. Whenever this happens, Aq_j will very nearly lie in the subspace $\langle q_1, \ldots, q_j \rangle$ and cancellation will occur during the computation of q_{j+1}. As a consequence, the computed q_{j+1} will be inaccurate, and it will not be truly orthogonal to $q_1, \ldots, q_j$. The computed B will be correspondingly inaccurate. The Arnoldi process can be reorganized so that it amounts to a modified Gram–Schmidt procedure. While this improves the stability somewhat, it cannot compensate for ill conditioning in the sequence $q_1, Aq_1, \ldots, Aq_{n-1}$.

The Arnoldi algorithm fails outright if at some step $b_{j+1,j} = 0$. This results from total cancellation in the computation of $\hat{q}_{j+1}$. The algorithm can be restarted by taking q_{j+1} to be any unit vector that is orthogonal to $\langle q_1, \ldots, q_j \rangle$. Such a q_{j+1} can be obtained by applying the Gram–Schmidt process to a vector chosen at random.

Exercise 5.3.12

(a)   Show that the Arnoldi algorithm fails ($b_{j+1,j} = 0$) if and only if the subspace $\langle q_1, \ldots, q_j \rangle$ is invariant under A.

(b)   How are a failure and subsequent restart reflected in the form of B?   □

The Lanczos Algorithm

The Arnoldi algorithm is most frequently applied to real, symmetric matrices, in which case it is called the *Lanczos algorithm*. If the initial vector q_1 is chosen to be real, then all subsequent computations remain within the real field. The resulting matrix B is not merely upper Hessenberg; it is symmetric and hence tridiagonal. The form of B allows for an important simplification in the computation: Recall that

$$\hat{q}_{j+1} = Aq_j - \sum_{i=1}^{j} q_i b_{ij}$$

Since $b_{ij} = 0$ for $i < j - 1$, all but two of the terms in the sum on the right-hand side are zero.

Since we are now working with a symmetric, tridiagonal matrix, it is convenient to alter the notation. Let

$$\alpha_j = b_{jj} \quad \text{and} \quad \beta_j = b_{j+1,j} = b_{j,j+1}$$

Then

$$\hat{q}_{j+1} = Aq_j - q_j\alpha_j - q_{j-1}\beta_{j-1}$$

The element α_j is defined by $\alpha_j = q_j^T Aq_j$, but a theoretically equivalent definition is $\alpha_j = q_j^T p$, where $p = Aq_j - q_{j-1}\beta_{j-1}$. These computations are the same in principle because q_{j-1} and q_j are (theoretically) orthogonal. In practice of course they yield slightly different results. The two ways of calculating α_j correspond to the classical and modified Gram–Schmidt procedures, respectively, so the latter is generally better (cf., Exercise 3.4.14 and the discussion that follows). Thus a step of the Lanczos algorithm is as follows:

$$\begin{aligned}
&p \leftarrow Aq_j - q_{j-1}\beta_{j-1} \qquad (q_0\beta_0 = 0) \\
&\alpha_j \leftarrow q_j^T p \\
&p \leftarrow p - q_j\alpha_j \\
&\beta_j \leftarrow \|p\|_2 \\
&q_{j+1} \leftarrow p/\beta_j \qquad \text{provided } \beta_j \neq 0
\end{aligned} \tag{5.3.8}$$

Starting with a unit vector q_1, we can perform $n - 1$ steps of (5.3.8) to obtain a symmetric, tridiagonal matrix T that is orthogonally similar to A. This was the original use of the algorithm proposed by Lanczos in 1950. Because of its instability and the subsequent development of the stable Householder reduction to tridiagonal form (Section 4.5), the Lanczos algorithm was largely ignored for many years. Around 1970 the method was resurrected for use with large, sparse matrices. In a typical application only a few of the extreme (largest or smallest) eigenvalues of A are desired, and the Lanczos process is carried only part way to completion. After $j - 1$ steps the leading j-by-j submatrix of T,

$$T_j = \begin{bmatrix} \alpha_1 & \beta_1 & & \\ \beta_1 & \alpha_2 & \ddots & \\ & \ddots & \ddots & \beta_{j-1} \\ & & \beta_{j-1} & \alpha_j \end{bmatrix}$$

will have been produced. In most cases the extreme eigenvalues of T_j provide surprisingly good approximations to the extreme eigenvalues of A, even when j is much smaller than n. The reasons for this are at least partly explained in Exercises 5.3.13 to 5.3.19. The eigenvalues of T_j can be calculated inexpensively by the QR algorithm, for example. Two other possibilities are the bisection method and Cuppen's divide-and-conquer method, both of which are discussed in Chapter 6.

The Lanczos procedure is well suited for sparse problems because the only use of A in (5.3.8) is to compute the matrix–vector products Aq_j. For most sparse matrices this computation is fairly inexpensive. The matrix A is not altered in any way during the computation. By contrast the Householder reduction, which also could be used to carry out partial reductions, performs a similarity transformation at each step. The resulting fill-in is severe; after only a few steps the matrix can no longer be regarded as sparse.

In order to use the Lanczos method successfully, one must cope effectively with its instability. Fortunately the errors that develop are not random; rather they are closely tied to the convergence of the algorithm. This is at least partly evident from what we have already observed. The "worst" thing that can happen is total cancellation in the calculation of q_{j+1}. This occurs when and only when $\langle q_1, \ldots, q_j \rangle$ is an invariant subspace. In this case $\beta_j = 0$, and every eigenvalue of T_j is an eigenvalue of A. A more careful analysis carried out by C. C. Paige showed what happens in less extreme situations: The errors in the computed vectors q_{j+1} tend to lie in the direction of approximate eigenvectors associated with eigenvalues of T_j that have nearly converged to eigenvalues of A. Most of the error can be removed from q_{j+1} by orthogonalizing it against a small number of these approximate eigenvectors or *Ritz vectors*, as they are usually called. For more details see [SEP], [MC], and the references cited therein.

Of course this additional orthogonalization makes the procedure more complicated and more expensive, and it requires the storage of a few (long) Ritz vectors. The main appeal of the original Lanczos process is its simplicity; the computational and storage requirements are low. For example, each iteration requires only the two most recent vectors; the others need not be saved.

In an effort to retain this simplicity, some workers have focused on Lanczos algorithms that do not reorthogonalize. The focus is shifted to an analysis of the eigenvalues of T_j and the identification of "spurious eigenvalues" caused by the nonorthogonality of the (q_j). Advocates of this approach suggest that the algorithm be used not only to calculate a few selected eigenvalues but also to calculate the complete spectrum if needed. In theory the process must terminate after $n - 1$ steps, but in practice, due to loss of orthogonality, it need never terminate; it can be carried on indefinitely as an iterative method. It has been observed that the resulting huge tridiagonal matrices T_j eventually have eigenvalues that approximate all eigenvalues of A well. Each T_j ($j \gg n$) has many more eigenvalues than A. The extra eigenvalues are concentrated in tight clusters that approximate eigenvalues of A. For more information on this approach see Cullum and Willoughby (1985) and the references cited therein.

The Lanczos procedure is also frequently used in conjunction with inverse iteration to approximate a few eigenvalues in the middle of the spectrum of A. Suppose, for example, we wish to find those eigenvalues of A that are closest to some specified real number μ. The eigenvalues of A that are closest to μ are mapped to the extreme eigenvalues of $(A - \mu I)^{-1}$. In order to apply the Lanczos algorithm to $(A - \mu I)^{-1}$, we need only be able to calculate $(A - \mu I)^{-1}q$ for any vector q. This may be feasible if, for example, A is sparse, since then a sparse LU decomposition of $A - \mu I$ would allow computation of $(A - \mu I)^{-1}q$ by forward and back substitution.

The following sequence of exercises shows that even for relatively small values of j, T_j has eigenvalues that approximate the extreme eigenvalues of A well. Questions of roundoff errors and stability are ignored.

Exercise 5.3.13   If $q_1, \ldots, q_j$ are the first j vectors produced by the Lanczos process and $Q_j = [q_1 \ \cdots \ q_j] \in \mathbb{R}^{n \times j}$, then $T_j = Q_j^T A Q_j$. Suppose A has an eigenvector v with associated eigenvalue λ, such that $v \in \langle q_1, \ldots, q_j \rangle$. Show that:

(a) There exists a unique $\hat{v} \in \mathbb{R}^j$ such that $v = Q_j \hat{v}$.

(b) $\hat{v} = Q_j^T v$.

(c) $\hat{v}$ is an eigenvector of T_j with associated eigenvalue λ. $\qquad\square$

Exercise 5.3.13 shows that if $\langle q_1, \ldots, q_j \rangle$ contains an eigenvector of A, then the associated eigenvalue is exactly an eigenvalue of T_j. This suggests the following approximation result: If $\langle q_1, \ldots, q_j \rangle$ contains vectors that are close to an eigenvector of A, then the eigenvalue associated with that eigenvector will be approximated by an eigenvalue of T_j. We will not take the time to prove this. The aim of the next few exercises is to show that $\langle q_1, \ldots, q_j \rangle$ contains vectors that approximate the eigenvectors associated with the extreme eigenvalues of A.

Exercise 5.3.14 Use induction on j to prove that $\langle q_1, \ldots, q_j \rangle = \langle q_1, A q_1, \ldots, A^{j-1} q_1 \rangle$, $j = 1, 2, 3, \ldots$. (This is valid for the Arnoldi algorithm as well.) $\qquad\square$

The power sequence $(A^i q_1)$ that surfaces in Exercise 5.3.14, already begins to show that $\langle q_1, \ldots, q_j \rangle$ contains good approximations to the dominant eigenvector(s) of A. However, the best approximations turn out to be much better than those given by the power sequence.

Exercise 5.3.15 Given a polynomial $p(x) = a_0 + a_1 x + a_2 x^2 + \cdots + a_m x^m$ with real coefficients, define $p(A) \in \mathbb{R}^{n \times n}$ by $p(A) = a_0 I + a_1 A + a_2 A^2 + \cdots + a_m A^m$. Let P_{j-1} denote the set of all polynomials of degree less than or equal to $j - 1$. Prove that $\langle q_1, \ldots, q_j \rangle = \{ p(A) q_1 \mid p \in P_{j-1} \}$. $\qquad\square$

Exercise 5.3.16 Show that if λ is an eigenvalue of A with associated eigenvector v, then $p(\lambda)$ is an eigenvalue of $p(A)$ with associated eigenvector v. $\qquad\square$

These two exercises, like Exercise 5.3.14, are also valid in the nonsymmetric case. From now on we will restrict our attention to symmetric matrices. Thus let $A \in \mathbb{R}^{n \times n}$ be symmetric. Then A has a complete orthonormal set of eigenvectors $v_1, \ldots, v_n$ associated with the real eigenvalues $\lambda_1, \ldots, \lambda_n$, respectively. Let us assume that the eigenvalues are ordered so that $\lambda_1 \geq \lambda_2 \geq \cdots \geq \lambda_n$. This is different from the ordering by absolute value assumed for the analysis of the power method. The starting vector q_1 can be expressed uniquely in terms of $v_1, \ldots, v_n$:

$$q_1 = c_1 v_1 + c_2 v_2 + \cdots + c_n v_n \qquad (5.3.9)$$

Exercise 5.3.17 Show that $\langle q_1, \ldots, q_j \rangle = \left\{ \sum_{i=1}^n p(\lambda_i) c_i v_i \mid p \in P_{j-1} \right\}$, where $c_1, \ldots, c_n$ are the constants determined by (5.3.9). $\qquad\square$

Thus a typical element of $\langle q_1, \ldots, q_j \rangle$ has the form

$$p(\lambda_1) c_1 v_1 + p(\lambda_2) c_2 v_2 + \cdots + p(\lambda_n) c_n v_n$$

Suppose we wish to approximate the eigenvector v_i. We will be able to do so from $\langle q_1, \ldots, q_j \rangle$ if there exists a polynomial $p \in P_{j-1}$ such that the coefficient $|p(\lambda_i) c_i|$

is much larger than $|p(\lambda_k)c_k|$ for $k \neq i$. The effect of the choice of starting vector is reflected in the coefficients $c_1, \ldots, c_n$. A choice of q_1 for which $|c_i|$ is either zero or extremely small would be very unlucky from the point of view of trying to approximate v_i. Let us assume that q_1 was not an unlucky choice, in the sense that none of the coefficients $|c_i|$ is exceptionally small. Then $\langle q_1, \ldots, q_j \rangle$ will contain good approximations to v_i if there exist polynomials $p \in P_{j-1}$ for which $|p(\lambda_i)|$ is much larger than $\max\{|p(\lambda_k)| \mid k \neq i\}$. With the help of the Chebychev polynomials introduced in the next exercise, we will be able to see that this is the case for the eigenvalues that lie near the ends of the spectrum.

Exercise 5.3.18   For $x \in [-1, 1]$ define an auxiliary variable $\theta \in [0, \pi]$ by $x = \cos\theta$. The value of θ is uniquely determined by x; in fact $\theta = \arccos x$. Now for $m = 0, 1, 2, \ldots$, define a function p_m on $[-1, 1]$ by

$$p_m(x) = \cos m\theta$$

(a)   Show that $|p_m(x)| \leq 1$ for all $x \in [-1, 1]$, $p_m(1) = 1$, $p_m(-1) = (-1)^m$, $|p_m(x)| = 1$ at $m + 1$ distinct points in $[-1, 1]$, and $p_m(x) = 0$ at m distinct points in $(-1, 1)$.

(b)   Use the trigonometric identities

$$\cos(\alpha + \beta) = \cos\alpha \cos\beta - \sin\alpha \sin\beta$$
$$\cos(\alpha - \beta) = \cos\alpha \cos\beta + \sin\alpha \sin\beta$$

with $\alpha = m\theta$ and $\beta = \theta$ to conclude that

$$p_{m+1}(x) + p_{m-1}(x) = 2x p_m(x) \qquad m = 1, 2, 3, \ldots$$

This yields the recursion

$$p_{m+1}(x) = 2x p_m(x) - p_{m-1}(x) \qquad m = 1, 2, 3, \ldots \qquad (5.3.10)$$

(c)   Determine $p_0(x)$ and $p_1(x)$ directly from the definition, and use (5.3.10) to calculate $p_2(x), p_3(x)$, and $p_4(x)$. Note that each is a polynomial in x and can therefore be extended in a natural way beyond the interval $[-1, 1]$. Graph the polynomials $p_2(x), p_3(x)$, and $p_4(x)$, focusing on the interval $[-1, 1]$, but notice that they grow rapidly once x leaves that interval.

(d)   Use the recursion (5.3.10) to prove that for all m, $p_m(x)$ is a polynomial of degree m. {From part (a) we know that p_m has all m of its zeros in $(-1, 1)$. Therefore it must grow rapidly once it leaves $[-1, 1]$.}

(e)   Use (5.3.10) to prove that $p_m'(1) = m^2$ and $p_m'(-1) = (-1)^{m+1}m^2$. This expresses quantitatively the rate of growth of p_m as x leaves the interval $[-1, 1]$.

(f)   Show that $|p_m'(x)| \geq m^2$ for all $x \notin [-1, 1]$. $\square$

Let $[a, b]$ be any closed interval with $a < b$. We can produce Chebychev polynomial behavior on $[a, b]$ by performing a transformation that maps $[a, b]$ onto $[-1, 1]$.

Exercise 5.3.19 Suppose $\lambda_1 > \lambda_2 > \lambda_n$.

(a) Show that the transformation

$$t = 1 + 2\frac{x - \lambda_2}{\lambda_2 - \lambda_n}$$

maps $[\lambda_n, \lambda_2]$ onto $[-1, 1]$.

(b) Define a polynomial $q_1 \in P_{j-1}$ by

$$q_1(x) = p_{j-1}(t) = p_{j-1}\left(1 + 2\frac{x - \lambda_2}{\lambda_2 - \lambda_n}\right)$$

where p_{j-1} is the Chebychev polynomial. Show that $|q_1(\lambda_i)| \le 1$ for $i = 2, \ldots, n$ and

$$|q_1(\lambda_1)| \ge 1 + 2(j - 1)^2 \frac{\lambda_1 - \lambda_2}{\lambda_2 - \lambda_n}$$

Thus for j sufficiently large, $|q_1(\lambda_1)| \gg \max_{i \ne 1} |q_1(\lambda_i)|$.

(c) Construct a polynomial $q_n \in P_{j-1}$ such that $|q_n(\lambda_i)| \le 1, i = 1, \ldots, n - 1$, and

$$|q_n(\lambda_n)| \ge 1 + 2(j - 1)^2 \frac{\lambda_{n-1} - \lambda_n}{\lambda_1 - \lambda_{n-1}}$$

(d) Suppose $\lambda_1 > \lambda_2 > \lambda_3 > \lambda_n$. Define $q_2 \in P_{j-1}$ by

$$q_2(x) = (x - \lambda_1)p_{j-2}\left(1 + 2\frac{x - \lambda_3}{\lambda_3 - \lambda_n}\right)$$

Show that $|q_2(\lambda_1)| = 0, |q_2(\lambda_i)| \le |\lambda_i - \lambda_1|$ for $i = 3, \ldots, n$, and

$$|q_2(\lambda_2)| \ge |\lambda_2 - \lambda_1|\left[1 + 2(j - 2)^2 \frac{\lambda_2 - \lambda_3}{\lambda_3 - \lambda_n}\right]$$

Show that for sufficiently large j, $|q_2(\lambda_2)| \gg \max_{i \ne 2} |q_2(\lambda_i)|$.

(e) Suppose $\lambda_2 > \lambda_3 > \lambda_4 > \lambda_n$. Construct a polynomial $q_3 \in P_{j-1}$ such that for sufficiently large j, $|q_3(\lambda_3)| \gg \max_{i \ne 3} |q_3(\lambda_i)|$. $\square$

Exercise 5.3.20 Show that the Lanczos equations

$$q_{i+1}\beta_i = Aq_i - q_i\alpha_i - q_{i-1}\beta_{i-1} \qquad i = 1, \ldots, j \ (\beta_0 = 0)$$

are equivalent to the single matrix equation

$$AQ_j = Q_jT_j + \beta_j q_{j+1}e_j^T$$

where $Q_j = \begin{bmatrix} q_1 & q_2 & \cdots & q_j \end{bmatrix} \in \mathbb{R}^{n \times j}$ and

$$T_j = \begin{bmatrix} \alpha_1 & \beta_1 & 0 & \cdots & & 0 \\ \beta_1 & \alpha_2 & \beta_2 & & & \vdots \\ 0 & \beta_2 & \alpha_3 & \ddots & & \\ \vdots & & \ddots & \ddots & & \beta_{j-1} \\ 0 & \cdots & & & \beta_{j-1} & \alpha_j \end{bmatrix} \in \mathbb{R}^{j \times j} \qquad \square$$

Exercise 5.3.21 This exercise demonstrates a simple test for convergence of eigenvalues. We will use the notation and the result established in the previous exercise. Let $\mu_1, \ldots, \mu_j$ be the eigenvalues of T_j with associated orthonormal eigenvectors $s_1, \ldots, s_j$. These can be computed inexpensively, especially the eigenvalues, if j is not too large. Let

$$M_j = \begin{bmatrix} \mu_1 & & & 0 \\ & \mu_2 & & \\ & & \ddots & \\ 0 & & & \mu_j \end{bmatrix}$$

and $S_j = \begin{bmatrix} s_1 & s_2 & \cdots & s_j \end{bmatrix} \in \mathbb{R}^{j \times j}$. Then $T_j = S_jM_jS_j^T$. Let $x_i = Q_js_i, i = 1, \ldots, j$. Show that

$$\| Ax_i - \mu_i x_i \|_2 = |\beta_j s_{ji}| \qquad i = 1, \ldots, j$$

where s_{ji} denotes the jth component of s_i. It is not hard to show (cf., Theorem 5.5.1 and Corollary 5.5.6) that A has an eigenvalue λ such that $|\lambda - \mu_i| \le \| Ax_i - \mu_i x_i \|_2$. Thus $|\beta_j s_{ji}|$ is an upper bound on the distance from μ_i to the nearest eigenvalue of A. Notice that if we are interested only in eigenvalues, we do not have to calculate all of S_j, just its last row. $\square$

Exercise 5.3.22 This exercise is for those who are familiar with the three-term recursion for generating orthogonal polynomials. First note that the Lanczos procedure is governed by a three-term recursion. Then show that the three-term recursion for orthogonal polynomials can be viewed as the Lanczos algorithm applied to a certain linear operator acting on an infinite-dimensional space. [*Note*: The recursion (5.3.10) for Chebychev polynomials is a special case.] $\square$

5.4
OTHER ALGORITHMS OF *QR* TYPE

The *QR* algorithm has a number of close relatives, some of which are quite useful. We will begin by mentioning several very slight variants. In Section 5.2 the *QL* algorithm arose in connection with duality. The *QL* algorithm differs from the *QR* algorithm only in that the triangular factor in the decomposition is lower instead of upper triangular. Another variant is the *RQ* algorithm, which appeared in Section 4.8 in connection with the computation of eigenvectors. Of course there is also an *LQ* algorithm. These four algorithms are identical in principle, but in practice there do exist situations in which one is preferable to the others. For example, the *RQ* algorithm is better than the *QR* for computing a partial set of eigenvectors of an upper Hessenberg matrix. For symmetric, tridiagonal matrices *QL* is more frequently used than *QR*. The most widely used codes stem from [HAC], contribution II/4. The reason for preferring *QL* is that in certain applications *graded* matrices appear. These are matrices whose elements increase dramatically in magnitude from top to bottom. For such matrices the *QL* algorithm determines the smallest eigenvalues with greater accuracy. Of course the same end can be achieved by reversing the rows and columns of the matrix and applying *QR*.

The *LR* Algorithm

The *LR* algorithm is similar in appearance to the *QR* algorithm, but it also has some significant differences. It is based on the *LU* decomposition, with which you became quite familiar in Chapter 1. In the context of the eigenvalue problem the *LU* decomposition is traditionally called the *LR* decomposition. Every matrix whose leading principal submatrices are nonsingular can be decomposed uniquely into a product *LR*, where *L* is unit lower triangular and *R* is upper triangular. The *LR* algorithm looks like the *QR* algorithm, except that the *LR* decomposition is used in place of the *QR* decomposition. Thus a typical *LR* step is

$$A_{m-1} - \sigma_m I = L_m R_m \qquad R_m L_m + \sigma_m I = A_m \qquad (5.4.1)$$

Exercise 5.4.1 (a) Show that $A_m = L_m^{-1} A_{m-1} L_m$.
(b) Show that if A_{m-1} has lower Hessenberg form, then so does A_m.
In the rest of the problem assume $A_{m-1} - \sigma_m I$ is nonsingular.
(c) Show that $A_m = R_m A_{m-1} R_m^{-1}$.
(d) Show that if A_{m-1} has upper Hessenberg form, then so does A_m.
(e) Show that if A_{m-1} is tridiagonal, then so is A_m.
[If $A_{m-1} - \sigma_m I$ is singular, the *LR* decomposition is not unique, and parts (d) and (e) do not necessarily hold. However it is always possible to choose the *LR* decomposition (if one exists) in such a way that parts (d) and (e) hold.] □

The *LR* algorithm breaks down for want of an *LR* decomposition if $A_{m-1} - \sigma_m I$ happens to have a singular leading principal submatrix. More importantly the algorithm is potentially unstable: if one of the leading principal submatrices is nearly singular, roundoff errors can destroy the computation. At least two different approaches to combating this problem have been taken. One is simply to allow row interchanges and use a partial-pivoting strategy. This gives

$$A_{m-1} - \sigma_m I = K_m R_m \qquad R_m K_m + \sigma_m I = A_m \qquad (5.4.2)$$

where K_m is a matrix obtained by permuting the rows of a unit-lower-triangular matrix. All entries of K_m have absolute value less than or equal to 1. Every matrix has a decomposition of this form, so breakdowns cannot occur. This algorithm preserves upper Hessenberg but not lower Hessenberg form.

Implicit versions of this algorithm exist, including a double-step LR algorithm. See [AEP] and [HAC], contribution II/16. The implicit LR algorithms are very similar to the implicit QR algorithms discussed in Section 4.7: each LR step creates a bulge in the Hessenberg form and chases it down the subdiagonal. The principal difference is that the bulge is chased not by rotators or reflectors but by Gaussian elimination operations with partial pivoting. The LR algorithm with partial pivoting can usually find the eigenvalues of a nonsymmetric matrix with the same accuracy as the QR algorithm at little more than half the cost. However, we do not recommend the LR algorithm for calculating eigenvectors.

The second approach to avoiding instability is simply to monitor the size of the pivots and multipliers generated during each LR decomposition. If dangerously large multipliers are encountered, the step is restarted with a different shift. This use of *exceptional shifts* also takes care of those cases in which $A_{m-1} - \sigma_m I$ fails to have an LR decomposition. This approach is used in conjunction with a preliminary reduction to tridiagonal form. It is therefore clearly necessary to forego row interchanges in order to preserve the form. The reduction to tridiagonal form is accomplished by an extension of the reduction to upper Hessenberg form by elimination discussed in Section 4.5. Once the zeros below the subdiagonal have been produced, the zeros above the superdiagonal are created by elimination using column operations. For details see [AEP] or Dax and Kaniel (1981).

Unfortunately the latter elimination must be done without the benefit of column pivoting because column interchanges would destroy the pattern of zeros below the subdiagonal. Therefore the process is potentially unstable. However it has been reported that the process can be carried out successfully for the vast majority of matrices of moderate size if extended-precision arithmetic is used. For those matrices that cannot be reduced stably to tridiagonal form, Dax and Kaniel (1981) have suggested that the QR algorithm be used. Another possibility is to apply the LR algorithm to the Hessenberg form or to a matrix that is intermediate between Hessenberg and tridiagonal forms.

The great advantage of reducing the matrix to tridiagonal form is that the LR steps are then very cheap. The number of arithmetic operations per implicit, tridiagonal LR step is $O(n)$, which is negligible compared to the cost of the reduction to tridiagonal form. Thus the tridiagonal LR algorithm, when it works, is quite inexpensive.

The decompositions $A = LR$ and $A = KR$, upon which (5.4.1) and (5.4.2) are based, both have an upper triangular factor R. It follows that the subspace spanned by the leading k columns of A is the same as that spanned by the leading

k columns of L and K, respectively, for $k = 1, \ldots, n-1$. Consequently all variants of the LR algorithm can be identified as forms of nested subspace iteration, just as the QR algorithm was in Section 5.2. Unfortunately the orthogonal similarity transformations $A_m = Q_m^T A_{m-1} Q_m$ of the QR algorithm are replaced by nonorthogonal similarity transformations $A_m = L_m^{-1} A_{m-1} L_m$ and $A_m = K_m^{-1} A_{m-1} K_m$ in the LR algorithm. As a consequence, convergence of an underlying sequence of subspaces to an invariant subspace does not unequivocally imply convergence of (A_m) to block triangular form.

The Cholesky *LR* Algorithm

This is an interesting variant of the LR algorithm, applicable to real symmetric or complex Hermitian matrices. If $A_{m-1} - \sigma_m I$ is positive definite, then it has a Cholesky decomposition $G_m G_m^*$, so a *Cholesky LR step* is possible:

$$A_{m-1} - \sigma_m I = G_m G_m^* \qquad G_m^* G_m + \sigma_m I = A_m$$

The next exercise demonstrates an interesting relationship between the Cholesky LR algorithm and the QR algorithm: one QR step equals two Cholesky LR steps with the same shift.

Exercise 5.4.2 Let $A_0 = A - \sigma I$ be positive definite. Let A_1, A_2 be the matrices generated by two Cholesky steps:

$$
\begin{aligned}
A_0 &= G_1 G_1^* & G_1^* G_1 &= A_1 \\
A_1 &= G_2 G_2^* & G_2^* G_2 &= A_2
\end{aligned}
\tag{5.4.3}
$$

(a) Show that A_1 is positive definite. [Thus the second line of (4.12.3) is meaningful, and A_2 is positive definite as well.]

(b) Show that $A_2 = (G_1 G_2)^* A_0 ((G_1 G_2)^*)^{-1}$ and $A_0^2 = (G_1 G_2)(G_1 G_2)^*$.

(c) Let B be the matrix produced by one QR step, starting from A_0:

$$A_0 = QR \qquad RQ = B$$

Show that $B = R A_0 R^{-1}$ and $A_0^2 = A_0^* A_0 = R^* R$.

(d) Show that $R = (G_1 G_2)^*$ and $B = A_2$. □

Because the Cholesky decomposition is stable, the Cholesky LR algorithm is also stable. It does however have one serious drawback that will probably prevent it from ever being widely used. The shifts have to be chosen so that $A_{m-1} - \sigma_m I$ is positive definite. This means that each shift must be smaller than the smallest eigenvalue. To get rapid convergence, we must choose shifts that approximate this eigenvalue well. The added requirement that the shifts be kept to the left of the eigenvalue rules out many good shifting strategies, for example the Rayleigh quotient and Wilkinson shifts. Shifts based on Newton's method have been tried with only modest success. No shift that rivals the Wilkinson shift has been found.

Hamiltonian Matrices and the SR Algorithm

In the quadratic regulator problem of optimal control theory, matrices of a special form arise:

$$B = \begin{bmatrix} A & N \\ K & -A^T \end{bmatrix} \in \mathbb{R}^{2n \times 2n}$$

where $A, K, N \in \mathbb{R}^{n \times n}$, $K = K^T$, and $N = N^T$. Such matrices are called *Hamiltonian*. The solution of the quadratic regulator problem requires that all the eigenvalues and a certain invariant subspace of a Hamiltonian matrix be found. This can be done satisfactorily by the QR algorithm, but it has the drawback that it does not exploit the Hamiltonian structure of the matrix. This structure is considerable; indeed it is comparable in extent to that of a symmetric matrix. It is therefore reasonable to seek a method that preserves and exploits the Hamiltonian structure. The SR algorithm is such a method. We begin by introducing the important matrix

$$J = \begin{bmatrix} 0 & I \\ -I & 0 \end{bmatrix} \in \mathbb{R}^{2n \times 2n}$$

Exercise 5.4.3   Show that the following three statements are equivalent: (i) B is Hamiltonian. (ii) JB is symmetric. (iii) $B^T J = -JB$. $\square$

Exercise 5.4.4   Let B be Hamiltonian.

  (a) Show that if z is an eigenvector of B with associated eigenvalue λ, then Jz is an eigenvector of B^T with eigenvalue $-\lambda$.

  (b) Conclude that if λ is an eigenvalue of B, then so is $-\lambda$. (Caution! We are discussing only *real* Hamiltonian matrices here. For complex Hamiltonian matrices the situation is more complicated.)

  (c) Show that if λ is a complex eigenvalue of B, then $\bar{\lambda}$, $-\lambda$, and $-\bar{\lambda}$ are also eigenvalues of B. $\square$

A matrix $S \in \mathbb{R}^{2n \times 2n}$ is called *symplectic* if $S^T J S = J$.

Exercise 5.4.5   (a) Show that if S_1 and S_2 are symplectic, then $S_1 S_2$ is symplectic.
(b) Show that if S is symplectic, then S is nonsingular and S^{-1} is symplectic. Show that $S^{-1} = -J S^T J$.
(c) Show that if S is symplectic, then S^{-T} and S^T are symplectic.
(d) Show that if $B, S \in \mathbb{R}^{2n \times 2n}$, B is Hamiltonian, S is symplectic, and $C = S^{-1} B S$, then C is Hamiltonian. $\square$

Almost every matrix $B \in \mathbb{R}^{2n \times 2n}$ has an *SR decomposition*

$$B = SR$$

where S is symplectic,

$$R = \begin{bmatrix} R_{11} & R_{12} \\ R_{21} & R_{22} \end{bmatrix} \tag{5.4.4}$$

each of the blocks R_{ij} is upper triangular, and R_{21} is strictly upper triangular. For details see [NLA], page 247. Matrices of this form are upper-triangular matrices in disguise. Indeed, let $\hat{R}$ be the matrix obtained by shuffling the rows and columns of R; that is, row (column) i is mapped to row (column)

$$2i - 1 \qquad \text{if } i \leq n$$

$$2i - 2n \qquad \text{if } i > n$$

Then $\hat{R}$ is upper triangular. We will refer to the form (5.4.4) as *deshuffled triangular form*.

Exercise 5.4.6 Show that $\hat{R}$ is upper triangular. Prove that the converse holds as well. □

The *SR algorithm* is just like the *QR* algorithm, except that it uses the *SR* decomposition in place of the *QR* decomposition. Given a matrix $B = B_0$, the *SR* algorithm forms a sequence (B_m) by

$$B_{m-1} - \sigma_m I = S_m R_m \qquad R_m S_m + \sigma_m I = B_m$$

with some suitably chosen shifts (σ_m).

Exercise 5.4.7 (a) Show that $B_m = S_m^{-1} B_{m-1} S_m$.
(b) Show that if B is Hamiltonian, then B_m is Hamiltonian for all m.
(c) Show that if $B_{m-1} - \sigma_m I$ is nonsingular, then $B_m = R_m B_{m-1} R_m^{-1}$. □

Since the factors R_m are disguised upper-triangular matrices, the *SR* algorithm can be identified as a form of nested subspace iteration. Thus we expect the iterates (B_m) to converge to (deshuffled) block triangular form. However, as in the case of the *LR* algorithm, convergence to certain invariant subspaces does not unconditionally guarantee convergence of the matrices. A matrix

$$B = \begin{bmatrix} B_{11} & B_{12} \\ B_{21} & B_{22} \end{bmatrix} \in \mathbb{R}^{2n \times 2n}$$

is said to be in *deshuffled Hessenberg Form* if B_{11}, B_{21}, and B_{22} are upper triangular and B_{12} is upper Hessenberg.

Exercise 5.4.8   Suppose B is in deshuffled upper Hessenberg form. Show that if B is "shuffled" in the manner described above, the resulting matrix $\hat{B}$ is in upper Hessenberg form, and conversely.   □

Exercise 5.4.9   Show that if $B = B_0$ is in deshuffled upper Hessenberg form and $B_{m-1} - \sigma_m I$ is nonsingular for all m, then all the SR iterates B_m are in deshuffled upper Hessenberg form. (It turns out that the nonsingularity condition is not essential.)   □

Exercise 5.4.10   Show that if

$$B = \begin{bmatrix} A & N \\ K & -A^T \end{bmatrix} \in \mathbb{R}^{2n \times 2n}$$

is Hamiltonian and deshuffled Hessenberg, then A and K are diagonal matrices and N is tridiagonal and symmetric.   □

The form described in Exercise 5.4.10 is called *pseudo-tridiagonal*. If we can reduce the Hamiltonian matrix B to pseudo-tridiagonal form to begin with, then this form will be preserved by the SR iterations. It turns out that every Hamiltonian matrix can be reduced to pseudo-tridiagonal form by symplectic similarity transformations. See Bunse-Gerstner and Mehrmann (1986). The algorithm is potentially unstable; indeed the basic algorithm can break down completely, so the size of certain pivots and multipliers must be monitored. If a dangerously large multiplier is encountered, the algorithm performs an "exceptional" symplectic similarity transformation whose only function is to try to make the multipliers smaller. This procedure works fairly well in practice if the matrices are not too large.

Once the matrix is in pseudo-tridiagonal form, the SR steps are very inexpensive, requiring only $O(n)$ arithmetic operations per step. Like LR steps without pivoting, the SR steps are potentially unstable, and certain pivots and multipliers must be monitored. If an unacceptably large multiplier is encountered, the step is restarted with an exceptional shift. Practical implementations of the SR algorithm use implicit rather than explicit SR steps. Since the real eigenvalues occur in $\pm$ pairs and the complex eigenvalues occur in sets of four, double and quadruple SR steps are used. See Bunse-Gerstner and Mehrmann (1986) for details.

General Theory of Algorithms of QR Type

The QR, LR, and SR algorithms have many similarities in their theoretical aspects and in their implementation. Watkins and Elsner (1990, 1991) have developed a general theory that includes all these algorithms as special cases. This work also encompasses hybrid algorithms that combine elements of, for example, the QR and LR algorithms.

5.5
SENSITIVITY OF EIGENVALUES
AND EIGENVECTORS

Since the matrix A whose eigensystem we wish to calculate is never known exactly, it is important to study how the eigenvalues and eigenvectors are affected by perturbations of A. Thus in this section we will ask and partially answer the question of how close the eigenvalues and eigenvectors of $A + \delta A$ are to those of A if $\|\delta A\|/\|A\|$ is small. This would be an important question even if the uncertainty in A were our only concern, but of course there is also another reason for asking it. We noted in Chapter 3 that any algorithm that transforms a matrix by rotators or reflectors constructed as prescribed in Section 3.2 is stable. The implicit QR algorithms are of this type. This means that an implicit QR algorithm determines the exact eigenvalues of a matrix $A + \delta A$ where $\|\delta A\|/\|A\|$ is small. If we can show that the eigenvalues of $A + \delta A$ are close to those of A, we will know that our answers are accurate. Of course it will turn out that we cannot always guarantee accurate eigenvalues; the accuracy depends upon certain condition numbers.

A related question is that of residuals. Suppose we have calculated an approximate eigenvalue λ and associated eigenvector v, and we wish to know whether they are accurate. It is natural to calculate the residual $r = Av - \lambda v$ and check whether it is small. Suppose $\|r\|$ *is* small. Does this guarantee that λ and v are accurate? As the following theorem shows, this question also reduces to that of the sensitivity of A.

THEOREM 5.5.1 Let $A \in \mathbb{C}^{n \times n}$, v an approximate eigenvector of A with $\|v\|_2 = 1$, λ an associated approximate eigenvalue, and r the residual: $r = Av - \lambda v$. Then λ and v are an exact eigenpair of some perturbed matrix $A + \delta A$, where $\|\delta A\|_2 = \|r\|_2$.

Proof Let $\delta A = -rv^*$. Then $\|\delta A\|_2 = \|r\|_2 \|v\|_2 = \|r\|_2$, and $(A + \delta A)v = Av - rv^*v = Av - r = \lambda v$.   ☐

Exercise 5.5.1 Verify that $\|rv^*\|_2 = \|r\|_2 \|v\|_2$ for all $r, v \in \mathbb{C}^n$.   ☐

The aim here is to give a brief overview of some of the most important results. For a detailed treatment see [AEP, Chap. 2]. A less thorough but more modern treatment is given in [MC].

Sensitivity of Eigenvalues

First of all, the eigenvalues of a matrix depend continuously on the entries of the matrix. This is so because the coefficients of the characteristic polynomial are continuous functions of the matrix entries, and the zeros of the characteristic polynomial, that is the eigenvalues, depend continuously on the coefficients. For more detailed information see [AEP] and the references cited therein. This means that we can put as small a bound as we please on how far the eigenvalues can wander, simply by making the perturbation of the matrix sufficiently small. But this information is too vague. It would be more useful to have a number κ such that if we perturb the matrix by ϵ, then the eigenvalues are perturbed by at most $\kappa\epsilon$, at least for sufficiently small ϵ. Then κ would serve as a condition number for the eigenvalues. It turns out that we can get such a κ if the matrix is simple. This is the content of our next theorem, which is due to F.L. Bauer and C.T. Fike. For the most part we will restrict our attention to simple matrices, since these are the ones that arise in practice. They are also much easier to analyze than the defective ones.

Theorem 5.5.2   Let $A \in \mathbb{C}^{n \times n}$ be a simple matrix, and suppose $V^{-1}AV = D$, where V is nonsingular and D is diagonal. Let $\delta A \in \mathbb{C}^{n \times n}$ be some perturbation of A, and let μ be an eigenvalue of $A + \delta A$. Then A has an eigenvalue λ such that

$$|\mu - \lambda| \leq \kappa_p(V)\|\delta A\|_p \tag{5.5.3}$$

for $1 \leq p \leq \infty$. $\square$

Remarks

1. This shows that $\kappa_p(V)$ is a condition number for the eigenvalues of A. [$\kappa_p(V)$ was defined in Section 2.3.]

2. We could make this theorem into a statement about relative errors by dividing both sides of (5.5.3) by $\|A\|_p$, but that would be an unnecessary complication.

Proof   Let $\delta D = V^{-1}(\delta A)V$. Then

$$\|\delta D\|_p \leq \|V^{-1}\|_p \|\delta A\|_p \|V\|_p = \kappa_p(V)\|\delta A\|_p \tag{5.5.4}$$

Since $D + \delta D$ is similar to $A + \delta A$, μ is an eigenvalue of $D + \delta D$. Let x be an associated eigenvector. If μ happens to be an eigenvalue of A, we are done, so suppose it is not. Then $\mu I - D$ is nonsingular, and the equation $(D + \delta D)x = \mu x$ can be rewritten as $x = (\mu I - D)^{-1}(\delta D)x$. Thus $\|x\|_p \leq \|(\mu I - D)^{-1}\|_p \|\delta D\|_p \|x\|_p$. Canceling out the factor $\|x\|_p$ and rearranging, we find that

$$\|(\mu I - D)^{-1}\|_p^{-1} \leq \|\delta D\|_p \tag{5.5.5}$$

The matrix $(\mu I - D)^{-1}$ is diagonal with main-diagonal entries $(\mu - \lambda_i)^{-1}$, where $\lambda_1, \ldots, \lambda_n$ are the eigenvalues of A. It follows easily that $\|(\mu I - D)^{-1}\|_2 = |\mu - \lambda|^{-1}$, where λ is the eigenvalue of A that is closest to μ. Thus (5.5.5) can be rewritten as

$$|\mu - \lambda| \leq \|\delta D\|_p$$

Combining this inequality with (5.5.4), we get our result. $\square$

Exercise 5.5.2

(a)   Let Δ be a diagonal matrix with main-diagonal entries $\delta_1, \ldots, \delta_n$. Verify that for $1 \le p \le \infty$, $\| \Delta \|_p = \max_{1 \le i \le n} |\delta_i|$.

(b)   Verify that $\| (\mu I - D)^{-1} \|_p = |\mu - \lambda|^{-1}$, where λ is the eigenvalue of the diagonal matrix D that is closest to μ. $\square$

The columns of the transforming matrix V are eigenvectors of A, so the condition number $\kappa_p(V)$ is a measure of how far from being linearly dependent the eigenvectors are: the larger the condition number, the closer they are to being dependent. From this viewpoint it would seem reasonable to assign an overall condition number of infinity to the eigenvalues of a defective matrix, since such matrices do not even have n linearly independent eigenvectors. Support for this viewpoint is given in Exercise 5.5.10 at the end of the section.

If A is Hermitian or even normal, V can be taken to be unitary (cf., Theorems 4.4.7 and 4.4.10). Unitary V satisfy $\kappa_2(V) = 1$, so the following corollary holds.

COROLLARY 5.5.6   Let $A \in \mathbb{C}^{n \times n}$ be normal, let δA be any perturbation of A, and let μ be any eigenvalue of $A + \delta A$. Then A has an eigenvalue λ such that

$$|\mu - \lambda| \le \| \delta A \|_2$$

Corollary 5.5.6 can be summarized by saying that the eigenvalues of a normal matrix are perfectly conditioned. If a normal matrix is perturbed slightly, the resulting perturbation of the eigenvalues is no greater than the perturbation of the matrix elements.

A weakness of Theorem 5.5.2 is that it gives a single condition number for all the eigenvalues. In fact it can happen that some of the eigenvalues are well conditioned while others are ill conditioned. This is true for both simple and defective matrices. It is therefore worthwhile to develop individual condition numbers for the eigenvalues. Again we will restrict our attention to the simple case; in fact, we will assume distinct eigenvalues.

Our discussion of individual condition numbers will depend on the notion of *left eigenvectors*, which we now introduce. Let λ be an eigenvalue of A. Then there exists a nonzero $x \in \mathbb{C}^n$ such that

$$Ax = \lambda x \qquad (5.5.7)$$

The eigenvalues of A^* are the complex conjugates of those of A, so there is a nonzero $y \in \mathbb{C}^n$ such that $A^* y = \bar{\lambda} y$. (In general $y \ne \bar{x}$.) This equation can also be written in the form

$$y^* A = \lambda y^* \qquad (5.5.8)$$

Any nonzero vector y^* that satisfies (5.5.8) is called a *left eigenvector* of A associated with the eigenvalue λ. Actually we will not be strict with the nomenclature and will refer to y itself as a left eigenvector of A, as if y and y^* were the same thing. In this context we will refer to any nonzero x satisfying (5.5.7) as a *right eigenvector*. A left eigenvector of A is just (the conjugate transpose of) a right eigenvector of A^*.

THEOREM 5.5.9   Let $A \in \mathbb{C}^{n \times n}$ have distinct eigenvalues $\lambda_1, \lambda_2, \ldots, \lambda_n$ with associated linearly independent right eigenvectors $x_1, x_2, \ldots, x_n$ and left eigenvectors $y_1, y_2, \ldots, y_n$. Then

$$y_j^* x_i \quad \begin{cases} = 0 & \text{if } i \neq j \\ \neq 0 & \text{if } i = j \end{cases}$$

(Two sequences of vectors that satisfy these relationships are said to be *biorthogonal*.)

Proof   Suppose $i \neq j$. From the equation $A x_i = \lambda_i x_i$ it follows that $y_j^* A x_i = \lambda_i y_j^* x_i$. On the other hand, $y_j^* A = \lambda_j y_j^*$, so $y_j^* A x_i = \lambda_j y_j^* x_i$. Thus $\lambda_i y_j^* x_i = \lambda_j y_j^* x_i$. Since $\lambda_i \neq \lambda_j$, it must be true that $y_j^* x_i = 0$.

It remains to be shown that $y_i^* x_i \neq 0$. Let's assume that $y_i^* x_i = 0$ and get a contradiction. From our assumption and the first part of the proof we see that $y_i^* x_k = 0$ for $k = 1, \ldots, n$. The vectors $x_1, \ldots, x_n$ are linearly independent, so they form a basis for $\mathbb{C}^n$. Notice that $y_i^* x_k$ is just the complex inner product (x_k, y_i), so y_i is orthogonal to $x_1, \ldots, x_n$, hence to every vector in $\mathbb{C}^n$. In particular it is orthogonal to itself, which implies $y_i = 0$, a contradiction.   □

The next theorem establishes the promised condition numbers for individual eigenvalues. Let's take a moment to set the scene. Suppose $A \in \mathbb{C}^{n \times n}$ has n distinct eigenvalues, and let λ be one of them. Let δA be a small perturbation satisfying $\| \delta A \|_2 = \epsilon$. Since the eigenvalues of A are distinct, and they depend continuously on the entries of A, we can assert that if ϵ is sufficiently small, $A + \delta A$ will have exactly one eigenvalue $\lambda + \delta \lambda$ that is close to λ. In Theorem 5.5.10 we will assume that all these conditions hold.

THEOREM 5.5.10   Let $A \in \mathbb{C}^{n \times n}$ have n distinct eigenvalues. Let λ be an eigenvalue with associated right and left eigenvectors x and y, respectively, normalized so that $\| x \|_2 = \| y \|_2 = 1$. Let $s = y^* x$. (Then by Theorem 5.5.9, $s \neq 0$.) Define

$$\kappa = \frac{1}{|s|} = \frac{1}{|y^* x|}$$

Let δA be a small perturbation satisfying $\| \delta A \|_2 = \epsilon$, and let $\lambda + \delta \lambda$ be the eigenvalue of $A + \delta A$ that approximates λ. Then

$$|\delta \lambda| \leq \kappa \epsilon + O(\epsilon^2)$$

Thus κ is a condition number for the eigenvalue λ.

Remarks

1. λ is a *simple* eigenvalue; that is, its algebraic multiplicity is 1. Therefore x and y are chosen from one-dimensional eigenspaces. This and the fact that they are chosen to have norm 1 guarantees that they are uniquely determined up to complex scalars of modulus 1. Hence s is determined up to a scalar of modulus 1, and κ is uniquely determined; that is, it is well defined.

2. The theorem is actually valid for any simple eigenvalue, regardless of whether or not the matrix is simple.

Proof From Theorem 5.5.2 we know that $|\delta\lambda| \leq \kappa_p(V)\,\epsilon$. This says more than that λ is merely continuous in A: It is Lipschitz continuous. This condition can be expressed briefly by the statement $|\delta\lambda| = O(\epsilon)$. It turns out that the same is true of the eigenvector as well: $A + \delta A$ has an eigenvector $x + \delta x$ associated with the eigenvalue $\lambda + \delta\lambda$, such that $\delta x = O(\epsilon)$. This depends on the fact that λ is a simple eigenvalue. For a proof see [AEP, page 67]. Expanding the equation $(A + \delta A)(x + \delta x) = (\lambda + \delta\lambda)(x + \delta x)$ and using the fact that $Ax = \lambda x$, we find that

$$(\delta A)x + A(\delta x) + O(\epsilon^2) = (\delta\lambda)x + \lambda(\delta x) + O(\epsilon^2)$$

Left multiplying by y^* and using the equation $y^*A = \lambda y^*$, we obtain

$$y^*(\delta A)x + O(\epsilon^2) = (\delta\lambda)y^*x + O(\epsilon^2)$$

hence

$$\delta\lambda = \frac{y^*(\delta A)x}{y^*x} + O(\epsilon^2)$$

Taking absolute values and noting that $|y^*(\delta A)x| \leq \|y\|_2\|\delta A\|_2\|x\|_2 = \epsilon$, we are done. $\square$

Exercise 5.5.3 Show that the condition number κ always satisfies $\kappa \geq 1$. $\square$

Exercise 5.5.4 Let $A \in \mathbb{C}^{n\times n}$ be Hermitian ($A^* = A$), and let x be a right eigenvector associated with the eigenvalue λ.

(a) Show that x is also a left eigenvector of A.

(b) Show that the condition number of the eigenvalue λ is $\kappa = 1$. $\square$

This exercise gives a second confirmation that the eigenvalues of a Hermitian matrix are perfectly conditioned. The results also hold for normal matrices. See also the following exercise.

Exercise 5.5.5 Let $A \in \mathbb{C}^{n \times n}$ have distinct eigenvalues $\lambda_1, \ldots, \lambda_n$ with associated right and left eigenvectors $v_1, \ldots, v_n$ and $w_1, \ldots, w_n$, respectively. By Theorem 5.5.9 we know that $w_i^* v_i \neq 0$ for all i, so we can assume without loss of generality that the vectors have been scaled so that $w_i^* v_i = 1$, $i = 1, \ldots, n$. This scaling does not determine the eigenvectors uniquely, but once v_i is chosen, then w_i^* is uniquely determined and vice versa. Let $V \in \mathbb{C}^{n \times n}$ be the matrix whose ith column is v_i, and let $W \in \mathbb{C}^{n \times n}$ be the matrix whose ith row is w_i^*.

    (a)   Show that $W = V^{-1}$.

    (b)   Show that the condition number κ_i associated with the ith eigenvalue is given by $\kappa_i = \| v_i \|_2 \| w_i \|_2$.

    (c)   Show that $\kappa_i \leq \kappa_2(V)$ for all i. Thus the overall condition number from Theorem 5.4.2 (with $p = 2$) always overestimates the individual condition numbers.

    (d)   Show that if A is normal, $\kappa_i = 1$ for all i. $\square$

Example 5.5.11 The 10-by-10 matrix

$$A = \begin{bmatrix} 10 & 10 & & & & \\ & 9 & 10 & & & \\ & & 8 & 10 & & \\ & & & \ddots & \ddots & \\ & & & & 2 & 10 \\ & & & & & 1 \end{bmatrix}$$

is a scaled-down version of an example from [AEP, page 90]. The entries below the main diagonal and those above the superdiagonal are all zero. The eigenvalues are obviously $10, 9, 8, \ldots, 2, 1$. We calculated the right and left eigenvectors, and thereby obtained the condition numbers, which are shown in Table 5.1. The eigenvalues are listed in pairs because they possess a certain symmetry: λ_i and λ_{11-i} have the same condition number. Notice that the condition numbers are fairly large, but the extreme eigenvalues are not as ill conditioned as the ones in the middle of the spectrum. Let A_ϵ be the matrix that is the same as A, except that the $(10, 1)$ entry is ϵ instead of 0. This perturbation of norm ϵ should cause a perturbation in λ_i for which $\kappa_i \epsilon$ is a rough bound. Table 5.2 gives the eigenvalues of A_ϵ for $\epsilon = 10^{-6}$, as calculated by the QR algorithm. It also shows how much the eigenvalues deviate from those of A and gives the numbers $\kappa_i \epsilon$ for comparison. As you can see, the numbers $\kappa_i \epsilon$ give good order-of-magnitude estimates of the actual perturbations.

TABLE 5.1

Condition Numbers of Eigenvalues of A

Eigenvalues (λ_i)	10, 1	9, 2	8, 3	7, 4	6, 5
Condition number (κ_i)	4.5×10^3	3.6×10^4	1.3×10^5	2.9×10^5	4.3×10^5

TABLE 5.2
Comparison of Eigenvalues of A and A_ϵ for $\epsilon = 10^{-6}$

Eigenvalues of A	10, 1	9, 2	8, 3	7, 4	6, 5
Eigenvalues of A_ϵ	10.0027	8.9740	8.0909	6.6614	6.4192
	0.9973	2.0260	2.9091	4.3386	4.5808
Perturbations of eigenvalues	0.0027	0.0260	0.0909	0.3386	0.4192
Scaled condition numbers $(\kappa_i \epsilon)$	0.0045	0.0361	0.1326	0.2931	0.4281

Notice also that the extreme eigenvalues are still quite close to the original values, while those in the middle have wandered quite far. We can expect that if ϵ is made much larger, some of the eigenvalues will lose their identities completely. Indeed, for $\epsilon = 10^{-5}$ the eigenvalues of A_ϵ, as computed by the QR algorithm, are

$$10.026 \qquad 8.68 \pm .29i \qquad 6.64 \pm .98i$$

$$0.974 \qquad 2.32 \pm .29i \qquad 4.36 \pm .98i$$

Only the two most extreme eigenvalues are recognizable. The others have collided in pairs to form complex conjugate eigenvalues.

We conclude this example with two remarks. (1) We also calculated the overall condition number $\kappa_2(V)$ given by Theorem 5.5.2, and found that it is about 2.5×10^6. This is clearly a gross overestimate of all the individual condition numbers. (2) The ill conditioning of the eigenvalues of A can be explained in terms of A's *departure from normality*. From Corollary 5.5.6 we know that the eigenvalues of a normal matrix are perfectly conditioned, and it is reasonable to expect that a matrix that is in some sense nearly normal would have well-conditioned eigenvalues. In Exercise 4.4.21 we found that a matrix that is upper triangular is normal if and only if it is a diagonal matrix. Our matrix A appears to be far from normal since it is upper triangular yet far from diagonal; it has very substantial superdiagonal entries. For more on departure from normality see [MC, Secs. 7.1 to 7.2.].

Sensitivity of Eigenvectors

In the proof of Theorem 5.5.10 we used the fact that if A has distinct eigenvalues, then the eigenvectors are Lipschitz continuous functions of the entries of A. This means that if x is an eigenvector of A, then any slightly perturbed matrix $A + \delta A$ will have an eigenvector $x + \delta x$ such that $\| \delta x \|_2 \approx \kappa \epsilon$. Here $\epsilon = \| \delta A \|_2$, and κ is a positive constant independent of δA. This is actually not hard to prove using notions from classical matrix theory and the theory of analytic functions. We will not take the time to work out the proof because it would take us too far from the main subject matter of this book. See Chapter 2 of [AEP]. Here we will take the existence of κ for granted and study its value, for κ is a condition number for the eigenvector x. We will consider the analysis of Wilkinson [AEP], which is also given in [MC], in a slightly different form. As you will see, this analysis yields results that are not completely satisfactory. For a newer, more precise (and also more involved) approach, see Meyer and Stewart (1988).

We suppose once again that $A \in \mathbb{C}^{n \times n}$ has distinct eigenvalues $\lambda_1, \ldots, \lambda_n$ associated with the right and left eigenvectors $x_1, \ldots, x_n$ and $y_1, \ldots, y_n$, respectively, all of which have Euclidean norm 1. Consider a slight perturbation δA with $\|\delta A\|_2 = \epsilon$, where ϵ is small enough that the perturbed matrix $A + \delta A$ has a unique eigenvalue $\lambda_k + \delta \lambda_k$ near λ_k associated with an eigenvector $x_k + \delta x_k$ near x_k. Since $x_1, \ldots, x_n$ is a basis for $\mathbb{C}^n$, the vector δx_k can be expressed as a linear combination

$$\delta x_k = \sum_{i=1}^{n} c_i x_i$$

where $c_1, \ldots, c_n$ are small coefficients. Thus $x_k + \delta x_k = (1 + c_k) x_k + \sum_{i \neq k} c_i x_i$.

Actually $x_k + \delta x_k$ is determined only up to a scalar multiple. We will normalize it by setting $c_k = 0$. This is the same as dividing $x_k + \delta x_k$ by $1 + c_k$. Thus

$$x_k + \delta x_k = x_k + \sum_{i \neq k} c_i x_i$$

Substituting this form into the equation $(A + \delta A)(x_k + \delta x_k) = (\lambda_k + \delta \lambda_k)(x_k + \delta x_k)$, using the equations $A x_i = \lambda_i x_i$, and performing some other simple manipulations, we find that

$$\sum_{i \neq k} c_i (\lambda_k - \lambda_i) x_i = (\delta A) x_k - (\delta \lambda_k) x_k + O(\epsilon^2)$$

Multiplying on the left by y_j^*, where $j \neq k$, and recalling that $y_j^* x_i = 0$ if $i \neq j$, we obtain

$$c_j (\lambda_k - \lambda_j) s_j = y_j^* (\delta A) x_k + O(\epsilon^2)$$

where $s_j = y_j^* x_j$, as before. Let $\beta_{jk} = y_j^* (\delta A) x_k$. Then

$$c_j = \frac{\beta_{jk}}{(\lambda_k - \lambda_j) s_j} + O(\epsilon^2)$$

We conclude that

$$x_k + \delta x_k = x_k + \sum_{i \neq k} \frac{\beta_{ik}}{(\lambda_k - \lambda_i) s_i} x_i + O(\epsilon^2) \tag{5.5.12}$$

Noting that each β_{ik} satisfies $|\beta_{ik}| \leq \epsilon$, we see that the size of δx_k is affected by the sensitivities ($\kappa_i = 1/|s_i|$) of the eigenvalues other than λ_k and the distance of λ_k from the other eigenvalues. As a corollary of (5.5.12) we have

$$\|\delta x_k\|_2 \leq \left(\sum_{i \neq k} \frac{\kappa_i}{|\lambda_k - \lambda_i|} \right) \epsilon + O(\epsilon^2) \tag{5.5.13}$$

Thus if λ_k is well separated from all other eigenvalues and all the other eigenvalues are well conditioned, the eigenvector x_k will be well conditioned. In particular, if a matrix has well-separated eigenvalues, all which are well conditioned, then all the eigenvectors will be well conditioned as well.

If A is normal, (5.5.12) yields a more precise result. In this case the eigenvectors $x_1, \ldots, x_n$ are orthonormal and all s_i satisfy $|s_i| = 1$ (Exercise 5.5.5). Thus

$$\|\delta x_k\|_2 = \left(\sum_{i \neq k} \frac{|\beta_{ik}|^2}{|\lambda_k - \lambda_i|^2} \right)^{1/2} + O(\epsilon^2) \leq \left(\sum_{i \neq k} \frac{1}{|\lambda_k - \lambda_i|^2} \right)^{1/2} \epsilon + O(\epsilon^2)$$

This shows that if A is normal, then x_k is well conditioned if λ_k is well separated from the other eigenvalues.

Example 5.5.14 For any $\epsilon \geq 0$, the matrix

$$A = \begin{bmatrix} 1 + \epsilon & 0 & 0 \\ 0 & 1 - \epsilon & 0 \\ 0 & 0 & 2 \end{bmatrix}$$

is Hermitian and hence normal. Its eigenvalues are obviously $1 + \epsilon$, $1 - \epsilon$, and 2. Associated with the poorly separated eigenvalues $1 + \epsilon$ and $1 - \epsilon$ are the orthonormal eigenvectors e_1 and e_2, respectively. Now consider the slightly perturbed matrix

$$A + \delta A = \begin{bmatrix} 1 + \epsilon & \sqrt{3}\epsilon & 0 \\ \sqrt{3}\epsilon & 1 - \epsilon & 0 \\ 0 & 0 & 2 \end{bmatrix}$$

which also happens to be Hermitian. You can easily check that for every $\epsilon \geq 0$, $A + \delta A$ has eigenvalues $1 + 2\epsilon$, $1 - 2\epsilon$, and 2. Associated with the clustered eigenvalues are the orthonormal eigenvectors $\sqrt{3}e_1/2 + e_2/2$ and $\sqrt{3}e_2/2 - e_1/2$. As you can see, the eigenvalues haven't moved far, but the eigenvectors have changed dramatically.

Notice that in the case $\epsilon = 0$, A has a two-dimensional eigenspace $\langle e_1, e_2 \rangle = \langle \sqrt{3}e_1/2 + e_2/2, \sqrt{3}e_2/2 - e_1/2 \rangle$ associated with the double eigenvalue 1. If ϵ is nonzero, $\langle e_1, e_2 \rangle$ is an invariant subspace but not an eigenspace. But notice that if ϵ is small, every vector $v \in \langle e_1, e_2 \rangle$ looks very much like an eigenvector, in that $Av \approx v$. Thus $\langle e_1, e_2 \rangle$ is almost an eigenspace; from a numerical standpoint we can view it as an eigenspace. From this practical viewpoint the invariant subspace $\langle e_1, e_2 \rangle$ is of greater interest than the two ill-conditioned eigenspaces.

Exercise 5.5.6 (a) Calculate the eigenvalues and eigenvectors of the matrix $A + \delta A$ of Example 5.5.14.
(b) Let $v \in \langle e_1, e_2 \rangle$ with $\| v \|_2 = 1$, and let A be as in Example 5.5.14. Show that the residual $r = Av - 1v$ satisfies $\| r \|_2 = \epsilon$. Thus v is almost an eigenvector associated with the "almost eigenvalue" 1. $\square$

Returning to the nonnormal case, let's take another look at (5.5.12) and (5.5.13). As in the normal case, clustered eigenvalues signal ill-conditioned eigenvectors. But (5.5.13) also suggests that if some of the eigenvalues other than λ_k are ill conditioned, then x_k will be ill conditioned, even if λ_k is well separated from the other eigenvalues. This is sometimes the case, but unfortunately (5.5.13) is not a reliable indicator. If some of the κ_i are large, then we know from Exercise 5.5.5, part (c), that the transforming matrix V, whose columns are $x_1, \ldots, x_n$, is ill conditioned. In other words, $x_1, \ldots, x_n$ are nearly linearly dependent. This implies that the sum in (5.5.12) can be small even though its coefficients are large. Thus the estimate (5.5.13) can be a gross overestimate. We conclude that this theory provides no definitive test for ill conditioning of eigenvectors of nonnormal matrices.

Example 5.5.14 shows that when two or more eigenvalues lie in a cluster, the invariant subspace associated with the entire cluster of eigenvalues is of greater interest than the individual eigenspaces. This suggests that we should study the stability of invariant subspaces under perturbations of A. This is indeed an important topic, but it is beyond the scope of this book. The general conclusions are that if A is normal, then an invariant subspace is stable if and only if the eigenvalues associated with the space are well separated from the other eigenvalues of A. If A is not normal, the separation of the eigenvalues continues to be important, but other factors come into play. For details see Stewart (1971, 1973).

Stability of Inverse Iteration

Inverse iteration is perhaps the most important technique for computing an eigenvector associated with an eigenvalue that has already been computed by some other means. In Section 4.3 we considered an heuristic explanation of why we should expect this technique to work well in spite of the fact that the computation involves a severely ill-conditioned matrix. Because of the importance of the technique, we will take another look at it here. Suppose $A \in \mathbb{C}^{n \times n}$ is a simple matrix with linearly independent eigenvectors $x_1, \ldots, x_n$ and associated eigenvalues $\lambda_1, \ldots, \lambda_n$. To keep the discussion uncluttered, we will assume that $\| A \| \approx 1$. Suppose further that we have calculated a number ρ that is a very good approximation to one of the eigenvalues, say λ_1, which will be assumed to be distinct from the other eigenvalues. Let γ be the distance from λ_1 to the nearest other eigenvalue. Then $|\lambda_1 - \rho| = \epsilon \gamma$ for some small ϵ.

To calculate a multiple of x_1 by inverse iteration, we choose an initial guess q_0 and solve the system $(A - \rho I)q_1 = q_0$. Then q_1 is very often a satisfactory approximation of the eigenvector. The reason for this is that if

$$q_0 = c_1 x_1 + c_2 x_2 + \cdots + c_n x_n$$

then

$$q_1 = (A - \rho I)^{-1} q_0 = \frac{c_1}{\lambda_1 - \rho} x_1 + \frac{c_2}{\lambda_2 - \rho} x_2 + \cdots + \frac{c_n}{\lambda_n - \rho} x_n$$

Assuming q_0 was not an unlucky guess, the coefficient c_1 will have about the same magnitude as the other coefficients of q_0. Then, since $|\lambda_1 - \rho|$ is small, the coefficient of x_1 in the expansion of q_1 must be much larger than the other coefficients. Thus q_1 points almost in the direction of x_1; that is, it is a good approximation to an eigenvector.

The possible defect to this argument is that we cannot calculate q_1 accurately because $A - \rho I$ is an ill-conditioned matrix (Exercise 4.3.13). How can we be sure that the computed q_1 is a good approximate eigenvector? Suppose we solve the system $(A - \rho I)q_1 = q_0$ by Gaussian elimination with partial pivoting. Then it is almost certain that the computed q_1 will be the exact solution of a perturbed system

$$(A - \rho I + \delta A)q_1 = q_0 \qquad (5.5.15)$$

where $\| \delta A \|$ is a small multiple of the unit roundoff error (cf., Theorem 2.5.5 and the accompanying discussion). This is so, regardless of how ill conditioned $A - \rho I$ is. In practice one would normalize q_1 at this point, but we will ignore that step. Rewriting (5.5.15), we see that the computed q_1 has the residual

$$r = (A - \rho I)q_1 = q_0 - (\delta A)q_1$$

This residual is typically small relative to q_1. First of all, the term $(\delta A)q_1$ is obviously small because δA is small. Furthermore q_0 is usually small in comparison with q_1 because of the large factor $1/(\lambda_1 - \rho)$ that multiplies the first coefficient in the eigenvector expansion of q_1. By Theorem 5.5.1 we know that the small residual implies that q_1 is an exact eigenvector of a slightly perturbed matrix $A + \delta A'$. Therefore, if x_1 is a well-conditioned eigenvector, q_1 will be a good approximation to (a multiple of) x_1. If the residual is not small enough, we can always perform another iteration or two.

Theorem 5.5.1 shows that it makes sense to stop iterating as soon as the residual is sufficiently small. Unfortunately this statement must be tempered with a word of warning: If λ_1 is an ill-conditioned eigenvalue, it can happen that the residuals never become satisfactorily small. This phenomenon is discussed in [HAC], contribution II/18. Exercise 5.5.8 may also provide some insight.

Exercise 5.5.7 At what points in the preceding discussion was the assumption $\| A \| \approx 1$ used? □

Exercise 5.5.8 Let $A \in \mathbb{C}^{n \times n}$ be a simple matrix with right and left eigenvectors $x_1, \ldots, x_n$ and $y_1, \ldots, y_n$, normalized so that each vector has Euclidean norm 1, with associated eigenvalues $\lambda_1, \ldots, \lambda_n$. Suppose λ_1 is ill conditioned. Thus $|y_1^* x_1|$ is small. Show that $\langle y_1 \rangle = \langle x_2, \ldots, x_n \rangle^\perp$. Let $z_2, \ldots, z_n$ be an orthonormal basis of $\langle x_2, \ldots, x_n \rangle$. Then $y_1, z_2, z_3, \ldots, z_n$ is an orthonormal basis for $\mathbb{C}^n$. The eigenvector x_1 can be expressed uniquely as a linear combination $x_1 = a_1 y_1 + a_2 z_2 + a_3 z_3 + \cdots + a_n z_n$. Show that $|a_1| = |y_1^* x_1| = 1/\kappa_1$. Let $z = a_2 z_2 + \cdots + a_n z_n$. Show that $\| x_1 - z \|_2 = 1/\kappa_1$. Thus z is very close to x_1, in spite of the fact that $z \in \langle x_2, \ldots, x_n \rangle$. Show that a step of inverse iteration $w = (A - \rho I)^{-1}z$, where $\rho \approx \lambda_1$, yields a w whose residual is not particularly small. □

Additional Exercises

Exercise 5.5.9 The Gerschgorin disk theorem is a powerful tool for perturbation theory that is used extensively in [AEP], though we have not used it here. Read the Gerschgorin disk theorem, Theorem 6.2.5, and apply it to the matrix $D + \delta D$ that appears in the proof of Theorem 5.5.2. Deduce a result similar to Theorem 5.5.2. For more delicate applications of this useful theorem see [AEP]. □

Exercise 5.5.10 For small $\epsilon \geq 0$ and $n \geq 2$ consider the n-by-n matrix

$$
J(\epsilon) = \begin{bmatrix}
0 & 1 & & & & \\
& 0 & 1 & & & \\
& & 0 & 1 & & \\
& & & \ddots & & \ddots \\
& & & & 0 & 1 \\
\epsilon & & & & & 0
\end{bmatrix}
$$

The entries that have not been indicated are zeros. Notice that $J(0)$ is a Jordan block, a severely defective matrix. It has only the eigenvalue 0, repeated n times, with a one-dimensional eigenspace (cf., Exercise 4.2.7).

(a) Show that the characteristic polynomial of $J(\epsilon)$ is $\lambda^n - \epsilon$. Show that every eigenvalue of $J(\epsilon)$ satisfies $|\lambda| = \epsilon^{1/n}$. Thus they all lie on the circle of radius $\epsilon^{1/n}$ centered on 0. (In fact it should be clear to you that the eigenvalues are the nth roots of ϵ: $\lambda_k = \epsilon^{1/n} e^{2\pi i k/n}$, $k = 1, \ldots, n$.)

(b) Sketch the graph of the function $f_n(\epsilon) = \epsilon^{1/n}$ for small $\epsilon \geq 0$ for a few values of n, for example $n = 2, 3$, and 4. Compare this with the graph of $g_\kappa(\epsilon) = \kappa\epsilon$ for a few values of κ. Show that $f_n'(\epsilon) \to \infty$ as $\epsilon \to 0$.

(c) Let $A = J(0)$, whose only eigenvalue is 0, and let $\delta A_\epsilon = J(\epsilon) - J(0)$. Let μ_ϵ be any eigenvalue of $A + \delta A_\epsilon$. Show that there is no real number κ such that $|\mu_\epsilon - 0| \leq \kappa \| \delta A \|_p$ for all $\epsilon > 0$. Thus there is no finite condition number for the eigenvalues of A. (*Remark*: The eigenvalues of $J(\epsilon)$ are continuous in ϵ but not Lipschitz continuous.)

(d) Consider the special case $n = 17$ and $\epsilon = 10^{-17}$. Observe that $\| \delta A \|_p = 10^{-17}$ but the eigenvalues of $A + \delta A$ all satisfy $|\lambda| = 1/10$. Thus the tiny perturbation 10^{-17} causes the relatively large perturbation $1/10$ in the eigenvalues. □

Exercise 5.5.11 Let A be the defective matrix

$$
A = \begin{bmatrix} 0 & 1 \\ 0 & 0 \end{bmatrix}
$$

Find left and right eigenvectors of A associated with the eigenvalue 0, and show that they satisfy $y^*x = 0$. Does this contradict Theorem 5.5.9? □

5.6
THE GENERALIZED EIGENVALUE PROBLEM

Numerous applications give rise to a more general form of eigenvalue problem

$$Ax = \lambda Bx \tag{5.6.1}$$

where A and $B \in \mathbb{C}^{n \times n}$. This problem has a rich theory, and numerous algorithms for solving it have been devised. This section provides only a brief overview.

A nonzero $x \in \mathbb{C}^n$ that satisfies (5.6.1) for some value of λ is called an *eigenvector* of the ordered pair (A, B), and λ is called the associated *eigenvalue*. Clearly (5.6.1) holds if and only if $(\lambda B - A)x = 0$, so λ is an eigenvalue of (A, B) if and only if the *characteristic equation*

$$\det(\lambda B - A) = 0$$

is satisfied. The expression $\lambda B - A$ is sometimes called a *matrix pencil*.

Most generalized eigenvalue problems can be reduced to standard eigenvalue problems. For example, if B is nonsingular, the eigenvalues of the pair (A, B) are just the eigenvalues of $B^{-1}A$.

Exercise 5.6.1 Let B be nonsingular.

  (a) Show that x is an eigenvector of (A, B) with associated eigenvalue λ if and only if x is an eigenvector of $B^{-1}A$ with eigenvalue λ.

  (b) Show that x is an eigenvector of (A, B) with associated eigenvalue λ if and only if Bx is an eigenvector of AB^{-1} with eigenvalue λ.

  (c) Show that $B^{-1}A$ and AB^{-1} are similar matrices.

  (d) Show that the characteristic equation of (A, B) is essentially the same as that of $B^{-1}A$ and AB^{-1}. $\square$

From Exercise 5.6.1 we see that if B is nonsingular, the pair (A, B) must have n eigenvalues, counting multiplicity. If B is singular, this need not be the case. Consider the following example.

Example 5.6.2 Let

$$A = \begin{bmatrix} 1 & 0 \\ 0 & -1 \end{bmatrix} \quad \text{and} \quad B = \begin{bmatrix} 1 & 0 \\ 0 & 0 \end{bmatrix}$$

and notice that B is singular. You can easily verify that the characteristic equation of (A, B) is $\lambda - 1 = 0$, whose degree is less than the order of the matrices. Evidently the pair (A, B) has only one eigenvalue, namely $\lambda = 1$. As we will see in a moment, it is reasonable to attribute to (A, B) a second eigenvalue $\lambda = \infty$.

Exercise 5.6.2 Let $A, B \in \mathbb{C}^{n \times n}$. (a) Prove that $\det(\lambda B - A)$ is a polynomial whose degree does not exceed n. (b) Prove that the degree of $\det(\lambda B - A)$ equals n if and only if B is nonsingular. $\square$

Suppose the pair (A, B) has a nonzero eigenvalue λ. Then, making the substitution $\mu = 1/\lambda$ in (5.6.1) and multiplying through by μ, we obtain

$$Bx = \mu A x$$

which is the equation of the generalized eigenvalue problem for the ordered pair (B, A). We conclude that the nonzero eigenvalues of (B, A) are the reciprocals of the nonzero eigenvalues of (A, B). Now suppose B is singular. Then zero is an eigenvalue of (B, A). In light of the reciprocal relationship we have just noted, it is reasonable to regard ∞ as an eigenvalue of (A, B). We will do just that. With this convention, most generalized eigenvalue problems have n eigenvalues. There is, however, a class of generalized eigenvalue problems that is not so well behaved. Let's look at an example.

Example 5.6.3 Let

$$A = \begin{bmatrix} 1 & 0 \\ 0 & 0 \end{bmatrix} \quad \text{and} \quad B = \begin{bmatrix} 1 & 0 \\ 0 & 0 \end{bmatrix}$$

Then $\det(\lambda B - A) = 0$ for all λ, so every complex number is an eigenvalue.

Exercise 5.6.3

(a) Let (A, B) be as in Example 5.6.3. Find the eigenvector associated with each eigenvalue. Find the null-spaces $\mathcal{N}(A)$ and $\mathcal{N}(B)$.

(b) Let A and B be any two matrices for which $\mathcal{N}(A) \cap \mathcal{N}(B) \neq \{0\}$. Show that every complex number is an eigenvalue of the pair (A, B), and every nonzero $x \in \mathcal{N}(A) \cap \mathcal{N}(B)$ is an eigenvector of (A, B) associated with every λ. $\square$

Any pair (A, B) for which the characteristic polynomial $\det(\lambda B - A)$ is identically zero is said to be *singular*. The term *singular pencil* is frequently used. If (A, B) is not singular, it is said to be *regular*. As we have just seen, (A, B) is singular whenever $\mathcal{N}(A)$ and $\mathcal{N}(B)$ have a nontrivial intersection. In most applications at least one of A and B is nonsingular, in which case (A, B) is regular. In such applications it is possible to reduce the generalized eigenvalue problem to the standard eigenvalue problem for one of the matrices AB^{-1}, $B^{-1}A$, BA^{-1}, or $A^{-1}B$.

Although this is a possibility that should not be ruled out, there are a number of reasons why it might not be the right course of action. Suppose we compute AB^{-1}. If B is ill conditioned (in the sense of Chapter 2), the eigenvalues of the computed AB^{-1} can be very far from the eigenvalues of (A, B), even though some or all of the eigenvalues of (A, B) are well conditioned. [For information about the sensitivity of the generalized eigenvalue problem see Stewart (1972).] A second reason not to compute AB^{-1} is that A and B might be symmetric. In many applications it is desirable to preserve the symmetry, and AB^{-1} will typically not be symmetric. Finally, A and B might be sparse, in which case we would certainly like to exploit the sparseness. But AB^{-1} will not be sparse, for the inverse of a sparse matrix is typically not sparse. For these reasons it is useful to develop algorithms that solve the generalized eigenvalue problem directly.

Equivalence Transformations

We know that two matrices have the same eigenvalues if they are similar, a fact that is exploited in numerous algorithms. If we wish to develop algorithms in the same spirit for the generalized problem, we need to know what sorts of transformations we can perform on pairs of matrices without altering the eigenvalues. It turns out that an even larger class than similarity transformations works. Two pairs of matrices (A, B) and $(\tilde{A}, \tilde{B})$ are said to be *equivalent* if there exist nonsingular matrices U and V such that $\tilde{A} = UAV$ and $\tilde{B} = UBV$. In the next exercise you will show that equivalent pairs have the same eigenvalues.

Exercise 5.6.4   Suppose (A, B) and $(\tilde{A}, \tilde{B})$ are equivalent pairs.

   (a)   Show that λ is an eigenvalue of (A, B) with associated eigenvector x if and only if λ is an eigenvalue of $(\tilde{A}, \tilde{B})$ with associated eigenvector $V^{-1}x$.

   (b)   Show that (A, B) and $(\tilde{A}, \tilde{B})$ have essentially the same characteristic equation. $\square$

Exercise 5.6.5   Investigate the equivalence transformations given by each of the following pairs of nonsingular matrices, assuming that the indicated inverses exist: (a) $U = B^{-1}$, $V = I$; (b) $U = I$, $V = B^{-1}$; (c) $U = A^{-1}$, $V = I$; (d) $U = I$, $V = A^{-1}$.

Computing Generalized Eigensystems: The Symmetric Case

If $A, B \in \mathbb{R}^{n \times n}$ are symmetric matrices and B is positive definite, the pair (A, B) is called a *symmetric* pair. Given a symmetric pair (A, B), B has a Cholesky decomposition $B = GG^T$, where G is lower triangular and has positive entries on the main diagonal. The equivalence transformation given by $U = G^{-1}$ and $V = G^{-T}$ transforms the generalized eigenvalue problem $Ax = \lambda Bx$ to the standard eigenvalue problem $G^{-1}AG^{-T}y = \lambda y$, where $y = G^T x$. Notice that the coefficient matrix $\tilde{A} = G^{-1}AG^{-T}$ is symmetric. This proves that a symmetric pair has real eigenvalues, and it has n linearly independent eigenvectors. The latter are not orthogonal in the conventional sense, but they are orthogonal with respect to an appropriately chosen inner product.

Exercise 5.6.6   The symmetric matrix $\tilde{A} = G^{-1}AG^{-T}$ has orthonormal eigenvectors $v_1, \ldots, v_n$, so $G^{-T}v_1, \ldots, G^{-T}v_n$ are eigenvectors of (A, B). Prove that these are orthonormal with respect to the *energy* inner product $(x, y)_B = y^T Bx$. $\square$

It is interesting that the requirement that B be positive definite is essential; it is not enough simply to specify that A and B be symmetric. It can be shown (cf., page 304 of [SEP]) that every $C \in \mathbb{R}^{n \times n}$ can be expressed as a product $C = AB^{-1}$, where A and B are symmetric. This means that the eigenvalue problem for any C can be reformulated as a generalized eigenvalue problem $Ax = \lambda Bx$, where A and B are symmetric. Thus there exist numerous examples of pairs (A, B) for which A and B are symmetric but the eigenvalues are complex.

One way to solve the eigenvalue problem for the symmetric pair (A, B) is to perform the transformation to $G^{-1}AG^{-T}$ and use one of the many available techniques for solving the symmetric eigenvalue problem. This has two of the same drawbacks as the transformation to AB^{-1}: (1) If B is ill conditioned, the eigenvalues will not be determined accurately and (2) $G^{-1}AG^{-T}$ does not inherit any sparseness that A and B might possess. The second drawback is not insurmountable. If B is sparse, its Cholesky factor G will often also be sparse. For example, if B is banded, G is also banded. Because of this it is possible to apply the Lanczos method (Section 5.3) or subspace iteration (Section 5.2) to $\tilde{A} = G^{-1}AG^{-T}$ at a reasonable cost. The only way in which these two methods use $\tilde{A}$ is to compute vectors $\tilde{A}x$ for a sequence of choices of x. For $\tilde{A}$ of the given form, we can compute $\tilde{A}x$ by first solving $G^{T}y = x$ for y, computing $z = Ay$, and then solving $Gw = z$ for w. Clearly $w = \tilde{A}x$. If G is sparse, the two systems $G^{T}y = x$ and $Gw = z$ can be solved cheaply. Even if G is somewhat ill conditioned, procedures of this type can generally compute the largest eigenvalues accurately.

In many applications both A and B are positive definite and only the smallest eigenvalues are sought. The Lanczos method produces good approximations to the largest and smallest eigenvalues, but the approximations to the small eigenvalues will be inaccurate if B is not well conditioned. The solution to this problem is to interchange the roles of A and B and solve the problem $Bx = \mu Ax$. The reciprocals of the smallest eigenvalues of (A, B) are the largest eigenvalues of (B, A), and these can often be computed accurately at reasonable cost.

There are numerous other possible algorithms for the symmetric generalized eigenvalue problem. A number of methods for the symmetric standard eigenvalue problem can be adapted to the generalized problem. For example, both the Jacobi method and the bisection method, described in Chapter 6, have been adapted to the generalized problem. See [SEP] for details and references.

Exercise 5.6.7   If B is symmetric and positive definite, it has a spectral decomposition $B = VDV^{T}$ where V is orthogonal and D is diagonal and positive definite. Show how this decomposition can be used to reduce the symmetric generalized eigenvalue problem to a standard symmetric problem.   □

Exercise 5.6.8   Two pairs (A, B) and $(\tilde{A}, \tilde{B})$ are said to be *congruent* if there exists a nonsingular $X \in \mathbb{R}^{n \times n}$ such that $\tilde{A} = XAX^{T}$ and $\tilde{B} = XBX^{T}$. Clearly any two congruent pairs are equivalent, but not conversely. Show that if (A, B) is a symmetric pair (and in particular B is positive definite), then $(\tilde{A}, \tilde{B})$ is also a symmetric pair. Thus we can retain the symmetric property by using only congruence transformations. The transformation to a standard eigenvalue problem using the Cholesky factor and the transformation suggested in Exercise 5.6.7 are both examples of congruence transformations.   □

Computing Generalized Eigensystems: The General Problem

We now consider the problem of finding the eigenvalues of a pair (A, B) that is not symmetric. We will examine an analog of the reduction to upper Hessenberg form and an extension of the QR algorithm. Unfortunately these algorithms do not preserve symmetry, so they are unsuitable for symmetric pairs.

The QR algorithm for the standard eigenvalue problem produces a sequence that converges to an upper-triangular form whose existence is guaranteed by Schur's theorem (Theorem 4.4.6). The analogous result for the generalized eigenvalue problem is the following.

Theorem 5.6.4 *(Generalized Schur Theorem)*   Let $A, B \in \mathbb{C}^{n \times n}$. Then there exist unitary $Q, Z \in \mathbb{C}^{n \times n}$ and upper-triangular $T, S \in \mathbb{C}^{n \times n}$ such that $Q^*AZ = T$ and $Q^*BZ = S$.

For a proof see, for example, [MC]. The pair (T, S) is equivalent to (A, B), so it has the same eigenvalues. But the eigenvalues of (T, S) are evident because

$$\det(\lambda S - T) = \prod_{i=1}^{n} (\lambda s_{ii} - t_{ii})$$

If there is an i for which both s_{ii} and t_{ii} are zero, the characteristic polynomial is identically zero; that is, the pair is singular. Otherwise the eigenvalues are t_{ii}/s_{ii}, $i = 1, \dots, n$. Some of these quotients may be ∞. For real (A, B) there is a real decomposition analogous to the real Schur theorem (Theorem 4.4.12). In this variant, Q and Z are real, orthogonal matrices, S is upper triangular, and T is block triangular with a 2-by-2 block for each complex conjugate pair of eigenvalues.

Reduction to Hessenberg-Triangular Form

Theorem 5.6.4 and its real analog give us simple forms to shoot for. Perhaps we can devise an algorithm that produces a sequence of equivalent pencils that converges to one of those forms. A big step in that direction is to transform the pair (A, B) to the Hessenberg-triangular form described in the next two theorems.

THEOREM 5.6.5   Let $A, B \in \mathbb{C}^{n \times n}$. Then there exist unitary $Q, Z \in \mathbb{C}^{n \times n}$, upper Hessenberg $H \in \mathbb{C}^{n \times n}$, and upper triangular $U \in \mathbb{C}^{n \times n}$ such that

$$Q^*AZ = H \qquad \text{and} \qquad Q^*BZ = U$$

There is a stable, direct algorithm to calculate Q, Z, H, and U in $O(n^3)$ flops. If A and B are real, Q, Z, H, and U can be taken to be real.

Proof   The proof will consist of an outline of an algorithm for transforming (A, B) to (H, U). This transformation will be effected by a sequence of reflectors and rotators applied on the left and right of A and B. Each transformation applied on the left (right) makes a contribution to Q^* (resp. Z). Every transformation applied to A must also be applied to B and vice versa. We know from Chapter 3 that B can be reduced to upper-triangular form by a sequence of reflectors applied on the left. Applying this sequence of reflectors to both B and A, we obtain a new (A, B), for which B is upper triangular. The rest of the algorithm consists of reducing A to upper Hessenberg form without destroying the upper-triangular form of B.

The first major step is to introduce zeros in the first column of A. This is done one element at a time, from bottom to top. First the $(n, 1)$ entry is set to zero by a rotator (or reflector) acting in the $(n, n - 1)$ plane, applied to A on the left. This alters rows $n - 1$ and n of A. The same transformation must also be applied to B. You can easily check that this operation will introduce one nonzero entry in the lower triangle of B, in the $(n, n - 1)$ position. This pollutant can be removed by the applying the appropriate rotation to columns $n - 1$ and n of B. That is, we apply a rotation in the $(n, n - 1)$ plane to B on the right. Applying the same rotation to A, we do not disturb the zero that we just introduced in column 1. We next apply a rotation to rows $n - 2$ and $n - 1$ of A in such a way that a zero is produced in position $(n - 1, 1)$. Applying the same rotation to B, we introduce a nonzero in position $(n - 1, n - 2)$. This can be removed by applying a rotator to columns $n - 1$ and $n - 2$. This operation does not disturb the zeros in the first column of A. Continuing in this manner, we can produce zeros in positions $(n - 2, 1), (n - 3, 1), \ldots, (3, 1)$.

This scheme cannot be used to produce a zero in position $(2, 1)$. In order to do that, we would have to apply a rotator to rows 1 and 2 of A. The same rotator applied to B would produce a nonzero in position $(2, 1)$. We could eliminate that entry by applying a rotator to columns 1 and 2 of B. Applying the same rotator to columns 1 and 2 of A, we would destroy all of the zeros that we had so painstakingly created in column 1. Thus we must leave the $(2, 1)$ entry of A as it is and move on to column 2.

The second major step is to introduce zeros in column 2 in positions $(n, 2)$, $(n - 1, 2), \ldots, (4, 2)$ by the same scheme as we used to create zeros in column 1. You can easily check that none of the rotators that are applied on the right operate on either column 1 or 2, so the zeros that have been created so far are maintained. We then proceed to column 3, and so on. After $n - 2$ major steps, we are done.

∎

Exercise 5.6.9 Count the flops required to execute the algorithm outlined in the proof of Theorem 5.6.5, (a) assuming Q and Z are not to be assembled, (b) assuming Q and Z are to be assembled. ∎

Exercise 5.6.10 Using the proof of Theorem 5.6.5 as a guide, derive an algorithm that uses Gaussian elimination with partial pivoting, instead of reflectors and rotators, to reduce a pair (A, B) to Hessenberg-triangular form. Show that this algorithm requires less than half as many flops as the version that uses reflectors and rotators. This algorithm is quite stable in practice, but the Q^{-1} and Z that it produces are not unitary. ∎

If B happens to be singular, this will be revealed, at least in principle, by the emergence of zeros on the main diagonal of the upper-triangular matrix U. It is convenient to gather these zeros at either the top or the bottom of the matrix. The next theorem, a refined version of Theorem 5.6.5, shows how this can be done.

THEOREM 5.6.6 Let $A, B \in \mathbb{C}^{n \times n}$, and let k be the dimension of the null-space of B. Then there exist unitary $Q, Z \in \mathbb{C}^{n \times n}$ such that $Q^*AZ = H$ and $Q^*BZ = U$, where H and U are of the form

$$H = \begin{bmatrix} H_{11} & H_{12} \\ 0 & H_{22} \end{bmatrix} \quad \text{and} \quad U = \begin{bmatrix} 0 & U_{12} \\ 0 & U_{22} \end{bmatrix}$$

with $H_{11} \in \mathbb{C}^{k \times k}$ upper *triangular*, H_{22} upper Hessenberg, and $U_{22} \in \mathbb{C}^{(n-k) \times (n-k)}$ upper triangular and nonsingular. There is a stable, direct algorithm to calculate Q, Z, H, and U in $O(n^3)$ flops. If A and B are real, Q, Z, H, and U can be taken to be real.

Proof In the case $k = 0$ this theorem follows immediately from Theorem 5.6.5, so assume $k > 0$. In Chapter 3 we discussed the QR decomposition with column pivoting. If we apply that algorithm to B, we obtain $\hat{B} = QR$, where $\hat{B}$ is a matrix obtained by permuting the columns of B, Q is unitary, and

$$R = \begin{bmatrix} R_{11} & R_{12} \\ 0 & 0 \end{bmatrix}$$

with R_{11} upper triangular and nonsingular and the block of zeros in the lower right-hand corner k by k. Clearly $\hat{B} = BP$, where P is a permutation matrix: P has exactly one 1 in each row and column, and zeros elsewhere. P is obviously orthogonal. Thus $R = Q^*BP$.

This decomposition does not quite meet our needs; we want the block of zeros to come out in the upper left-hand corner. Instead of the QR decomposition with column pivoting, we need an RQ decomposition with row pivoting. This yields $PB = RQ$; that is $R = PBQ^*$, where P is a permutation matrix, Q is unitary, and

$$R = \begin{bmatrix} 0 & R_{12} \\ 0 & R_{22} \end{bmatrix}$$

with R_{22} nonsingular and upper triangular. In Exercise 5.6.11 you will show how to construct such a decomposition. Assume now that B has been reduced to this special form (the form of R), which is the desired form. Applying the same operations to A, we get a new A. The rest of the algorithm consists of reducing this A to the form H specified in the statement of the theorem, without ruining the structure of B. We proceed just as in the proof of Theorem 5.6.5. That is, we begin by applying a rotator to rows $n - 1$ and n of A to set the $(n, 1)$ entry to zero. This creates a nonzero in the $(n, n - 1)$ position of B, which can be eliminated by a rotator acting on columns $n - 1$ and n. Then by similar means we set the $(n - 1, 1)$ entry of A to zero, and so on up the first column.

But notice that as we approach the top of the column, the algorithm suddenly becomes simpler: We set the $(k + 1, 1)$ entry of A to zero by a rotator acting on rows k and $k + 1$. We then apply the same rotator to B. Because of the k-by-k block of zeros in the upper left-hand corner, this rotator introduces no new unwanted nonzeros into B (Draw a picture!), so there is no need to apply a correcting rotation to the columns. The same situation holds for all of the rest of the zeros that are created in the first column of A. In fact it is now even possible to set the $(2, 1)$ entry of A to zero, which was not possible in the proof of Theorem 5.6.5. Recall that we were prevented from doing this because it would have created a nonzero in the $(2, 1)$ position of B. To set this to zero would necessitate applying a rotator to columns 1 and 2, which would destroy the zeros that had been created in column 1 of A. But in the present situation the rotator that sets the $(2, 1)$ entry of A to zero does not create a nonzero in the $(2, 1)$ position of B, so the correcting rotator on columns 1 and 2 is not needed. Thus we can zero out the entire first column of A, except for the main diagonal entry.

We then turn to the second column, in which we create zeros in positions $(n, 2), (n - 1, 2), \ldots, (k + 2, 2)$, just as in Theorem 5.6.5. Furthermore, if $k \geq 2$, we can create zeros in positions $(k + 1, 2), \ldots, (3, 2)$ without having to perform any rotations on the columns. If $k = 1$, we cannot set the $(3, 2)$ entry to zero. In general we can reduce the first k columns of A to upper-triangular form. On subsequent columns we must settle for upper Hessenberg form, as in Theorem 5.6.5. Thus A can be reduced to the form H given in the statement of the theorem.

$\square$

Exercise 5.6.11 Let $B \in \mathcal{C}^{n \times n}$. Using the QR decomposition algorithm with column pivoting as a guide, show how to carry out an RQ decomposition: $\hat{B} = RQ$, where $\hat{B}$ is a matrix obtained from B by permuting rows, Q is a product of reflectors, and

$$R = \begin{bmatrix} 0 & R_{12} \\ 0 & R_{22} \end{bmatrix}$$

with R_{22} nonsingular and upper triangular. *Hint:* Introduce zeros into B one row at a time, starting with the bottom row. $\square$

The form given by Theorem 5.6.6 is quite useful. Clearly

$$\det(\lambda U - H) = \det(-H_{11}) \det(\lambda U_{22} - H_{22}) = (-1)^k h_{11} h_{22} \cdots h_{kk} \det(\lambda U_{22} - H_{22})$$

If any of $h_{11}, \ldots, h_{kk}$ equals zero, (H, U) [and hence also (A, B)] is a singular pair. Otherwise the k nonzero entries $h_{11}, \ldots, h_{kk}$ indicate an infinite eigenvalue of multiplicity k. The other eigenvalues are eigenvalues of the smaller Hessenberg-triangular pair (H_{22}, U_{22}). These eigenvalues are all finite since U_{22} is nonsingular.

In Theorem 5.6.6 we act as if we know the value of k, the dimension of the null-space of B. In practice we do not, and finding it is not a trivial matter. If we could work in exact arithmetic, we could carry out the RQ decomposition with row pivoting and quit as soon as we run out of nonzero entries. The size of the remaining block of zeros is k. In practice the arithmetic is inexact, and numbers that should have been zero will be nonzero but (hopefully) small. Thus we must set a threshold and declare that all numbers below this threshold are effectively zero. In the course of the RQ decomposition, once we reach a block of effectively zero numbers, we declare the block to be zero, thus determining k numerically. As we noted in Chapter 3, this approach is not completely reliable. (See the example given in [SLS, page 31] and [MC, page 245]). However, it can be expected to work well in the vast majority of cases. For an absolutely reliable numerical determination of k, the singular value decomposition, discussed in Chapter 6, must be used in place of the RQ decomposition.

The deflation strategy outlined here is not the only one, nor is it necessarily the best. A different strategy is outlined in [MC, Sec. 7.7.5].

The QZ Algorithm

Thanks to Theorem 5.6.6 we can now restrict our attention to pairs (A, B) for which A is upper Hessenberg and B is upper triangular and nonsingular. We can even assume that A is properly upper Hessenberg, since otherwise the eigenvalue problem can be split into two or more subproblems in the obvious manner. A form of the implicit QR algorithm called the QZ algorithm can be used to find the eigenvalues of such pairs. Each step of the QZ algorithm effectively performs a step of the QR algorithm on AB^{-1} and $B^{-1}A$ simultaneously. Let $C = AB^{-1}$, and note that C is a properly upper Hessenberg matrix. Recall that if we wished to perform a QR step on C, we would first choose a shift σ, then perform a QR decomposition of the shifted matrix:

$$C - \sigma I = QR \tag{5.6.7}$$

then calculate the next iterate $\hat{C}$ by $\hat{C} = RQ + \sigma I$ or, equivalently,

$$\hat{C} = Q^*CQ \tag{5.6.8}$$

It turns out that we can perform this step implicitly, without ever forming AB^{-1}. You can easily check that the first column of $C - \sigma I$ has the form $x = [x_1 \ x_2 \ 0 \ \cdots \ 0]^T$, where $x_1 = a_{11}b_{11}^{-1} - \sigma$ and $x_2 = a_{21}b_{11}^{-1}$. Let Q_{12} be a rotator (or reflector) in the $(1, 2)$ plane whose first column is proportional to x. Notice that x_2 is certainly nonzero, which implies that Q_{12} is a nontrivial rotator. Transform (A, B) to (Q_{12}^*A, Q_{12}^*B). This transformation recombines the first two rows of each matrix. It does not disturb the upper Hessenberg form of A, but it does disturb the upper-triangular form of B, creating a bulge in the $(2, 1)$ position. (Draw a picture!) Because $b_{11} \neq 0$ and Q_{12} is a nontrivial rotator, this bulge is certainly nonzero.

The rest of the algorithm consists of returning the pair to Hessenberg-triangular form by chasing the bulge. Let Z_{12} be a (nontrivial) rotator acting on columns 1 and 2 that eliminates the bulge; that is, for which $Q_{12}^*BZ_{12}$ is again upper triangular. Applying the same transformation to A, we obtain $Q_{12}^*AZ_{12}$, which has a bulge in the $(3, 1)$ position. Because $a_{32} \neq 0$ and Z_{12} is a nontrivial rotator, this new bulge is certainly nonzero. It can be eliminated by left multiplication by a nontrivial rotator Q_{23}^* acting in the $(2, 3)$ plane. This operation deposits a nonzero entry in the $(2, 1)$ position of $Q_{23}^*Q_{12}^*AZ_{12}$. This entry will not be altered by subsequent transformations. Applying Q_{23}^* to B as well, we find that $Q_{23}^*Q_{12}^*BZ_{12}$ has a bulge (which is certainly nonzero) in the $(3, 2)$ position. We have now chased the bulge from the $(2, 1)$ position of B over to A and back to the $(3, 2)$ position of B. We can now chase it to the $(4, 2)$ position of A by applying a rotator Z_{23} on the right, and so on. It is finally chased away completely by $Z_{n-1,n}$, which acts on columns $n - 1$ and n to remove the bulge from the $(n, n - 1)$ position of the transformed B matrix without introducing a bulge in A. This completes the implicit QZ step. We have transformed (A, B) to $(\tilde{A}, \tilde{B})$, given by

$$\tilde{A} = \tilde{Q}^* A \tilde{Z} \qquad \tilde{B} = \tilde{Q}^* B \tilde{Z} \tag{5.6.9}$$

where

$$\tilde{Q} = Q_{12} Q_{23} \cdots Q_{n-1,n} \quad \text{and} \quad \tilde{Z} = Z_{12} Z_{23} \cdots Z_{n-1,n} \tag{5.6.10}$$

$\tilde{Q}$ and $\tilde{Z}$ are unitary matrices. $\tilde{A}$ is upper Hessenberg, and $\tilde{B}$ is upper triangular.

Exercise 5.6.12 Verify that the subdiagonal entries of $\tilde{A}$ satisfy $\tilde{a}_{k+1,k} \neq 0$ for $k = 1, 2, \ldots, n - 2$. [You might also like to show that $\tilde{a}_{n,n-1} \neq 0$ if and only if the shift σ is not an eigenvalue of (A, B).] $\square$

Exercise 5.6.13 Verify that the flop count for an implicit QZ step is $O(n^2)$. [It is $4n^2$ if fast rotators (discussed in Appendix A) are used and the transforming matrices $\tilde{Q}$ and $\tilde{Z}$ are not accumulated.] $\square$

So far we have described a QZ step, but we have not yet shown that it effects a QR step implicitly. Let $\tilde{C} = \tilde{A}\tilde{B}^{-1}$. We will use one of the uniqueness theorems from Section 5.3 to show that $\tilde{C}$ is essentially the same as the matrix $\hat{C}$ obtained from one QR step applied to $C = AB^{-1}$. First of all, $\tilde{C}$ is in upper Hessenberg form, and it follows easily from Exercise 5.6.12 that $\tilde{c}_{k+1,k} \neq 0$ for $k = 1, \ldots, n - 2$. Bearing in mind that $\tilde{Q}$ and $\tilde{Z}$ are unitary, we see from (5.6.9) that

$$\tilde{C} = \tilde{Q}^* A \tilde{Q} \tag{5.6.11}$$

Next we note that the first columns of $\tilde{Q}$ and Q are proportional, where Q is the unitary transforming matrix in (5.6.8). This is because Q_{12} was chosen so that its first column is proportional to x, the first column of $C - \sigma I$. Since each of the transformations $Q_{23}, \ldots, Q_{n-1,n}$ leaves the first column unchanged, $\tilde{Q} = Q_{12} Q_{23} \cdots Q_{n-1,n}$ has the same first column as Q_{12}. On the other hand, the first column of Q is also proportional to x by (5.6.7), since R is upper triangular. In light of (5.6.8) and (5.6.11), we can apply Theorem 5.3.5 with the roles of A, B, and $\tilde{B}$ played by C, $\tilde{C}$, and $\hat{C}$, respectively, to conclude that Q and $\tilde{Q}$ are essentially the same, as are $\hat{C}$ and $\tilde{C}$. (You may find that the paragraph following the proof of Theorem 5.3.5 provides additional insight.) Thus the QZ step essentially carries out a step of the QR algorithm on AB^{-1}.

To see that the QZ step is also a QR step on $B^{-1}A$, consider how a QR step acts on $E = B^{-1}A$. First a QR decomposition is performed:

$$E - \sigma I = ZU \tag{5.6.12}$$

where Z is unitary and U is upper triangular. Then the similarity transformation

$$\hat{E} = Z^*EZ$$

is carried out. On the other hand, letting $\tilde{E} = \tilde{B}^{-1}\tilde{A}$, we have

$$\tilde{E} = \tilde{Z}^*E\tilde{Z}$$

by (5.6.9). If we can show that Z and $\tilde{Z}$ have proportional first columns, it will follow from Theorem 5.3.5 that $\tilde{E}$ is essentially $\hat{E}$.

By (5.6.12) the first column of Z is proportional to the first column of $B^{-1}A - \sigma I$. From the second equation in (5.6.9) we deduce $\tilde{Z} = B^{-1}\tilde{Q}\tilde{B}$. Thus the first column of $\tilde{Z}$, which can be expressed as $\tilde{Z}e_1$, is given by $\tilde{Z}e_1 = B^{-1}\tilde{Q}\tilde{B}e_1$. Since $\tilde{B}$ is upper triangular, $\tilde{B}e_1 = \alpha e_1$ for some constant α. Now $\tilde{Q}e_1$ is the first column of $\tilde{Q}$, which has already been shown to be proportional to the first column of $AB^{-1} - \sigma I$. Thus

$$\tilde{Z}e_1 = \beta B^{-1}(AB^{-1} - \sigma I)e_1 = \beta(B^{-1}A - \sigma I)B^{-1}e_1$$

for some constant β. Since B^{-1} is also upper triangular, $B^{-1}e_1 = \gamma e_1$ for some constant γ. Thus $\tilde{Z}e_1 = \gamma\beta(B^{-1}A - \sigma I)e_1$, which shows that the first column of $\tilde{Z}$ is proportional to the first column of $B^{-1}A - \sigma I$. We conclude that Z and $\tilde{Z}$ have proportional first columns. Thus, by Theorem 5.3.5, they are essentially the same, and so are $\hat{E}$ and $\tilde{E}$.

Now suppose we perform a sequence of QZ steps to generate a sequence of equivalent pairs (A_k, B_k). Then if the shifts are chosen reasonably, the matrices $C_k = A_k B_k^{-1}$ and $E_k = B_k^{-1}A_k$ will converge to upper-triangular or block triangular form. Thus $A_k = C_k B_k = B_k E_k$ will also converge to block triangular form, from which the generalized eigenvalues are either evident or can be determined easily.

Another question that needs to be addressed is that of how the shifts are determined. From what we know about the QR algorithm, we realize it makes sense to take the shift for the kth step to be an eigenvalue of the lower right-hand 2-by-2 submatrix of either $A_{k-1}B_{k-1}^{-1}$ or $B_{k-1}^{-1}A_{k-1}$. In the following exercise you will show how to determine these submatrices without calculating B_{k-1}^{-1} explicitly.

Exercise 5.6.14   (a) Suppose B is a nonsingular block triangular matrix

$$B = \begin{bmatrix} B_{11} & B_{12} \\ 0 & B_{22} \end{bmatrix}$$

Show that B^{-1} has the block triangular form

$$B^{-1} = \begin{bmatrix} B_{11}^{-1} & X \\ 0 & B_{22}^{-1} \end{bmatrix}$$

Determine what X must be.

(b) Let B be nonsingular and upper triangular. Show how to calculate the lower right-hand k-by-k submatrix of B^{-1} cheaply for $k = 1, 2, 3$. Give explicit formulas.
(c) Let (A, B) have Hessenberg-triangular form. Let M denote the lower right-hand 2-by-2 submatrix of $B^{-1}A$. By making the appropriate partition of $B^{-1}A$, show that M depends only on the lower right-hand 2-by-2 submatrix of B^{-1}. Derive explicit formulas for the entries of M.
(d) Let (A, B) have Hessenberg-triangular form. Let N denote the lower right-hand 2-by-2 submatrix of AB^{-1}. Show that N depends only on the lower right-hand 3-by-3 submatrix of B^{-1}. □

We have derived a single-step QZ algorithm. It is also natural to consider a double-step QZ algorithm analogous to the double-step QR algorithm, to be used in order to avoid complex arithmetic when A and B are real. Such an algorithm is indeed possible. It is just like the single-step algorithm, except that the bulges are fatter. The single and double-step QZ algorithms are unconditionally stable.

Exercise 5.6.15   Derive a double-step implicit QZ algorithm. □

Exercise 5.6.16   Derive an algorithm that is like the QZ algorithm, except that it uses Gaussian elimination operations, instead of rotators or reflectors, to chase the bulge. This is called the LZ algorithm. It is an alternative to the QZ algorithm that is well worth considering. While it is somewhat less stable, its flop count is lower by a factor of at least 2. □

6

Other Methods for the Symmetric Eigenvalue Problem

Let $A \in \mathbb{R}^{n \times n}$ be real and symmetric. Then the eigensystem of A has special properties: the eigenvalues are real, and there exists a set of n orthonormal real eigenvectors. Furthermore, the eigenvalues are well conditioned. Because of these and other special properties, the symmetric eigenvalue problem is easier to handle than the general problem, and there is a correspondingly wider variety of techniques available. In the previous two chapters we examined a number of methods, most of which can be applied effectively to symmetric matrices. Two noteworthy examples are the QR and Lanczos algorithms. In this chapter we will study three more methods. Each is very different from the others; they illustrate amply the rich variety of approaches to solving a single problem. One feature they have in common is that all can be implemented efficiently in parallel. As preparation for this chapter you need only have read through Section 4.5.

6.1
THE JACOBI METHOD

The Jacobi method is one of the oldest numerical methods for the eigenvalue problem. It is older than matrix theory itself, dating back to a paper by Jacobi (1846). Like most numerical methods it was little used in the precomputer era. It enjoyed a brief renaissance in the 1950s, but in the 1960s it was supplanted as the method of choice by the QR algorithm. In recent years it has made a comeback because of its adaptability to parallel computers.

The Basic Ideas

Consider a 2-by-2, real symmetric matrix

$$A = \begin{bmatrix} a & b \\ b & d \end{bmatrix}$$

One way to find the eigenvalues of A is to find a rotator

$$Q = \begin{bmatrix} c & -s \\ s & c \end{bmatrix}$$

such that $Q^T A Q$ is a diagonal matrix. Then the main-diagonal entries of $Q^T A Q$ are the eigenvalues of A. It turns out to be easy to find such a Q. A direct computation shows that

$$Q^T A Q = \begin{bmatrix} c^2 a + s^2 d + 2csb & (c^2 - s^2)b + cs(d - a) \\ (c^2 - s^2)b + cs(d - a) & c^2 d + s^2 a - 2csb \end{bmatrix} \qquad (6.1.1)$$

Notice that if $d = a$, then $Q^T A Q$ can be made diagonal by setting $c = s = 1/\sqrt{2}$. Otherwise, letting

$$\hat{t} = \frac{2b}{a - d} \qquad (6.1.2)$$

we see that $Q^T A Q$ is diagonal if and only if

$$\frac{2cs}{c^2 - s^2} = \hat{t} \qquad (6.1.3)$$

Recall that $c = \cos \theta$ and $s = \sin \theta$ for some angle θ. The double-angle formulas $\cos 2\theta = \cos^2 \theta - \sin^2 \theta$ and $\sin 2\theta = 2 \sin \theta \cos \theta$ allow (6.1.3) to be rewritten as

$$\tan 2\theta = \hat{t} \qquad (6.1.4)$$

For each real number $\hat{t}$ there is a unique angle $\theta \in (-\pi/4, \pi/4)$ for which (6.1.4) holds. There is no need to calculate θ, for c and s can be found directly from $\tan 2\theta$ with the help of trigonometric identities. First of all, letting $t = \tan \theta$, we have the half-angle formula

$$t = \frac{\hat{t}}{1 + \sqrt{1 + \hat{t}^2}} \qquad (6.1.5)$$

Then, since $1 = c^2 + s^2 = c^2(1 + t^2)$, c and s can be obtained from t by

$$c = \frac{1}{\sqrt{1 + t^2}} \qquad \text{and} \qquad s = ct \qquad (6.1.6)$$

We know that $c > 0$ because $\theta \in (-\pi/4, \pi/4)$. Actually c and s are not even needed for the computation of $Q^T A Q$. You can easily verify that

$$Q^T A Q = \begin{bmatrix} a + tb & 0 \\ 0 & d - tb \end{bmatrix} \qquad (6.1.7)$$

Exercise 6.1.1   Verify the computation (6.1.1), and show that $Q^T A Q$ is diagonal if and only if (6.1.3) holds, where $\hat{t}$ is defined by (6.1.2).   □

Exercise 6.1.2   [Verification of (6.1.5)]

(a)   Show that the formula

$$\tan \theta = \frac{\sin 2\theta}{1 + \cos 2\theta}$$

holds for all θ in the domain of the tangent function.

(b)   Verify that for all $\theta \in (-\pi/4, \pi/4)$,

$$\cos 2\theta = \frac{1}{\sqrt{1 + \tan^2 2\theta}} \quad \text{and} \quad \sin 2\theta = \cos 2\theta \tan 2\theta$$

These are essentially identical to (6.1.6).

(c)   Deduce (6.1.5) from the equations derived in parts (a) and (b).   □

Exercise 6.1.3   Verify (6.1.7), where $t = s/c$, provided that (6.1.3) holds, with $\hat{t}$ given by (6.1.2).   □

If $b = 0$, then A is already a diagonal matrix, so we can skip the computation completely. You can easily check that in this case we get $\theta = 0$. If b is nonzero but sufficiently small, the error introduced by simply setting b to zero will be negligible. Suppose $|b|$ is very small (on the order of roundoff error) compared with $\min\{|a|, |d|\}$. Then, since it is always the case that $|t| \leq 1$, we have $a + tb \approx a$ and $d - tb \approx d$ in (6.1.7). That is, a and b undergo insignificant modifications, regardless of the value of θ. In this case the computation can and should be skipped.

If $a - d$ is very small, there is a chance of overflow in the computation (6.1.2) and also in (6.1.5), in which $\hat{t}$ is squared. We can eliminate this danger by working with the reciprocal of (6.1.2) whenever $|a - d| < |2b|$. We compute

$$\hat{k} = \frac{a - d}{2b}$$

instead of (6.1.2). Equation (6.1.4) is replaced by

$$\cot 2\theta = \hat{k}$$

Instead of (6.1.5) we have,

$$t = \frac{\text{sign}(\hat{k})}{|\hat{k}| + \sqrt{1 + \hat{k}^2}} \tag{6.1.8}$$

Then c and s are given by (6.1.6), as before.

Exercise 6.1.4 [Verification of (6.1.8)]

(a) Show that if $\theta \in (-\pi/4, \pi/4)$ and $\theta \neq 0$, then

$$
\sin 2\theta = \begin{cases} \dfrac{1}{\sqrt{1 + \cot^2 2\theta}} & \text{if } \theta > 0 \\[3mm] \dfrac{-1}{\sqrt{1 + \cot^2 2\theta}} & \text{if } \theta < 0 \end{cases}
$$

and $\cos 2\theta = \sin 2\theta \cot 2\theta$.

(b) Use these formulas, together with the formula from part (a) of Exercise 6.1.2, to deduce (6.1.8). ☐

Exercise 6.1.5 Use (6.1.2), (6.1.5), and (6.1.7) to determine the eigenvalues of each of the following two matrices. Perform the computations in exact arithmetic, and check your answers by calculating the eigenvalues by some other means. Notice that there is no need to calculate c and s.

$$
\text{(a)} \quad \begin{bmatrix} 2 & 1 \\ 1 & 3 \end{bmatrix} \qquad \text{(b)} \quad \begin{bmatrix} 5 & 6 \\ 6 & 1 \end{bmatrix}
$$

(c) Repeat the computations, using (6.1.8) instead of (6.1.5). ☐

Exercise 6.1.6 Suppose we choose to calculate c and s using (6.1.6). Explain how these values can be used to determine a complete orthonormal set of eigenvectors of A. ☐

Finding the eigenvalues of a 2-by-2 matrix is not an impressive feat. Now let's see what we can do with n-by-n matrices. Let $A \in \mathbb{R}^{n \times n}$ be symmetric. It is reasonably clear that we can use the formulas developed above to create a plane rotator Q such that any prescribed off-diagonal entry of $Q^T A Q$ is zero. (Rotators of this form are called *Jacobi rotators*.) Indeed, suppose that for some specified i and j with $i > j$ we wish to set the (i, j) and (j, i) entries to zero.

We can accomplish this with a rotator that operates in the (i, j) plane:

$$
Q = \begin{bmatrix} I & & & & \\ & c & & -s & \\ & & I & & \\ & s & & c & \\ & & & & I \end{bmatrix}
$$

where c and s are defined by (6.1.6), via (6.1.5) or (6.1.8). The numbers a, b, and d used to generate $\hat{t} = \tan 2\theta$ or $\hat{k} = \cot 2\theta$ are, of course, given by

$$
\begin{bmatrix} a & b \\ b & d \end{bmatrix} = \begin{bmatrix} a_{jj} & a_{ij} \\ a_{ij} & a_{ii} \end{bmatrix}
$$

If we overwrite A by $Q^T A Q$, as is usually done, we must make the following substitutions. First of all,

$$a_{jj} \leftarrow a_{jj} + t a_{ij}$$
$$a_{ii} \leftarrow a_{ii} - t a_{ij}$$
$$a_{ij} \leftarrow 0$$

Since premultiplication by Q^T affects rows j and i, we have

$$a_{jk} \leftarrow c a_{jk}^{(old)} + s a_{ik}^{(old)}$$
$$a_{ik} \leftarrow -s a_{jk}^{(old)} + c a_{ik}^{(old)} \qquad k = 1, 2, \dots, n; \; k \neq i, j \qquad (6.1.9)$$

Since postmultication by Q affects columns j and i, we have also

$$a_{kj} \leftarrow c a_{kj}^{(old)} + s a_{ki}^{(old)}$$
$$a_{ki} \leftarrow -s a_{kj}^{(old)} + c a_{ki}^{(old)} \qquad k = 1, 2, \dots, n; \; k \neq i, j \qquad (6.1.10)$$

Because A is symmetric, there is redundancy in Eqs. 6.1.9 and 6.1.10. If these operations are carried out by computer, only about half of A will be stored, and only half of the operations in (6.1.9) and (6.1.10) will be performed.

Exercise 6.1.7 (a) Verify that Eqs. 6.1.9 and 6.1.10 are correct.
(b) Suppose only the lower half of A is being stored and updated. Describe exactly how the operations in (6.1.9) and (6.1.10) should be carried out.
(c) In what way are the operations simpler in the case $i = j + 1$? □

As you may already suspect, our plan is to carry out a succession of similarity transformations by Jacobi rotators. It follows that additional savings in arithmetic can be realized by implementing the rotators as fast, scaled rotators, as described in Appendix A.

The Classical Jacobi Method

The similarity transformation that sets a_{ij} to zero does not by itself yield any information about the eigenvalues of A, but it does bring the matrix closer to diagonal form, in a sense that will be made precise below. The original procedure employed by Jacobi, now known as the *classical Jacobi method*, was to search the matrix for the largest (in magnitude) off-diagonal entry and perform a rotation to set that entry to zero. The resulting matrix is then searched for its largest off-diagonal entry, and another similarity transformation is performed to set that entry to zero. This process is carried out repeatedly. Of course each zero that is created can be destroyed by a subsequent rotation, so the process does not terminate after finitely many steps; rather, it is an iterative process, as it must be.

It is not hard to prove that the sequence of matrices so produced converges to diagonal form yielding the eigenvalues of A. Indeed the convergence becomes quite swift, once the matrix is sufficiently close to diagonal form. This method was employed by Jacobi (1846) in the solution of an eigenvalue problem that arose in a study of perturbations of planetary orbits. The system was of order seven because there were then seven known planets. In the paper Jacobi stressed that the computations (by hand!) are quite easy. That was easy for him to say, for the actual computations were done by a student.

Finding the Eigenvectors

The classical Jacobi method can be used to determine the eigenvectors as well as the eigenvalues. In fact, the following remarks apply not only to the classical method, but also to all of the variants that will be discussed shortly. Suppose that after m Jacobi rotations the matrix is essentially in diagonal form. Then we have essentially

$$D = Q_m^T \cdots Q_1^T A Q_1 \cdots Q_m$$

where D is diagonal and each Q_i is a Jacobi rotator. Let

$$Q = Q_1 Q_2 \cdots Q_m$$

Then Q is orthogonal and $D = Q^T A Q$, so the columns of Q are n orthonormal eigenvectors of A.

It is an easy matter to accumulate Q. One starts with an array named Q, which is initially set equal to the identity matrix. Each time a Jacobi rotation $A \leftarrow Q_i^T A Q_i$ is performed, the update $Q \leftarrow Q Q_i$ is carried out as well. In the end the columns of Q are the eigenvectors.

Cyclic Jacobi Methods

The classical Jacobi procedure is quite appropriate for hand calculation but inefficient for computer implementation. For a human being working on a small matrix, it is a simple matter to identify the largest off-diagonal entry. The hard part is the arithmetic. However, for a computer operating on a larger matrix, the arithmetic is easy; the search for the largest entry is the expensive part.

Exercise 6.1.8 (a) Show that determining the largest off-diagonal entry of an n-by-n symmetric matrix requires that $O(n^2)$ numbers be examined. (b) Show that each similarity transformation of an n-by-n symmetric matrix by a Jacobi rotator requires $O(n)$ arithmetic operations. (c) Conclude that for large n, the expensive part of a classical Jacobi step is the search for the largest entry. Does your conclusion change if the operations are to be done in parallel?   □

Because the search for the largest entry is so time consuming, a class of variants of Jacobi's procedure, called *cyclic Jacobi methods*, was introduced. A cyclic Jacobi method sweeps through the matrix, setting entries to zero in some prespecified order and paying no attention to the magnitudes of the entries. In each complete *sweep*, each off-diagonal entry is set to zero once. For example, one could perform a *sweep by columns*, which would create zeros in the order

$$(2, 1), (3, 1), \ldots, (n, 1), (3, 2), (4, 2), \ldots, (n, 2), \ldots, (n, n - 1) \qquad (6.1.11)$$

Alternatively one could sweep by rows or by diagonals, for example. We will refer to the sweep-by-columns Jacobi method defined by the ordering (6.1.11) as the *special cyclic Jacobi method*. Cyclic Jacobi methods are obviously an improvement from an organizational standpoint, since the search is eliminated and the entries of A are accessed in a systematic manner.

Unfortunately it is more difficult to prove convergence. Indeed, Hansen (1963) gave an example that shows that for some orderings there are matrices for which the method does not converge. Fortunately such cases are rare; in fact, Hansen's example is quite bizarre. Experimental evidence indicates that for practically any symmetric matrix and any ordering, the cyclic Jacobi method will converge. Moreover, the special cyclic method (6.1.11) does always converge. A proof was given by Forsythe and Henrici (1960). Finally, when a cyclic Jacobi method does converge, it converges quadratically, regardless of the order of elimination. We will take a brief look at convergence, after which we will study some schemes for implementing cyclic Jacobi methods in parallel.

Convergence of Jacobi Methods

Every symmetric matrix $A \in \mathbb{R}^{n \times n}$ can be expressed uniquely as a sum $A = D + E$, where D is a diagonal matrix and E is a symmetric matrix whose main-diagonal entries are all zero. Recall that the Frobenius norm $\| E \|_F$ is defined to be the square root of the sum of the squares of the entries of E. Clearly $\| E \|_F$ is a measure of how far A deviates from being a diagonal matrix. The object of the various versions of Jacobi's method is to drive $\| E \|_F$ to zero. The most fundamental fact about the convergence of Jacobi methods is that each Jacobi rotation makes progress in that direction, as the following theorem shows.

THEOREM 6.1.12   Let $A \in \mathbb{R}^{n \times n}$ be symmetric, and let $\hat{A} = Q^T A Q$, where Q is a Jacobi rotator that sets a_{ij} to zero. Let D and $\hat{D}$ be diagonal matrices, and E and $\hat{E}$ symmetric matrices with zeros on the main diagonal, uniquely determined by the equations $A = D + E$ and $\hat{A} = \hat{D} + \hat{E}$. Then

$$\| \hat{E} \|_F^2 = \| E \|_F^2 - 2 a_{ij}^2$$

You will prove Theorem 6.1.12 by working the following exercise.

Exercise 6.1.9   (a) Recall that if $U \in \mathbb{R}^{n \times n}$ is orthogonal, then $\| Ux \|_2 = \| x \|_2$ for all $x \in \mathbb{R}^n$. Use this fact to prove that $\| UB \|_F = \| B \|_F$ for all $B \in \mathbb{R}^{n \times n}$.
(b) For the rest of the exercise we will use the notation introduced in the statement of Theorem 6.1.12. Use the result of part (a) to prove that $\| \hat{A} \|_F = \| A \|_F$.

(c) Since A and $\hat{A}$ are identical in all but rows and columns i and j, we can focus on those two rows and columns. Prove that $\hat{a}_{ik}^2 + \hat{a}_{jk}^2 = a_{ik}^2 + a_{jk}^2$ if $k \neq i, j$. By symmetry the same result holds with the order of the subscripts reversed.
(d) Conclude from the previous two parts that $\| \hat{E} \|_F^2 = \| E \|_F^2 - 2a_{ij}^2$.   □

Exercise 6.1.10   (a) Prove that $\| A \|_F^2 = \| D \|_F^2 + \| E \|_F^2$. Obviously the same result holds for $\hat{A}$.
(b) Show that $\hat{a}_{ii}^2 + \hat{a}_{jj}^2 = a_{ii}^2 + a_{jj}^2 + 2a_{ij}^2$.   □

Theorem 6.1.12 does not quite imply convergence of Jacobi methods in general. It shows that $\| E \|_F$ gets smaller with each step, but it does not show that $\| E \|_F \to 0$. In the next two exercises you will prove the convergence of the classical Jacobi method and of a class of variants of the cyclic Jacobi method called threshold Jacobi methods.

Exercise 6.1.11   (a) In the classical Jacobi method we always annihilate the off-diagonal element of greatest magnitude. Suppose that at some step we choose to annihilate a_{ij}. Prove that $|a_{ij}|^2 \geq \| E \|_F^2 / N$, where $N = n(n-1)$.
(b) Conclude that $\| \hat{E} \|_F^2 \leq (1 - 2/N) \| E \|_F^2$. Thus after m steps of the classical Jacobi method, $\| E \|_F^2$ will be reduced at least by a factor of $(1 - 2/N)^m$. In other words, the classical Jacobi method converges at least linearly.   □

Exercise 6.1.12   A *threshold Jacobi method* is a cyclic Jacobi method for which the (i, j) rotation is performed only if $|a_{ij}|$ exceeds some threshold τ. Otherwise the rotation is skipped. Usually a new threshold is computed before each sweep. For example, one can compute the threshold $\tau = \| E \|_F / \sqrt{N}$.

(a)   Prove that any threshold Jacobi method with the threshold $\tau = \| E \|_F / \sqrt{N}$ converges.

(b)   Prove that any threshold Jacobi method that satisfies the following conditions converges: (i) At each sweep there is at least one off-diagonal entry that exceeds the threshold for that sweep. (ii) There is a constant $K > 0$ such that on every sweep the threshold satisfies $\tau \geq \| E \|_F / K$.   □

Theorem 6.1.12 does not imply convergence for cyclic Jacobi methods in general because of the possibility that each rotation annihilates a very small element, while at the same time large off-diagonal elements are being pushed around the matrix, but never being annihilated. This is essentially what happens in the example of Hansen (1963). Fortunately this cannot happen when the special cyclic Jacobi method is used.

The results we have seen so far tell nothing about the asymptotic convergence rate of Jacobi methods. It turns out that all variants of Jacobi's method converge quadratically in some sense. We will restrict our discussion to cyclic Jacobi methods. Henrici (1958) proved that if a cyclic Jacobi method is applied to a symmetric matrix having distinct eigenvalues, and the method converges, then it converges quadratically. For an exact statement and rigorous proof of the result, see the original work. The following exercise shows clearly, if not rigorously, why quadratic convergence takes place.

Exercise 6.1.13   Let $A \in \mathbb{R}^{n \times n}$ be a symmetric matrix with distinct eigenvalues λ_1, λ_2, ..., λ_n, and let $\delta = \min\{|\lambda_i - \lambda_j| : i \neq j\}$. We will continue to use the notation $A = D + E$ established above. Suppose that at the beginning of a sweep $\|E\|_F = \epsilon$, where ϵ is small compared with δ. Then $|a_{ij}| \leq O(\epsilon)$ for all $i \neq j$. The elements must stay this small since $\|E\|_F$ is nonincreasing. Suppose further that $\|E\|_F$ is small enough that the main diagonal entries of A are fairly close to the eigenvalues.

(a)   Show that each rotator

$$Q = \begin{bmatrix} c & -s \\ s & c \end{bmatrix}$$

generated during the sweep must satisfy $|s| \leq O(\epsilon)$ and $c \approx 1$.

(b)   Using the result of part (a) and Eqs. 6.1.9 and 6.1.10, show that once each a_{ik} is set to zero, subsequent rotations can make it no bigger than $O(\epsilon^2)$. Thus, at the end of a complete sweep, every off-diagonal entry is $O(\epsilon^2)$; the convergence is quadratic. $\square$

The argument in Exercise 6.1.13 relies on the separation of the eigenvalues. This is needed to ensure that the rotation angles tend to zero as $\|E\|_F \to 0$: The angle of the rotator used to annihilate a_{ij} is given by

$$\tan 2\theta = \frac{2a_{ij}}{a_{jj} - a_{ii}}$$

It is assumed that a_{ij} is small. Furthermore $a_{jj} - a_{ii}$ cannot be small because a_{ii} and a_{jj} are converging to different eigenvalues. Therefore θ must be small. This argument breaks down for matrices with repeated eigenvalues, since a_{ii} and a_{jj} could be converging to the same eigenvalue. One might therefore suspect that the convergence fails to be quadratic in this case. It turns out that this fear is groundless. A theorem in [SEP, pp. 182–183] shows that if a_{ii} and a_{jj} are converging to the same eigenvalue, and $\|E\|_F = O(\epsilon)$, where ϵ is sufficiently small, then $|a_{ij}| = O(\epsilon^2)$; that is, it is tiny. Early in the section it was suggested that a rotation be skipped whenever $|a_{ij}|$ (then called $|b|$) is very small. If this advice is followed, there will be no danger that the presence of repeated eigenvalues will cause the generation of rotators with large angles.

Mobile Parallel Jacobi Schemes

If you gave a reasonable amount of thought to part (c) of Exercise 6.1.8, you have already noticed that most of the arithmetic performed during a similarity transformation by a Jacobi rotator can be done in parallel. Now let us note that there is also an opportunity for parallelism on a higher level: several Jacobi rotations can be performed simultaneously. The reason for this is that each Jacobi rotation affects only two rows and columns. Suppose, for example, we perform a rotation in the $(1, 2)$ plane, setting a_{21} to zero. In this operation only the first two rows and columns are altered. In particular, a_{43}, a_{33}, and a_{44} are untouched. These are exactly the elements that are used to determine a Jacobi rotation in the $(3, 4)$ plane, so we can simultaneously determine a rotator that sets a_{43} to zero. This operation affects only rows and columns 3 and 4. In particular it does not affect a_{21}. It should now be obvious that we can also set $a_{65}, a_{87}, \ldots$ to zero at the same time. Thus we can perform p simultaneous rotations, where

$$p = \begin{cases} n/2 & \text{if } n \text{ is even} \\ (n-1)/2 & \text{if } n \text{ is odd} \end{cases}$$

We have grouped the indices $1, 2, 3, \ldots, n$ into the pairs

$$(1, 2), (3, 4), (5, 6), \ldots, (m - 1, m)$$

where $m = n$ if n is even and $m = n - 1$ if n is odd. We will refer to a set of independent rotations performed simultaneously as a *batch*. After carrying out this batch of rotations, we naturally want to pair off the indices in some other way and carry out another batch. For example, we could try a pairing that starts like this: $(1, 3), (2, 4), (5, 7), (6, 8), \ldots$. It is natural to ask whether we can perform a complete cyclic Jacobi sweep by performing a relatively small number of large batches of rotations. The answer turns out to be yes.

Let us first determine a lower bound on the number of batches per sweep. Since an n-by-n matrix has $n(n - 1)/2$ entries below the main diagonal, a complete sweep consists of $n(n - 1)/2$ rotations. If n is even, there are at most $n/2$ rotations in a batch, so a complete sweep can be done in not less than $n - 1$ batches. If n is odd, there are at most $(n - 1)/2$ rotations in a batch, so a sweep can be done in not less than n batches. We will consider a scheme that performs a complete sweep in exactly n batches, regardless of whether n is odd or even. Thus the scheme is optimal when n is odd and nearly optimal when n is even. Furthermore, we will see that it is almost equivalent to the special cyclic Jacobi method. This scheme was proposed independently in several different guises by a number of researchers. See Modi and Pryce (1985) and Whiteside, Ostlund, and Hibbard (1984). The observation that all of these schemes are the same was made by Luk and Park (1987, 1989).

As motivation for the scheme, let us consider the problem of performing batches of simultaneous Jacobi rotations on a large multiprocessor whose computing power and memory are distributed over a large, two-dimensional array of small processors. Each processor handles a small portion of the matrix, and neighboring matrix elements are handled by neighboring processors. In order to perform a rotation involving indices i and j, the processors handling these rows and columns must communicate. This is obviously easiest if the numbers i and j are not far apart. In this environment the pairing

$$(1, 2), (3, 4), (5, 6), \ldots$$

by which the elements $a_{21}, a_{43}, a_{65}, \ldots$ can be set to zero simultaneously, is obviously very desirable. [See also Exercise 6.1.7, part (c).] Let's call this *pairing A*.

Another desirable pairing is

$$(2, 3), (4, 5), (6, 7), \ldots$$

which can be used to set $a_{32}, a_{54}, a_{76}, \ldots$ to zero simultaneously. We'll call this *pairing B*. In this pairing the index 1 is not paired off with any other index. If n is even, index n is not paired off either. In this case one is tempted to pair off indices 1 and n, but that would be bad from the communication standpoint. It turns out that pairings A and B are all we need. This may seem surprising, given that these pairings allow only the elements immediately below the main diagonal to be set to zero. The trick is to allow the rows and columns to migrate in such a way that every element below the main diagonal eventually finds its way into a subdiagonal position and gets set to zero.

Exercise 6.1.14 Show that the operation of interchanging any two rows and the same two columns of a matrix is an orthogonal similarity transformation on the matrix. Thus the eigenvalues are unchanged by this operation. □

Consider the following *mobile Jacobi scheme*, which can be described in two sentences:

1. Perform batches of Jacobi rotations using pairings A and B alternately.
2. After each rotation, interchange the participating rows and columns. (6.1.13)

Table 6.1 shows what this scheme does when $n = 5$. The vertical lines indicate the pairings and interchanges. Each column represents a batch of two rotations. For the first batch, pairing A is used: a_{21} and a_{43} are set to zero and rows and columns 1 and 2 are interchanged, as are rows and columns 3 and 4. For the second batch, pairing B is used. The pairing is (2,3) and (4,5); however, the numbers have been scrambled to reflect the interchanges that took place during the first batch, so the effective pairing is (1,4) and (3,5). For the third batch, pairing A is used again, but this time the effective pairing is (2,4) and (1,5).

TABLE 6.1
Mobile Parallel Jacobi scheme for $n = 5$

	Batch										
	1	2	3	4	5	6	7	8	9	10	
1	1	2	2	4	4	5	5	3	3	1	1
2	2	1	4	2	5	4	3	5	1	3	2
3	3	4	1	5	2	3	4	1	5	2	3
4	4	3	5	1	3	2	1	4	2	5	4
5	5	5	3	3	1	1	2	2	4	4	5

You can easily check that in the 10 rotations that are carried out in the first five batches, each pair from the set $\{1, 2, 3, 4, 5\}$ occurs exactly once. Thus these five batches constitute one Jacobi sweep. Notice also that after five batches, the order of the indices has been reversed. The next five batches, starting with pairing B, are the same as the first five turned upside down. At the end of these five batches, the indices have been returned to their original locations. It will be shown below that this is the general pattern. Each set of n consecutive batches constitutes a sweep that reverses the order of the rows and columns. Thus each $2n$ batches is a double sweep that restores the order of the rows and columns.

You probably suspect that the cost of the interchanges is small. As a matter of fact, they can be done for free. Consider a rotation that modifies rows i and j as in (6.1.9). If we also wish to interchange rows i and j, we can do so simply by computing the number that was destined for a_{jk} and storing it in position a_{ik} instead, and vice versa. This costs nothing.

Exercise 6.1.15 In this exercise we use matrix terminology to derive the same conclusion as in the previous paragraph. Consider a rotation that modifies rows and columns j and $j + 1$ to change A to $Q^T A Q$. (Here we have taken $i = j + 1$ for convenience.) If we then interchange the two rows and columns that participated in the rotation, $Q^T A Q$ is changed to $P^T Q^T A Q P$, where P is a certain simple permutation matrix. The two operations can be performed at once by changing A to $\Pi^T A \Pi$, where $\Pi = QP$. What do Q, P, and Π look like? Conclude that the similarity transformation with Π costs no more than the similarity transformation with Q alone. □

Exercise 6.1.16 Construct analogs of Table 6.1 for the cases $n = 6$ and $n = 7$. Check that in each case, n batches of rotations constitute a complete sweep that reverses the order of the indices. □

Exercise 6.1.17 While the mobile Jacobi scheme (6.1.13) is best suited for a certain type of parallel computer architecture, it can be programmed on any computer, parallel or sequential. Write an algorithm that performs a double parallel mobile sweep. It is convenient to perform double sweeps since these restore the rows and columns to their original order. Your algorithm should exploit symmetry by operating only on the lower half of A. [See Exercise 6.1.7, parts (b) and (c).] □

Exercise 6.1.18

(a) Using your algorithm from Exercise 6.1.17 as a core, write a Fortran subroutine to calculate the eigenvalues of a real, symmetric matrix A by the mobile Jacobi method. Write clear, structured, well-documented code, in conformity with the Fortran programming tips at the end of Section 1.6. At the very beginning, before performing any sweeps, your subroutine should calculate the number $b = \max\{|a_{ij}| : 1 \le j \le i \le n\}$. (If $b = 0$, quit. All eigenvalues are zero.) Before each double sweep calculate $s = \max\{|a_{ij}| : i \ne j\}$.

Normally a subroutine should not do any printing. However, since we want this assignment to be a learning experience, print out s each time you calculate it. This way you will be able to observe the rapid convergence. (What convergence rate do you observe? Why?) Print the main-diagonal entries of A as well, so you can see how well they approximate the eigenvalues. (Display as many decimal places as is justified by the precision of your computer.) Use double precision (real*8) throughout so that you can best appreciate the swift convergence. (Use D-format when you print out s. You need not display many digits; the exponent is what is important in this case.)

If s is sufficiently small, you can quit. A good time to stop is when s/b is smaller than the unit roundoff error of the machine. This happens as soon as the condition $(b + s).EQ.(b)$ is TRUE.[§] (You should also quit if you have performed many sweeps (say 100) without convergence. Set a flag to indicate failure. This measure eliminates the possibility of infinite looping. It could prove particularly useful while you are debugging your program.)

(b) In order to test your subroutine, write a driver program and another subroutine that generates random symmetric matrices with known eigenvalues. This can be done as follows: Start with a diagonal matrix with the desired eigenvalues. Perform mobile Jacobi-like double sweeps, but choose the angle of each rotator at random. After one double sweep with random angles, you will have a full, random matrix with known eigenvalues. Perform two or three double sweeps for good measure. You may use any convenient pseudo–random number generator to generate your angles. Many Fortran compilers provide one that gives uniformly distributed random numbers in the interval $(0,1)$. (See your system's Fortran manual for details.) In order to get random angles in the interval $(0, 2\pi)$, just multiply your random numbers by $2\pi \approx 6.28318$.

(c) Use your parallel Jacobi subroutine to calculate the eigenvalues of randomly generated n-by-n matrices with eigenvalues $1, 2, \ldots, n$, for $n = 7, 8, 9,$ and 10.

(d) If your computer is a parallel machine, you might like to try some larger test problems and see what kind of speedup you can achieve by running your code in parallel. □

Exercise 6.1.19

(a) Given $i \neq j$, let $Q_{ij} \in \mathbb{R}^{n \times n}$ denote a plane rotator acting in the (i,j) plane. Show that $Q_{ij}Q_{kl} = Q_{kl}Q_{ij}$ if $k, l, i,$ and j are all different. The commutativity of Q_{ij} and Q_{kl} reflects the fact that they could be applied to a matrix A one at a time in either order or both at the same time. Let $Q_{ijkl} = Q_{ij}Q_{kl}$. What does Q_{ijkl} look like? What does a product $Q_{12}Q_{34}Q_{56} \cdots Q_{m-1,m}$ look like? This is an example of a *compound*

[§] Recently Demmel and Veselić have proposed a more stringent stopping criterion that guarantees that the tiny eigenvalues of A (if any) will be computed to full precision: Do not stop until each a_{ij} is tiny relative to a_{ii} and a_{jj}.

rotator. The batches of rotations performed in the mobile Jacobi method are just compound rotations with interchanges.

(b) Consider the product $Q_{34}^T Q_{12}^T A Q_{12} Q_{34} = Q_{12}^T Q_{34}^T A Q_{12} Q_{34}$. Q_{34}^T acts on rows 3 and 4 of A, and Q_{12} acts on columns 1 and 2. These rows and columns intersect in the positions

$$(3, 1) \qquad (3, 2)$$
$$(4, 1) \qquad (4, 2)$$

so these four elements are affected by both operations. What basic theorem of matrix algebra can be invoked here to show that it does not matter which operation is carried out first? □

We still need to justify the claim that the mobile Jacobi scheme is essentially the special cyclic Jacobi method. We will begin by modeling the scheme as a conveyor belt, from which it will be clear that any n consecutive batches constitute a Jacobi sweep. Consider a conveyor belt with $2n$ positions, as shown in Figure 6.1 for the case $n = 7$. Numbered positions alternate with blank positions around the belt. In the belt's initial state (numbered 0) the odd numbers are on the left side of the belt, poised to move downward, and the even numbers are on the right side, set to move up. The belt can be used to define a mobile Jacobi scheme as follows:

Each forward step of the belt corresponds to a batch of rotations. Numbers that pass one another during the step are paired off, a Jacobi rotation between those two indices is performed, and the rows and columns are interchanged. For

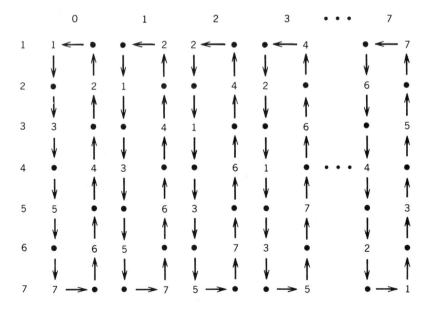

Figure 6.1 Conveyor belt scheme for $n = 7$.

example, during the step from state 0 to state 1 the number 1 moves downward, passing the upward moving number 2. Similarly 3 passes 4, and 5 passes 6. The number 7 moves to the right, passing nothing. Thus the pairing for this batch is $(1, 2), (3, 4)$, and $(5, 6)$. This is what we called pairing A.

The movement of the numbers on the belt corresponds to the physical movement of the rows and columns in the matrix. Think of each row (and column) as being labeled by a number that represents its initial position in the matrix. As the row moves around the matrix, it retains its label while changing its physical position. The numbers on the conveyor belt correspond to the labels. The current physical position of a row (or column) is determined by its height on the belt. For example, in state 1 the label 2 is at the top, the next highest label is 1, followed by 4, 3, 6, 5, and 7. This means that the current physical ordering of rows and columns in the array is 2, 1, 4, 3, 6, 5, 7. In general the physical location of a row (column) can be determined by checking its height on the belt, as determined by the number that stands at the same level in the left margin of the figure.

As the belt moves from state 1 to state 2, the second batch of rotations and interchanges is performed. The number 2 stays at the top and passes nothing, numbers 1, 3, and 5 move downward, passing the upward moving numbers 4, 6, and 7, respectively, so the pairing, stated in terms of the labels, is $(1, 4), (3, 6)$, and $(5, 7)$. However, it is clear from the figure that if we express the same pairing in terms of the current physical locations of the rows and columns, it is $(2, 3), (4, 5)$, and $(6, 7)$. This is what we have called pairing B. In the next batch, performed as the belt moves from state 2 to state 3, the effective pairing is $(2,4), (1,6)$, and $(3,7)$, but in terms of the current physical locations of the rows and columns, it is just pairing A again.

By now it should be clear that the scheme defined by the conveyor belt is exactly the same as the mobile Jacobi scheme (6.1.13). The beauty of the conveyor belt model is that it shows the flow of the rows and columns through the array very clearly. Specifically, notice that in n steps each number moves exactly half way around the belt. For example, the number 1 moves from the top left corner to the bottom right corner. In the process it passes each other number exactly once. In fact, every number passes each other number once. Therefore, in a set of n consecutive batches, every possible pairing of indices occurs exactly once. In other words, a Jacobi sweep takes place. At the end of the sweep, the ordering of the rows and columns has been reversed. Another n steps on the conveyor belt effect another sweep and bring the rows and columns back to their original positions.

Exercise 6.1.20

    (a) Write down all of the pairings that occur in the first seven batches in Figure 6.1, and check that each pairing occurs exactly once. Of course the schedule of pairings should be identical to that obtained for the case $n = 7$ in Exercise 6.1.16.

    (b) Draw the analog of Figure 6.1 for the case $n = 6$. Write down all of the pairings that occur in six steps, and note that each possible pairing occurs exactly once. Compare your schedule of pairings with that which you obtained for the case $n = 6$ in Exercise 6.1.16.   □

To see that the sweeps performed by the mobile Jacobi scheme (6.1.13) are essentially special cyclic Jacobi sweeps, we will adopt yet another point of view. Consider the problem of trying to perform a special cyclic Jacobi sweep (6.1.11) on a computer that has as many processors as we need. To what extent can the rotations be performed in parallel? The first n pairings are $(1, 2), (1, 3), \ldots, (1, n)$, all of which involve row 1. Thus none can be performed until the ones before it have been carried out. For example, the $(1, 3)$ rotation cannot be performed until after the $(1, 2)$ rotation. Once the $(1, 3)$ rotation has been done, the $(1, 4)$ rotation can be carried out. But notice that this rotation and the subsequent rotations $(1, 5), \ldots, (1, n)$ do not involve either of the indices 2 and 3. Therefore the $(2, 3)$ rotation can be carried out at any time after the $(1, 3)$ rotation, even though it does not appear in the ordering until after the $(1, n)$ rotation (cf., Exercise 6.1.19). Thus the rotations $(1, 4)$ and $(2, 3)$ can be performed simultaneously.

After the $(2, 3)$ rotation come $(2, 4), (2, 5), \ldots, (2, n)$, followed by $(3, 4)$, $(3, 5), \ldots, (3, n)$, and so on. The $(1, 5)$ and $(2, 4)$ rotations can be performed simultaneously, as can $(1, 6)$ and $(2, 5)$. But notice that now $(3, 4)$ can be done as well, so in fact $(1, 6)$, $(2, 5)$, and $(3, 4)$ can be done together. By now we have established a pattern. For example, the next two batches of rotations will be $(1, 7)$, $(2, 6)$, and $(3, 5)$, followed by $(1, 8)$, $(2, 7)$, $(3, 6)$, and $(4, 5)$. As you can see, the parallelism increases as we move further into the sweep. Eventually it peaks and then steadily declines, as shown in Table 6.2, which gives the schedule for an entire sweep in the case $n = 7$.

Exercise 6.1.21

(a) Compile tables analogous to Table 6.2 for the cases $n = 5, 6$, and 8.

(b) Show that in general $2n - 3$ batches (some consisting of a single rotation) are needed to perform a complete sweep. This is almost twice the optimum number. $\square$

TABLE 6.2
Schedule of pairings for parallel execution of a special cyclic Jacobi sweep in the case $n = 7$

1	(1,2)		
2	(1,3)		
3	(1,4)	(2,3)	
4	(1,5)	(2,4)	
5	(1,6)	(2,5)	(3,4)
6	(1,7)	(2,6)	(3,5)
7	(2,7)	(3,6)	(4,5)
8	(3,7)	(4,6)	
9	(4,7)	(5,6)	
10	(5,7)		
11	(6,7)		

The schedule depicted in Table 6.2 is clearly quite inefficient, and one is tempted to give up on the special cyclic ordering. However, one must not overlook the fact that the second sweep can begin before the first is finished. Referring to Table 6.2, we notice that from step 8 on, indices 1 and 2 are not involved in any subsequent rotations. This means that the $(1, 2)$ rotation of the second sweep can be done on step 8. Similarly, the $(1, 3)$ rotation can be done on step 9, since from that point on the index 3 is not involved in any rotations of the first sweep. On step 10, rotations $(1, 4)$ and $(2, 3)$ can be performed, and on step 11 rotations $(1, 5)$ and $(2, 4)$ can be done. To see the complete picture, we can depict the second sweep, see Table 6.3, as the mirror image of the first sweep. The two sweeps can then be integrated as shown in Table 6.4. Clearly we can join any number of sweeps in this way to produce a very efficient scheme.

Exercise 6.1.22   Generate the analogs of Tables 6.2, 6.3, and 6.4 for the case $n = 8$. Be careful! The two sweeps do not fit together quite so compactly as in the odd case. ☐

As our next task let us see how we can reformulate the special cyclic Jacobi scheme so that it can be implemented efficiently on a multiprocessor array. In the interest of minimizing difficulties in communication, we would like to perform rotations between adjacent rows and columns only. The first pairing is $(1, 2)$, for which the indices are adjacent. Notice that if we interchange these rows and columns after the rotation, index 1 will lie between 2 and 3. In particular, 1 and 3 will be adjacent. We can then carry out the $(1, 3)$ rotation and, following the pattern of the first step, interchange the rows and columns. Now 1 is adjacent to 4, and we can perform the $(1, 4)$ rotation. But notice that the interchange of 1 and 3 also caused rows and columns 2 and 3 to become adjacent once again. Thus the $(2, 3)$ rotation can be performed concurrently with the $(1, 4)$ rotation. Naturally we interchange the rows and columns also. Continuing in this manner we find that we can carry out the entire sweep.

This is illustrated in Table 6.5 for the case $n = 7$. The vertical bars indicate the pairings. At every step the indices are in just the right position for the rotations

TABLE 6.3
Second special cyclic Jacobi
sweep for $n = 7$

1			(1,2)
2			(1,3)
3		(2,3)	(1,4)
4		(2,4)	(1,5)
5	(3,4)	(2,5)	(1,6)
6	(3,5)	(2,6)	(1,7)
7	(4,5)	(3,6)	(2,7)
8		(4,6)	(3,7)
9		(5,6)	(4,7)
10			(5,7)
11			(6,7)

TABLE 6.4
Two overlapping sweeps in the case $n = 7$

1	(1,2)		
2	(1,3)		
3	(1,4)	(2,3)	
4	(1,5)	(2,4)	
5	(1,6)	(2,5)	(3,4)
6	(1,7)	(2,6)	(3,5)
7	(2,7)	(3,6)	(4,5)
8	(3,7)	(4,6)	(1,2)
9	(4,7)	(5,6)	(1,3)
10	(5,7)	(2,3)	(1,4)
11	(6,7)	(2,4)	(1,5)
12	(3,4)	(2,5)	(1,6)
13	(3,5)	(2,6)	(1,7)
14	(4,5)	(3,6)	(2,7)
15		(4,6)	(3,7)
16		(5,6)	(4,7)
17			(5,7)
18			(6,7)

to be performed in accordance with the schedule in Table 6.2. At the end of the sweep the order of the indices has been reversed. A second sweep would be as shown in Table 6.6; this is just the first sweep turned upside down. It is now a simple matter to combine the two sweeps so that they are performed in accordance with the schedule in Table 6.4. As we noted above, any number of sweeps can

TABLE 6.5
Special cyclic Jacobi sweep with interchanges

	Batch											
	1	2	3	4	5	6	7	8	9	10	11	
1	1	2	2	3	3	4	4	5	5	6	6	7
2	2	1	3	2	4	3	5	4	6	5	7	6
3	3	3	1	4	2	5	3	6	4	7	5	5
4	4	4	4	1	5	2	6	3	7	4	4	4
5	5	5	5	5	1	6	2	7	3	3	3	3
6	6	6	6	6	6	1	7	2	2	2	2	2
7	7	7	7	7	7	7	1	1	1	1	1	1

TABLE 6.6
The second special cyclic Jacobi sweep with interchanges

	Batch 1	2	3	4	5	6	7	8	9	10	11	
7	7	7	7	7	7	7	1	1	1	1	1	1
6	6	6	6	6	6	1	7	2	2	2	2	2
5	5	5	5	5	1	6	2	7	3	3	3	3
4	4	4	4	1	5	2	6	3	7	4	4	4
3	3	3	1	4	2	5	3	6	4	7	5	5
2	2	1	3	2	4	3	5	4	6	5	7	6
1	1	2	2	3	3	4	4	5	5	6	6	7

be joined in this way. This is illustrated in Table 6.7, which shows two complete sweeps and the beginning of a third.

Exercise 6.1.23   Generate the analogs of Tables 6.5, 6.6, and 6.7 for the case $n = 8$. ☐

Referring to Table 6.7, we see that after the first four batches, which serve to start up the process, three rotations are carried out in each batch. This is for the case $n = 7$. In Exercise 6.1.23 you worked out the case $n = 8$. From your analog of Table 6.7 you can see that after an initial start-up period of five batches, a steady-state situation is reached, in which there are alternating batches of four and three rotations. The important thing to notice now is that in either case, once

TABLE 6.7
Overlapping special cyclic Jacobi sweeps with interchanges

	Batch 1	2	3	4	5	6	7	8	9	10	11	12	13	14	15	16	17	18	19
1	1	2	2	3	3	4	4	5	5	6	6	7	7	1	1	2	2	3	3
2	2	1	3	2	4	3	5	4	6	5	7	6	1	7	2	1	3	2	4
3	3	3	1	4	2	5	3	6	4	7	5	1	6	2	7	3	1	4	2
4	4	4	4	1	5	2	6	3	7	4	1	5	2	6	3	7	4	1	5
5	5	5	5	5	1	6	2	7	3	1	4	2	5	3	6	4	7	5	1
6	6	6	6	6	6	1	7	2	1	3	2	4	3	5	4	6	5	7	6
7	7	7	7	7	7	7	1	1	2	2	3	3	4	4	5	5	6	6	7

this steady state is reached, the batches consist of pairings A and B alternately with interchanges.

In other words, the special cyclic Jacobi method, as we have just formulated it, is essentially the same as the mobile Jacobi scheme (6.1.13). The only difference is that the special cyclic Jacobi method has a start-up period. Since we know that the special cyclic Jacobi method always converges, we can guarantee that the mobile Jacobi scheme will always converge, simply by including the start-up period at the beginning of the first sweep. According to Luk and Park (1987), the start-up is unnecessary.

Exercise 6.1.24   For even values of n the mobile Jacobi scheme (6.1.13) is not optimal. It completes a sweep in n batches, whereas an optimal scheme would take only $n - 1$. Brent and Luk (1985) have proposed an optimal scheme based on the scheduling algorithm that has been used for many years in round-robin tournaments of all kinds. This scheme is easily modeled by a conveyor belt that is somewhat different from the one that was considered earlier. Figure 6.2 shows the belt in its initial state for the case $n = 8$. Perhaps it is best to imagine that the numbers represent chess players sitting at a long table with $n/2$ chessboards on it. Each player plays a game with the player sitting directly opposite. Once all games are finished, each player gets up and follows the arrow to the next seat. That is, all except player 1, who stays in the same seat all night. Once all players are seated, the next round of games commences. In our case we are interested in pairings for rotations, not chess games, but the principle is the same.

(a)   It is obvious that in one complete trip around the loop, each player will sit opposite player 1 exactly once. Show that in the case $n = 8$, each player sits opposite each other player exactly once in one complete cycle.

(b)   Show that for arbitrary even n, each player is paired with each other player exactly once in one cycle.   □

Each cycle consists of $n - 1$ batches of $n/2$ pairings, so this is an optimal schedule of pairings. Unfortunately the cyclic Jacobi method based on this schedule does not always converge; Luk and Park (1987) have given a counterexample that is just an extension of the bizarre example of Hansen (1963) cited earlier. However, the method works well in practice.

Exercise 6.1.25   The round-robin idea can also be used with odd values of n, as shown in Figure 6.3. The stationary chair remains vacant, and the player sitting opposite the vacant chair sits out the round. It turns out that this scheme is equivalent to the mobile Jacobi scheme (6.1.13). You can prove this by showing that this scheme

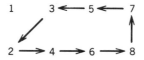

Figure 6.2   Conveyor belt model of a round-robin chess tournament.

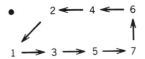

Figure 6.3   Round-robin chess tournament for odd n.

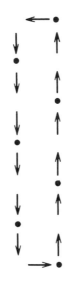

Figure 6.4   Fill in the blanks.

is the same as the conveyor belt scheme of Figure 6.1, except that the labels must be permuted.

(a)   Replace the blank spots in Figure 6.4 with the numbers 1 through 7 in such a way that the scheme defined by this conveyor belt is the same as the scheme defined by Figure 6.3. Notice that the two schemes define pairings differently. In one the pairings are determined by who sits opposite whom, whereas in the other they are determined by who passes whom.

(b)   Explain how to replace the dots (in addition to the blank spots) in Figure 6.4 by the numbers 1 through 7 in such as way as to improve your understanding of the equivalence of the two schemes.

(c)   Prove that for any odd n the two conveyor belt schemes are equivalent.

□

6.2
THE SLICING OR BISECTION METHOD

In Section 4.5 we saw how to reduce an arbitrary real symmetric matrix to tridiagonal form. The method of slicing, also known as the bisection or Sturm sequence method, is a simple, inexpensive procedure for calculating the eigenvalues of a real, symmetric, tridiagonal matrix. As we shall see, it is best suited for finding a few specified eigenvalues or all eigenvalues in some specified interval.

Our approach to the method of slicing is based on the notions of congruence and inertia. We begin by noting that if $A \in \mathbb{R}^{n \times n}$ is symmetric, then SAS^T is also symmetric, for any $S \in \mathbb{R}^{n \times n}$. A symmetric matrix $B \in \mathbb{R}^{n \times n}$ is said to be *congruent* to A if there exists a nonsingular matrix $S \in \mathbb{R}^{n \times n}$ such that $B = SAS^T$. It is important that S is nonsingular.

Exercise 6.2.1   Show that (a) A is congruent to A; (b) if B is congruent to A, then A is congruent to B; (c) if B is congruent to A and C is congruent to B, then C is congruent to A. (In other words, congruence is an equivalence relation.) ☐

The eigenvalues of a symmetric $A \in \mathbb{R}^{n \times n}$ are all real. Let $\nu(A)$, $\zeta(A)$, and $\pi(A)$ denote the number of negative, zero, and positive eigenvalues of A, respectively. The order triple $(\nu(A), \zeta(A), \pi(A))$ is called the *inertia* of A. The key result for our development is Sylvester's law of inertia, which states that congruent matrices have the same inertia.

THEOREM 6.2.1 *(Sylvester's Law of Inertia)*   Let $A, B \in \mathbb{R}^{n \times n}$ be symmetric, and suppose B is congruent to A. Then $\nu(B) = \nu(A)$, $\zeta(B) = \zeta(A)$, and $\pi(B) = \pi(A)$.

The proof is deferred to the end of the section.

We will use Sylvester's law to slice the set of eigenvalues into subsets. Let $\lambda_1, \ldots, \lambda_n$ be the eigenvalues of A, ordered so that $\lambda_1 \geq \lambda_2 \geq \ldots \geq \lambda_n$. Suppose that for any given $\sigma \in \mathbb{R}$, we can determine the inertia of $A - \sigma I$. The number $\pi(A - \sigma I)$ equals the number of eigenvalues of A that are greater than σ. If $\pi(A - \sigma I) = i$, with $0 < i < n$, then

$$\lambda_n \leq \cdots \leq \lambda_{i+1} \leq \sigma < \lambda_i \leq \cdots \leq \lambda_1$$

This splits the set of eigenvalues of A into two subsets. We will see that by repeatedly slicing the set of eigenvalues with systematically chosen values of σ, we can determine all the eigenvalues of A with great precision.

Exercise 6.2.2   Suppose you have a subroutine that can calculate $\pi(A - \sigma I)$ for any value of σ. Devise an algorithm that calculates all the eigenvalues of A in some specified interval $(a, b]$ with error less than some specified tolerance $\epsilon > 0$. (A solution will be provided later in this section.) ☐

Let us now see how we can compute $\pi(A - \sigma I)$. In Theorem 1.7.25 we found that $A - \sigma I$ can be expressed as a product

$$A - \sigma I = L_\sigma D_\sigma L_\sigma^T \tag{6.2.2}$$

where L_σ is unit lower triangular and D_σ is diagonal, provided that the leading principal minors of $A - \sigma I$ are all nonzero. Assuming that this is the case, the decomposition (6.2.2) yields $\pi(A - \sigma I)$ very easily: D_σ is congruent to $A - \sigma I$, so it has the same inertia. Since D_σ is diagonal, $\pi(D_\sigma)$ equals the number of positive elements on the main diagonal of D_σ.

It is a routine matter to calculate the LDL^T decomposition of a symmetric matrix by Gaussian elimination. In the present context $A - \sigma I$ is tridiagonal, so the computation is especially simple and inexpensive. It is simple enough that we will derive it here in a few lines, even though it is a special case of an algorithm that we already discussed in Chapter 1. By Theorem 1.9.3 the band structure of $A - \sigma I$ is inherited by the triangular factors in the LU decomposition. Since the LDL^T decomposition is just a variant of the LU decomposition, the matrix L_σ in (6.2.2) must be bidiagonal. Thus (6.2.2) can be written more explicitly as

$$
\begin{bmatrix}
\alpha_1 - \sigma & \beta_1 & & 0 \\
\beta_1 & \alpha_2 - \sigma & \ddots & \\
& \ddots & \ddots & \beta_{n-1} \\
0 & & \beta_{n-1} & \alpha_n - \sigma
\end{bmatrix}
$$

$$
=
\begin{bmatrix}
1 & & & 0 \\
l_1 & 1 & & \\
& \ddots & \ddots & \\
0 & & l_{n-1} & 1
\end{bmatrix}
\begin{bmatrix}
\delta_1 & & & 0 \\
& \delta_2 & & \\
& & \ddots & \\
0 & & & \delta_n
\end{bmatrix}
\begin{bmatrix}
1 & l_1 & & 0 \\
& 1 & \ddots & \\
& & \ddots & l_{n-1} \\
0 & & & 1
\end{bmatrix}
$$

It is a simple matter to carry out the matrix multiplications on the right-hand side and find that

$$
\alpha_1 - \sigma = \delta_1
$$
$$
\left.
\begin{aligned}
\beta_i &= l_i \delta_i \\
\alpha_{i+1} - \sigma &= \delta_{i+1} - l_i^2 \delta_i
\end{aligned}
\right\} \qquad i = 1, 2, \ldots, n-1
$$

Combining equations and solving for δ_{i+1} we find that

$$
\delta_1 = \alpha_1 - \sigma
$$
$$
\delta_{i+1} = \alpha_{i+1} - \sigma - \frac{\beta_i^2}{\delta_i} \qquad i = 1, \ldots, n-1 \tag{6.2.3}
$$

Thus $\delta_1, \ldots, \delta_n$ can be calculated very cheaply. The number of positive δ_i is $\pi(A - \sigma I)$.

Algorithm (6.2.3) breaks down whenever $\delta_i = 0$ for some i. In fact this occurs when and only when one of the leading principal minors of $A - \sigma I$ is zero.

Exercise 6.2.3 Suppose $\det(A_k - \sigma I) \neq 0$ for $k = 1, \ldots, j - 1$.

    (a) Show that $\delta_k \neq 0$ for $k = 1, \ldots, j - 1$.

    (b) Show that $\delta_j = 0$ if and only if $\det(A_j - \sigma I) = 0$. $\square$

A zero value of one of the δ_i is an extremely rare event. The ith leading principal minor $\det(A_i - \sigma I)$ is a polynomial in σ of degree i, so it has at most i distinct zeros. Thus the number of distinct values of σ for which some leading principal minor of $(A - \sigma I)$ is zero is at most $\sum_{i=1}^{n} i = \frac{1}{2}n(n + 1)$. There is little likelihood of hitting one of these numbers exactly. Nevertheless we must be prepared to take some corrective action in those rare cases when some δ_i is zero.

The simplest imaginable remedy turns out to work very well in practice. Simply replace the offending zero by a small number ϵ whose magnitude is comparable to that of the roundoff errors in the problems. The effect of this action is the same as that of replacing α_i by $\alpha_i + \epsilon$ in the matrix A. In Section 5.5 we saw that the symmetric eigenvalue problem is well behaved with respect to small perturbations of the data. A perturbation of magnitude ϵ causes a pertubation of the same magnitude in the eigenvalues. Thus the error introduced by this maneuver is insignificant. For those who are dissatisfied with this remedy, a second remedy is given in Exercise 6.2.9.

Now that we know how to calculate $\pi(A - \sigma I)$ for any value of σ, let us see how we can systematically determine the eigenvalues of A. For convenience we will introduce the shorthand notation

$$\pi(\sigma) = \pi(A - \sigma I)$$

A straightforward bisection approach works very well. Suppose we wish to find all eigenvalues in the interval $(a, b]$. We begin by calculating $\pi(a)$ and $\pi(b)$ to find out how many eigenvalues lie in the interval. If $\pi(a) = i$ and $\pi(b) = j$, then

$$a < \lambda_i \leq \cdots \leq \lambda_{j+1} \leq b$$

so there are $i - j$ eigenvalues in the interval. Now let $\sigma = (a+b)/2$, the midpoint of the interval, and calculate $\pi(\sigma)$. From this we can deduce how many eigenvalues lie in each of the intervals $(a, \sigma]$ and $(\sigma, b]$. More precise information can be obtained by bisecting each of these intervals, and so on. Any interval that is found to contain no eigenvalues can be removed from further consideration. An interval that contains a single eigenvalue can be bisected repeatedly until the eigenvalue has been located with sufficient precision. If we know that $\lambda_k \in (\sigma_1, \sigma_2]$, where $\sigma_2 - \sigma_1 = \epsilon$, then the approximation

$$\lambda_k \approx \frac{\sigma_1 + \sigma_2}{2}$$

is in error by at most $\epsilon/2$.

EXAMPLE 6.2.4   Consider the matrix

$$A = \begin{bmatrix} 2 & 1 & & & & \\ 1 & 2 & 1 & & \text{\Large 0} & \\ & 1 & 2 & 1 & & \\ & & 1 & 2 & 1 & \\ \text{\Large 0} & & & 1 & 2 & 1 \\ & & & & 1 & 2 \end{bmatrix} \in \mathbb{R}^{6 \times 6}$$

We will determine how many eigenvalues of A lie in the interval $(0, 1.5]$ and locate each of them roughly. To find $\pi(0)$ we set $\sigma = 0$ in (6.2.3) to obtain

$$\delta_1 = 2 \qquad\qquad \delta_4 = 2 - \frac{3}{4} = \frac{5}{4}$$

$$\delta_2 = 2 - \frac{1}{2} = \frac{3}{2} \qquad \delta_5 = 2 - \frac{4}{5} = \frac{6}{5}$$

$$\delta_3 = 2 - \frac{2}{3} = \frac{4}{3} \qquad \delta_6 = 2 - \frac{5}{6} = \frac{7}{6}$$

Thus $\pi(0) = 6$; all six eigenvalues of A are greater than zero. Now setting $\sigma = 1.5$ in (6.2.3), we find that

$$\delta_1 = \frac{1}{2} \qquad\qquad \delta_4 = \frac{1}{2} - \frac{6}{7} = -\frac{5}{14}$$

$$\delta_2 = \frac{1}{2} - 2 = -\frac{3}{2} \qquad \delta_5 = \frac{1}{2} + \frac{14}{5} = \frac{33}{10}$$

$$\delta_3 = \frac{1}{2} + \frac{2}{3} = \frac{7}{6} \qquad \delta_6 = \frac{1}{2} - \frac{10}{33} = \frac{13}{66}$$

Since four of the δ_i are positive, $\pi(1.5) = 4$.

   Thus four eigenvalues of A are greater than 1.5, and we conclude that A has two eigenvalues in the interval $(0, 1.5]$. We bisect the interval by calculating $\pi(0.75)$. This is tedious by hand but poses no problem for a calculator. We find that $\pi(0.75) = 5$. Thus A has one eigenvalue in each of the intervals $(0, 0.75]$ and $(0.75, 1.5]$. Bisecting the first interval several times, we get the following results:

σ	$\pi(\sigma)$
0.375	5
0.1875	6
0.28125	5

Thus the eigenvalue λ_6 lies in the interval $(0.1875, 0.28125]$. Bisecting the interval $(0.75, 1.5]$ several times, we obtain the following:

σ	$\pi(\sigma)$
1.125	4
0.9375	4
0.84375	4

Thus the eigenvalue λ_5 lies in the interval $(0.75, 0.84375]$. We can compute these eigenvalues with any desired accuracy, limited only by the precision of the computation, by carrying out further bisections.

Exercise 6.2.4   Let A be the matrix of Example 6.2.4. Determine how many eigenvalues A has in the interval $(2.3, 3.9]$. For each of these eigenvalues find an interval of length 0.1 that contains the eigenvalue.   □

The previous example and exercise illustrate a strength of the bisection method. If only rough bounds on the eigenvalues are needed, the bisection method can get them very cheaply. It is also clear, however, that if very accurate eigenvalues are needed, the bisection method can get them very cheaply. It is also clear, however, that if very accurate eigenvalues are required, many bisections will be needed. If an eigenvalue is known to lie in an interval of length 1, s bisections are needed to confine it to an interval of length 2^{-s}. Fortunately bisections are very inexpensive. Examining (6.2.3), we see that each bisection requires only some $2n$ subtractions and n divisions if the squares β_i^2 have been calculated in advance. In addition each of the n numbers δ_i must be compared with zero.

It is clear that the slicing method could be run very efficiently on a parallel computer. If p processors are available, then p subintervals can be searched simultaneously. See, for example, the multisection algorithm of Lo, Philippe, and Sameh (1987).

Exercise 6.2.5   Write an algorithm that uses bisection to calculate all eigenvalues of a given symmetric, tridiagonal matrix A that lie in a specified interval $(a, b]$. Each eigenvalue should be determined with an error not exceeding some specified tolerance $\epsilon > 0$.   □

Exercise 6.2.6   Write a Fortran subroutine that implements the algorithm of Exercise 6.2.5. Be sure to write clear, structured code, document it with a reasonable number of comments, and document clearly all of the variables that are passed to and from the subroutine. Try out your subroutine on the matrix

$$
\begin{bmatrix}
16 & 1 &   &   &   \\
 1 & 8 & 1 &   &   \\
   & 1 & 4 & 1 &   \\
   &   & 1 & 2 & 1 \\
   &   &   & 1 & 1
\end{bmatrix}
$$

whose eigenvalues are approximately 16.124, 8.126, 4.244, 2.208, and 0.2972. Now try

$$
\begin{bmatrix}
 2 & -1 &   &   &   &   &   \\
-1 & 2 & -1 &   &   &   &   \\
   & -1 & 2 & -1 &   &   &   \\
   &   & -1 & 2 & -1 &   &   \\
   &   &   & -1 & 2 & -1 &   \\
   &   &   &   & -1 & 2 & -1 \\
   &   &   &   &   & -1 & 2
\end{bmatrix} \in \mathbb{R}^{7 \times 7}
$$

whose eigenvalues are $4 \sin^2 (k\pi/16)$, $k = 1, \ldots, 7$.

If you would like to experiment with larger versions of this matrix, the n-by-n version has eigenvalues $4\sin^2[k\pi/2(n+1)]$, $k = 1,\ldots,n$. {The eigenvectors are $v^{(k)}$, $k = 1,\ldots,n$, where $v_i^{(k)} = \sin[ik\pi/(n+1)]$, $i,k = 1,\ldots,n$.} ☐

If we wish to use the slicing method to find the complete set of eigenvalues of a matrix, we need only find an interval $(a, b]$ that is guaranteed to contain all the eigenvalues. This is not difficult.

Exercise 6.2.7 Let $A \in \mathbb{R}^{n \times n}$ be symmetric.

(a) Let $r = \|A\|$, where $\|\cdot\|$ denotes any induced matrix norm. Show that all eigenvalues of A lie in the interval $[-r, r]$. Thus for any $\epsilon > 0$, no matter how small, the half-open interval $(-r - \epsilon, r]$ contains all eigenvalues of A.

(b) Let

$$\hat{r} = \max_{1 \le i \le n} \sum_{j=1}^{n} |a_{ij}|$$

Show that all eigenvalues of A lie in $[-\hat{r}, \hat{r}]$. ☐

A better bound is provided by the Gerschgorin disk theorem, which states that the eigenvalues of A lie within certain disks in the complex plane. Let $A \in \mathbb{C}^{n \times n}$, and for $i = 1, \ldots, n$, let

$$r_i = \sum_{\substack{j=1 \\ j \ne i}}^{n} |a_{ij}|$$

The ith *Gerschgorin disk* D_i associated with A is the set of $\lambda \in \mathbb{C}$ such that $|\lambda - a_{ii}| \le r_i$. This is the closed disk of radius r_i centered at a_{ii}.

Theorem 6.2.5 *(Gerschgorin Disk Theorem)* Let $A \in \mathbb{C}^{n \times n}$. Then each eigenvalue of A lies within one of the Gerschgorin disks of A.

Proof Let λ be an eigenvalue of A, and let $v = [v_1, \ldots, v_n]^T \ne 0$ be an associated eigenvector. Let i be an index such that

$$|v_i| = \max_{1 \le j \le n} |v_j|$$

Examining the ith component of the vector equation $Av = \lambda v$, we see that

$$\lambda v_i = a_{ii} v_i + \sum_{\substack{j=1 \\ j \ne i}}^{n} a_{ij} v_j$$

Subtracting $a_{ii}v_i$ from both sides and taking absolute values, we obtain

$$|\lambda - a_{ii}||v_i| \le \sum_{\substack{j=1 \\ j \ne i}}^{n} |a_{ij}||v_j| \le \left(\sum_{\substack{j=1 \\ j \ne i}}^{n} |a_{ij}| \right) |v_i| = r_i|v_i|$$

Dividing both sides by the positive constant $|v_i|$, we find that $|\lambda - a_{ii}| \le r_i$; that is, λ is contained in the ith Gerschgorin disk of A. ☐

Exercise 6.2.8 Suppose $A \in \mathbb{R}^{n \times n}$ is symmetric.

(a) Show that each eigenvalue of A must lie in one of the "Gerschgorin intervals" $[a_{ii} - r_i, a_{ii} + r_i]$. Conclude that all eigenvalues of A lie in the interval $[c, b]$, where

$$c = \min_{i=1,\dots,n} (a_{ii} - r_i) \qquad b = \max_{i=1,\dots,n} (a_{ii} + r_i)$$

(b) Let $[-\hat{r}, \hat{r}]$ be the interval defined in Exercise 6.2.7, part (b). Show that $[c, b] \subseteq [-\hat{r}, \hat{r}]$. ☐

Exercise 6.2.9 This exercise develops a second way of dealing with the case $\delta_i = 0$. In order to keep the notation simple, we will consider the case $\delta_1 = \alpha_1 - \sigma = 0$, which establishes the general pattern.

(a) Explain how to proceed if $\beta_1 = 0$.

(b) For this and subsequent parts of the problem assume $\beta_1 \ne 0$. Let

$$A_1 = \begin{bmatrix} 0 & \beta_1 \\ \beta_1 & \alpha_2 - \sigma \end{bmatrix}$$

Show that

$$A - \sigma I = \left[\begin{array}{cc|ccc} & & 0 & \cdots & 0 \\ \multicolumn{2}{c|}{A_1} & \beta_2 & \cdots & 0 \\ \hline 0 & \beta_2 & & & \\ \vdots & \vdots & & A_2 & \\ 0 & 0 & & & \end{array} \right]$$

$$= \left[\begin{array}{cc|ccc} & & 0 & \cdots & 0 \\ \multicolumn{2}{c|}{I} & 0 & \cdots & 0 \\ \hline \beta_2/\beta_1 & 0 & & & \\ \vdots & \vdots & & I & \\ 0 & 0 & & & \end{array} \right] \left[\begin{array}{cc|ccc} & & 0 & \cdots & 0 \\ \multicolumn{2}{c|}{A_1} & 0 & \cdots & 0 \\ \hline 0 & 0 & & & \\ \vdots & \vdots & & A_2 & \\ 0 & 0 & & & \end{array} \right] \left[\begin{array}{cc|ccc} & & \beta_2/\beta_1 & \cdots & 0 \\ \multicolumn{2}{c|}{I} & 0 & \cdots & 0 \\ \hline 0 & 0 & & & \\ \vdots & \vdots & & I & \\ 0 & 0 & & & \end{array} \right]$$

(c)   Show that A_1 has one positive and one negative eigenvalue. Don't forget that $\beta_1 \neq 0$. [Can you do this one two different ways? The easiest way is to examine $\det(A_1)$.]

(d)   Explain how to use the results of parts (b) and (c) to determine $\pi(\sigma)$.

☐

In Chapter 2 we discussed the dangers of using small pivots in Gaussian elimination. Since the LDL^T decomposition algorithm is just a form of Gaussian elimination, we might expect that the slicing algorithm would yield poor results whenever a small (pivot) δ_i occurs. The next exercise shows, at least informally, that this fear is unjustified. For this we can thank the fact that the outcome of the algorithm depends only on the signs of the δ_i, not on their exact values.

Exercise 6.2.10   Suppose $\beta_i \neq 0$. Show that if $|\delta_i|$ is sufficiently small, then:

(a) δ_i and δ_{i+1} have opposite signs. Thus if a small perturbation in δ_i occurs and causes it to change sign, this will be offset by a simultaneous sign change in δ_{i+1}.

(b) $|\delta_{i+1}|$ is large, and δ_{i+1} is approximately proportional to δ_i^{-1}.

(c)   The difference between δ_{i+2} and $\alpha_{i+2} - \sigma$ is small, and it is approximately proportional to δ_i. Thus the smaller $|\delta_i|$ is (and the larger $|\delta_{i+1}|$ is), the less impact δ_{i+1} has on the subsequent calculations.

(d)   For comparison with Exercise 6.2.9 consider the case $i = 1$. Show that if $\alpha_3 - \sigma \neq 0$, then for sufficiently small $|\delta_1|$ the slicing method yields the same result as the method proposed in Exercise 6.2.9 for case $\delta_i = 0$.

(e)   The conclusion of part (a) is valid for $i = 1, \ldots, n - 1$. Discuss the case $i = n$. What happens when a small δ_n is perturbed slightly so that it changes sign? Do you think that this will cause any problems?

(f)   Discuss the case $\beta_i = 0$.   ☐

A rigorous justification of the slicing method requires a backward error analysis. See [AEP, pages 299ff.] and [HAC, contribution II/5].

Sturm Sequence Viewpoint

As before, we assume that $A \in \mathbb{R}^{n \times n}$ is symmetric and tridiagonal. For $k = 1, 2, \ldots, n$, let A_k denote the k-by-k leading principal submatrix of A, and let $p_k(\lambda) = \det(A_k - \lambda I)$. Thus $p_k(\lambda)$ is a polynomial in λ of degree k. The sequence of polynomials $p_1, p_2, \ldots, p_n$ is called a *Sturm sequence*; it satisfies a number of interesting properties. First, it can be shown [SEP, page 186] that the zeros are interlaced: if $\lambda_1 \geq \lambda_2 \geq \cdots \geq \lambda_{k+1}$ are the eigenvalues of A_{k+1}, and $\mu_1 \geq \mu_2 \geq \cdots \geq \mu_k$ are the eigenvalues of A_k, then

$$\lambda_1 \geq \mu_1 \geq \lambda_2 \geq \mu_2 \geq \cdots \geq \mu_k \geq \lambda_{k+1}$$

Exercise 6.2.11   The interlacing property can be used to determine how many positive eigenvalues $A - \sigma I$ has, for any σ.

  (a) Show that if $p_1(0) > 0$ [$p_1(0) < 0$] then A has at least one positive (negative) eigenvalue.

  (b) For simplicity assume for the remainder of the problem that $p_1(0), p_2(0), \ldots, p_n(0)$ are all nonzero; that is, none of $A_1, \ldots, A_n$ has a zero eigenvalue. [Recall that $p_k(0) = \det(A_k)$ equals the product of the eigenvalues of A_k.] Let π_k and ν_k denote the number of positive and negative eigenvalues of A_k, respectively. Thus $\pi_k + \nu_k = k$. Show that $\pi_{k+1} \geq \pi_k$ and $\nu_{k+1} \geq \nu_k$. Show that if $p_k(0)$ and $p_{k+1}(0)$ have the same sign, then $\pi_{k+1} = \pi_k + 1$ and $\nu_{k+1} = \nu_k$, whereas if $p_k(0)$ and $p_{k+1}(0)$ have opposite signs, then $\pi_{k+1} = \pi_k$ and $\nu_{k+1} = \nu_k + 1$.

  (c) Letting $p_0 = 1$, show that the number of negative eigenvalues of A is equal to the number of sign changes in the sequence

$$p_0(0), p_1(0), p_2(0), \ldots, p_n(0) \tag{6.2.6}$$

  and the number of positive eigenvalues is equal to the number of "non–sign changes" in (6.2.6).

  (d) Show how the sequence $p_0(\sigma), p_1(\sigma), \ldots, p_n(\sigma)$ can be used to determine the number of eigenvalues of A that are greater than σ.   □

In order to implement the results of Exercise 6.2.11, we need to be able to calculate the sequence $p_0(\sigma), p_1(\sigma), \ldots, p_n(\sigma)$ for any given σ. The next exercise provides a recursion that allows us to calculate these numbers inexpensively.

Exercise 6.2.12   Calculate $p_k(\sigma) = \det(A_k - \sigma I)$ by performing a column expansion on the kth column.   □

Solution:

$$p_k(\sigma) = (\alpha_k - \sigma)p_{k-1}(\sigma) - \beta_{k-1}^2 p_{k-2}(\sigma) \tag{6.2.7}$$

This equation is valid for $k = 1, 2, \ldots, n$ provided we define $p_0 = 1$ and $p_1 = 0$.
  Equation (6.2.7) allows the calculation of $p_0(\sigma), \ldots, p_n(\sigma)$ for a given σ in $O(n)$ arithmetic operations. In practice there is some danger of overflow. Since $p_n(\sigma)$ has degree n (which could be large), even modest values of σ can give rise to very large values of $p_n(\sigma)$. It was found that this problem can be remedied by working with the ratios

$$\delta_k(\sigma) = \frac{p_k(\sigma)}{p_{k-1}(\sigma)}$$

instead of the numbers $p_k(\sigma)$. These ratios do not have such a tendency to grow.

Exercise 6.2.13 (a) Show that the number of eigenvalues of A that are greater (less) than σ is equal to the number of positive (negative) numbers in the sequence $\delta_1(\sigma), \delta_2(\sigma), \ldots, \delta_n(\sigma)$.
(b) Clearly $\delta_1(\sigma) = \alpha_1 - \sigma$. Use (6.2.7) to determine a recursion for $\delta_k(\sigma)$, $k = 2, \ldots, n$. □

If you worked Exercise 6.2.12 correctly, you got a formula identical to (6.2.3). Thus the Sturm sequence method using the ratios $\delta_k(\sigma)$ is identical to the slicing method using the LDL^T decomposition.

Proof of Sylvester's Law of Inertia

Exercise 6.2.14 (a) Show that the dimension of the null space of a diagonal matrix is equal to the number of zeros on the main diagonal.
(b) Show that the dimension of the null space of a symmetric matrix $A \in \mathbb{R}^{n \times n}$ is equal to the number of eigenvalues of A that are zero.
(c) Let

$$A = \begin{bmatrix} 0 & 1 \\ 0 & 0 \end{bmatrix}$$

Show that the dimension of the null-space of A does not equal the number of zero eigenvalues. □

Let $A \in \mathbb{R}^{n \times n}$ and $B \in \mathbb{R}^{n \times n}$ be two congruent, symmetric matrices. Thus there exists a nonsingular $S \in \mathbb{R}^{n \times n}$ such that $B = S^T A S$. We wish to show that $\pi(A) = \pi(B)$, $\nu(A) = \nu(B)$, and $\zeta(A) = \zeta(B)$. The null-spaces of A and B obviously have the same dimension, so by Exercise 6.2.14, $\zeta(A) = \zeta(B)$. It follows that $\pi(A) + \nu(A) = \pi(B) + \nu(B)$. If we can show that $\pi(A) = \pi(B)$, it will follow that $\nu(A) = \nu(B)$ also. By Spectral Theorem 4.4.12 there exist orthogonal matrices $P, Q \in \mathbb{R}^{n \times n}$ and diagonal matrices $D, E \in \mathbb{R}^{n \times n}$ such that

$$D = P^T A P \qquad \text{and} \qquad E = Q^T B Q$$

The main-diagonal entries of D and E are the eigenvalues of A and B, respectively. Thus we have to show that D and E have the same number of positive entries on the main diagonal. Suppose this is not the case. Then, without loss of generality, E has more positive main-diagonal entries than D has. Letting $C = P^T S Q$, we have

$$E = C^T D C$$

Let $x \in \mathbb{R}^n$ be a nonzero vector that satisfies

$$\begin{aligned} x_i &= 0 && \text{if } e_{ii} \leq 0 \\ (Cx)_i &= 0 && \text{if } d_{ii} > 0 \end{aligned} \qquad (6.2.8)$$

We know that such an x exists because (6.2.8) is a homogeneous linear system with n unknowns and fewer than n equations. Then

$$x^T E x = \sum_{e_{ii}>0} e_{ii} x_i^2 \ge 0$$

In fact $x^T E x > 0$ because $x_i \ne 0$ for some i. Let $z = Cx$. Then $z_i = 0$ whenever $d_{ii} > 0$, so $z^T D z \le 0$. Thus $0 < x^T E x = z^T D z \le 0$. This is a contradiction. $\qquad\qquad\qquad\qquad\qquad\qquad\qquad\qquad\qquad\qquad\qquad\qquad$ □

6.3
CUPPEN'S DIVIDE-AND-CONQUER METHOD

In this section we will examine a relatively recent method for calculating the complete eigensystem of a symmetric tridiagonal matrix by a divide-and-conquer technique. This method works well in parallel computing environments because it breaks the task into a large number of subtasks that can be performed independently. Tests have shown that the divide-and-conquer technique is much faster than standard programs based on the QR algorithm on parallel machines. Surprisingly it also seems to be faster for large matrices on sequential machines. This success has been attributed to deflation effects. We will look at the general ideas without going into complete detail. The interested reader can consult the original works of Bunch, Nielsen, and Sorensen (1978), Cuppen (1981), and Dongarra and Sorensen (1987). The method is based on a procedure for updating the eigenvalues and eigenvectors of a matrix that has been subjected to a perturbation of rank one. We will look at this procedure first.

Rank-One Updates

Let $\tilde{A} \in I\!\!R^{n \times n}$ be a symmetric matrix whose eigenvalues and eigenvectors are known. Thus we have an orthogonal $\tilde{Q} \in I\!\!R^{n \times n}$ and diagonal $\tilde{D} \in I\!\!R^{n \times n}$ such that $\tilde{A} = \tilde{Q}\tilde{D}\tilde{Q}^T$ (Theorem 4.4.12). The columns of $\tilde{Q}$ are orthonormal eigenvectors of $\tilde{A}$, and main-diagonal entries of $\tilde{D}$ are the eigenvalues. Suppose we would like to know the eigenvalues and eigenvectors of a perturbed matrix

$$A = \tilde{A} + H$$

where the perturbation $H \in I\!\!R^{n \times n}$ is a symmetric matrix of rank one. It would be nice if we could somehow use the eigensystem of $\tilde{A}$ to compute the eigensystem of A cheaply, rather than starting over from scratch. It turns out that this can be done.

Exercise 6.3.1   (a) Let $B \in I\!\!R^{n \times n}$ be a matrix of rank one. Show that there exist nonzero vectors u and $v \in I\!\!R^n$ such that $B = uv^T$. (b) Let $H \in I\!\!R^{n \times n}$ be a symmetric matrix of rank one. Show that there exist nonzero $\rho \in \mathbb{R}$ and $w \in \mathbb{R}^n$ such that $H = \rho w w^T$. Show that w can be chosen so that $\|w\|_2 = 1$, in which case ρ is unique. To what extent is w nonunique? (c) Let $H = \rho w w^T$, where $\|w\|_2 = 1$. Determine the complete eigensystem of H. □

From Exercise 6.3.1 part (b), we see that $A = \tilde{A} + \rho w w^T$ for some nonzero $\rho \in \mathbb{R}$ and $w \in \mathbb{R}^n$ with $\|w\|_2 = 1$. Since $\tilde{A} = \tilde{Q} \tilde{D} \tilde{Q}^T$,

$$A = \tilde{Q}(\tilde{D} + \rho z z^T)\tilde{Q}^T$$

where $z = \tilde{Q}^T w$. We seek an orthogonal $Q \in \mathbb{R}^{n \times n}$ and a diagonal $D \in \mathbb{R}^{n \times n}$ such that $A = QDQ^T$. If we can find an orthogonal $\hat{Q}$ and a diagonal D such that $\tilde{D} + \rho z z^T = \hat{Q} D \hat{Q}^T$, then $A = QDQ^T$, where

$$Q = \tilde{Q} \hat{Q}$$

Thus is suffices to consider the matrix $\tilde{D} + \rho z z^T$.

The eigenvalues of $\tilde{D}$ are $\tilde{d}_1, \ldots, \tilde{d}_n$, its main-diagonal entries, and the associated eigenvectors are $e_1, e_2, \ldots, e_n$, the standard basis vectors of $\mathbb{R}^n$. The next exercise shows that it can happen that some of these are also eigenvalues and eigenvectors of $\tilde{D} + \rho z z^T$.

Exercise 6.3.2 Let z_i denote the ith component of z. Suppose $z_i = 0$ for some i. (a) Show that the ith column of $\tilde{D} + \rho z z^T$ is $\tilde{d}_i e_i$ and the ith row is $\tilde{d}_i e_i^T$. (b) Show that $\tilde{d}_i$ is an eigenvalue of $\tilde{D} + \rho z z^T$ with associated eigenvector e_i. (c) Show that if $q = [q_1, q_2, \ldots, q_n]^T \in \mathbb{R}^n$ is an eigenvector of $\tilde{D} + \rho z z^T$, associated with some eigenvalue $\lambda \neq \tilde{d}_i$, then $q_i = 0$. □

From this exercise we see that if $z_i = 0$, we get an eigenvalue and eigenvector for free. Furthermore, the ith row and column of $\tilde{D} + \rho z z^T$ can be ignored during the computation of the other eigenvalues and eigenvectors. Thus we effectively work with a submatrix; that is, we deflate the problem.

Although exact zero components of z are unusual, it is not at all unusual that some components are small enough that we might consider ignoring them. The next exercise helps us decide when to set a small z_i to zero.

Exercise 6.3.3 Given λ and x with $\|x\|_2 = 1$, an approximate eigenvalue and eigenvector for the matrix B, one way to test the quality of the approximation is to calculate the norm of the residual $Bx - \lambda x$. If $\|Bx - \lambda x\|$ is small, then λ and x are in some sense a good approximate eigenpair. (For symmetric matrices this criterion is very good, as we know from Section 5.5) Calculate the norm of the residual in the case $B = \tilde{D} + \rho z z^T$, $\lambda = \tilde{d}_i$, and $x = e_i$. Show that

$$\|(\tilde{D} + \rho z z^T)e_i - \tilde{d}_i e_i\|_2 = |\rho z_i|$$

□

Thus if $|\rho z_i|$ is sufficiently small, we can set z_i to zero and deflate the ith row and column from the problem. The usual criterion is $|\rho z_i| < \epsilon$, where ϵ is some accepted noise level for the problem.

A second opportunity for deflation presents itself whenever $\tilde{D}$ has two eigenvalues that are very close together. First consider the extreme case in which two eigenvalues are exactly equal, say $\tilde{d}_i = \tilde{d}_j$. Let M denote the 2-by-2 submatrix of $\tilde{D} + \rho z z^T$ consisting of the intersection of rows i and j with columns i and j. Then

$$M = d \begin{bmatrix} 1 & 0 \\ 0 & 1 \end{bmatrix} + \rho \begin{bmatrix} z_i \\ z_j \end{bmatrix} [z_i \ z_j]$$

where $d = \tilde{d}_i = \tilde{d}_j$. Let $U = \begin{bmatrix} c & -s \\ s & c \end{bmatrix}$ be a rotator that creates a zero in the second component of $[z_i \ z_j]^T$, say

$$\begin{bmatrix} c & -s \\ s & c \end{bmatrix} \begin{bmatrix} z_i \\ z_j \end{bmatrix} = \begin{bmatrix} \tilde{z}_i \\ 0 \end{bmatrix}$$

Then, since the identity matrix commutes with U,

$$UMU^T = d \begin{bmatrix} 1 & 0 \\ 0 & 1 \end{bmatrix} + \rho \begin{bmatrix} \tilde{z}_i \\ 0 \end{bmatrix} [\tilde{z}_i \ 0]$$

Let $\tilde{U} \in \mathbb{R}^{n \times n}$ be the plane rotator that has the rotator U embedded in rows and columns i and j. Then

$$\tilde{U}(\tilde{D} + \rho z z^T)\tilde{U}^T = \tilde{D} + \rho \tilde{z}\tilde{z}^T$$

where $\tilde{z}$ has a zero in the jth position. Thus the jth row and column can be deleted from the transformed matrix.

Exercise 6.3.4 Suppose $\tilde{D}$ has k eigenvalues that are equal. Show how to transform the problem so that $k - 1$ rows and columns can be deleted.   □

Now suppose eigenvalues $\tilde{d}_i$ and $\tilde{d}_j$ are close but not necessarily equal. Then

$$M = \begin{bmatrix} \tilde{d}_i & 0 \\ 0 & \tilde{d}_j \end{bmatrix} + \rho \begin{bmatrix} z_i \\ z_j \end{bmatrix} [z_i \ z_j]$$

Let $\delta = \tilde{d}_i - \tilde{d}_j$, and let U be the rotator defined above. Then

$$UMU^T = \begin{bmatrix} c^2\tilde{d}_i + s^2\tilde{d}_j & 0 \\ 0 & s^2\tilde{d}_i + c^2\tilde{d}_j \end{bmatrix} + \rho \begin{bmatrix} \tilde{z}_i \\ 0 \end{bmatrix} [\tilde{z}_i \ 0] + cs\delta \begin{bmatrix} 0 & 1 \\ 1 & 0 \end{bmatrix} \quad (6.3.1)$$

If $|cs\delta| < \epsilon$, the last term is negligible. If we set it to zero, we can delete row and column j from the big problem.

Exercise 6.3.5 (a) Verify equation (6.3.1). (b) How small does $|\delta|$ have to be to guarantee that deflation will always occur, regardless of what U turns out to be? (c) If $|c|$ or $|s|$ is small, deflation can take place even if $|\delta|$ is much larger than ϵ. What condition is signaled by a small value of $|c|$ or $|s|$?   □

After making all possible deflations, we can assume that $\tilde{D}$ has distinct eigen-values and all components of z are nonzero. Without loss of generality we will assume that $\tilde{d}_1 < \tilde{d}_2 < \ldots < \tilde{d}_n$. It is not hard [following Cuppen (1981)] to see what the eigenvalues and eigenvectors of $\tilde{D} + \rho z z^T$ must look like. Let λ be an eigenvalue with associated eigenvector q. Then $(\tilde{D} + \rho z z^T)q = \lambda q$; that is

$$(\tilde{D} - \lambda I)q + \rho(z^T q)z = 0 \tag{6.3.2}$$

We were able to interchange $z^T q$ with z because $z^T q$ is a scalar. Before going on, we need to know that $z^T q \neq 0$ and that λ is distinct from $\tilde{d}_1, \tilde{d}_2, \ldots, \tilde{d}_n$. These facts are not hard to verify.

Exercise 6.3.6 (a) Use (6.3.2) to show that if $z^T q = 0$, then $\lambda = \tilde{d}_i$, for some i, and $q = c e_i$, for some nonzero scalar c. Conclude that $z_i = 0$; this contradicts one of our assumptions. (b) Use the ith row of (6.3.2) to show that if $\lambda = \tilde{d}_i$, then either $z^T q = 0$ or $z_i = 0$. Thus $\lambda \neq \tilde{d}_i$. $\qquad\square$

Since $\lambda \neq \tilde{d}_i$ for $i = 1, \ldots, n$, $(\tilde{D} - \lambda I)^{-1}$ exists. Multiplying equation (6.3.2) by $(\tilde{D} - \lambda I)^{-1}$, we obtain

$$q + \rho(z^T q)(\tilde{D} - \lambda I)^{-1}z = 0 \tag{6.3.3}$$

Multiplying this equation by z^T on the left and then dividing through by the nonzero scalar $z^T q$, we see that

$$1 + \rho z^T(\tilde{D} - \lambda I)^{-1}z = 0 \tag{6.3.4}$$

Since $(\tilde{D} - \lambda I)^{-1}$ is a diagonal matrix, (6.3.4) can also be written as

$$1 + \rho \sum_{i=1}^{n} \frac{z_i^2}{\tilde{d}_i - \lambda} = 0 \tag{6.3.5}$$

Every eigenvalue of $\tilde{D} + \rho z z^T$ must satisfy (6.3.5), which is known as the *secular equation*.

A great deal can be learned about the solutions of (6.3.5) by studying the behavior of the function

$$f(\lambda) = 1 + \rho \sum_{i=1}^{n} \frac{z_i^2}{\tilde{d}_i - \lambda}$$

f is a rational function with the n distinct poles $\tilde{d}_1, \tilde{d}_2, \ldots, \tilde{d}_n$. Since

$$f'(\lambda) = \rho \sum_{i=1}^{n} \frac{z_i^2}{(\tilde{d}_i - \lambda)^2}$$

we see that if $\rho > 0$, then $f'(\lambda) > 0$ for all λ a which f is defined. Thus each continuous piece of f is a strictly increasing function. As λ approaches the pole

$\tilde{d}_i$, the function is dominated by the ith term $z_i^2/(\tilde{d}_i - \lambda)$. Thus

$$\lim_{\lambda \to \tilde{d}_i -} f(\lambda) = +\infty \qquad \lim_{\lambda \to \tilde{d}_i +} f(\lambda) = -\infty$$

Each of the terms $z_i^2/(\tilde{d}_i - \lambda)$ tends to zero as $\lambda \to \pm\infty$, so

$$\lim_{\lambda \to \pm\infty} f(\lambda) = 1$$

Thus if $\rho > 0$, the graph of f is as in Figure 6.5. From this graph it is immediately obvious that the secular equation $f(\lambda) = 0$ has exactly one solution between each pair of poles and one additional solution to the right of $\tilde{d}_n$. Denoting these solutions $d_1 < d_2 < \cdots < d_n$, we have

$$\tilde{d}_1 < d_1 < \tilde{d}_2 < d_2 < \tilde{d}_3 < \cdots < d_{n-1} < \tilde{d}_n < d_n$$

In addition it is not hard to get the upper bound $d_n < \tilde{d}_n + \rho$.

Exercise 6.3.7   (a) Recall from Exercise 4.4.32 that the trace of a matrix equals the sum of its eigenvalues. Using this fact show that $d_n < \tilde{d}_n + \rho$ if $\rho > 0$. (b) More generally, show that if $\rho > 0$, then $d_i = \tilde{d}_i + c_i\rho$, where $c_i > 0, i = 1, \ldots, n$, and $\sum_{i=1}^{n} c_i = 1$. Thus $d_i < \tilde{d}_i + \rho$ for all i.   □

If $\rho < 0$, then $f'(\lambda) < 0$ for all λ at which f is defined, so each continuous piece of f is strictly decreasing. the graph of f is then essentially the reverse of what is depicted in Figure 6.5.

Exercise 6.3.8   (a) Graph f in the case $\rho < o$. Conclude that if $\rho < 0$, the eigenvalues of $\tilde{D} + \rho z z^T$ satisfy

$$\tilde{d}_1 + \rho < d_1 < \tilde{d}_1 < d_2 < \tilde{d}_2 < \cdots < \tilde{d}_{n-1} < d_n < \tilde{d}_n$$

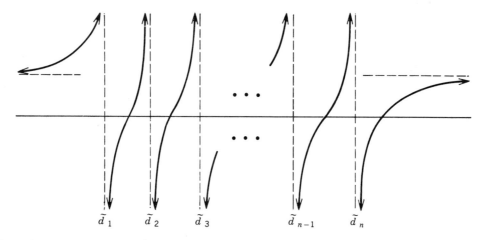

Figure 6.5   Graph of $f(\lambda) = 1 + \rho \sum_{i=1}^{n} z_i^2/(\tilde{d}_i - \lambda)$ in the case $\rho > 0$.

(b) Show that if $\rho < 0$, then $d_i = \tilde{d}_i + c_i\rho$, where $c_i > 0$, $i = 1, \ldots, n$, and $\sum_{i=1}^{n} c_i = 1$. Thus $\tilde{d}_i + \rho < d_i$ for all i. □

The following theorem summarizes our findings.

THEOREM 6.3.6 Let $\tilde{D}$ be a diagonal matrix with main-diagonal entries $\tilde{d}_1 < \tilde{d}_2 < \cdots < \tilde{d}_n$. Let $\rho \in \mathbb{R}$ and $z = [z_1, z_2, \ldots, z_n]^T \in \mathbb{R}^n$ with $\| z \|_2 = 1$. Assume $z_i \neq 0$, $i = 1, \ldots, n$. Let $d_1 \leq d_2 \leq \cdots \leq d_n$ denote the eigenvalues of $\tilde{D} + \rho z z^T$. Then if $\rho > 0$, $\tilde{d}_1 < d_1 < \tilde{d}_2 < d_2 < \tilde{d}_3 < \cdots < \tilde{d}_n < d_n < \tilde{d}_n + \rho$; if $\rho < 0$, $\tilde{d}_1 + \rho < d_1 < \tilde{d}_1 < d_2 < \tilde{d}_2 < \cdots < \tilde{d}_{n-1} < d_n < \tilde{d}_n$. For all $\rho \in \mathbb{R}$, there exist positive constants $c_1, \ldots, c_n$ such that $c_1 + c_2 + \cdots + c_n = 1$ and $d_i = \tilde{d}_i + \rho c_i$, $i = 1, \ldots, n$.

Since we know that each interval $(\tilde{d}_i, \tilde{d}_{i+1})$ contains exactly one eigenvalue of $\tilde{D} + \rho z z^T$, it is easy to solve the secular equation (6.3.5) numerically and thereby determine the eigenvalues of $\tilde{D} + \rho z z^T$ with as much accuracy as we please (subject, of course, to the precision of the machine and the accuracy of the data). The bisection method, which we employed in the previous section, could also be used here. A faster method is Newton's method, with which you may be familiar. Since $f(\lambda)$ is a rational function, Bunch, Nielsen, and Sorensen proposed a method that uses rational function approximations. They proved that their method always converges monotonically and quadratically. It has also been shown to work very well in practice. See Bunch, Nielsen, and Sorenson (1978) and Dongarra and Sorenson (1987) for details.

Once the eigenvalues of $\tilde{D} + \rho z z^T$ have been found, it is a simple matter to compute the eigenvectors. By (6.3.3) an eigenvector q associated with the eigenvalue λ is given by

$$q = c(\tilde{D} - \lambda I)^{-1} z$$

where c is any nonzero scalar. Thus the components of q are given by

$$q_i = \frac{c z_i}{\tilde{d}_i - \lambda} \qquad i = 1, \ldots, n$$

Exercise 6.3.9 Show that the choices of c that give $\| q \|_2 = 1$ are $c = \pm \sqrt{\rho / f'(\lambda)}$. □

Exercise 6.3.10 (a) Write the matrix

$$a = \begin{bmatrix} 5 & 2 \\ 2 & 2 \end{bmatrix}$$

in the form $\tilde{D} + \rho z z^T$, where $\tilde{D}$ is diagonal. (There are many ways to do this. A simple choice is $z = (1/\sqrt{2})\,[1\ \ 1]^T$.)
(b) Graph the function $f(\lambda)$ given by this choice of ρ and z.
(c) Calculate the eigenvalues and eigenvectors of A by performing a rank-one update. □

Exercise 6.3.11   Our derivation of the secular equation followed Cuppen (1981). Earlier approaches were based on the characteristic polynomial of $\tilde{D} + \rho z z^T$. Let ψ_n denote the characteristic polynomial. More generally let ψ_k be the characteristic polynomial of the leading k-by-k submatrix.

(a)   Show that

$$\psi_n(\lambda) = (\lambda - \tilde{d}_n)\psi_{n-1}(\lambda) - \rho_n^2 \prod_{k=1}^{n-1}(\lambda - \tilde{d}_k)$$

[*Hint:* Making use of the fact that the determinant function is linear in each of its rows, break the determinant $\det(\lambda I - (\tilde{D} + \rho z z^T))$ into a sum of two determinants with bottom rows $[0, \ldots, 0, \lambda - \tilde{d}_{nn}]$ and $\rho z_n[z_1, z_2, \ldots, z_n]$, respectively. The second determinant can be simplified by subtracting the appropriate multiple of the last row from each of the other rows. These row operations of type 1 do not change the determinant.]

(b)   Prove by induction that

$$\psi_n(\lambda) = \prod_{k=1}^{n}(\lambda - \tilde{d}_k) - \rho \sum_{j=1}^{n} z_j^2 \prod_{k \neq j}(\lambda - \tilde{d}_k)$$

(c)   Show that if $z_i \neq 0$, then $\tilde{d}_i$ is not a root of the characteristic equation.

(d)   Show that for $\lambda \neq \tilde{d}_1, \ldots, \tilde{d}_n$,

$$\psi_n(\lambda) = \left[\prod_{k=1}^{n}(\lambda - \tilde{d}_k)\right]\left[1 + \rho \sum_{j=1}^{n} \frac{z_j^2}{\tilde{d}_j - \lambda}\right]$$

Conclude that λ is an eigenvalue of $\tilde{D} + \rho z z^T$ if and only if λ is a root of the secular equation (6.3.5).   □

The Divide-and-Conquer Scheme

The scheme operates on symmetric, tridiagonal matrices, so we must begin by tridiagonalizing the matrix using the algorithm described in Section 4.6. Having done this, we can assume that our matrix has the form

$$A = \begin{bmatrix} \alpha_1 & \beta_1 & & & & 0 \\ \beta_1 & \alpha_2 & \beta_2 & & & \\ & \beta_2 & \alpha_3 & \ddots & & \\ & & \ddots & \ddots & \beta_{n-1} \\ 0 & & & \beta_{n-1} & \alpha_n \end{bmatrix}$$

This can be rewritten as

$$A = \tilde{A} + \rho w w^T$$

where $\rho = 2\beta_i$,

$$\tilde{A} = \begin{bmatrix} \tilde{A}_1 & 0 \\ 0 & \tilde{A}_2 \end{bmatrix} = \left[\begin{array}{cccc|cccc} \alpha_1 & \beta_1 & & & & & & \\ \beta_1 & \alpha_2 & \ddots & & & & & \\ & \ddots & \ddots & \beta_{i-1} & & & 0 & \\ & & \beta_{i-1} & (\alpha_i - \beta_i) & & & & \\ \hline & & & & (\alpha_{i+1} - \beta_i) & \beta_{i+1} & & \\ & & 0 & & \beta_{i+1} & \ddots & \ddots & \\ & & & & & \ddots & \ddots & \beta_{n-1} \\ & & & & & & \beta_{n-1} & \alpha_n \end{array} \right]$$

and

$$w = \frac{1}{\sqrt{2}} \begin{bmatrix} 0 \\ \vdots \\ 0 \\ 1 \\ \hline 1 \\ 0 \\ \vdots \\ 0 \end{bmatrix} \tag{6.3.7}$$

This can be done for any choice of i. If $i \approx n/2$ is chosen, then $\tilde{A}_1$ and $\tilde{B}_1$ will be of about the same size. Suppose we calculate the eigenvalues and eigenvectors of $\tilde{A}_1$ and $\tilde{A}_2$ by any method to get $\tilde{A}_1 = \tilde{Q}_1 \tilde{D}_1 \tilde{Q}_1^T$ and $\tilde{A}_2 = \tilde{Q}_2 \tilde{D}_2 \tilde{Q}_2^T$, where $\tilde{Q}_1$ and $\tilde{Q}_2$ are orthogonal and $\tilde{D}_1$ and $\tilde{D}_2$ are diagonal. Then $\tilde{A} = \tilde{Q} \tilde{D} \tilde{Q}^T$, where

$$\tilde{Q} = \begin{bmatrix} \tilde{Q}_1 & 0 \\ 0 & \tilde{Q}_2 \end{bmatrix}, \quad \text{and} \quad \tilde{D} = \begin{bmatrix} \tilde{D}_1 & 0 \\ 0 & \tilde{D}_2 \end{bmatrix} \tag{6.3.8}$$

$\tilde{Q}$ is orthogonal and $\tilde{D}$ is diagonal. Thus

$$A = \tilde{Q} \tilde{D} \tilde{Q}^T + \rho w w^T = \tilde{Q}(\tilde{D} + \rho z z^T)\tilde{Q}^T$$

where

$$z = \tilde{Q}^T w \tag{6.3.9}$$

We can determine the eigensystem of $\tilde{D} + \rho z z^T$ by the rank-one update scheme outlined above. This gives $\tilde{D} + \rho z z^T = \hat{Q} D \hat{Q}^T$, where $\hat{Q}$ is orthogonal and D is diagonal. Thus $A = Q D Q^T$, where

$$Q = \tilde{Q} \hat{Q} \tag{6.3.10}$$

Exercise 6.3.12   Summarize all calculations required for the computation of the eigensystem of A, given the eigensystems of $\tilde{A}_1$ and $\tilde{A}_2$. Count the flops required for each step, including all of the operations involved in the rank-one update. [It is fair to assume that the cost of calculating each eigenvalue is $O(n)$ flops. Why?] Which step is the most expensive? Which operations can be done in parallel? □

Exercise 6.3.13   The choice of w and ρ that we used to split A is far from the only possibility. Determine all choices of ρ and w such that $w = [0 \; \cdots \; 0 \; c \, | \, s \; 0 \; \cdots \; 0]^T$, $c^2 + s^2 = 1$, and $A = \tilde{A} + \rho w w^T$, where the $(i+1, \, i)$ entry of $\tilde{A}$ is zero. □

So far we have not specified how to calculate the eigensystems of $\tilde{A}_1$ and $\tilde{A}_2$. The divide-and-conquer approach is to divide each of these problems into two subproblems, just as the original problem was. The resulting four subproblems can be broken into eight smaller problems, and so on. If the subdivision process is carried out as far as possible, trivial 1-by-1 and 2-by-2 problems are eventually reached. Alternatively we can stop the subdivision process at some point and solve the many subproblems by the QR algorithm, for example. This is the approach taken by Dongarra and Sorenson (1987). We then obtain the solution of the original problem by a sequence of rank-one updates.

The divide-and-conquer algorithm is highly parallelizable. The original problem is rapidly divided into many small subproblems that can be solved independently. If k subdivisions are carried out, we get 2^k subproblems. Once these are solved, 2^{k-1} rank-one updates can be performed on pairs of subproblems independently to reduce their number from 2^k to 2^{k-1}. In the next step 2^{k-2} updates are done to reduce their number from 2^{k-1} to 2^{k-2}, and so on. While the number of simultaneous rank-one updates decreases at each step as we move back up toward the original problem, the opportunity for parallelism within each rank-one update increases. An update of a j-by-j submatrix requires that j eigenvalues be calculated. These can be calculated simultaneously, as can the associated eigenvectors. In addition the eigenvector matrix $\hat{Q}$ must be multiplied by $\tilde{Q}$ to yield Q (6.3.10). This can also be done in parallel. The extent to which all of these operations can be parallelized depends directly on the size of the matrix.

It is also important to discuss the parallelizability of the reduction to tridiagonal form. This is, after all, an expensive part of the algorithm.

Exercise 6.3.14   Review the algorithm for reducing a symmetric matrix to tridiagonal form. Show that the first step is highly parallelizable. Subsequent steps are like the first step, but they act on progressively smaller matrices. Show that the parallelizability decreases as the algorithm progresses. □

Exercise 6.3.14 shows that the parallelizability of the reduction to tridiagonal form decreases with each step of the algorithm. This can be offset by the fact that the operations on the tridiagonal matrix can be begun before the reduction to tridiagonal form is complete. Once j steps of the reduction have been carried out, the upper j-by-j submatrix of the tridiagonal matrix is available. If any processors are idle, they can begin operations on this submatrix. By the end of the reduction to tridiagonal form, the eigenvalue calculation can be well under way.

Calculation of Eigenvalues Only If only eigenvalues are wanted, the number of calculations can be reduced substantially. This is far from evident at first. In order to calculate $z = \tilde{Q}^T w$ as in (6.3.9), we need the matrix $\tilde{Q}$ from (6.3.8), so it appears that we need the complete sets of eigenvectors $\tilde{Q}_1$ and $\tilde{Q}_2$ from the subproblems. However the vector w given by (6.3.7) has nonzero entries only in positions i and $i + 1$. It follows immediately that only the last row of $\tilde{Q}_1$ and the first row of $\tilde{Q}_2$ are needed.

Exercise 6.3.15 Suppose $Q = \tilde{Q}\hat{Q}$. Show that the jth row of Q depends only on the jth row of $\tilde{Q}$. Show that the divide-and-conquer algorithm to calculate eigenvalues requires only that the first and last row of each intermediate eigenvector matrix [e.g., Q in (6.3.10) or $\tilde{Q}_1, \tilde{Q}_2$ in (6.3.8)] be calculated. How much impact does this have on the overall cost of the calculations? $\qquad$ $\square$

7

The Singular Value Decomposition

7.1
INTRODUCTION

In order to read this chapter, you will have to have read the first four sections of Chapter 4. In Section 7.2 we will study a variant of the QR algorithm, for which you will need to have read most of the material in Sections 4.5 to 4.8 and Section 5.3 as well. This chapter is largely independent of Chapter 6.

Throughout the chapter we will restrict our attention to real matrices. This is done solely to simplify the exposition; the generalization to the complex setting is routine.

Let $A \in \mathbb{R}^{n \times n}$ be symmetric. Then by Corollary 4.4.14 there exists an orthonormal basis $v_1, \ldots, v_n$ of $\mathbb{R}^n$, consisting of eigenvectors of A. Each v_i satisfies $Av_i = \lambda_i v_i$, where λ_i is the (real) eigenvalue associated with v_i. These relationships can be expressed by the following diagram,

$$
\begin{array}{ccc}
 & A & \\
 & \lambda_1 & \\
v_1 & \longrightarrow & v_1 \\
 & \lambda_2 & \\
v_2 & \longrightarrow & v_2 \\
\vdots & & \vdots \\
 & \lambda_n & \\
v_n & \longrightarrow & v_n
\end{array}
\qquad (7.1.1)
$$

which portrays the action of A as a linear transformation mapping $\mathbb{R}^n$ into $\mathbb{R}^n$. The diagram describes the action of A completely, since A is completely determined by its action on a basis of $\mathbb{R}^n$. Diagram (7.1.1) is equivalent to the statement (Theorem 4.4.13) that there exists an orthogonal matrix V and a diagonal matrix D such that $A = VDV^T$. The columns of V are $v_1, v_2, \ldots, v_n$, and the main-diagonal entries of D are $\lambda_1, \lambda_2, \ldots, \lambda_n$.

It is reasonable to ask to what extent (7.1.1) can be generalized to nonsymmetric matrices. If A is nonsymmetric but normal, (7.1.1) continues to be true, except that some of $\lambda_1, \ldots, \lambda_n$ and $v_1, \ldots, v_n$ are complex. If A is not normal but simple, we must give up the orthogonality of the basis. If A is not simple, (7.1.1) ceases to hold. One well-known generalization is the Jordan canonical form. See, for example, Lancaster and Tisminetsky (1985). If A is not square, say $A \in \mathbb{R}^{n \times m}$ with $n \ne m$, then even the Jordan canonical form does not exist. In this section we will develop an extension of (7.1.1), valid for all $A \in \mathbb{R}^{n \times m}$, called the singular value decomposition (SVD).

Only a moment's thought reveals a significant change that will have to be made in (7.1.1) if we wish to extend it to nonsquare matrices. Every $A \in \mathbb{R}^{n \times m}$ can be viewed as a linear transformation $A : \mathbb{R}^m \rightarrow \mathbb{R}^n$, mapping $\mathbb{R}^m$ into $\mathbb{R}^n$. The domain consists of m-tuples, while the range consists of n-tuples. Thus our extension of (7.1.1) will have to have different sets of vectors on left and right.

For the rest of this section A will denote a matrix in $\mathbb{R}^{n \times m}$. Recall from Section 3.5 that A has two important spaces associated with it—the *null space* and the *range*—given by

$$\mathcal{N}(A) = \{x \in \mathbb{R}^m \mid Ax = 0\}$$

$$\mathcal{R}(A) = \{Ax \mid x \in \mathbb{R}^m\}$$

The null space is a subspace of $\mathbb{R}^m$, and the range is a subspace of $\mathbb{R}^n$. Recall that the range is also called the column space of A (Exercise 3.5.7), and its dimension is called the *rank* of A, denoted rank(A).

THEOREM 7.1.2 If $A \in \mathbb{R}^{n \times m}$, then $m = \dim(\mathcal{N}(A)) + \dim(\mathcal{R}(A))$.

For a proof see any elementary linear algebra text. Better yet, prove it yourself.

Exercise 7.1.1 Prove Theorem 7.1.2. [*Outline:* Start with a basis for $\mathcal{N}(A)$. Extend it to a basis of $\mathbb{R}^m$. Show that the image under A of the appropriate subset of this basis is a basis of $\mathcal{R}(A)$. Count the vectors in each basis.] □

From Theorem 7.1.2 we see that rank(A) $= m - \dim(\mathcal{N}(A))$.

Our development of the SVD will be based on the matrices $A^T A \in \mathbb{R}^{m \times m}$ and $AA^T \in \mathbb{R}^{n \times n}$. Let us therefore explore the properties of these matrices and their relationships to A and A^T.

Exercise 7.1.2 (Review) Prove that $A^T A$ and AA^T are (a) symmetric, (b) positive semidefinite. □

THEOREM 7.1.3 $\mathcal{N}(A^T A) = \mathcal{N}(A)$.

Proof It is obvious that $\mathcal{N}(A) \subseteq \mathcal{N}(A^T A)$. To prove that $\mathcal{N}(A^T A) \subseteq \mathcal{N}(A)$, suppose $x \in \mathcal{N}(A^T A)$. Then $A^T Ax = 0$. Recall that the standard inner product in $\mathbb{R}^n$ is given by $(w, z) = z^T w$. An easy computation shows that $(A^T Ax, x) = (Ax, Ax)$ (cf., Lemma 3.5.2). Thus $0 = (A^T Ax, x) = (Ax, Ax) = \| Ax \|_2^2$. Therefore $Ax = 0$; that is $x \in \mathcal{N}(A)$. □

COROLLARY 7.1.4 Rank($A^T A$) = rank(A) = rank(A^T) = rank(AA^T).

Proof Rank($A^T A$) $= m - \dim(\mathcal{N}(A^T A)) = m - \dim(\mathcal{N}(A)) = \text{rank}(A)$. The second equation is a basic result of linear algebra. The third equation is the same as the first equation, except that the roles of A and A^T are reversed. □

THEOREM 7.1.5 If v is an eigenvector of $A^T A$ associated with a nonzero eigenvalue λ, then Av is an eigenvector of AA^T associated with the same eigenvalue.

Proof $AA^T(Av) = A(A^T A)v = A(\lambda v) = \lambda(Av)$. □

COROLLARY 7.1.6 $A^T A$ and AA^T have the same nonzero eigenvalues, counting multiplicity.

Exercise 7.1.3 Let v_1 and v_2 be eigenvectors of $A^T A$. Show that if v_1 and v_2 are orthogonal, then Av_1 and Av_2 are orthogonal. □

Exercise 7.1.4 Let $B \in \mathbb{R}^{m \times m}$ be a simple matrix with linearly independent eigenvectors $v_1, v_2, \ldots, v_m \in \mathbb{R}^m$, associated with eigenvalues $\lambda_1, \lambda_2, \ldots, \lambda_m \in \mathbb{R}$. Suppose $\lambda_1, \ldots, \lambda_r$ are nonzero and $\lambda_{r+1}, \ldots, \lambda_m$ are zero. Show that $\mathcal{R}(B) = \langle v_1, \ldots, v_r \rangle$. Therefore rank($B$) $= r$, the number of nonzero eigenvalues. □

The matrices $A^T A$ and AA^T are both symmetric and hence simple. Thus Corollary 7.1.6 and Exercise 7.1.4 yield a second proof that they have the same rank, which equals the number of nonzero eigenvalues. Since $A^T A$ and AA^T generally have different dimensions, they cannot have exactly the same eigenvalues. The difference is made up by a zero eigenvalue of the appropriate multiplicity. If rank ($A^T A$) $= \text{rank}(AA^T) = r$, and $r < m$, then $A^T A$ has zero as an eigenvalue of multiplicity $m - r$. If $r < n$, then AA^T has zero as an eigenvalue of multiplicity $n - r$.

Exercise 7.1.5 (a) Give an example of a matrix $A \in \mathbb{R}^{1 \times 2}$ such that $A^T A$ has zero as an eigenvalue and AA^T does not. (b) Where does the proof of Theorem 7.1.5 break down in the case of $\lambda = 0$? □

THEOREM 7.1.7 *(SVD Theorem)* Let $A \in \mathbb{R}^{n \times m}$ have rank r. Then there exist real numbers $\sigma_1 \geq \sigma_2 \geq \cdots \geq \sigma_r > 0$, an orthonormal basis $v_1, \ldots, v_m$ of $\mathbb{R}^m$, and an orthonormal basis $u_1, \ldots, u_n$ of $\mathbb{R}^n$, such that

$$
\begin{array}{llll}
Av_i = \sigma_i u_i & i = 1, \ldots, r & A^T u_i = \sigma_i v_i & i = 1, \ldots, r \\
Av_i = 0 & i = r+1, \ldots, m & A^T u_i = 0 & i = r+1, \ldots, n
\end{array}
$$
$$(7.1.8)$$

Equations 7.1.8 imply that $v_1, \ldots, v_m$ are eigenvectors of $A^T A$, $u_1, \ldots, u_n$ are eigenvectors of AA^T, and $\sigma_1^2, \ldots, \sigma_r^2$ are the nonzero eigenvalues of $A^T A$ and AA^T.

Proof You can easily verify that the assertions in the final sentence are true. This determines how $v_1, \ldots, v_m$ must be chosen. Let $v_1, \ldots, v_m$ be an orthonormal basis of $\mathbb{R}^m$ consisting of eigenvectors of A^TA, and let $\lambda_1, \ldots, \lambda_m$ be the associated eigenvalues. Since A^TA is positive semidefinite, all of its eigenvalues are nonnegative. Assume $v_1, \ldots, v_m$ are ordered so that $\lambda_1 \geq \lambda_2 \geq \cdots \geq \lambda_m$. Since $r = \text{rank}(A) = \text{rank}(A^TA)$, it must be that $\lambda_r > 0$ and $\lambda_{r+1} = \lambda_{r+2} = \cdots = \lambda_m = 0$. For $i = 1, \ldots, r$, define σ_i and u_i by

$$\sigma_i = \| Av_i \|_2 \qquad \text{and} \qquad u_i = \frac{1}{\sigma_i} Av_i$$

These definitions imply that $Av_i = \sigma_i u_i$ and $\| u_i \|_2 = 1$, $i = 1, \ldots, r$. The result of Exercise 7.1.3 implies that $u_1, \ldots, u_r$ are orthogonal and hence orthonormal. It is easy to show that $\sigma_i^2 = \lambda_i$, $i = 1, \ldots, r$. Indeed $\sigma_i^2 = \| Av_i \|_2^2 = (Av_i, Av_i) = (A^TAv_i, v_i) = (\lambda_i v_i, v_i) = \lambda_i$. It now follows easily that $A^Tu_i = \sigma_i v_i$, $i = 1, \ldots, r$, for $A^Tu_i = (1/\sigma_i)A^TAv_i = (\lambda_i/\sigma_i)v_i = \sigma_i v_i$.

The proof is now complete, except that we have not defined $u_{r+1}, \ldots, u_n$, assuming $r < n$. By Theorem 7.1.5 the vectors $u_1, \ldots, u_r$ are eigenvectors of AA^T associated with nonzero eigenvalues. Since $AA^T \in \mathbb{R}^{n \times n}$ and $\text{rank}(AA^T) = r$, AA^T must have a null-space of dimensions $n - r$. Let $u_{r+1}, \ldots, u_n$ be any orthonormal basis of $\mathcal{N}(AA^T)$. Noting that $u_{r+1}, \ldots, u_n$ are eigenvectors of AA^T associated with the eigenvalue zero, we see that $u_{r+1}, \ldots, u_n$ are orthogonal to $u_1, \ldots, u_r$. Thus $u_1, \ldots, u_n$ is an orthonormal basis of $\mathbb{R}^n$ consisting of eigenvectors of AA^T. Since $\mathcal{N}(AA^T) = \mathcal{N}(A^T)$, we have $A^Tu_i = 0$ for $i = r + 1, \ldots, n$. This completes the proof. $\qquad\square$

The numbers $\sigma_1, \ldots, \sigma_r$ are called the *singular values* of A. Let $k = \min\{n, m\}$. If $r < k$, it is usual to adjoin $k - r$ zero singular values $\sigma_{r+1} = \cdots = \sigma_k = 0$. The vectors $v_1, v_2, \ldots, v_m$ are called *right singular vectors* of A, and $u_1, u_2, \ldots, u_n$ are called *left singular vectors* of A. Singular vectors are not uniquely determined; they are no more uniquely determined than any eigenvectors of length 1. Any singular vector can be replaced by its opposite, and if A^TA or AA^T happens to have an eigenspace of dimension ≥ 2, an even greater loss of uniqueness occurs.

A^T has the same singular values as A. The right (left) singular vectors of A^T are the left (right) singular vectors of A.

Theorem 7.1.7 allows us to draw, for any $A \in \mathbb{R}^{n \times m}$, a diagram in the spirit of (7.1.1):

$$
\begin{array}{ccc}
 & A & \\
 & \sigma_1 & \\
v_1 & \longrightarrow & u_1 \\
 & \sigma_2 & \\
v_2 & \longrightarrow & u_2 \\
\vdots & & \vdots \\
 & \sigma_r & \\
v_r & \longrightarrow & u_r \\
\left.\begin{array}{c} v_{r+1} \\ \vdots \\ v_m \end{array}\right\} & \longrightarrow & 0
\end{array}
$$

An analogous diagram holds for A^T. Drawing the two diagrams side by side, we have

$$
\begin{array}{ccc}
& A & A^T \\
\end{array}
$$

$$(7.1.9)$$

which serves as a pictorial representation of the SVD theorem.

Exercise 7.1.6   What form does (7.1.9) take when $A \in \mathbb{R}^{n \times n}$ is symmetric? How does this compare with (7.1.1)? (Remember that singular values are nonnegative.) $\square$

The singular value decomposition displays orthonormal bases of the four spaces $\mathcal{R}(A)$, $\mathcal{N}(A)$, $\mathcal{R}(A^T)$, and $\mathcal{N}(A^T)$. It is clear from (7.1.9) that

$$
\begin{array}{ll}
\mathcal{R}(A^T) = \langle v_1, \ldots, v_r \rangle & \mathcal{R}(A) = \langle u_1, \ldots, u_r \rangle \\
\mathcal{N}(A) = \langle v_{r+1}, \ldots, v_m \rangle & \mathcal{N}(A^T) = \langle u_{r+1}, \ldots, u_n \rangle
\end{array}
$$

From these representations we see that $\mathcal{R}(A^T) = \mathcal{N}(A)^\perp$ and $\mathcal{R}(A) = \mathcal{N}(A^T)^\perp$; we proved these equalities in Theorem 3.5.3 by other means.

The singular value decomposition is usually expressed as a matrix decomposition, as follows:

THEOREM 7.1.10 *(SVD Theorem)*   Let $A \in \mathbb{R}^{n \times m}$ have rank r. Then there exist $U \in \mathbb{R}^{n \times n}$, $\Sigma \in \mathbb{R}^{n \times m}$, and $V \in \mathbb{R}^{m \times m}$, such that U and V are orthogonal, Σ has the form

$$\sigma_1 \geq \sigma_2 \geq \cdots \geq \sigma_r > 0$$

$$(7.1.11)$$

and

$$A = U \Sigma V^T$$

Proof Let $v_1, \ldots, v_m$ and $u_1, \ldots, u_n$ be right and left singular vectors, and let $\sigma_1, \ldots, \sigma_r$ be the nonzero singular values of A. Let $V = [v_1, \ldots, v_m] \in \mathbb{R}^{m \times m}$ and $U = [u_1, \ldots, u_n] \in \mathbb{R}^{n \times n}$. Then U and V are orthogonal. The equations

$$Av_i = \begin{cases} \sigma_i u_i & i = 1, \ldots, r \\ 0 & i = r+1, \ldots, m \end{cases}$$

can be combined into a single matrix equation

$$A[v_1, \ldots, v_r \mid v_{r+1}, \ldots, v_m] = [u_1, \ldots, u_r \mid u_{r+1}, \ldots, u_n] \begin{bmatrix} \sigma_1 & & & 0 \\ & \ddots & & \\ 0 & & \sigma_r & 0 \\ & 0 & & 0 \end{bmatrix}$$

that is, $AV = U\Sigma$. Since $VV^T = I$, we see immediately that $A = U\Sigma V^T$. $\qquad\square$

Exercise 7.1.7 Suppose $A = U\Sigma V^T$, where $U = [u_1, \ldots, u_n] \in \mathbb{R}^{n \times n}$ and $V = [v_1, \ldots, v_m] \in \mathbb{R}^{m \times m}$ are orthogonal and $\Sigma \in \mathbb{R}^{n \times m}$ has the form (7.1.11). (a) Express A^T, $A^T A$, and AA^T in terms of U, Σ, and V. (b) Show that $v_1, \ldots, v_m$ are eigenvectors of $A^T A$, $u_1, \ldots, u_n$ are eigenvectors of AA^T, and $\sigma_1^2, \ldots, \sigma_r^2$ are the nonzero eigenvalues of $A^T A$ and AA^T. $\qquad\square$

In the product $A = U\Sigma V^T$, the last $n - r$ columns of U and $m - r$ columns of V are superfluous because they interact only with blocks of zeros in Σ. This observation leads to the following variant of the SVD theorem.

THEOREM 7.1.12 Let $A \in \mathbb{R}^{n \times m}$ have rank r. Then there exist $\hat{U} \in \mathbb{R}^{n \times r}$, $\hat{\Sigma} \in \mathbb{R}^{r \times r}$, and $\hat{V} \in \mathbb{R}^{m \times r}$ such that $\hat{U}$ and $\hat{V}$ are isometries (cf., Section 3.4), $\hat{\Sigma}$ is a diagonal matrix with main-diagonal entries $\sigma_1 \geq \sigma_2 \geq \cdots \geq \sigma_r > 0$, and

$$A = \hat{U}\hat{\Sigma}\hat{V}^T$$

Exercise 7.1.8 Prove Theorem 7.1.12. $\qquad\square$

The following theorem gives yet another useful form of the SVD.

THEOREM 7.1.13 Let $A \in \mathbb{R}^{n \times m}$ have rank r. Let $\sigma_1, \ldots, \sigma_r$ be the nonzero singular values of A, with associated right and left singular vectors $v_1, \ldots, v_r$ and $u_1, \ldots, u_r$, respectively. Then

$$A = \sum_{j=1}^{r} \sigma_j u_j v_j^T$$

Proof Let $B = \sum_{j=1}^{r} \sigma_j u_j v_j^T \in \mathbb{R}^{n \times m}$. To show that $A = B$, it suffices to show that $Av_i = Bv_i$, $i = 1, \ldots, m$, since $v_1, \ldots, v_m$ is a basis of $\mathbb{R}^m$. If $i \leq r$, we have $Av_i = \sigma_i u_i$ and $Bv_i = \sum_{j=1}^{r} \sigma_j u_j (v_j^T v_i)$. Since $v_1, \ldots, v_m$ is an orthonormal set, $v_j^T v_i = 0$ unless $j = i$, in which case $v_i^T v_i = 1$. Thus all terms in the sum are zero, except the ith term, and $Bv_i = \sigma_i u_i$. If $i > r$, then $Av_i = 0$ and $Bv_i = \sum_{j=1}^r \sigma_j u_j (v_j^T v_i) = \sum_{j=1}^r \sigma_j u_j (0) = 0$.   □

Exercise 7.1.9 Obtain a second proof of Theorem 7.1.13 by making the appropriate partition of the equation $A = \hat{U}\hat{\Sigma}\hat{V}^T$ from Theorem 7.1.12.   □

Exercise 7.1.10   Given $A \in \mathbb{R}^{n \times m}$, consider the symmetric matrix

$$M = \begin{bmatrix} 0 & A \\ A^T & 0 \end{bmatrix} \in \mathbb{R}^{(n+m) \times (n+m)}$$

Show that a simple relationship exists between the singular vectors of A and the eigenvectors of M. Show how to build an orthogonal basis of $\mathbb{R}^{n+m}$ consisting of eigenvectors of M, given the singular vectors of A.   □

It is clear that we can calculate the singular value decomposition of any matrix A by calculating the eigenvalues and eigenvectors of $A^T A$ and AA^T. This approach is illustrated in the next example and the exercises that follow. In the next section we will discuss a different approach, in which the SVD is computed without forming $A^T A$ or AA^T explicitly.

Example 7.1.14 Find the singular values and right and left singular vectors of the matrix

$$A = \begin{bmatrix} 1 & 2 & 0 \\ 2 & 0 & 2 \end{bmatrix}$$

Since $A^T A$ is 3 by 3 and AA^T is 2 by 2, it seems reasonable to work with the latter. Since

$$AA^T = \begin{bmatrix} 5 & 2 \\ 2 & 8 \end{bmatrix}$$

the characteristic polynomial is $(\lambda-5)(\lambda-8)-4 = \lambda^2 - 13\lambda + 36 = (\lambda-9)(\lambda-4)$, and the eigenvalues of AA^T are $\lambda_1 = 9$ and $\lambda_2 = 4$. The singular values of A are therefore

$$\sigma_1 = 3 \qquad \text{and} \qquad \sigma_2 = 2$$

The left singular vectors of A are eigenvectors of AA^T. Solving $(9I - AA^T)u = 0$, we find that every multiple of $[1 \; 2]^T$ is an eigenvector of AA^T associated with the eigenvalue $\lambda_1 = 9$. Then solving $(4I - AA^T)u = 0$, we find that the other eigenspace of AA^T consists of multiples of $[2 \; -1]^T$.

Since we want representatives with unit Euclidean norm, we take

$$u_1 = \frac{1}{\sqrt{5}} \begin{bmatrix} 1 \\ 2 \end{bmatrix} \qquad \text{and} \qquad u_2 = \frac{1}{\sqrt{5}} \begin{bmatrix} 2 \\ -1 \end{bmatrix}$$

(What other choices for u_1 and u_2 could we have made?) These are the left singular vectors of A. Notice that they are orthogonal, as they must be. We can find the right singular vectors v_1, v_2, and v_3 by calculating the eigenvectors of $A^T A$. However v_1 and v_2 are more easily found by the formula

$$v_i = \frac{1}{\sigma_i} A^T u_i \qquad i = 1, 2$$

which is a trivial variation of one of the equations in (7.1.8). Thus

$$v_1 = \frac{1}{3\sqrt{5}} \begin{bmatrix} 5 \\ 2 \\ 4 \end{bmatrix} \qquad \text{and} \qquad v_2 = \frac{1}{\sqrt{5}} \begin{bmatrix} 0 \\ 2 \\ -1 \end{bmatrix}$$

Notice that these vectors are orthonormal. v_3 must satisfy $Av_3 = 0$. Solving the equation $Av = 0$ and normalizing the solution, we get

$$v_3 = \frac{1}{3} \begin{bmatrix} -2 \\ 1 \\ 2 \end{bmatrix}$$

We could have found v_3 without reference to A by applying the Gram–Schmidt process, for example, to find a vector orthogonal to both v_1 and v_2. Normalizing that vector, we would get $\pm v_3$.

Now that we have the singular values and singular vectors of A, we can easily construct the matrices U, Σ, and V of Theorem 7.1.10. We have

$$U = [u_1 \, u_2] = \frac{1}{\sqrt{5}} \begin{bmatrix} 1 & 2 \\ 2 & -1 \end{bmatrix}$$

$$\Sigma = \begin{bmatrix} \sigma_1 & 0 & 0 \\ 0 & \sigma_2 & 0 \end{bmatrix} = \begin{bmatrix} 3 & 0 & 0 \\ 0 & 2 & 0 \end{bmatrix}$$

$$V = [v_1 \, v_2 \, v_3] = \frac{1}{3\sqrt{5}} \begin{bmatrix} 5 & 0 & -2\sqrt{5} \\ 2 & 6 & \sqrt{5} \\ 4 & -3 & 2\sqrt{5} \end{bmatrix}$$

You can easily check that $A = U\Sigma V^T$. In so doing you will notice that v_3 plays no role in the computation. This is an instance of the remark made just prior to Theorem 7.1.12. It is an easy exercise for you to write down matrices $\hat{U}$, $\hat{\Sigma}$, and $\hat{V}$ satisfying the hypotheses of Theorem 7.1.12.

Exercise 7.1.11 Let A be the matrix of Example 7.1.14. Calculate the eigenvalues and eigenvectors of $A^T A$. Compare them with the σ_1, σ_2, v_1, v_2, and v_3 calculated in Example 7.1.14.   ☐

Exercise 7.1.12 Let $A = [3 \; 4] \in \mathbb{R}^{1 \times 2}$. (a) Find the singular values and singular vectors of A. (b) Express A as a product $A = U\Sigma V^T$, where U, Σ, and V satisfy the hypotheses of Theorem 7.1.10.   ☐

Exercise 7.1.13   Repeat Exercise 7.1.12 with the matrix

$$A = \begin{bmatrix} 3 & 2 \\ 2 & 3 \\ 2 & -2 \end{bmatrix}$$ □

Exercise 7.1.14   (a) Repeat Exercise 7.1.12 with the matrix

$$A = \begin{bmatrix} 3 & 1 \\ 6 & 2 \end{bmatrix}$$

(b) Find matrices $\hat{U}$, $\hat{\Sigma}$, and $\hat{V}$ satisfying the hypotheses of Theorem 7.1.12. How does the product $A = \hat{U}\hat{\Sigma}\hat{V}^T$ relate to Theorem 7.1.13 in this case?   □

7.2
COMPUTING THE SVD

Diagram (7.1.9) gives a clear, complete picture of the action of A and of A^T. It is therefore reasonable to expect that the singular value decomposition will be very useful. This is indeed the case, and we will examine some applications in subsequent sections. Because the SVD employs orthogonal matrices (orthonormal bases), we can expect it to be not only an important theoretical device, but also a powerful computational tool. If this expectation is to be realized, we must have accurate, efficient means of calculating singular values and singular vectors.

The option of forming $A^T A$ (or AA^T) and calculating its eigenvalues and (possibly) eigenvectors should not be overlooked. This approach has the advantage of being relatively inexpensive; we studied several good algorithms for calculating eigenvalues and eigenvectors of symmetric matrices in Chapters 4, 5, and 6. The disadvantage of this approach is that the smaller singular values will be calculated inaccurately. This is a consequence of the "loss of information through squaring" phenomenon (see Example 3.5.11), which occurs when we compute $A^T A$ from A.

We can get some idea why this information loss occurs by considering an example. Suppose the entries of the matrix A are known to be correct to about six decimal places. If A has, say, $\sigma_1 \approx 1$ and $\sigma_{17} \approx 10^{-3}$, then σ_{17} is fairly small compared with σ_1, but it is still well above the error level $\epsilon \approx 10^{-5}$ or 10^{-6}. We ought to be able to calculate σ_{17} with some precision, perhaps to two or three decimal places. The entries of $A^T A$ also have about six-digits accuracy. Associated with the singular values σ_1 and σ_{17}, $A^T A$ has the eigenvalues $\lambda_1 = \sigma_1^2 \approx 1$ and $\lambda_{17} = \sigma_{17}^2 \approx 10^{-6}$. Notice that λ_{17} is of about the same magnitude as the errors in the entries of $A^T A$. Therefore we cannot expect to calculate λ_{17} accurately.

The larger singular values and associated singular vectors of A *can* be determined accurately from the eigenvalues and eigenvectors of $A^T A$. As we shall see, there are some applications that require only this information. For such applications this may be the best approach. Other applications require the smaller singular values. For these we must find a procedure that avoids calculating $A^T A$ and $A A^T$. In this section we will study one such procedure, in which A is reduced to a condensed form and then an implicit version of the QR algorithm is applied to the condensed form to extract its singular values and (optionally) vectors. This is just one of several possible approaches. For example, each of the three methods discussed in Chapter 6 has implicit variants for calculating singular values.

Exercise 7.2.1 Let $A \in \mathbb{R}^{n \times m}$. How many flops are required to calculate some or all of the singular values of A by a method that first calculates $A^T A$ and then reduces $A^T A$ to tridiagonal form (a) when $n \gg m$, (b) when $n = m$? Exploit symmetry at each step. □

Reduction to Bidiagonal Form

In Chapter 4 we found that the eigenvalue problem can be made much easier if we first reduce the matrix to a simpler form, such as tridiagonal or Hessenberg form. The same turns out to be true of the singular value decomposition. The eigenvalue problem requires that the reduction be done via similarity transformations. For the singular value decomposition $A = U \Sigma V^T$, it is clear that similarity transformations are not necessary, but the transforming matrices should be orthogonal. We will see that we can reduce any matrix to a bidiagonal form by applying reflectors on both left and right. The algorithms that we are about to discuss work well for dense matrices. We will not cover the sparse case.

We continue to assume that we are dealing with a matrix $A \in \mathbb{R}^{n \times m}$, but we will now make the additional assumption that $n \geq m$. This does not imply any loss of generality, for if $n < m$, we can operate on A^T instead of A. If the SVD of A^T is $A^T = U \Sigma V^T$, then the SVD of A is $A = V \Sigma^T U^T$.

A matrix $B \in \mathbb{R}^{n \times m}$ ($n \geq m$) is said to be *bidiagonal* if $b_{ij} = 0$ whenever $i > j$ or $i < j - 1$. This means that B has the form

$$
\begin{bmatrix}
* & * & & & & \\
 & * & * & & 0 & \\
 & & * & \ddots & & \\
 & & & \ddots & \ddots & * \\
 & & & & & * \\
 & 0 & & & &
\end{bmatrix}
$$

THEOREM 7.2.1 Let $A \in \mathbb{R}^{n \times m}$ with $n \geq m$. Then there exist orthogonal $\hat{U} \in \mathbb{R}^{n \times n}$ and $\hat{V} \in \mathbb{R}^{m \times m}$, both products of a finite number of reflectors, and a bidiagonal $B \in \mathbb{R}^{n \times m}$, such that

$$
A = \hat{U} B \hat{V}^T
$$

There is a finite algorithm to calculate $\hat{U}$, $\hat{V}$, and B.

We will prove Theorem 7.2.1 by describing the construction. It is very similar to the QR decomposition by reflectors (Section 3.2) and the reduction to upper Hessenberg form by reflectors (Section 4.5), so let's just sketch the procedure. The first step creates zeros in the first column and row of A. Let $\hat{U}_1 \in \mathbb{R}^{n \times n}$ be a reflector such that

$$
\hat{U}_1 \begin{bmatrix} a_{11} \\ a_{21} \\ \vdots \\ a_{n1} \end{bmatrix} = \begin{bmatrix} * \\ 0 \\ \vdots \\ 0 \end{bmatrix}
$$

Then the first column of $\hat{U}_1 A$ consists of zeros, except for the $(1, 1)$ entry. Now let $[\hat{a}_{11}\ \hat{a}_{12}\ \cdots\ \hat{a}_{1m}]$ denote the first column of $\hat{U}_1 A$, and let $\hat{V}_1 \in \mathbb{R}^{m \times m}$ be a reflector of the form

$$
\hat{V}_1 = \left[\begin{array}{c|ccc} 1 & 0 & \cdots & 0 \\ \hline 0 & & & \\ \vdots & & \overline{V}_1 & \\ 0 & & & \end{array} \right]
$$

such that

$$
[a_{12}\ \cdots\ a_{1n}]\, \overline{V}_1 = [*\ 0\ \cdots\ 0]
$$

Then the first row of $\hat{U}_1 A \hat{V}_1$ consists of zeros, except for the first two entries. Because the first column of $\hat{V}_1$ is e_1, the first column of $\hat{U}_1 A$ is unaltered by right multiplication by $\hat{V}_1$. Thus $\hat{U}_1 A \hat{V}_1$ has the form

$$
\hat{U}_1 A \hat{V}_1 = \left[\begin{array}{c|cccc} * & * & 0 & \cdots & 0 \\ \hline 0 & & & & \\ \vdots & & & \hat{A} & \\ 0 & & & & \end{array} \right]
$$

The second step of the construction is identical to the first, except that it acts on the submatrix $\hat{A}$. It is easy to show that the rotators used on the second step do not destroy the zeros created on the first step. After two steps we have

$$
\hat{U}_2 \hat{U}_1 A \hat{V}_1 \hat{V}_2 = \left[\begin{array}{c|c|cccc} * & * & 0 & 0 & \cdots & 0 \\ \hline 0 & * & * & 0 & \cdots & 0 \\ \hline 0 & 0 & & & & \\ \vdots & \vdots & & & \hat{\hat{A}} & \\ 0 & 0 & & & & \end{array} \right]
$$

The third step acts on the submatrix $\hat{\hat{A}}$, and so on. After m steps we have

$$\hat{U}_m \cdots \hat{U}_2 \hat{U}_1 A \hat{V}_1 \hat{V}_2 \cdots \hat{V}_{m-2} = \begin{bmatrix} * & * & & & 0 \\ & \ddots & \ddots & & \\ & & & * & \\ 0 & & & & * \\ \hline & & 0 & & \end{bmatrix} = B$$

Notice that steps $m - 1$ and m require multiplications on the left only. Let $\hat{U} = \hat{U}_1 \hat{U}_2 \cdots \hat{U}_m$ and $\hat{V} = \hat{V}_1 \hat{V}_2 \cdots \hat{V}_{m-2}$. Then $\hat{U}^T A \hat{V} = B$; that is $A = \hat{U} B \hat{V}^T$.

Exercise 7.2.2 Carry out a flop count for this algorithm. Assume that $\hat{U}$ and $\hat{V}$ are not to be assembled explicitly. (Recall from Section 3.2 that if $x \in \mathbb{R}^k$ and $U \in \mathbb{R}^{k \times k}$ is a reflector, the operations $x \rightarrow Ux$ and $x^T \rightarrow x^T U$ each cost about $2k$ flops.) Show that the cost of the right multiplications is slightly less than that of the left multiplications, but for large n and m, the difference is negligible. Notice that the total flop count is about twice that of a QR decomposition by reflectors. ☐

Exercise 7.2.3 Explain how the reflectors can be stored efficiently. How much storage space is required, in addition to the array that contains A initially? Assume that A is to be destroyed. ☐

Exercise 7.2.4 Write a Fortran subroutine that performs the reduction to bidiagonal form. ☐

Exercise 7.2.5 (a) Show that $B^T B = \hat{V}^T A^T A \hat{V}$ and $B^T B$ is tridiagonal. (b) Which takes more flops (i) calculating $A^T A$ and then reducing it to tridiagonal form or (ii) reducing A to bidiagonal form and then calculating $B^T B$? ☐

In many applications (e.g., least-squares problems) n is much larger than m. In this case it is sometimes more efficient to perform the reduction to bidiagonal form in two stages. In the first stage a QR decomposition of A is performed:

$$A = QR = [Q_1 \ Q_2] \begin{bmatrix} \hat{R} \\ 0 \end{bmatrix}$$

where $\hat{R} \in \mathbb{R}^{m \times m}$ is upper triangular. This involves multiplications by reflectors on the left side of A only. In the second stage the relatively small matrix $\hat{R}$ is reduced to bidiagonal form $\hat{R} = \tilde{U} \tilde{B} \tilde{V}^T$. All of these matrices are m by m. It is easy to show that

$$A = [Q_1 \ Q_2] \begin{bmatrix} \tilde{U} & 0 \\ 0 & I \end{bmatrix} \begin{bmatrix} \tilde{B} \\ 0 \end{bmatrix} \tilde{V}^T$$

Letting

$$\hat{U} = [Q_1 \; Q_2] \begin{bmatrix} \tilde{U} & 0 \\ 0 & I \end{bmatrix} = [Q_1 \tilde{U}, Q_2] \in \mathbb{R}^{n \times n}$$

$$B = \begin{bmatrix} \tilde{B} \\ 0 \end{bmatrix} \in \mathbb{R}^{n \times m}$$

and $\hat{V} = \tilde{V} \in \mathbb{R}^{m \times m}$, we have $A = \hat{U} B \hat{V}^T$.

The advantage of this arrangement is that the right multiplications are applied to the small matrix $\hat{R}$ instead of the large matrix A. They therefore cost a lot less. The disadvantage is that the right multiplications destroy the upper-triangular form of R. Thus most of the left multiplications must be repeated on the small matrix $\hat{R}$. If the ratio n/m is sufficiently large, the added cost of the extra left multiplications will be more than offset by the savings in the right multiplications.

Exercise 7.2.6 (a) Count the flops for the reduction to bidiagonal form, starting with a QR decomposition. Assume that the orthogonal matrices Q, $\tilde{U}$, $\hat{U}$, and $\hat{V}$ are not formed explicitly. (b) Show that for large n and m this procedure requires fewer flops than the other procedure as long as $n/m > 5/3$. (c) Show that if n/m is large, the cost of this procedure is only slightly more than half that of the other procedure. ☐

It is tempting to look for a revised algorithm that exploits the special structure of $\hat{R}$ rather than destroying it. Such a procedure, using plane rotators, was mentioned by Chan (1982), but as he pointed out, nothing is saved unless fast, scaled rotators are used. Using fast rotators, one can devise an algorithm whose asymptotic flop count is less than that of the original procedure for all ratios $n/m > 1$. The catch is that fast rotators have a large overhead, and they will not prove cost effective unless m is fairly large.

As we have already noted, the various applications of the SVD have different requirements. Some require only the singular values, while others require the right or left singular vectors, or both. If any of the singular vectors are needed, then the matrices $\hat{U}$ and/or $\hat{V}$ have to be computed explicitly. Usually it is possible to avoid calculating $\hat{U}$, but there are numerous applications for which $\hat{V}$ is needed. The flop counts in Exercises 7.2.2 and 7.2.6 do not include the cost of computing $\hat{U}$ or $\hat{V}$.

Exercise 7.2.7 (a) Determine the flop count for assembling $\hat{V}$ (cf., Section 3.4). (b) Determine the flop count for calculating the first m columns of $\hat{U}$ if a preliminary QR decomposition (i) has not been done and (ii) has been done. Notice that the latter takes more than twice as many flops as the former. (We calculate only the first m columns of $\hat{U}$ because that is all that is needed in a typical application requiring $\hat{U}$.) (c) Show that if a preliminary QR decomposition has been done, the increased cost of accumulating the first m columns of $\hat{U}$ more than offsets the savings accrued earlier. Thus there is no point in performing the preliminary QR decomposition in those problems in which the left singular vectors are needed. ☐

Since B is bidiagonal, it has the form

$$B = \begin{bmatrix} \tilde{B} \\ 0 \end{bmatrix}$$

where $\tilde{B} \in \mathbb{R}^{m \times m}$ is bidiagonal. The problem of finding the SVD of A can be reduced to that of finding the SVD of the small, bidiagonal matrix B. First of all, the equation $A = \hat{U} B \hat{V}^T$ can be rewritten as $A = \hat{U}_1 \tilde{B} \hat{V}^T$, where $\hat{U}_1 \in \mathbb{R}^{n \times m}$ consists of the first m columns of $\hat{U}$. If $\tilde{B} = \tilde{U} \Sigma \tilde{V}^T$ is the SVD of $\tilde{B}$, then $A = U_1 \Sigma V^T$, where $U_1 = \hat{U}_1 \tilde{U}$ and $V = \hat{V} \tilde{V}$. This is not exactly the SVD in the sense of Theorem 7.1.10. That form can be obtained by adjoining $n - m$ rows of zeros to Σ and $n - m$ columns U_2 to U_1 in such a way that the resulting matrix $U = [U_1 \ U_2]$ is orthogonal. The choice $U_2 = \hat{U}_2$ (last $n - m$ columns of $\hat{U}$) works. In practice this extension is seldom necessary, since the decomposition $A = U_1 \Sigma V^T$ usually suffices.

Properly Bidiagonal Matrices

For notational convenience we now drop the tilde from $\tilde{B}$ and let $B \in \mathbb{R}^{m \times m}$ be a bidiagonal matrix, say

$$B = \begin{bmatrix} \beta_1 & \gamma_1 & & & 0 \\ & \beta_2 & \gamma_2 & & \\ & & \beta_3 & \ddots & \\ & & & \ddots & \gamma_{m-1} \\ 0 & & & & \beta_m \end{bmatrix}$$

We will say that B is a *properly* bidiagonal matrix if $\beta_i \neq 0$ and $\gamma_i \neq 0$ for all i.

Exercise 7.2.8   (a) Show that both BB^T and $B^T B$ are tridiagonal. (b) Show that B is properly bidiagonal if and only if both BB^T and $B^T B$ are properly tridiagonal (Hessenberg) matrices. $\square$

If B is not properly bidiagonal, the problem of finding the SVD of B can be reduced to two or more smaller subproblems. First of all, if some $\gamma_k = 0$, then

$$B = \begin{bmatrix} B_1 & 0 \\ 0 & B_2 \end{bmatrix} \tag{7.2.2}$$

where $B_1 \in \mathbb{R}^{k \times k}$ and $B_2 \in \mathbb{R}^{(m-k) \times (m-k)}$ are both bidiagonal. The SVDs of B_1 and B_2 can be found separately and combined to yield the SVD of B. If some $\beta_k = 0$, a small amount of work transforms B to a form that can be reduced. This is left as an exercise.

Exercise 7.2.9

(a)   Show that $\beta_k = 0$ for some k if and only if zero is a singular value of B.

(b)   Suppose $\beta_m = 0$. If $\gamma_{m-1} = 0$ as well, B has the form (7.2.2) with $B_2 = [0] \in \mathbb{R}^{1 \times 1}$. This is the zero singular value. We can deflate the problem and operate on B_1. Now suppose $\gamma_{m-1} \neq 0$. Show that γ_{m-1} can be eliminated by a sequence of rotators applied to B from the right. The first rotator acts on columns $m - 1$ and m and transforms γ_{m-1} to zero. At the same time it creates a new nonzero in position $(m - 2, m)$. The second rotator acts on columns $m - 2$ and m, annihilates the $(m - 2, m)$ entry, and creates a new nonzero entry in position $(m - 3, m)$. Subsequent rotators push the nonzero entry up the mth column until it falls off the top of the matrix. The resulting revised bidiagonal matrix can be deflated because it has $\gamma_{m-1} = 0$.

(c)   Suppose $k \neq m$, $\beta_k = 0$, and $\beta_{k+1} \neq 0$. If $\gamma_k = 0$, B has the form (7.2.2) and can be reduced. Suppose $\gamma_k \neq 0$. Show that γ_k can be eliminated by a sequence of rotators applied to B on the left. The first acts on rows k and $k + 1$, annihilates γ_k, and creates a new nonzero in position $(k, k + 2)$. Subsequent rotators push the nonzero entry across the kth row and eventually off the edge of the matrix. The resulting bidiagonal matrix has the form (7.2.2) because $\gamma_k = 0$. Since $\beta_k = 0$ also, the matrix B in (7.2.2) has the form discussed in part (b), so the reduction described there should be performed immediately.   □

The Implicit QR Algorithm for the SVD

As a result of Exercise 7.2.9, we see that we can always assume, without loss of generality, that B is a properly bidiagonal matrix. A form of the implicit QR algorithm can be used to find the SVD of any such matrix. We will describe the algorithm first and then justify it.

Suppose $B \in \mathbb{R}^{m \times m}$ is a properly bidiagonal matrix. Then both BB^T and $B^T B$ are properly tridiagonal matrices, so we could find their eigenvalues inexpensively by the QR algorithm. The algorithm that we are about to develop is equivalent to the QR algorithm on both BB^T and $B^T B$, but it is carried out without ever forming these matrices explicitly. We begin a QR step by choosing a shift. The lower right-hand 2-by-2 submatrix of BB^T is

$$\begin{bmatrix} \beta_{m-1}^2 + \gamma_{m-1}^2 & \beta_m \gamma_{m-1} \\ \beta_m \gamma_{m-1} & \beta_m^2 \end{bmatrix}$$

Calculate the eigenvalues of this submatrix and take the shift σ to be that eigenvalue which is closer to β_m^2. This is the Wilkinson shift on BB^T. It is a good choice because it guarantees convergence, and the convergence is rapid in practice. We could have chosen the shift from $B^T B$ instead of BB^T. We chose the latter because its lower right-hand 2-by-2 submatrix has a slightly simpler form.

A QR step on $B^T B$ with shift σ would perform the similarity transformation $B^T B \to Q^T B^T B Q$, where Q is the orthogonal factor from the QR decomposition:

$$B^T B - \sigma I = QR \tag{7.2.3}$$

Since we plan to take an implicit step, all we need is the first column of Q. Because R is upper triangular, the first column of Q is proportional to the first column of $B^T B - \sigma I$, which is

$$\begin{bmatrix} \beta_1^2 - \sigma \\ \gamma_1 \beta_1 \\ 0 \\ \vdots \\ 0 \end{bmatrix} \tag{7.2.4}$$

Let V_{12} be a rotator (or reflector) in the $(1, 2)$ plane whose first column is proportional to (7.2.4). Multiply B by V_{12} on the right. The operation $B \to BV_{12}$ alters only the first two columns of B, and as you can easily check, it creates a new nonzero entry (a "bulge") in the $(2, 1)$ position. (Draw a picture!)

Now find a rotator U_{12}^T in the $(1, 2)$ plane such that $U_{12}^T B V_{12}$ has a zero in the $(2, 1)$ position. This operation acts on rows 1 and 2 and creates a new bulge in the $(1, 3)$ position. Let V_{23} be a rotator acting on columns 2 and 3 such that $U_{12}^T B V_{12} V_{23}$ has a zero in the $(1, 3)$ position. This creates a bulge in the $(3, 2)$ position. Applying additional rotators U_{23}^T, V_{34}, U_{34}^T, ..., we chase the bulge through positions $(2, 4)$, $(4, 3)$, $(3, 5)$, $(5, 4)$, ..., $(m, m - 1)$, and finally off of the matrix completely. The result is a bidiagonal matrix

$$\hat{B} = U_{m-1,m}^T \cdots U_{23}^T U_{12}^T B V_{12} V_{23} \cdots V_{m-1,m} \tag{7.2.5}$$

that is nearly proper.

Exercise 7.2.10   (a) Show that each of the rotators $V_{12}, \ldots, V_{n-1,n}, U_{12}, \ldots, U_{n-1,n}$ is a nontrivial rotator (i.e., rotates through an angle other than $0°$ or $180°$), and at the same time deduce that the entries of $\hat{B}$ satisfy $\hat{\beta}_i \neq 0$, $i = 1, \ldots, m - 1$, and $\hat{\gamma}_i \neq 0$, $i = 1, \ldots, m - 2$. (It can happen that $\hat{\beta}_m = \hat{\gamma}_{m-1} = 0$. In fact, it can be shown that this happens when and only when σ is exactly an eigenvalue of BB^T and $B^T B$.) (b) Show that the off-diagonal entries of $\hat{B}\hat{B}^T$ and $\hat{B}^T \hat{B}$ are all nonzero, except possibly the $(m, m - 1)$ entry.   □

Letting

$$U = U_{12} U_{23} \cdots U_{m-1,m} \quad \text{and} \quad V = V_{12} V_{23} \cdots V_{m-1,m} \tag{7.2.6}$$

we can rewrite (7.2.5) as

$$\hat{B} = U^T B V \tag{7.2.7}$$

In addition we have $\hat{B}\hat{B}^T = U^T BB^T U$ and $\hat{B}^T\hat{B} = V^T B^T BV$. As we shall see, $\hat{B}\hat{B}^T$ and $\hat{B}^T\hat{B}$ are essentially the same matrices as we would have obtained by taking one shifted QR step with shift σ, starting with BB^T and $B^T B$, respectively. If we set $\hat{B} \to B$ and perform repeated QR steps, both BB^T and $B^T B$ will tend to diagonal form. The main-diagonal entries will converge to the eigenvalues. If the Wilkinson shift is used, the $(m, m-1)$ and (m, m) entries of both BB^T and $B^T B$ will converge very rapidly, the former to zero and the latter to an eigenvalue. Of course we do not deal with BB^T or $B^T B$ directly; we deal with B. The rapid convergence of BB^T and $B^T B$ translates into convergence of γ_{m-1} to zero and $|\beta_m|$ to a singular value of B.

Exercise 7.2.11   Prove that if the (m, m) entries of both BB^T and $B^T B$ tend to an eigenvalue λ, then $\gamma_{m-1} \to 0$, and $|\beta_m|$ converges to a singular value of B.   □

Once γ_{m-1} becomes negligible, it can be considered to be zero, and the problem can be deflated. Performing shifted QR steps on the remaining $(m-1)$-by-$(m-1)$ submatrix, we can force γ_{m-2} quickly to zero, exposing another singular value in the β_{m-1} position. Continuing in this manner, we soon find all the singular values of B.

During the whole procedure, all the γ_k tend slowly toward zero. If at any point one of them becomes negligible, the problem should be reduced to two smaller subproblems.

Exercise 7.2.12   Write an algorithm to calculate the singular values of a bidiagonal matrix B. Include the mechanisms for reduction developed in Exercise 7.2.9.   □

Exercise 7.2.13   Write a Fortran program that implements the algorithm you wrote in Exercise 7.2.12. Test it on the matrix $B \in \mathbb{R}^{m \times m}$ given by $\beta_i = 1$ and $\gamma_i = 1$ for all i. Try various values of m. The singular values of B are

$$\sigma_j = 2\cos\frac{j\pi}{2m+1} \qquad j = 1, 2, \ldots, m \qquad\qquad □$$

Calculating the Singular Vectors

If only singular values are needed, there is no need to keep a record of the many rotators used during the QR steps. However, if the right (or left) singular vectors are desired, we must keep track of the rotators $V_{i,i+1}$ (or $U_{i,i+1}$). Let us suppose we wish to compute the right singular vectors of $A \in \mathbb{R}^{n \times m}$ and we have already calculated a decomposition

$$A = \hat{U}_1 B \hat{V}^T \qquad\qquad (7.2.8)$$

where $B \in \mathbb{R}^{m \times m}$ is bidiagonal, $\hat{U}_1 \in \mathbb{R}^{n \times m}$ has orthonormal columns, and $\hat{V} \in \mathbb{R}^{m \times m}$ is orthogonal. Needing the right singular vectors, we have calculated $\hat{V}$ explicitly and saved it. As we perform the QR steps on B, we need to take into account each rotator V_{ij} that multiplies B on the right. This can be done by making the update $\hat{V}V_{ij} \rightarrow \hat{V}$ along with the update $BV_{ij} \rightarrow B$. Since $(BV_{ij})(\hat{V}V_{ij})^T = B\hat{V}^T$, we see that this update preserves the overall product in (7.2.8). Of course this procedure should also be followed for the right rotators used in the reduction procedure described in Exercise 7.2.9. Once B has been reduced to diagonal form, the singular values lie on the main diagonal of B, and the right singular vectors of A are the columns of $\hat{V}$. The singular values do not necessarily appear in descending order. If left singular vectors are needed, then $\hat{U}_1$ must be calculated explicitly and saved. Then for each rotator U_{ij}^T that is applied to B on the left, the update $\hat{U}_1 U_{ij} \rightarrow U_1$ should be made along with the update $U_{ij}^T B \rightarrow B$. In the end the m columns of $\hat{U}_1$ are the left singular vectors of A.

The updates of B are inexpensive because B is very sparse. By contrast the updates of the full matrices $\hat{V}$ and $\hat{U}_1$ are relatively expensive. While the cost of an entire QR step without updating $\hat{U}_1$ or $\hat{V}$ is $O(m)$ flops, the additional costs of updating $\hat{V}$ and $\hat{U}_1$ are $O(m^2)$ and $O(nm)$ flops per QR step, respectively. It follows that if the right or left singular vectors are needed, the QR steps become much more expensive. The added cost can usually be decreased by employing the ultimate shift strategy suggested in Section 4.8.

Justification of the Implicit QR Step

Although we are mainly interested in bidiagonal matrices, we will begin in a more general setting. B need not even be square. Let $B \in \mathbb{R}^{n \times m}$. If we wish to perform QR steps on BB^T and $B^T B$ with a shift σ, we begin by taking QR decompositions

$$BB^T - \sigma I = PS \qquad \text{and} \qquad B^T B - \sigma I = QR \qquad (7.2.9)$$

where $P \in \mathbb{R}^{n \times n}$ and $Q \in \mathbb{R}^{m \times m}$ are orthogonal and $S \in \mathbb{R}^{n \times n}$ and $R \in \mathbb{R}^{m \times m}$ are upper triangular. If we now define $\overline{B} \in \mathbb{R}^{n \times m}$ by

$$\overline{B} = P^T B Q \qquad (7.2.10)$$

we see immediately from (7.2.9) and (7.2.10) that

$$\overline{B}\,\overline{B}^T = P^T BB^T P = SP + \sigma I$$
$$\overline{B}^T \overline{B} = Q^T B^T B Q = RQ + \sigma I$$

This means that the transformations $BB^T \rightarrow \overline{B}\,\overline{B}^T$ and $B^T B \rightarrow \overline{B}^T \overline{B}$ are both shifted QR steps.

Now suppose B is bidiagonal. Then the bidiagonal form is inherited by $\overline{B}$, as the following exercises show. As in Chapter 1, we will say that B is *lower k-banded* (*upper k-banded*) if $b_{ij} = 0$ whenever $i - j > k$ ($j - i > k$). A bidiagonal matrix is one that is both lower 0-banded and upper 1-banded.

Exercise 7.2.14 Show that if $B \in \mathbb{R}^{n \times m}$ is lower k-banded and $S \in \mathbb{R}^{n \times n}$ and $R \in \mathbb{R}^{m \times m}$ are upper triangular, then both SB and BR are lower k-banded. $\square$

If σ is not an eigenvalue of BB^T or B^TB, then the upper-triangular matrices S and R in (7.2.9) are both nonsingular.

Exercise 7.2.15   Show that (a) if R and S in (7.2.9) are both nonsingular, then $\overline{B} = SBR^{-1}$ and $\overline{B}^T = RB^TS^{-1}$; (b) if B in part (a) is lower k-banded, then so is $\overline{B}$; (c) if B is upper k-banded, then so is $\overline{B}$; (d) if B is bidiagonal, then so is $\overline{B}$; (e) if B is a properly bidiagonal matrix, then so is $\overline{B}$.   □

If $B \in \mathbb{R}^{m \times m}$ is a properly bidiagonal matrix, then part (d) of Exercise 7.2.15 remains valid even if σ is an eigenvalue of BB^T and B^TB. In this case part (e) is almost true as well, except that $\overline{B}$ has a zero in its $(m-1, m)$ position; that is, $\overline{\gamma}_{m-1} = 0$. A careful discussion of this case requires some extra effort, so we will omit it. The result $\overline{\gamma}_{m-1} = 0$ is, of course, a theoretical result. In practice, roundoff errors will cause $\overline{\gamma}_{m-1}$ to be nonzero.

We wish to show that the matrix $\overline{B}$ given by (7.2.10) is essentially the same as $\hat{B}$ given by (7.2.5) and (7.2.7). The next theorem will make that possible.

THEOREM 7.2.11   Let $B \in \mathbb{R}^{m \times m}$ be nonsingular. Suppose $\overline{B}$, $\hat{B} \in \mathbb{R}^{m \times m}$ are properly bidiagonal matrices, P, Q, U, $V \in \mathbb{R}^{m \times m}$ are orthogonal matrices,

$$\overline{B} = P^TBQ \qquad\qquad (7.2.12)$$

and

$$\hat{B} = U^TBV \qquad\qquad (7.2.13)$$

Suppose further that Q and V have essentially the same first column; that is, $q_1 = v_1d_1$, where $d_1 = \pm 1$. Then there exist orthogonal diagonal matrices D and E such that $V = QD$, $U = PE$, and

$$\hat{B} = E\overline{B}D$$

In other words, $\overline{B}$ and $\hat{B}$ are essentially the same.

Proof   We note first that $\overline{B}^T\overline{B} = Q^T(B^TB)Q$ and $\hat{B}^T\hat{B} = V^T(B^TB)V$. By Exercise 7.2.8, $\overline{B}^T\overline{B}$ and $\hat{B}^T\hat{B}$ are properly tridiagonal matrices. Applying Theorem 5.3.1 (or 5.3.5) with $\mathbb{F} = \mathbb{R}$ and with B^TB, $\overline{B}^T\overline{B}$, and $\hat{B}^T\hat{B}$ playing the roles of A, B, and $\tilde{B}$, respectively, we find that Q and V are essentially equal. More precisely, there exists an orthogonal diagonal matrix D such that $V = QD$. Using this equation, we can rewrite (7.2.12) and (7.2.13) to obtain

$$U\hat{B} = BV = BQD = P(\overline{B}D)$$

Defining $C = U\hat{B} = P(\overline{B}D)$, we note that $U\hat{B}$ and $P(\overline{B}D)$ are both QR decompositions of C, since U and P are orthogonal and $\hat{B}$ and $\overline{B}D$ are upper triangular. Since $\hat{B}$ and $\overline{B}D$ are not assumed to have positive main-diagonal entries, we cannot conclude that they are equal. Instead we can draw the weaker conclusion (Exercise 3.2.21) that there exists an orthogonal diagonal matrix E such that $U = PE$ and $\hat{B} = E\overline{B}D$. This completes the proof.   □

To apply this theorem to our present situation, note that the first column of the matrix $V = V_{12}V_{23} \cdots V_{m-1,m}$ of (7.2.6) is the same as the first column of V_{12}. This is easy to see if we think of accumulating V by starting with V_{12}, then multiplying on the right by $V_{23}, V_{34}, \ldots, V_{m-1,m}$, successively. Since each of these is a rotator acting on columns other than column 1, the first column of V must be the same as the first column of V_{12}. The first column of V_{12} was chosen to be essentially the same as the first column of the orthogonal matrix Q in the QR decomposition of $B^T B - \sigma I$. This is exactly the Q that appears in (7.2.3) and (7.2.10). We can now apply Theorem 7.2.11 to conclude that the matrix $\hat{B}$ defined by (7.2.5) and (7.2.7) is essentially the same as the matrix $\overline{B}$ defined by (7.2.10). This completes the justification of the implicit QR step, at least in the case that the shift is not exactly equal to an eigenvalue.

Exercise 7.2.16 Work out the details for the case that σ is an eigenvalue of BB^T and $B^T B$. You may find the ideas in Exercise 4.6.13 useful. □

7.3
SOME BASIC APPLICATIONS
OF SINGULAR VALUES

Computation of Norms and Condition Numbers

In Chapter 2 we defined the *spectral* matrix norm to be the norm induced by the Euclidean vector norm:

$$\| A \|_2 = \max_{x \neq 0} \frac{\| Ax \|_2}{\| x \|_2}$$

The discussion in Chapter 2 was restricted to square matrices, but this definition makes sense for nonsquare matrices as well. Geometrically $\| A \|_2$ represents the maximum magnification that can be undergone by any vector $x \in \mathbb{R}^m$ when acted on by A. In light of (7.1.9), it should not be surprising that $\| A \|_2$ equals the maximum singular value of A.

THEOREM 7.3.1 Let $A \in \mathbb{R}^{n \times m}$ have singular values $\sigma_1 \geq \sigma_2 \geq \cdots \geq 0$. Then

$$\| A \|_2 = \sigma_1$$

Proof We must show that $\max_{x \neq 0} \| Ax \|_2 / \| x \|_2 = \sigma_1$. First notice that if v_1 is the right singular vector of A associated with σ_1, then

$$\frac{\| Av_1 \|}{\| v_1 \|} = \sigma_1 \frac{\| u_1 \|_2}{\| v_1 \|_2} = \sigma_1$$

so $\max_{x \neq 0} \| Ax \|_2 / \| x \|_2 \geq \sigma_1$. Now we must show that no other vector is magnified by more than σ_1.

Let $x \in \mathbb{R}^m$. Then x can be expressed as a linear combination of the right singular vectors of A: $x = c_1 v_1 + c_2 v_2 + \cdots + c_m v_m$. Since $v_1, \ldots, v_m$ are orthonormal, $\| x \|_2^2 = |c_1|^2 + \cdots + |c_m|^2$. Now $Ax = c_1 A v_1 + \cdots + c_r A v_r + \cdots + c_m A v_m = \sigma_1 c_1 u_1 + \cdots + \sigma_r c_r u_r + 0 + \cdots + 0$, where r is the rank of A. Since $u_1, \ldots, u_r$ are also orthonormal, $\| Ax \|_2^2 = |\sigma_1 c_1|^2 + \cdots + |\sigma_r c_r|^2$. Thus $\| Ax \|_2^2 \leq \sigma_1^2 (|c_1|^2 + \cdots + |c_r|^2) \leq \sigma_1^2 \| x \|_2^2$; that is, $\| Ax \|_2 / \| x \|_2 \leq \sigma_1$. This completes the proof. $\square$

Since A and A^T have the same singular values, we have the following corollary.

COROLLARY 7.3.2 $\| A \|_2 = \| A^T \|_2$.

Exercise 7.3.1 Recall that the *Frobenius* matrix norm is defined by

$$\| A \|_F = \left(\sum_{i=1}^{n} \sum_{j=1}^{m} |a_{ij}|^2 \right)^{1/2}$$

Show that $\| A \|_F = (\sigma_1^2 + \sigma_2^2 + \cdots + \sigma_r^2)^{1/2}$
(*Hint:* Show that if $B = UC$ where U is orthogonal, then $\| B \|_F = \| C \|_F$.) $\square$

Now suppose A is square, say $A \in \mathbb{R}^{n \times n}$, and nonsingular. The spectral condition number of A is defined by

$$\kappa_2(A) = \| A \|_2 \| A^{-1} \|_2$$

Let us see how $\kappa_2(A)$ can be expressed in terms of the singular values of A. Since A has rank n, it has n strictly positive singular values, and its action is described completely by the following diagram:

$$
\begin{array}{ccc}
 & A & \\
v_1 & \xrightarrow{\ \sigma_1\ } & u_1 \\
v_2 & \xrightarrow{\ \sigma_2\ } & u_2 \\
\vdots & & \vdots \\
v_n & \xrightarrow{\ \sigma_n\ } & u_n
\end{array}
$$

It follows that the corresponding diagram for A^{-1} is

$$
\begin{array}{ccc}
 & A^{-1} & \\
u_1 & \xrightarrow{\ \sigma_1^{-1}\ } & v_1 \\
u_2 & \xrightarrow{\ \sigma_2^{-1}\ } & v_2 \\
\vdots & & \vdots \\
u_n & \xrightarrow{\ \sigma_n^{-1}\ } & v_n
\end{array}
$$

In terms of matrices we have $A = U\Sigma V^T$ and $A^{-1} = V^{-T}\Sigma^{-1}U^{-1} = V\Sigma^{-1}U^T$. Either way we see that the singular values of A^{-1}, in descending order, are $\sigma_n^{-1} \geq \sigma_{n-1}^{-1} \geq \cdots \geq \sigma_1^{-1} > 0$. Applying Theorem 7.3.1 to A^{-1}, we conclude that $\|A^{-1}\|_2 = \sigma_n^{-1}$. These observations imply the following theorem.

THEOREM 7.3.3 Let $A \in \mathbb{R}^{n \times n}$ be a nonsingular matrix with singular values $\sigma_1 \geq \sigma_2 \geq \cdots \geq \sigma_n > 0$. Then

$$\kappa_2(A) = \frac{\sigma_1}{\sigma_n}$$

Exercise 7.3.2 Formulate and prove an expression for the Frobenius condition number $\kappa_F(A)$ in terms of the singular values of A. $\square$

Another expression for the condition number that was given in Chapter 2 is

$$\kappa_2(A) = \frac{\text{maxmag}(A)}{\text{minmag}(A)}$$

where

$$\text{maxmag}(A) = \max_{x \neq 0} \frac{\|Ax\|_2}{\|x\|_2}$$

$$\text{minmag}(A) = \min_{x \neq 0} \frac{\|Ax\|_2}{\|x\|_2}$$

This gives a slightly different view of the condition number. From Theorem 7.3.1 we know that $\text{maxmag}(A) = \sigma_1$. It must therefore be true that $\text{minmag}(A) = \sigma_n$.

Exercise 7.3.3 Prove that $\text{minmag}(A) = \sigma_n$. Show that the minimum magnification is obtained by taking $x = v_n$. $\square$

In Chapter 3 we observed that the equation

$$\kappa_2(A) = \frac{\text{maxmag}(A)}{\text{minmag}(A)} \tag{7.3.4}$$

can be used to extend the definition of κ_2 to certain nonsquare matrices. Specifically, if $A \in \mathbb{R}^{n \times m}$, $n \geq m$, and $\text{rank}(A) = m$, then $\text{minmag}(A) > 0$, and we can take (7.3.4) as the definition of the condition number of A. If A is nonzero but does not have full rank, then (still assuming $n \geq m$) $\text{minmag}(A) = 0$, and it is reasonable to define $\kappa_2(A) = \infty$. With this convention the following theorem holds, regardless of whether or not A has full rank.

THEOREM 7.3.5 Let $A \in \mathbb{R}^{n \times m}$, $n \geq m$, be a nonzero matrix with singular values $\sigma_1 \geq \sigma_2 \geq \cdots \geq \sigma_m \geq 0$. Then $\text{maxmag}(A) = \sigma_1$, $\text{minmag}(A) = \sigma_m$, and $\kappa_2(A) = \sigma_1/\sigma_m$.

The proof is left as an easy exercise for you.

Exercise 7.3.4   (a) Let $A \in \mathbb{R}^{n \times m}$ with $n \geq m$. Show that $\| A^T A \|_2 = \| A \|_2^2$ and $\kappa_2(A^T A) = \kappa_2(A)^2$.
(b) Let $M \in \mathbb{R}^{n \times n}$ be positive definite, and let G be the Cholesky factor of M, so that $M = GG^T$. Show that $\| M \|_2 = \| G \|_2^2$ and $\kappa_2(M) = \kappa_2(G)^2$. $\square$

Exercise 7.3.5   Let $A \in \mathbb{R}^{n \times m}$ with singular values $\sigma_1 \geq \sigma_2 \geq \cdots \geq \sigma_m$ and right singular vectors $v_1, v_2, \ldots, v_m$. We have seen that $\| Ax \|_2 / \| x \|_2$ is maximized when $x = v_1$ and minimized when $x = v_m$. More generally, show that for $k = 1, 2, \ldots, m$,

$$\sigma_k = \frac{\| Av_k \|_2}{\| v_k \|_2} = \max \left\{ \frac{\| Ax \|_2}{\| x \|_2} \;\middle|\; x \neq 0, \; x \in \langle v_1, \ldots, v_{k-1} \rangle^\perp \right\}$$

$$= \min \left\{ \frac{\| Ax \|_2}{\| x \|_2} \;\middle|\; x \neq 0, \; x \in \langle v_{k+1}, \ldots, v_m \rangle^\perp \right\} \qquad \square$$

Exercise 7.3.6   (a) Let $A \in \mathbb{R}^{2 \times 2}$ with singular values $\sigma_1 \geq \sigma_2 > 0$. Show that the set $\{ Ax \mid \| x \|_2 = 1 \}$ (the image of the unit circle) is an ellipse in $\mathbb{R}^2$ whose major and minor semiaxes have lengths σ_1 and σ_2, respectively.
(b) Let $A \in \mathbb{R}^{n \times m}$, $n \geq m$, rank$(A) = m$. Show that the set $\{ Ax \mid \| x \|_2 = 1 \}$ is an m-dimensional hyperellipsoid with semiaxes $\sigma_1, \sigma_2, \ldots, \sigma_m$. [Notice that the lengths of the longest and shortest semiaxes are maxmag(A) and minmag(A), respectively.] $\square$

Exercise 7.3.7   In this exercise you will prove some results that were used in the analysis of the sensitivity of the least-squares problem in Chapter 3. Let $A \in \mathbb{R}^{n \times m}$, $n \geq m$, rank$(A) = m$, with singular values $\sigma_1 \geq \sigma_2 \geq \cdots \geq \sigma_m > 0$.

    (a)   Determine the singular value decompositions of the matrices $(A^T A)^{-1}$, $(A^T A)^{-1} A^T$, $A(A^T A)^{-1}$, and $A(A^T A)^{-1} A^T$ in terms of the SVD of A. (Draw diagrams or use matrices, whichever you prefer.)

    (b)   Infer from part (a) that $\| (A^T A)^{-1} \|_2 = \sigma_m^{-2}$, $\| (A^T A)^{-1} A^T \|_2 = \sigma_m^{-1}$, $\| A(A^T A)^{-1} \|_2 = \sigma_m^{-1}$, and $\| A(A^T A)^{-1} A^T \|_2 = 1$. $(A^T A)^{-1} A^T$ is called the *pseudoinverse* of A. $A(A^T A)^{-1}$ is the pseudoinverse of A^T. Pseudo-inverses will be discussed in greater generality in the following section. $\square$

Numerical Rank Determination

In the absence of roundoff errors and uncertainties in the data, the singular value decomposition reveals the rank of a matrix. Unfortunately the presence of errors and uncertainties makes the question of rank meaningless. As we shall see, a small perturbation in a matrix that is not of full rank can and typically will increase the rank.

Exercise 7.3.8 Let $A \in \mathbb{R}^{n \times m}$ with $\text{rank}(A) = r < \min\{n, m\}$. Use the SVD of A to show that for every $\epsilon > 0$, there exists a full-rank matrix $A_\epsilon \in \mathbb{R}^{n \times m}$ such that $\| A - A_\epsilon \|_2 < \epsilon$.   □

The nonnegative number $\| A - A_\epsilon \|_2$ is a measure of the distance between the matrices A and A_ϵ. Exercise 7.3.8 shows that every rank-deficient matrix has full-rank matrices arbitrarily close to it; this suggests that matrices of full rank are abundant. This impression is strengthened by the next theorem and its corollary.

THEOREM 7.3.6 Let $A \in \mathbb{R}^{n \times m}$ with $\text{rank}(A) = r > 0$. Let $A = U \Sigma V^T$ be the singular value decomposition of A. For $k = 1, \ldots, r - 1$ define $A_k = U \Sigma_k V^T$, where $\Sigma_k \in \mathbb{R}^{n \times m}$ is defined by

$$
\Sigma_k = \begin{bmatrix}
\sigma_1 & & & & & \\
 & \sigma_2 & & & 0 & \\
 & & \ddots & & & 0 \\
0 & & & \sigma_k & & \\
\hline
 & & 0 & & & 0
\end{bmatrix}
$$

(We assume as usual that $\sigma_1 \geq \sigma_2 \geq \cdots \geq \sigma_r$.) Then $\text{rank}(A_k) = k$, and

$$
\sigma_{k+1} = \| A - A_k \|_2 = \min\{\| A - B \|_2 \mid \text{rank}(B) = k\}
$$

That is, of all the matrices of rank k, A_k is closest to A.

Proof It is obvious that $\text{rank}(A_k) = k$. Since $A - A_k = U(\Sigma - \Sigma_k)V^T$, it is clear that the largest singular value of $A - A_k$ is σ_{k+1}. Therefore $\| A - A_k \|_2 = \sigma_{k+1}$. It remains to be shown only that for any other matrix B of rank k, $\| A - B \|_2 \geq \sigma_{k+1}$.

Given such a B, note first that $\mathcal{N}(B)$ has dimension $m - k$, for $\dim(\mathcal{N}(B)) = \dim(\mathbb{R}^m) - \dim(\mathcal{R}(B)) = m - \text{rank}(B) = m - k$. Also, the space $\langle v_1, \ldots, v_{k+1} \rangle$ has dimension $k + 1$. (As usual, $v_1, \ldots, v_m$ denote the columns of V.) Since $\mathcal{N}(B)$ and $\langle v_1, \ldots, v_{k+1} \rangle$ are two subspaces of $\mathbb{R}^m$, the sum of whose dimensions exceeds m, they must have a nontrivial intersection. Let $\hat{x}$ be a nonzero vector in $\mathcal{N}(B) \cap \langle v_1, \ldots, v_{k+1} \rangle$. We can and will assume that $\| \hat{x} \|_2 = 1$. Since $\hat{x} \in \langle v_1, \ldots, v_{k+1} \rangle$, there exist scalars $c_1, \ldots, c_{k+1}$ such that $\hat{x} = c_1 v_1 + \cdots + c_{k+1} v_{k+1}$. Because $v_1, \ldots, v_{k+1}$ are orthonormal, $|c_1|^2 + \cdots + |c_{k+1}|^2 = \| \hat{x} \|_2^2 = 1$. Since $\hat{x} \in \mathcal{N}(B)$, $B\hat{x} = 0$. Thus

$$
(A - B)\hat{x} = A\hat{x} = \sum_{i=1}^{k+1} c_i A v_i = \sum_{i=1}^{k+1} \sigma_i c_i u_i
$$

Since $u_1, \ldots, u_{k+1}$ are also orthonormal,

$$
\| (A - B)\hat{x} \|_2^2 = \sum_{i=1}^{k+1} |\sigma_i c_i|^2 \geq \sigma_{k+1}^2 \sum_{i=1}^{k+1} |c_i|^2 = \sigma_{k+1}^2
$$

Therefore

$$\| A - B \|_2 \geq \frac{\| (A - B)\hat{x} \|_2}{\| \hat{x} \|_2} \geq \sigma_{k+1}$$

This completes the proof. $\square$

Exercise 7.3.9 In the first part of the proof of Theorem 7.3.6, we used the SVD of A, A_k, and $A - A_k$ in the form given in Theorem 7.1.10. Write down the other forms of the SVD of A_k and $A - A_k$: (a) Diagram (7.1.9), (b) Theorem 7.1.12, (c) Theorem 7.1.13. $\square$

COROLLARY 7.3.7 Suppose $A \in {I\!\!R}^{n \times m}$ has full rank. Thus rank$(A) = r = \min\{n, m\}$. Let $\sigma_1 \geq \sigma_2 \geq \cdots \geq \sigma_r > 0$ be the singular values of A. Let $B \in {I\!\!R}^{n \times m}$ satisfy $\| A - B \|_2 < \sigma_r$. Then B also has full rank.

This result is an immediate consequence of Theorem 7.3.6. From Corollary 7.3.7 we see that if A has full rank, then all matrices sufficiently close to A also have full rank. From Exercise 7.3.8 we know that every rank-deficient matrix has full-rank matrices arbitrarily close to it. By Corollary 7.3.7, each of these full-rank matrices is surrounded by other matrices of full rank. In topological language, the set of matrices of full rank is an open dense subset of ${I\!\!R}^{n \times m}$. Its complement, the set of rank-deficient matrices, is therefore closed and nowhere dense. This discussion is meant to convince you that almost all matrices have full rank.

If a matrix does not have full rank, any small perturbation is almost certain to transform it to a matrix that does have full rank. It follows that in the presence of uncertainty in the data, it is impossible to calculate the rank of a matrix or even detect that it is rank deficient. (This is a generalization of the assertion, made in Chapters 1 and 2, that it is impossible to detect whether a square matrix is singular.) Nevertheless in certain applications it is reasonable to call a matrix *numerically rank deficient* if it is close to a rank-deficient matrix.

Let ϵ be some positive number that represents the magnitude of the data uncertainties in the matrix A. If there exist matrices B of rank k such that $\| A - B \|_2 < \epsilon$ and, on the other hand, for every matrix C of rank $\leq k - 1$ we have $\| A - C \|_2 \gg \epsilon$, then we will say that the *numerical rank* of A is k. From Theorem 7.3.6 we know that this condition is satisfied if and only if the singular values of A satisfy

$$\sigma_1 \geq \sigma_2 \geq \cdots \geq \sigma_k \gg \epsilon > \sigma_{k+1} \geq \cdots$$

Thus the numerical rank can be determined by examining the singular values. A matrix that has k "large" singular values, the others being "tiny," has numerical rank k. However, if the set of singular values has no convenient gap, it may be impossible to assign a meaningful numerical rank to the matrix.

Example 7.3.8 The matrix

$$A = \begin{bmatrix} 1/3 & 1/3 & 2/3 \\ 2/3 & 2/3 & 4/3 \\ 1/3 & 2/3 & 3/3 \\ 2/5 & 2/5 & 4/5 \\ 3/5 & 1/5 & 4/5 \end{bmatrix}$$

is obviously of rank 2, since its third column is the sum of the first two columns. We calculated the singular values of A using a canned program that employs an algorithm of the type described in the previous section. The numbers are stored to about 16 decimal digits accuracy in the machine, so a perturbation of magnitude about 10^{-16} is present in the machine representation of A. Additional roundoff errors are made in the calculation of the singular values. Therefore a reasonable choice of ϵ would be, say, $\epsilon = 10^{-15}$. The computed singular values were

$$\sigma_1 = 2.599 \qquad \sigma_2 = .368 \qquad \text{and} \qquad \sigma_3 = .866 \times 10^{-16}$$

Since $\sigma_1 \geq \sigma_2 \gg \epsilon > \sigma_3$, the matrix has numerical rank 2.

Example 7.3.9 Imagine a 2000×1000 matrix with singular values $\sigma_n = (.9)^n$, $n = 1, 2, \ldots, 1000$. Then $\sigma_1 = .9$ and $\sigma_{1000} = 1.75 \times 10^{-46}$. It is clear that σ_1 is "large" and σ_{1000} is "tiny" by any standards, so the numerical rank of the matrix lies somewhere between 1 and 1000. However, it is impossible to specify the numerical rank exactly because there are no gaps in the singular values. For example

$$\sigma_{261} = 1.14 \times 10^{-12} \qquad \sigma_{262} = 1.03 \times 10^{-12}$$

$$\sigma_{263} = 9.24 \times 10^{-13} \qquad \sigma_{264} = 8.31 \times 10^{-13}$$

If $\epsilon = 10^{-12}$, say, it might be reasonable to say that the numerical rank is approximately 260, but it is certainly not possible to specify it exactly.

Distance to Nearest Singular Matrix

We conclude this section by considering the implications of Theorem 7.3.6 for square, nonsingular matrices. Let $A \in \mathbb{R}^{n \times n}$ be nonsingular, and let A_s denote the singular matrix that is closest to A, in the sense that $\| A - A_s \|_2$ is as small as possible. In Theorem 2.3.17 we showed that

$$\frac{\| A - A_s \|}{\| A \|} \geq \frac{1}{\kappa(A)}$$

for any induced matrix norm, and we mentioned that for the 2-norm, equality holds. We now have the tools to prove this.

COROLLARY 7.3.10   Let $A \in \mathbb{R}^{n \times n}$ be nonsingular. (Thus A has singular values $\sigma_1 \geq \sigma_2 \geq \cdots \geq \sigma_n > 0$.) Let A_s be the singular matrix that is closest to A, in the sense that $\| A - A_s \|_2$ is as small as possible. Then $\| A - A_s \|_2 = \sigma_n$, and

$$\frac{\| A - A_s \|_2}{\| A \|_2} = \frac{1}{\kappa_2(A)}$$

These results are immediate consequences of Theorems 7.3.1, 7.3.3, and 7.3.6. In words, the distance from A to the nearest singular matrix is equal to the smallest singular value of A, and the "relative distance" to the nearest singular matrix is equal to the reciprocal of the condition number.

7.4
THE SVD AND THE
LEAST-SQUARES PROBLEM

Let $A \in \mathbb{R}^{n \times m}$, $r = \text{rank}(A)$, and $b \in \mathbb{R}^n$, and consider the system of equations

$$Ax = b \tag{7.4.1}$$

with unknown $x \in \mathbb{R}^m$. If it happens that $n > m$, then the system is overdetermined, and we cannot expect to find an exact solution. Thus we will seek an x such that $\| b - Ax \|_2$ is minimized. This is exactly the least-squares problem, which we studied in Chapter 3. There we found that if $n \geq m$ and $\text{rank}(A) = m$, the least-squares problem has a unique solution. If $\text{rank}(A) < m$, the solution is not unique; there are many $x \in \mathbb{R}^m$ for which $\| b - Ax \|_2$ is minimized. Even if $n < m$, it can happen that (7.4.1) does not have a solution, so we might as well include that case as well. Thus we will make no assumption about the relative sizes of n and m.

Because the solution of the least-squares problem is sometimes not unique, we will consider the following additional problem: of all the $x \in \mathbb{R}^n$ that minimize $\| b - Ax \|_2$, find one for which $\| x \|_2$ is as small as possible. As we shall see, this problem always has a unique solution. Initially we shall assume A and b are known exactly, and all computations are carried out exactly. Once we have settled the theoretical issues, we will discuss the practical questions.

Suppose we have the exact singular value decomposition: $A = U \Sigma V^T$, where $U \in \mathbb{R}^{n \times n}$ and $V \in \mathbb{R}^{m \times m}$ are orthogonal and

$$\Sigma = \begin{bmatrix} \hat{\Sigma} & 0 \\ 0 & 0 \end{bmatrix} = \begin{bmatrix} \sigma_1 & & & & 0 & \\ & \sigma_2 & & & & 0 \\ & & \ddots & & & \\ 0 & & & \sigma_r & & \\ \hline & & 0 & & & 0 \end{bmatrix} \in \mathbb{R}^{n \times m}$$

with $\sigma_1 \geq \sigma_2 \geq \cdots \geq \sigma_r > 0$. Because U is orthogonal, $\| b - Ax \|_2 = \| U^T(b - Ax) \|_2 = \| U^T b - \Sigma(V^T x) \|_2$. Letting $c = U^T b$ and $y = V^T x$, we have

$$\| b - Ax \|_2^2 = \| c - \Sigma y \|_2^2 = \sum_{i=1}^{r} |c_i - \sigma_i y_i|^2 + \sum_{i=r+1}^{m} |c_i|^2 \qquad (7.4.2)$$

It is clear that this expression is minimized when and only when

$$y_i = \frac{c_i}{\sigma_i} \qquad i = 1, \ldots, r$$

Notice that when $r < m$, $y_{r+1}, \ldots, y_m$ do not appear in (7.4.2). Thus they have no effect on the residual and can be chosen arbitrarily. Among all the solutions so obtained, $\| y \|_2$ is clearly minimized when and only when $y_{r+1} = \cdots = y_m = 0$. Since $x = Vy$ and V is orthogonal, $\| x \|_2 = \| y \|_2$. Thus $\| x \|_2$ is minimized when and only when $\| y \|_2$ is. This proves that the least-squares problem has exactly one minimum norm solution.

It is useful to repeat the development, using partitioned matrices. Let

$$c = \begin{bmatrix} \hat{c} \\ d \end{bmatrix} \qquad \text{and} \qquad y = \begin{bmatrix} \hat{y} \\ z \end{bmatrix}$$

where $\hat{c}, \hat{y} \in \mathbb{R}^r$. Then (7.4.2) can be rewritten as

$$\| b - Ax \|_2^2 = \left\| \begin{bmatrix} \hat{c} \\ d \end{bmatrix} - \begin{bmatrix} \hat{\Sigma} & 0 \\ 0 & 0 \end{bmatrix} \begin{bmatrix} \hat{y} \\ z \end{bmatrix} \right\|_2^2 = \left\| \begin{bmatrix} \hat{c} - \hat{\Sigma}\hat{y} \\ d \end{bmatrix} \right\|_2^2 = \| \hat{c} - \hat{\Sigma}\hat{y} \|_2^2 + \| d \|_2^2$$

$$(7.4.3)$$

This is minimized when and only when $\hat{y} = \hat{\Sigma}^{-1}\hat{c}$; that is, $y_i = c_i/\sigma_i$, $i = 1, \ldots, r$. We can choose z arbitrarily, but we get the minimum norm solution by taking $z = 0$. The norm of the minimal residual is $\| d \|_2$. This solves the problem completely in principle. We summarize the procedure:

1. Calculate $\begin{bmatrix} \hat{c} \\ d \end{bmatrix} = c = U^T b$.

2. Let $\hat{y} = \hat{\Sigma}^{-1}\hat{c}$.

3. Let $y = \begin{bmatrix} \hat{y} \\ z \end{bmatrix} \in \mathbb{R}^n$, where z can be chosen arbitrarily. To get the   (7.4.4)
    minimum norm solution, take $z = 0$.

4. Let $x = Vy$.

Practical Considerations

In practice we do not know the exact rank of A. It is best to use the numerical rank, discussed in Section 7.3, instead. All "tiny" singular values should be set to zero.

We have solved the least-squares problem under the assumption that we have the whole matrices U and V at hand. However, you can easily check that the calculation of $\hat{c}$ uses only the first r columns of U, where, in practice, r is the numerical rank. If only the minimum norm solution is wanted, only the first r columns of V are used. While the numerical rank is usually not known in advance, it can never exceed $\min\{n, m\}$, so at most $\min\{n, m\}$ columns of U and V are needed.

If $n \gg m$, the computation of U can be expensive, even if we compute only the first m columns. In fact the computation of U can be avoided completely. U is the product of many reflectors and rotators that are generated during the reduction to bidiagonal form and the subsequent QR steps. Since U is needed only so that we can calculate $c = U^T b$, we can simply update b instead of assembling U. As each rotator or reflector U_i is generated, we make the update $U_i^T b \rightarrow b$. In the end b will have been transformed into c. This is much less expensive than calculating U explicitly just to get $c = U^T b$. In the process we get not only $\hat{c}$, but also d, so we can compute the norm of the residual $\| d \|_2$. If several least-squares problems with the same A but different right-hand sides $b^{(1)}, b^{(2)}, \ldots$ are to be solved, the updates must be applied to all of the $b^{(i)}$ at once, since the U_i will not be saved.

No matter how the calculations are organized, the SVD is an expensive way to solve the least-squares problem. Its principal advantage is that it gives a completely reliable means of determining the numerical rank for rank-deficient least-squares problems.

Exercise 7.4.1   Estimate the total number of flops needed to transform b to c by repeated updates. How does this compare with the flop count for calculating the first m columns of U?   □

Exercise 7.4.2   Study the effects of treating very small nonzero singular values as nonzeros in the solution of the least-squares problem.   □

Exercise 7.4.3   Suppose $n < m$ and rank(A) = n.

(a)   Show that in this case the minimum norm least-squares problem is actually a constrained minimization problem.

(b)   This problem can be solved by an SVD. Show that it can also be solved by an LQ decomposition: $A = LQ$, where $L = [\tilde{L}\ 0] \in \mathbb{R}^{n \times m}$, $\tilde{L} \in \mathbb{R}^{n \times n}$ is nonsingular and lower triangular, and $Q \in \mathbb{R}^{m \times m}$ is orthogonal.

(c)   How does one calculate an LQ decomposition? Sketch an algorithm for calculating the LQ decomposition and solving the constrained minimization problem.   □

The Pseudoinverse

The pseudoinverse, also known as the Moore–Penrose generalized inverse, is an interesting generalization of the ordinary inverse. Although only square, nonsingular matrices have inverses in the ordinary sense, every $A \in \mathbb{R}^{n \times m}$ has a pseudoinverse. Just as the solution of a square, nonsingular linear system $Ax = b$ can be expressed in terms of A^{-1} as $x = A^{-1}b$, the minimum norm solution to a least-squares problem with a possibly nonsquare coefficient matrix A can be expressed in terms of the pseudoinverse $A^{\dagger}$ as $x = A^{\dagger}b$.

Given $A \in \mathbb{R}^{n \times m}$ with rank r, the action of A is completely described by the diagram

$$
\begin{array}{ccc}
v_1 & \xrightarrow{\;\sigma_1\;} & u_1 \\
v_2 & \xrightarrow{\;\sigma_2\;} & u_2 \\
\vdots & & \vdots \\
v_r & \xrightarrow{\;\sigma_r\;} & u_r \\
\left.\begin{array}{c} v_{r+1} \\ \vdots \\ v_m \end{array}\right\} & \longrightarrow & 0
\end{array}
$$

where $v_1, \ldots, v_m$ and $u_1, \ldots, u_n$ are complete orthonormal sets of right and left singular vectors, respectively, and $\sigma_1 \geq \sigma_2 \geq \cdots \geq \sigma_r > 0$ are the nonzero singular values of A. In matrix form,

$$ A = U\Sigma V^T $$

We wish to define the pseudoinverse $A^{\dagger} \in \mathbb{R}^{m \times n}$ so that it is as much like a true inverse as possible. Therefore we must certainly require $A^{\dagger}u_i = \sigma_i^{-1}v_i$ for $i = 1, \ldots, r$. There is no reasonable choice for $A^{\dagger}u_{r+1}, \ldots, A^{\dagger}u_n$ other than to make them zero. This assures, among other things, that A is the pseudoinverse of $A^{\dagger}$. Thus we define the *pseudoinverse* of A to be the matrix $A^{\dagger} \in \mathbb{R}^{m \times n}$ that is uniquely specified by the diagram

$$
\begin{array}{ccc}
u_1 & \xrightarrow{\;\sigma_1^{-1}\;} & v_1 \\
u_2 & \xrightarrow{\;\sigma_2^{-1}\;} & v_2 \\
\vdots & & \vdots \\
u_r & \xrightarrow{\;\sigma_r^{-1}\;} & v_r \\
\left.\begin{array}{c} u_{r+1} \\ \vdots \\ u_n \end{array}\right\} & \longrightarrow & 0
\end{array}
$$

We see immediately that $\mathrm{rank}(A^{\dagger}) = \mathrm{rank}(A)$, $u_1, \ldots, u_n$ and $v_1, \ldots, v_m$ are right and left singular vectors of $A^{\dagger}$, respectively, and $\sigma_1^{-1}, \ldots, \sigma_r^{-1}$ are the nonzero singular values. The restricted operators $A: \langle v_1, \ldots, v_r \rangle \rightarrow \langle u_1, \ldots, u_r \rangle$ and $A^{\dagger}: \langle u_1, \ldots, u_r \rangle \rightarrow \langle v_1, \ldots, v_r \rangle$ are true inverses of one another.

What does $A^{\dagger}$ look like as a matrix? You can answer this question in the simplest case by working the following exercise.

Exercise 7.4.4   Show that if

$$\Sigma = \begin{bmatrix} \sigma_1 & & 0 & \\ & \ddots & & 0 \\ 0 & & \sigma_r & \\ \hline & 0 & & 0 \end{bmatrix} \in \mathbb{R}^{n \times m}$$

then

$$\Sigma^{\dagger} = \begin{bmatrix} \sigma_1^{-1} & & 0 & \\ & \ddots & & 0 \\ 0 & & \sigma_r^{-1} & \\ \hline & 0 & & 0 \end{bmatrix} \in \mathbb{R}^{m \times n} \qquad\qquad \square$$

To see what $A^{\dagger}$ looks like in general, note that the equations

$$A^{\dagger} u_i = \begin{cases} v_i \sigma_i^{-1} & i = 1, \ldots, r \\ 0 & i = r + 1, \ldots, n \end{cases}$$

can be expressed as a single matrix equation

$$A^{\dagger}[u_1, u_2, \ldots, u_r \mid u_{r+1}, \ldots, u_n] =$$

$$[v_1, v_2, \ldots, v_r \mid v_{r+1}, \ldots, v_m] \begin{bmatrix} \sigma_1^{-1} & 0 & \cdots & 0 & \\ 0 & \sigma_2^{-1} & \cdots & 0 & 0 \\ \vdots & \vdots & & & \\ 0 & 0 & \cdots & \sigma_r^{-1} & \\ \hline & 0 & & & 0 \end{bmatrix}$$

or $A^{\dagger} U = V\Sigma^{\dagger}$. Thus

$$A^{\dagger} = V\Sigma^{\dagger} U^T \qquad\qquad (7.4.5)$$

This is the SVD of $A^{\dagger}$ in matrix form, and it gives us a means of calculating $A^{\dagger}$ by computing the SVD of A. However, there is seldom any reason to compute the pseudoinverse; it is mainly a theoretical tool. In this respect the pseudoinverse plays a role much like that of the ordinary inverse.

It is easy to make the claimed connection between the pseudoinverse and the least-squares problem.

THEOREM 7.4.6   Let $A \in \mathbb{R}^{n \times m}$ and $b \in \mathbb{R}^n$, and let $x \in \mathbb{R}^m$ be the minimum norm solution of

$$\| b - Ax \|_2 = \min_{w \in \mathbb{R}^m} \| b - Aw \|_2$$

Then $x = A^{\dagger} b$.

Proof   By (7.4.4), $x = Vy = V \begin{bmatrix} \hat{y} \\ 0 \end{bmatrix} = V \begin{bmatrix} \hat{\Sigma}^{-1}\hat{c} \\ 0 \end{bmatrix} = V \begin{bmatrix} \hat{\Sigma}^{-1} & 0 \\ 0 & 0 \end{bmatrix} \begin{bmatrix} \hat{c} \\ d \end{bmatrix} = V\Sigma^{\dagger}c = V\Sigma^{\dagger}U^{T}b = A^{\dagger}b.$ □

The pseudoinverse is used in the study of the sensitivity of the rank-deficient least-squares problem. See [SLS] or Stewart (1977).

Exercise 7.4.5   Show that if $A = \sum_{i=1}^{r} \sigma_i u_i v_i^T$, then $A^{\dagger} = \sum_{i=1}^{r} \sigma_i^{-1} v_i u_i^T$.   □

Exercise 7.4.6 (Pseudoinverses of full-rank matrices)

(a)   Show that if $A \in \mathbb{R}^{n \times m}$, $n \geq m$, rank$(A) = m$, then $A^{\dagger} = (A^T A)^{-1} A^T$ (cf., Exercise 7.3.7). What is the connection between this result and the normal equations (3.5.10)?

(b)   Show that if $A \in \mathbb{R}^{n \times m}$, $n \leq m$, and rank$(A) = n$, then $A^{\dagger} = A^T (AA^T)^{-1}$.   □

Exercise 7.4.7   (a) Let $A \in \mathbb{R}^{n \times m}$ and $B = A^{\dagger} \in \mathbb{R}^{m \times n}$. Show that the following four relationships hold:

$$BAB = B \qquad ABA = A$$
$$(BA)^T = BA \qquad (AB)^T = AB \tag{7.4.7}$$

(b) Conversely, show that if A and B satisfy (7.4.7), then $B = A^{\dagger}$. Thus equations (7.4.7) characterize the pseudoinverse. In many books they are used to define the pseudoinverse.   □

7.5
ANGLES AND DISTANCES
BETWEEN SUBSPACES

Angles between Subspaces

In this section we will see how to describe the relative orientation of two subspaces of $\mathbb{R}^n$ in terms of certain *principal angles* between them. These are the angles formed by certain *principal vectors* in the spaces. The study of principal angles and vectors has numerous applications, notably canonical correlation analysis.[§] At the end of the section we will study the closely related notion of distances between subspaces.

Let $\mathcal{S}$ and $\mathcal{T}$ be two subspaces of $\mathbb{R}^n$. We could allow them to have different dimensions, but for simplicity we will assume dim$(\mathcal{S}) =$ dim$(\mathcal{T}) = k > 0$. The *smallest angle* between $\mathcal{S}$ and $\mathcal{T}$ is defined to be the smallest angle that can be

---

[§] See, for example, Johnson and Wichern (1988). The connection between the notions of angle and correlation is easily seen: The correlation of two zero-mean unit vectors is just the cosine of the angle between them.

formed between a vector in $\mathcal{S}$ and a vector in $\mathcal{T}$. Recall that the angle between nonzero vectors $u \in \mathcal{S}$ and $v \in \mathcal{T}$ is the unique $\theta \in [0, \pi]$ defined by

$$\cos \theta = \frac{(u, v)}{\| u \|_2 \| v \|_2}$$

Obviously the angle is unaltered when u and v are multiplied by positive constants, so it suffices to consider vectors of length 1. Since the angle is minimized when the cosine is maximized, the smallest angle satisfies

$$\cos \theta_1 = \max \left\{ (u, v) \mid u \in \mathcal{S}, \| u \|_2 = 1, \ v \in \mathcal{T}, \| v \|_2 = 1 \right\} \qquad (7.5.1)$$

Exercise 7.5.1   Show that (a) θ_1 always lies in $[0, \pi/2]$; (b) $\theta_1 = \pi/2$ if and only if $\mathcal{S}$ and $\mathcal{T}$ are mutually orthogonal; (c) $\theta_1 = 0$ if and only if $\mathcal{S} \cap \mathcal{T} \neq \{0\}$; (d) if $\mathcal{S}$, $\mathcal{T} \subseteq \mathbb{R}^3$ and $\dim(\mathcal{S}) = \dim(\mathcal{T}) = 2$, then $\theta_1 = 0$. $\qquad \square$

The smallest angle is also called the *first principal angle* between $\mathcal{S}$ and $\mathcal{T}$. Suppose the maximum in (7.5.1) is attained when $u = u_1$ and $v = v_1$. The *second principal angle* θ_2 is then defined to be the smallest angle that can be formed between a vector in $\mathcal{S}$ that is orthogonal to u_1 and a vector in $\mathcal{T}$ that is orthogonal to v_1. Thus

$$\cos \theta_2 = \max \left\{ (u, v) \mid u \in \mathcal{S} \cap \langle u_1 \rangle^{\perp}, \| u \|_2 = 1, \ v \in \mathcal{T} \cap \langle v_1 \rangle^{\perp}, \| v \|_2 = 1 \right\}$$

Let u_2, v_2 be a pair for which the maximum is attained. In general the principal angles are defined recursively for $i = 1, \ldots, k$ by

$$\cos \theta_i = \max \left\{ (u, v) \mid u \in \mathcal{S} \cap \langle u_1, \ldots, u_{i-1} \rangle^{\perp}, \right.$$

$$\left. \| u \|_2 = 1, \ v \in \mathcal{T} \cap \langle v_1, \ldots, v_{i-1} \rangle^{\perp}, \| v \|_2 = 1 \right\}$$

The vectors $u_i \in \mathcal{S} \cap \langle u_1, \ldots, u_{i-1} \rangle^{\perp}$ and $v_i \in \mathcal{T} \cap \langle v_1, \ldots, v_{i-1} \rangle^{\perp}$ are chosen to be a pair for which the maximum is attained. Obviously $0 \leq \theta_1 \leq \theta_2 \leq \cdots \leq \theta_k \leq \pi/2$. The orthonormal vectors $u_1, \ldots, u_k \in \mathcal{S}$ and $v_1, \ldots, v_k \in \mathcal{T}$ are called *principal vectors*. Clearly $\mathcal{S} = \langle u_1, \ldots, u_k \rangle$ and $\mathcal{T} = \langle v_1, \ldots, v_k \rangle$. The principal vectors are not uniquely determined but, as we shall see, the principal angles are.

Example 7.5.2   A pair of principal vectors u_i and v_i can always be exchanged for $-u_i$ and $-v_i$.

Example 7.5.3   Let $\mathcal{S} = \mathcal{T}$. Then $\theta_1 = \cdots = \theta_k = 0$. The principal vectors $u_1, \ldots, u_k \in \mathcal{S}$ can be chosen arbitrarily, subject only to orthonormality, and $v_i = u_i$, $i = 1, \ldots, k$.

Exercise 7.5.2

   (a)   Show that if $\dim(\mathcal{S} \cap \mathcal{T}) = i$, then $\theta_1 = \theta_2 = \cdots = \theta_i = 0$ and (if $i < k$) $\theta_{i+1} \neq 0$.

   (b)   Show that $\mathcal{S} = \mathcal{T}$ if and only if $\theta_k = 0$. $\qquad \square$

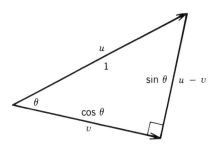

Figure 7.1

The principal angles and vectors satisfy a number of simple and interesting geometric relationships, in addition to those that are obvious consequences of the definition. We will begin with two elementary lemmas.

LEMMA 7.5.4 Let $u, v \in R^n$ with $\| u \|_2 = 1$ and $(u - v, v) = 0$. Let θ denote the angle between u and v. (If $v = 0$, we define $\theta = \pi/2$.) Then

$$\cos \theta = \| v \|_2 \qquad \text{and} \qquad \sin \theta = \| u - v \|_2$$

Proof From Figure 7.1 we see that this lemma states a geometrically obvious fact. Here is a proof anyway. By definition, $\cos \theta = (u, v)/\| u \|_2 \| v \|_2$. Since $(u - v, v) = 0$, we have $(u, v) = (v, v) = \| v \|_2^2$. Since also $\| u \|_2 = 1$, we conclude that $\cos \theta = \| v \|_2$. By the Pythagorean theorem (Lemma 3.5.5), $\| u \|_2^2 = \| v \|_2^2 + \| u - v \|_2^2$, so $\| u - v \|_2^2 = 1 - \cos^2 \theta = \sin^2 \theta$. Thus $\| u - v \|_2 = \sin \theta$. $\qquad \square$

LEMMA 7.5.5 Let $u, v \in R^n$ with $\| u \|_2 = \| v \|_2 = 1$. Let θ be the angle between u and v. Then

$$\| u - v \cos \theta \|_2 = \sin \theta$$

$v \cos \theta$ is the multiple of v that is closest to u.

Proof Figure 7.2 shows that this lemma is also obvious. Let us first show that $v \cos \theta$ is the multiple of v that is closest to u. Applying the projection theorem (Theorem 3.5.6) with $\mathcal{S} = \langle v \rangle$, we find that the multiple cv that is closest to u

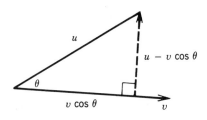

Figure 7.2

is characterized by $(u - cv, v) = 0$. Solving this equation for c, we find that $c = (u, v)/(v, v) = (u, v) = \cos\theta$. Now applying Lemma 7.5.4 with v replaced by $v\cos\theta$, we have $\| u - v\cos\theta \| = \sin\theta$. $\square$

The result of the next exercise will be used in the proof of the theorem that follows.

Exercise 7.5.3   Let $u, v \in \mathbb{R}^n$ be nonzero vectors, and let θ be the angle between them. (a) Show that the angle between u and $-v$ is $\pi - \theta$. (b) Show that if $\theta > \pi/2$, then $\| u - (-v) \|_2 < \| u - v \|_2$. [*Hint:* Use the identity $\| w \|_2^2 = (w, w)$.] (c) Show that if v is the best approximation of u from $\mathcal{T}$, then $\theta \leq \pi/2$. $\square$

THEOREM 7.5.6   The first principal angle θ_1 and principal vectors $u_1 \in \mathcal{S}$ and $v_1 \in \mathcal{T}$ satisfy

$$\sin\theta_1 = \| u_1 - v_1\cos\theta_1 \|_2 = \min_{v \in \mathcal{T}} \| u_1 - v \|_2 = \min_{\substack{u \in \mathcal{S} \\ \| u \|_2 = 1}} \min_{v \in \mathcal{T}} \| u - v \|_2$$

$$\sin\theta_1 = \| v_1 - u_1\cos\theta_1 \|_2 = \min_{u \in \mathcal{S}} \| v_1 - u \|_2 = \min_{\substack{v \in \mathcal{T} \\ \| v \|_2 = 1}} \min_{u \in \mathcal{S}} \| v - u \|_2$$

Proof   The equation $\sin\theta_1 = \| u_1 - v_1\cos\theta_1 \|_2$ is a consequence of Lemma 7.5.5. As for the other two equations, it is obvious that

$$\| u_1 - v_1\cos\theta_1 \|_2 \geq \min_{v \in \mathcal{T}} \| u_1 - v \|_2 \geq \min_{\substack{u \in \mathcal{S} \\ \| u \|_2 = 1}} \min_{v \in \mathcal{T}} \| u - v \|_2$$

To get equality, it suffices to prove that for every $u \in \mathcal{S}$ and $v \in \mathcal{T}$ with $\| u \|_2 = 1$, $\| u - v \|_2 \geq \| u_1 - v_1\cos\theta_1 \|_2$. Given such a u and v, let $\hat{v}$ denote the best approximation of u from $\mathcal{T}$. Then certainly $\| u - \hat{v} \|_2 \leq \| u - v \|_2$. Let θ denote the angle between u and $\hat{v}$. (If $\hat{v} = 0$, we set $\theta = \pi/2$.) By Exercise 7.5.3, $\theta \leq \pi/2$. Since $(u - \hat{v}, \hat{v}) = 0$ by Theorem 3.5.6, we have $\| u - \hat{v} \|_2 = \sin\theta$ from Lemma 7.5.4. Since $\theta_1 \leq \theta$ and $\sin$ is an increasing function on $[0, \pi/2]$, $\| u_1 - v_1\cos\theta_1 \|_2 = \sin\theta_1 \leq \sin\theta = \| u - \hat{v} \|_2 \leq \| u - v \|_2$. This proves the first string of equations. The second string is proved by reversing the roles of $\mathcal{S}$ and $\mathcal{T}$. $\square$

Theorem 7.5.6 is easily generalized to yield statements about the other principal angles and vectors. For this purpose it is convenient to introduce some new notation. For $i = 1, \ldots, k$ let

$$\mathcal{S}_i = \mathcal{S} \cap \langle u_1, \ldots, u_{i-1} \rangle^\perp = \langle u_i, \ldots, u_k \rangle$$
$$\mathcal{T}_i = \mathcal{T} \cap \langle v_1, \ldots, v_{i-1} \rangle^\perp = \langle v_i, \ldots, v_k \rangle$$

Roughly speaking, $\mathcal{S}_i$ and $\mathcal{T}_i$ are just $\mathcal{S}$ and $\mathcal{T}$ with the first $i - 1$ principal vectors removed.

THEOREM 7.5.7 For $i = 1, \ldots, k$,

$$\sin \theta_i = \| u_i - v_i \cos \theta_i \|_2 = \min_{v \in \mathcal{T}_i} \| u_i - v \|_2 = \min_{\substack{u \in \mathcal{S}_i \\ \| u \|_2 = 1}} \min_{v \in \mathcal{T}_i} \| u - v \|_2$$

$$\sin \theta_i = \| v_i - u_i \cos \theta_i \|_2 = \min_{u \in \mathcal{S}_i} \| v_i - u \|_2 = \min_{\substack{v \in \mathcal{T}_i \\ \| v \|_2 = 1}} \min_{u \in \mathcal{S}_i} \| v - u \|_2$$

Theorem 7.5.7 is an immediate consequence of Theorem 7.5.6, because θ_i, u_i, and v_i are the first principal angles and vectors for the subspaces $\mathcal{S}_i$ and $\mathcal{T}_i$.

Not only are $u_1, \ldots, u_k$ and $v_1, \ldots, v_k$ orthonormal, but additional orthogonality relations hold between the u_i and v_j as well.

THEOREM 7.5.8 If $i \neq j$, then $(u_i, v_j) = 0$.

Proof First assume $i < j$. From Theorem 7.5.7,

$$\| u_i - v_i \cos \theta_i \|_2 = \min_{v \in \mathcal{T}_i} \| u_i - v \|_2$$

so by the projection theorem, $(u_i - v_i \cos \theta_i, v) = 0$ for all $v \in \mathcal{T}_i$. Since $i < j$, $v_j \in \mathcal{T}_i$, so $(u_i - v_i \cos \theta_i, v_j) = 0$. Since $(v_i, v_j) = 0$ as well, we conclude that $(u_i, v_j) = 0$. To get the result for $i > j$, just reverse the roles of $\mathcal{S}$ and $\mathcal{T}$. $\square$

Theorem 7.5.8 can also be expressed in matrix form as follows.

COROLLARY 7.5.9 Let $U_1 = [u_1, u_2, \ldots, u_k] \in \mathbb{R}^{n \times k}$, $V_1 = [v_1, \ldots, v_k] \in \mathbb{R}^{n \times k}$, and

$$\Gamma_1 = \begin{bmatrix} \cos \theta_1 & & & 0 \\ & \cos \theta_2 & & \\ & & \ddots & \\ 0 & & & \cos \theta_k \end{bmatrix} \in \mathbb{R}^{k \times k}$$

Then $U_1^T V_1 = V_1^T U_1 = \Gamma_1$.

Exercise 7.5.4 Verify Corollary 7.5.9. $\square$

Exercise 7.5.5 (a) Let $u, v \in \mathbb{R}^n$ with $\| u \|_2 = \| v \|_2 = 1$, and let θ be the angle between u and v. Show that $\| u - v \|_2 = 2 \sin(\theta/2)$. [You might find the half-angle formula $\sin^2(\theta/2) = (1 - \cos \theta)/2$ useful.] (b) Draw a picture that shows that the result of part (a) is geometrically obvious. (c) Show that the principal angles and vectors satisfy

$$2 \sin(\theta_i/2) = \| u_i - v_i \|_2 = \min \{ \| u - v \|_2 \mid u \in \mathcal{S}_i, \| u \|_2 = 1, \ v \in \mathcal{T}_i, \| v \|_2 = 1 \}$$

for $i = 1, \ldots, k$, where $\mathcal{S}_i = \langle u_i, \ldots, u_k \rangle$ and $\mathcal{T}_i = \langle v_i, \ldots, v_k \rangle$, as above. $\square$

Computing Principal Angles and Vectors

In applications the subspaces $\mathcal{S}$ and $\mathcal{T}$ are usually provided in the form of a basis for each subspace. If the bases are not orthonormal, they can be orthonormalized by one of the techniques from Chapter 3. Let us therefore assume that we have orthonormal bases $p_1, p_2, \ldots, p_k$ and $q_1, q_2, \ldots, q_k$ of $\mathcal{S}$ and $\mathcal{T}$, respectively. Let $P_1 = [p_1, p_2, \ldots, p_k] \in \mathbb{R}^{n \times k}$ and $Q_1 = [q_1, q_2, \ldots, q_k] \in \mathbb{R}^{n \times k}$. We wish to determine the principal angles and vectors between the spaces. This is equivalent to determining the matrices U_1, V_1, and Γ_1, defined in Corollary 7.5.9. Since $u_1, \ldots, u_k$ and $p_1, \ldots, p_k$ are both bases of the same space, $\mathcal{R}(U_1) = \mathcal{R}(P_1) = \mathcal{S}$. Similarly $\mathcal{R}(V_1) = \mathcal{R}(Q_1) = \mathcal{T}$. The matrices P_1, Q_1, U_1, and V_1 are all isometries (cf., Section 3.4) because they have orthonormal columns.

Exercise 7.5.6 Let $U, P \in \mathbb{R}^{n \times k}$ have linearly independent columns. (a) Show that $\mathcal{R}(U) = \mathcal{R}(P)$ if and only if there exists a nonsingular $M \in \mathbb{R}^{k \times k}$ such that $U = PM$. (b) Suppose $\mathcal{R}(U) = \mathcal{R}(P)$ and U and P have orthonormal columns. Show that the matrix M of part (a) has orthonormal columns. $\square$

By Exercise 7.5.6, there exist orthogonal matrices $M_1, N_1 \in \mathbb{R}^{k \times k}$ such that

$$U_1 = P_1 M_1 \qquad \text{and} \qquad V_1 = Q_1 N_1$$

If we can figure out how to calculate M_1 and N_1, we can use them to determine U_1 and V_1. Recalling from Corollary 7.5.9 that $U_1^T V_1 = \Gamma_1$, we have

$$P_1^T Q_1 = M_1 U_1^T V_1 N_1^T = M_1 \Gamma_1 N_1^T$$

Since M_1 and N_1 are orthogonal and Γ_1 is diagonal, $M_1 \Gamma_1 N_1^T$ is the SVD of $P_1^T Q_1$. This gives us a means of calculating the principal angles and vectors:

Calculate

1. $P_1^T Q_1$.
2. the SVD $P_1^T Q_1 = M_1 \Gamma_1 N_1^T$. Let $\gamma_1 \geq \gamma_2 \geq \cdots \geq \gamma_k$ denote
    the singular values.   (7.5.10)
3. $\theta_i = \arccos \gamma_i$, $i = 1, \ldots, k$ (principal angles).
4. $U_1 = P_1 M_1$ and $V_1 = Q_1 N_1$ (principal vectors).

This also settles a theoretical question. The principal angles are uniquely determined: They are determined by the singular values of a matrix that is independent of $u_1, \ldots, u_k$ and $v_1, \ldots, v_k$, so they do not depend on how the principal vectors are chosen. By step 4 of (7.5.10), the principal vectors have exactly as much arbitrariness as singular vectors do.

Exercise 7.5.7 Show that if $\theta_1 > \theta_2 > \cdots > \theta_k$, the principal vectors are almost uniquely determined. $\square$

As the following exercise shows, the computation $\theta_i = \arccos \gamma_i$ cannot deliver accurate values for angles near zero.

Exercise 7.5.8   (a) Use the mean-value theorem to show that the function $\theta = \arccos \gamma$ is ill conditioned for γ near 1 (and θ near 0). That is, small perturbations in γ can cause relatively large perturbations in θ. (b) Show that for γ bounded away from 1, $\arccos \gamma$ is well conditioned. Specifically, if γ_1, $\gamma_2 < 0.8$, then $|\arccos \gamma_1 - \arccos \gamma_2| < \frac{5}{3}|\gamma_1 - \gamma_2|$.   □

If we wish to calculate small principal angles accurately, we must find another method. The following lemma is a start in that direction.

LEMMA 7.5.11   Let $W_1 \in \mathbb{R}^{n \times k}$ be an isometry, and consider the partitioned form

$$W_1 = \begin{bmatrix} W_{11} \\ W_{21} \end{bmatrix} \qquad W_{11} \in \mathbb{R}^{n_1 \times k}, \; W_{21} \in \mathbb{R}^{n_2 \times k}, \; n_1 + n_2 = n$$

Let $\gamma_1 \geq \gamma_2 \geq \cdots \geq \gamma_k$ be the singular values of W_{11}, and let $\sigma_1 \leq \sigma_2 \leq \cdots \leq \sigma_k$ be the singular values of W_{21} in ascending order. Then

$$\gamma_i^2 + \sigma_i^2 = 1 \qquad i = 1, \ldots, k$$

Proof   Since W_1 has orthonormal columns,

$$I = W_1^T W_1 = W_{11}^T W_{11} + W_{21}^T W_{21}$$

It follows immediately that $W_{11}^T W_{11}$ and $W_{21}^T W_{21}$ have common eigenvectors: If $W_{11}^T W_{11} v = \lambda v$, then $W_{21}^T W_{21} v = \mu v$, where

$$\lambda + \mu = 1 \tag{7.5.12}$$

and vice versa. Since the eigenvalues of $W_{11}^T W_{11}$ and $W_{21}^T W_{21}$ are $\gamma_1^2 \geq \gamma_2^2 \geq \cdots \geq \gamma_k^2$ and $\sigma_1^2 \leq \sigma_2^2 \leq \cdots \leq \sigma_k^2$, respectively, (7.5.12) implies that $\gamma_i^2 + \sigma_i^2 = 1$ for $i = 1, 2, \ldots, k$.   □

Exercise 7.5.9   Using the notation of Lemma 7.5.11, show that:

   (a)   If $n_1 < k$, then $\gamma_i = 0$ and $\sigma_i = 1$ for $i = n_1 + 1, \ldots, k$.
   (b)   If $n_2 < k$, then $\gamma_i = 1$ and $\sigma_i = 0$ for $i = 1, \ldots, k - n_2$.   □

THEOREM 7.5.13   Let $\mathcal{S}$ and $\mathcal{T}$ be k-dimensional subspaces of $\mathbb{R}^n$ with principal angles $\theta_1 \geq \theta_2 \geq \cdots \geq \theta_k$. Let $p_1, \ldots, p_k$ and $p_{k+1}, \ldots, p_n$ be orthonormal bases for $\mathcal{S}$ and $\mathcal{S}^\perp$, respectively, and let $q_1, \ldots, q_k$ and $q_{k+1}, \ldots, q_n$ be orthonormal bases for $\mathcal{T}$ and $\mathcal{T}^\perp$, respectively. Let

$$P_1 = [p_1, \ldots, p_k] \qquad P_2 = [p_{k+1}, \ldots, p_n]$$
$$Q_1 = [q_1, \ldots, q_k] \qquad Q_2 = [q_{k+1}, \ldots, q_n]$$

Then the singular values of $P_2^T Q_1$ and $P_1^T Q_2$ are

$$\sin \theta_1 \leq \sin \theta_2 \leq \cdots \leq \sin \theta_k$$

Proof Let $P = [P_1 \ P_2] \in \mathbb{R}^{n \times n}$. Then P is an orthogonal matrix. Since Q_1 has orthonormal columns, the matrix

$$W_1 = P^T Q_1$$

must also have orthonormal columns. W_1 can be written in the partitioned form

$$W_1 = \begin{bmatrix} P_1^T Q_1 \\ P_2^T Q_1 \end{bmatrix}$$

The singular values of $P_1^T Q_1$ are $\gamma_i = \cos \theta_i$, $i = 1, \ldots, k$, so by Lemma 7.5.11, $P_2^T Q_1$ has singular values

$$\sigma_i = \sqrt{1 - \gamma_i^2} = \sqrt{1 - \cos^2 \theta_i} = \sin \theta_i \qquad i = 1, \ldots, k$$

Reversing the roles of P and Q, we find that $Q_2^T P_1$ also has singular values $\sin \theta_i$, $i = 1, \ldots, k$. $P_1^T Q_2$ is the transpose of $Q_2^T P_1$, so it has the same singular values. $\qquad\square$

Exercise 7.5.10

(a) Referring to the notation of Theorem 7.5.13, show that if $k > n/2$, then $\sin \theta_i = 0$ for $i = 1, \ldots, 2k - n$.

(b) How does this result relate to the geometry of k-dimensional subspaces of $\mathbb{R}^n$? $\qquad\square$

Exercise 7.5.11 Formulate and prove statements about the function $\theta = \arcsin \sigma$ analogous to the results of Exercise 7.5.8. In particular, show that arcsin is well conditioned when θ is small. $\qquad\square$

It is now clear how to calculate small principal angles accurately.

Calculate:

1. $p_{k+1}, \ldots, p_n$, an orthonormal basis of $\mathcal{S}^\perp$. Let $P_2 = [p_{k+1}, \ldots, p_n]$.
2. $P_2^T Q_1$.
3. the singular values $\sigma_1 \leq \sigma_2 \leq \cdots \leq \sigma_k$ of $P_2^T Q_1$.   (7.5.14)
4. $\theta_i = \arcsin \sigma_i$, $i = 1, \ldots, k$.

This procedure gives accurate values of the small θ_i and inaccurate values of the θ_i that are close to $\pi/2$.

Exercise 7.5.12 (Review) Suppose the space $\mathcal{S}$ is given in the form of a nonorthonormal basis $s_1, s_2, \ldots, s_k$. How can a QR decomposition by reflectors be used to calculate both $p_1, \ldots, p_k$ and $p_{k+1}, \ldots, p_n$, orthonormal bases for $\mathcal{S}$ and $\mathcal{S}^\perp$, respectively? $\qquad\square$

The matrices $P_1^T Q_1$, $P_2^T Q_1$, and $P_1^T Q_2$ are all submatrices of the orthogonal matrix

$$P^T Q = \begin{bmatrix} P_1^T Q_1 & P_1^T Q_2 \\ P_2^T Q_1 & P_2^T Q_2 \end{bmatrix}$$

Theorem 7.5.13 shows that there are simple relationships between the singular values of these submatrices. A more detailed statement of the relationships between the singular values and singular vectors of the blocks of an orthogonal or unitary matrix is given by Stewart (1977), Theorem A1.

The Distance between Subspaces

We continue to let $\mathscr{S}$ and $\mathscr{T}$ denote the k-dimensional subspaces of $\mathbb{R}^n$. In Section 5.2 we defined the *distance* between $\mathscr{S}$ and $\mathscr{T}$ by

$$d(\mathscr{S}, \mathscr{T}) = \max_{\substack{u \in \mathscr{S} \\ \| u \|_2 = 1}} d(u, \mathscr{T})$$

where

$$d(u, \mathscr{T}) = \min_{v \in \mathscr{T}} \| u - v \|_2$$

The following theorem makes it very easy to work with this distance function.

THEOREM 7.5.15 In the notation of Theorem 7.5.13,

$$d(\mathscr{S}, \mathscr{T}) = \| Q_2^T P_1 \|_2 = \| Q_1^T P_2 \|_2 = \| P_1^T Q_2 \|_2 = \| P_2^T Q_1 \|_2$$

Proof Let $u \in \mathscr{S}$ with $\| u \|_2 = 1$. Since $\mathbb{R}^n = \mathscr{T} \oplus \mathscr{T}^\perp$, u can be expressed uniquely as a sum, $u = v + v^\perp$, where $v \in \mathscr{T}$ and $v^\perp \in \mathscr{T}^\perp$. By the projection theorem (Theorem 3.5.6), $d(u, \mathscr{T}) = \| v^\perp \|_2$. Letting Q be the orthogonal matrix $Q = [Q_1 \ Q_2]$, we have

$$Q^T v^\perp = \begin{bmatrix} Q_1^T v^\perp \\ Q_2^T v^\perp \end{bmatrix}$$

so $\| v^\perp \|_2^2 = \| Q^T v^\perp \|_2^2 = \| Q_1^T v^\perp \|_2^2 + \| Q_2^T v^\perp \|_2^2$. Since $v^\perp \in \mathscr{T}^\perp$, $Q_1^T v^\perp = 0$, so $\| v^\perp \|_2 = \| Q_2^T v^\perp \|_2$. Notice also that $Q_2^T v = 0$. Thus

$$d(u, \mathscr{T}) = \| v^\perp \|_2 = \| Q_2^T v^\perp \|_2 = \| Q_2^T v^\perp + Q_2^T v \|_2 = \| Q_2^T u \|_2$$

and

$$d(\mathscr{S}, \mathscr{T}) = \max_{\substack{u \in \mathscr{S} \\ \| u \|_2 = 1}} \| Q_2^T u \|_2$$

Since $u \in \mathcal{G} = \mathcal{R}(P_1)$, there is a (unique) $x \in \mathbb{R}^k$ such that $u = P_1 x$. Since P_1 is an isometry, $\| x \|_2 = \| u \|_2 = 1$. Conversely, given $x \in \mathbb{R}^k$ with $\| x \|_2 = 1$, $P_1 x$ is a member of $\mathcal{G}$ satisfying $\| P_1 x \|_2 = 1$. Therefore

$$d(\mathcal{G}, \mathcal{T}) = \max_{\substack{x \in \mathbb{R}^k \\ \| x \|_2 = 1}} \| Q_2^T P_1 x \|_2 = \| Q_2^T P_1 \|_2$$

By Theorem 7.3.1, the spectral norm of a matrix is equal to the largest singular value. Since $Q_1^T P_2$ has the same singular values as $Q_2^T P_1$ (Theorem 7.5.13), we have $d(\mathcal{G}, \mathcal{T}) = \| Q_2^T P_1 \|_2$ as well. Since $P_1^T Q_2$ and $P_2^T Q_1$ are the transposes of $Q_2^T P_1$ and $Q_1^T P_2$, respectively, it is also true that $d(\mathcal{G}, \mathcal{T}) = \| Q_2^T P_1 \|_2 = \| Q_1^T P_2 \|_2$. $\square$

COROLLARY 7.5.16   Let $\mathcal{G}$ and $\mathcal{T}$ be two k-dimensional subspaces of $\mathbb{R}^n$. Then $d(\mathcal{G}, \mathcal{T})$ equals the sine of the largest principal angle between $\mathcal{G}$ and $\mathcal{T}$.

Proof   From Theorem 7.5.13 we know that the largest singular value of $P_2^T Q_1$ is $\sin \theta_k$. Therefore $d(\mathcal{G}, \mathcal{T}) = \| P^T Q_1 \|_2 = \sin \theta_k$. $\square$

Exercise 7.5.13   Let $\mathcal{G}$ and $\mathcal{T}$ be two k-dimensional subspaces of $\mathbb{R}^n$. Then $\mathcal{G}^\perp$ and $\mathcal{T}^\perp$ both have dimension $n - k$, so $d(\mathcal{G}^\perp, \mathcal{T}^\perp)$ is defined. Show that $d(\mathcal{G}^\perp, \mathcal{T}^\perp) = d(\mathcal{G}, \mathcal{T})$. $\square$

THEOREM 7.5.17   The function $d(\cdot, \cdot)$ is a metric on the set of k-dimensional subspaces of $\mathbb{R}^n$. That is, for any k-dimensional subspaces $\mathcal{G}$, $\mathcal{T}$, and $\mathcal{U}$ of $\mathbb{R}^n$,

(i) $d(\mathcal{G}, \mathcal{T}) > 0$ if $\mathcal{G} \neq \mathcal{T}$

(ii) $d(\mathcal{G}, \mathcal{G}) = 0$

(iii) $d(\mathcal{G}, \mathcal{T}) = d(\mathcal{T}, \mathcal{G})$

(iv) $d(\mathcal{G}, \mathcal{U}) \leq d(\mathcal{G}, \mathcal{T}) + d(\mathcal{T}, \mathcal{U})$

Proof   Properties (i) and (ii) follow immediately from the definition of $d(\cdot, \cdot)$. Property (iii) is obviously a consequence of Theorem 7.5.15. To prove (iv), let $r_1, \ldots, r_k$ denote an orthonormal basis for $\mathcal{U}$ and let $R_1 = [r_1, \ldots, r_k] \in \mathbb{R}^{n \times k}$. Then, by Theorem 7.5.15,

$$d(\mathcal{G}, \mathcal{U}) = \| P_2^T R_1 \| = \| P_2^T I R_1 \|_2 = \| P_2^T Q Q^T R_1 \|_2$$

$$= \left\| P_2^T [Q_1 \ Q_2] \begin{bmatrix} Q_1^T \\ Q_2^T \end{bmatrix} R_1 \right\|_2 = \left\| [P_2^T Q_1 \ P_2^T Q_2] \begin{bmatrix} Q_1^T R_1 \\ Q_2^T R_1 \end{bmatrix} \right\|_2$$

$$= \| (P_2^T Q_1)(Q_1^T R_1) + (P_2^T Q_2)(Q_2^T R_1) \|_2$$

$$\leq \| P_2^T Q_1 \|_2 \| Q_1^T \|_2 \| R_1 \|_2 + \| P_2^T \|_2 \| Q_2 \|_2 \| Q_2^T R_1 \|_2$$

$$= \| P_2^T Q_1 \|_2 + \| Q_2^T R_1 \|_2 = d(\mathcal{G}, \mathcal{T}) + d(\mathcal{T}, \mathcal{U})$$

This completes the proof. $\square$

Fast Rotators

Algorithms that are primarily based on rotators generally perform twice as many multiplications as additions. The reason for this is easily seen when one examines the basic operation of multiplying a rotator by a vector:

$$\begin{bmatrix} c & s \\ -s & c \end{bmatrix} \begin{bmatrix} a \\ b \end{bmatrix} = \begin{bmatrix} ca + sb \\ -sa + cb \end{bmatrix} \tag{A.1}$$

This operation requires four multiplications and only two additions, so an algorithm that performs k such operations will require $4k$ multiplications and $2k$ additions. If two of the multiplications could somehow be eliminated from (A.1), the multiplication count would be halved. Let us see what can be done in this direction.

Consider a factorization of the rotator:

$$\begin{bmatrix} c & s \\ -s & c \end{bmatrix} = \begin{bmatrix} c & 0 \\ 0 & c \end{bmatrix} \begin{bmatrix} 1 & t \\ -t & 1 \end{bmatrix} \tag{A.2}$$

where $t = s/c$. Recall that s and c stand for sine and cosine, respectively, so t stands for tangent. The operation

$$\begin{bmatrix} 1 & t \\ -t & 1 \end{bmatrix} \begin{bmatrix} a \\ b \end{bmatrix} = \begin{bmatrix} a + tb \\ -ta + b \end{bmatrix}$$

requires only two multiplications because of the ones in the matrix. Of course we have not really saved anything, because we still have to multiply by the diagonal matrix $\begin{bmatrix} c & 0 \\ 0 & c \end{bmatrix}$. However, a typical algorithm multiplies a single matrix by a large number of rotators. If the scaling factors $\begin{bmatrix} c & 0 \\ 0 & c \end{bmatrix}$ from these rotators can in some way be accumulated inexpensively and multiplied into the matrix all at once at the end, considerable savings can be made.

Consider an n-by-n array A that is continually being updated by multiplication by rotators on the left. If the rotators are factored in some scheme similar to (A.2), then A must be represented as a product $A = DB$, where

$$D = \begin{bmatrix} d_1 & 0 & \cdots & 0 \\ 0 & d_2 & \cdots & 0 \\ \vdots & \vdots & \ddots & \vdots \\ 0 & 0 & \cdots & d_n \end{bmatrix}$$

is a diagonal matrix in which the scaling factors are accumulated. (Initially $D = I$.) A can be recovered from B and D at any time by multiplying the ith row of B by d_i for $i = 1, \ldots, n$. The factorization (A.2) turns out to be too simple for the task at hand, for we are not allowed to work with A directly; we must work with B and D.

Suppose we wish to multiply A by the rotator Q^T to obtain $\hat{A} = Q^T A$. How should D and B be modified so that we obtain $\hat{D}$ and $\hat{B}$ for which $\hat{A} = \hat{D}\hat{B}$? Since Q^T operates on only two rows of A, we can restrict our attention to these two rows. Thus we have

$$\begin{bmatrix} \hat{a}_{i1} & \hat{a}_{i2} & \cdots & \hat{a}_{in} \\ \hat{a}_{j1} & \hat{a}_{j2} & \cdots & \hat{a}_{jn} \end{bmatrix} = \begin{bmatrix} c & s \\ -s & c \end{bmatrix} \begin{bmatrix} a_{i1} & a_{i2} & \cdots & a_{in} \\ a_{j1} & a_{j2} & \cdots & a_{jn} \end{bmatrix} \tag{A.3}$$

If we were to carry out this operation in the conventional manner, we would calculate

$$\left. \begin{array}{l} \hat{a}_{ik} = c a_{ik} + s a_{jk} \\ \hat{a}_{jk} = -s a_{ik} + c a_{jk} \end{array} \right\} \quad k = 1, 2, \ldots, n$$

which requires $4n$ multiplications and $2n$ additions. Writing (A.3) in factored form, we have

$$\begin{bmatrix} \hat{d}_i & 0 \\ 0 & \hat{d}_j \end{bmatrix} \begin{bmatrix} \hat{b}_{i1} & \hat{b}_{i2} & \cdots & \hat{b}_{in} \\ \hat{b}_{j1} & \hat{b}_{j2} & \cdots & \hat{b}_{jn} \end{bmatrix} = \begin{bmatrix} c & s \\ -s & c \end{bmatrix} \begin{bmatrix} d_i & 0 \\ 0 & d_j \end{bmatrix} \begin{bmatrix} b_{i1} & b_{i2} & \cdots & b_{in} \\ b_{j1} & b_{j2} & \cdots & b_{jn} \end{bmatrix}$$
$$\tag{A.4}$$

Our task is to find $\hat{D}$ and $\hat{B}$ such that this equation holds and $\hat{B}$ can be obtained from B inexpensively. The latter requirement will be satisfied if for some p and q,

$$\begin{bmatrix} \hat{b}_{i1} & \hat{b}_{i2} & \cdots & \hat{b}_{in} \\ \hat{b}_{j1} & \hat{b}_{j2} & \cdots & \hat{b}_{jn} \end{bmatrix} = \begin{bmatrix} 1 & p \\ -q & 1 \end{bmatrix} \begin{bmatrix} b_{i1} & b_{i2} & \cdots & b_{in} \\ b_{j1} & b_{j2} & \cdots & b_{jn} \end{bmatrix} \tag{A.5}$$

since the calculations

$$\left. \begin{array}{l} \hat{b}_{ik} = b_{ik} + p b_{jk} \\ \hat{b}_{jk} = -q b_{ik} + b_{jk} \end{array} \right\} \quad k = 1, 2, \ldots, n$$

require only $2n$ multiplications and $2n$ additions. Suppose we take (A.5) as the definition of $\hat{B}$, where p and q are still to be determined. Then (A.4) holds provided that

$$\begin{bmatrix} \hat{d}_i & 0 \\ 0 & \hat{d}_j \end{bmatrix} \begin{bmatrix} 1 & p \\ -q & 1 \end{bmatrix} = \begin{bmatrix} c & s \\ -s & c \end{bmatrix} \begin{bmatrix} d_i & 0 \\ 0 & d_j \end{bmatrix}$$

Carrying out the indicated matrix products, we find that it suffices to take

$$\begin{aligned} \hat{d}_i &= c d_i \\ \hat{d}_j &= c d_j \\ p &= (s/c)(d_j/d_i) \\ q &= (s/c)(d_i/d_j) \end{aligned} \tag{A.6}$$

This scheme breaks down when $c = 0$. In fact there are also dangers when c is close to zero. In (A.6) the scale factors d_i and d_j are multiplied by c. Indeed, for each rotator Q_m^T that alters the ith row, d_i will be multiplied by such a c_m. If some of these c_m are very close to zero, d_i can underflow after only a few rotations. At the same time, some of the entries of the ith row of B might also overflow. In order to minimize this underflow/overflow threat, we need an alternate scheme to use in those cases where $|c|$ is small. Since $c^2 + s^2 = 1$, we see that when $|c|$ is small, $|s|$ is not small.

A factorization analogous to (A.2) that uses s as the scale factor instead of c is

$$\begin{bmatrix} c & s \\ -s & c \end{bmatrix} = \begin{bmatrix} s & 0 \\ 0 & s \end{bmatrix} \begin{bmatrix} k & 1 \\ -1 & k \end{bmatrix}$$

where $k = c/s$ (= kotangent). This factorization is too simple to use in our scheme, but it suggests that we try replacing (A.5) by

$$\begin{bmatrix} \hat{b}_{i1} & \hat{b}_{i2} & \cdots & \hat{b}_{in} \\ \hat{b}_{j1} & \hat{b}_{j2} & \cdots & \hat{b}_{jn} \end{bmatrix} = \begin{bmatrix} p & 1 \\ -1 & q \end{bmatrix} \begin{bmatrix} b_{i1} & b_{i2} & \cdots & b_{in} \\ b_{j1} & b_{j2} & \cdots & b_{jn} \end{bmatrix} \tag{A.7}$$

for some p and q. Just as in (A.5), these operations require only $2n$ flops. Taking (A.7) to be the definition of $\hat{B}$, we see that (A.4) holds provided that

$$\begin{bmatrix} \hat{d}_i & 0 \\ 0 & \hat{d}_j \end{bmatrix} \begin{bmatrix} p & 1 \\ -1 & q \end{bmatrix} = \begin{bmatrix} c & s \\ -s & c \end{bmatrix} \begin{bmatrix} d_i & 0 \\ 0 & d_j \end{bmatrix}$$

Carrying out these matrix products, we see that it suffices to take

$$\begin{aligned} \hat{d}_i &= s d_j \\ \hat{d}_j &= s d_i \\ p &= (c/s)(d_i/d_j) \\ q &= (c/s)(d_j/d_i) \end{aligned} \tag{A.8}$$

Shrinkage of the scale factors d_i and d_j can be minimized by using (A.5) and (A.6) whenever $|c| \geq |s|$ and (A.7) and (A.8) whenever $|c| < |s|$.

Rotators That Set an Entry to Zero

In many applications the rotators are determined in such a way that they set a certain entry to zero:

$$\begin{bmatrix} c & s \\ -s & c \end{bmatrix} \begin{bmatrix} a_i \\ a_j \end{bmatrix} = \begin{bmatrix} * \\ 0 \end{bmatrix}$$

For this we choose s and c so that $sa_i = ca_j$ and $s^2 + c^2 = 1$. Of course a_i and a_j are not available directly but in the factored form

$$\begin{bmatrix} a_i \\ a_j \end{bmatrix} = \begin{bmatrix} d_i & 0 \\ 0 & d_j \end{bmatrix} \begin{bmatrix} b_i \\ b_j \end{bmatrix}$$

Thus we require s and c such that

$$sd_ib_i = cd_jb_j \quad \text{and} \quad c^2 + s^2 = 1 \tag{A.9}$$

We also need the numbers p and q from either (A.6) or (A.8). As we shall soon see, we can calculate p and q directly and avoid calculating s and c altogether.

Let us suppose tentatively that we are going to use (A.5) and (A.6). In the unlikely event that $b_i = 0$, a zero can be placed in the desired spot simply by interchanging rows i and j. (This operation is actually a reflection rather than a rotation, but it doesn't make any difference.) If $b_i \neq 0$, then by (A.9), $c \neq 0$ and

$$\frac{s}{c} = \frac{d_j}{d_i} \frac{b_j}{b_i} \tag{A.10}$$

Thus, by (A.6)

$$p = \frac{b_j}{b_i} \frac{d_j^2}{d_i^2} \quad \text{and} \quad q = \frac{b_j}{b_i} \tag{A.11}$$

This shows that for the calculation of p and q we need only b_i, b_j, and the squares of d_i and d_j. Therefore it makes more sense to store the numbers d_k^2 rather than the d_k themselves. In order to update d_i^2 and d_j^2 by (A.6), we need c^2, not c. Letting $t = s/c$, we have $1 = c^2 + s^2 = c^2(1 + t^2)$, so $c^2 = 1/(1 + t^2)$. But by (A.6), $t^2 = pq$, so t^2 is easy to compute. To summarize, we compute p and q by (A.11). Then $t^2 = pq$, $c^2 = 1/(1 + t^2)$, $\hat{d}_i^2 = c^2d_i^2$, $\hat{d}_j^2 = c^2d_j^2$, and B is updated by (A.5).

However, if it turns out that $t^2 > 1$, then $|s| > |c|$, and we should use (A.7) and (A.8) instead of (A.5) and (A.6). In this case the arithmetic that has been done up to the point of computing t^2 has not been completely wasted. Notice that the p and q of (A.8) are just the reciprocals of the p and q of (A.6), so the new values can be obtained by inverting the p and q that have already been computed. Then the new p and q satisfy $pq = k^2$, the square of the cotangent, and s^2 is given by $s^2 = 1/(1 + k^2)$. Finally $\hat{d}_i^2 = s^2 d_j^2$, $\hat{d}_j^2 = s^2 d_i^2$, and B is updated by (A.7). The entire computation is summarized by the following algorithm. We write g_k for d_k^2.

Algorithm to Set Up a Fast Rotator Q^T and Compute $Q^T A$

The rotator operates on rows i and j and sets a_{jm} to zero.
A is stored as a product DB, as explained above.

if $(b_{jm} \neq 0)$ then
 if $(b_{jm} + b_{im} = b_{jm})$ then $(b_{im}$ is effectively zero)
  interchange rows i and j (including scale factors)
 $q \leftarrow b_{jm}/b_{im}$
 $p \leftarrow q(g_j/g_i)$
 $r \leftarrow pq$ (tangent squared)
 if $(r \leq 1)$ then
  $r \leftarrow 1/(1 + r)$ (cosine squared)
  $g_i = rg_i$
  $g_j = rg_j$ (A.12)
  update B by (A.5)
 else
  $p \leftarrow 1/p$
  $q \leftarrow 1/q$
  $r \leftarrow pq$ (cotangent squared)
  $r \leftarrow 1/(1 + r)$ (sine squared)
  $\sigma \leftarrow g_i$
  $g_i \leftarrow rg_i$
  $g_j \leftarrow r\sigma$
  update B by (A.7)

Except for the update of B, the number of operations is small and independent of the size of the matrix. Therefore the total cost is roughly equal to that of the update of B.

So far we have considered only multiplication on the left. Many algorithms perform both left and right multiplications by reflectors. This can be accomplished by factoring A in the form $A = D_1 B D_2$, where both D_1 and D_2 are diagonal matrices. Left multiplications are performed as already described and do not involve D_2. The procedure for right multiplication is just the transpose of the procedure for left multiplication. Each right multiplication affects two columns of B and the two associated scale factors in D_2. D_1 is unaffected. A can be recovered from D_1, B, and D_2 at any time by multiplying the rows of B by the scale factors in D_1 and the columns of B by the scale factors in D_2.

In an algorithm that performs similarity transformations, each rotator that is applied on the right is the transpose of a rotator that has just been applied on the left. It follows immediately that $D_1 = D_2$. Thus the factorization of A has the form $A = DBD$, and only one set of scale factors has to be retained. Each similarity transformation by a rotator can be performed just as in (A.12), except that the update of B involves the modification of both rows and columns.

Exercise A.1   (a) Write an expanded version of (A.12), in which the modifications of B are made explicit and a complete similarity transformation $A \rightarrow Q^T A Q$ is performed.

(b) Show that the execution of this similarity transformation requires about $4n$ flops.   □

Exercise A.2   (a) In a single or double QR step (Section 4.7) the similarity transformations are applied to matrices that are nearly in upper Hessenberg form. Taking the zeros into account, show that each similarity transformation by a fast rotator requires just over $2n$ flops.

(b) Show that a single implicit QR step with fast rotators requires about $2n^2$ flops, and a double step requires about $4n^2$ flops.   □

Jacobi Rotators

Fast rotators can also be used in conjunction with all variants of Jacobi's method (Section 6.1). A real, symmetric matrix A can be represented in the form $A = DBD$, where B is symmetric and D is a diagonal matrix of scale factors. A Jacobi rotation to annihilate a_{ij} is set in motion by calculating the number

$$\hat{t} = \tan 2\theta = \frac{2a_{ij}}{a_{jj} - a_{ii}}$$

[cf., (6.1.2)] or its reciprocal. Since A is now stored in the form DBD, the computation of $\hat{t}$ becomes slightly more complicated:

$$\hat{t} = \frac{2d_i b_{ij} d_j}{d_j^2 b_{jj} - d_i^2 b_{ii}}$$

The numbers $t = \tan \theta$ and $c = \cos \theta$ can then be calculated as in Section 6.1 and used to calculate the numbers p and q of (A.5) and (A.6). We always use (A.5) and (A.6) rather than (A.7) and (A.8) because θ always lies in the interval $[-\frac{\pi}{4}, \frac{\pi}{4}]$. Thus $p = t(d_j/d_i)$, $q = t(d_i/d_j)$, $\hat{d}_i = cd_i$, and $\hat{d}_j = cd_j$. In this case it makes sense to store the d_i rather than their squares.

Guarding against Underflow by Rescaling

The procedure that we have outlined is designed to minimize the likelihood of underflow of the scale factors. We must nevertheless recognize that an underflow could eventually occur. In order to avoid this catastrophe, we must periodically rescale the arrays and reset the scale factors to 1. Since this costs a number of multiplications, we would rather not do it too frequently. It is easy to determine a simple strategy that guarantees that the scale factors will never underflow.

    Suppose we are using a scheme that stores the squared scale factors $g_i = d_i^2$. Each time the ith row is modified by a rotator, g_i is multiplied by a number, c^2 or s^2, that is not less than $1/2$. Thus after row i has been modified k times, $g_i \geq 2^{-k}$.

If we are using a computer on which 2^{-m} is the smallest positive number that can be represented by the floating point number system, then underflow cannot occur as long as $k \le m$. The value of m varies from computer to computer; on many machines it also depends upon whether single- or double-precision arithmetic is being used. The smallest value of m on any commmonly used machine is 128. Many scientific computers have $m > 1000$.

Exercise A.3   Check that in a single (double) step of the QR algorithm each g_i is modified at most twice (four times). Thus underflow of the scale factors cannot occur as long as the arrays are rescaled after every $m/2$ steps, counting double steps as two steps. $\square$

Exercise A.4   If the mobile Jacobi method of Section 6.1 is programmed using fast rotators, how frequently must the arrays be rescaled to guarantee that the scale factors will not underflow? Remember that in this case we store the scale factors themselves, not their squares. $\square$

Connected with the threat that the g_i might underflow is the possibility that some of the entries of B might overflow. The growth of these entries is rigidly bound to the decay of the g_i (why?), so the precautions that one takes against underflow also generally guard against overflow in B. However, if the entries of A happen to be large to begin with, there can be problems. Suppose, for example, the average entry of A has magnitude 10^{25}. Then there is considerably less room for growth than if the average magnitude were 1. This problem is easily remedied. Before the rotations are begun, sweep once through the matrix and determine $m = \max |a_{ij}|$. Then work with the matrix $m^{-1}A$, whose entries are not large.

APPENDIX **B**

Software for Matrix Computations

LINPACK and EISPACK programs and other software are available from NETLIB. You can access NETLIB by sending electronic mail messages to either

netlib@research.att.com

or

netlib@ornl.gov

The one-line message

send index

will cause an index of NETLIB, along with brief instructions for using NETLIB, to be sent to you by electronic mail. The index has a brief description of each package. If you would like a more detailed description of LINPACK, for example, you can get one by sending the message

send index for linpack

To obtain the actual software, send a message such as

send dgeco from linpack

In response you will receive the LINPACK program DGECO and all of the auxiliary subroutines that DGECO calls. If you already have the auxiliary routines and just want DGECO, send the message

send only dgeco from linpack

Your requests will be processed by a computer, not a human being, so you should not deviate much from the syntax shown here. Messages to NETLIB can contain more than one request, but separate requests should be on separate lines. For more information about NETLIB, see Dongarra and Grosse (1987).

If you wish to obtain the entire contents of some large package such as LIN-PACK or EISPACK, you should not attempt to get them from NETLIB. Instead you should order a tape from

National Energy Software Center
Argonne National Laboratory
9700 South Cass Avenue
Argonne, IL   60439

phone   312-972-7250

The superb matrix manipulation programs MATLAB and GAUSS are not in the public domain. You can buy a copy of MATLAB from

The Math Works
21 Eliot Street
South Natlick, MA   01760

phone   508-653-1415

MATLAB is the favorite of numerical analysts, but GAUSS has its fans. GAUSS is available from

Aptech Systems, Inc.
26250 196th Place SE
Kent, WA   98042

phone   206-631-6679

The information in this appendix is correct as of 1990. It is of course subject to change in the future.

Bibliography

Part I
REFERENCE BOOKS
FOR MATRIX COMPUTATIONS

[AEP] J. H. Wilkinson, *The Algebraic Eigenvalue Problem*, Oxford Univ. Press, London/New York, 1965.

[CSLAS] G. Forsythe and C. B. Moler, *Computer Solution of Linear Algebraic Systems*, Prentice–Hall, Englewood Cliffs, N.J., 1967.

[EG] B. T. Smith et al., *Matrix Eigensystem Routines—EISPACK Guide*, 2nd ed., Springer-Verlag, New York, 1976. B. S. Garbow et al., *Eispack Guide Extension*, Springer-Verlag, New York, 1972.

[HAC] J. H. Wilkinson and C. Reinsch, *Handbook for Automatic Computation, vol. II, Linear Algebra*, Springer-Verlag, New York, 1971.

[IMC] G. W. Stewart, *Introduction to Matrix Computations*, Academic Press, New York, 1973.

[LUG] J. J. Dongarra, J. R. Bunch, C. B. Moler, and G. W. Stewart, *LINPACK Users' Guide*, SIAM, Philadelphia, 1979.

[MC] G. H. Golub and C. F. Van Loan, *Matrix Computations*, 2nd ed., Johns Hopkins Univ. Press, Baltimore, 1989.

[NLA] W. Bunse and A. Bunse-Gerstner, *Numerische Lineare Algebra*, Teubner, Stuttgart, 1985.

[SEP] B. N. Parlett, *The Symmetric Eigenvalue Problem*, Prentice–Hall, Englewood Cliffs, N.J., 1980.

[SLS] C. L. Lawson and R. J. Hanson, *Solving Least Squares Problems*, Prentice–Hall, Englewood Cliffs, N.J., 1974.

[SPDS] A. George and J. W. Liu, *Computer Solution of Large Sparse Positive Definite Systems*, Prentice–Hall, Englewood Cliffs, N.J., 1981.

Part II
OTHER REFERENCES

Batterson, S., and J. Smillie (1990), "Rayleigh quotient iteration for nonsymmetric matrices," *Math. Comp.*, **55**, pp. 169–178.

Björck, A. (1967), "Solving least squares problems by Gram–Schmidt orthogonalization," *BIT*, **7**, pp. 1–21.

Björck, A., and G. H. Golub (1973), "Numerical methods for computing angles between linear subspaces," *Math. Comp.*, **27**, pp. 579–594.

Brent, R. P., and F. T. Luk (1985), "The solution of singular-value and eigenvalue problems on multiprocessor arrays," *SIAM J. Sci. Stat. Comput.*, **6**, pp. 69–84.

Bunch, J. R., C. P. Nielsen, and D. C. Sorensen (1978), "Rank-one modification of the symmetric eigenproblem," *Numer. Math.*, **31**, pp. 31–48.

Bunse-Gerstner, A., and V. Mehrmann (1986), "A symplectic QR-like algorithm for the solution of the real algebraic Riccati equation," *IEEE Trans. Auto. Control*, **31**, pp. 1104–1113.

Chan, T. F. (1982), "An improved algorithm for computing the singular value decomposition," *ACM Trans. Math. Software*, **8**, pp. 72–83. (algorithm on pp. 84–88.)

Chan, T. F. (1987), *Rank-revealing QR factorizations*, Linear Algebra Appl. 88/89, pp. 67–82.

Cline, A. K., C. B. Moler, G. W. Stewart, and J. H. Wilkinson (1979), "An estimate for the condition number of a matrix," *SIAM J. Numer. Anal.*, **16**, pp. 368–375.

Cullum, J. K., and R. A. Willoughby (1985), *Lanczos Algorithms for Large Symmetric Eigenvalue Computations*, Birkhäuser, Boston.

Cuppen, J. J. M. (1981), "A divide and conquer method for the symmetric tridiagonal eigenproblem," *Numer. Math.*, **36**, pp. 177–195.

Dax, A., and S. Kaniel (1981), "The *ELR* method for computing the eigenvalues of a general matrix," *SIAM J. Numer. Anal.*, **18**, pp. 597–605.

Dongarra, J. J., J. Du Croz, S. Hammarling, and I. Duff (1990), "A set of level 3 basic linear algebra subprograms," *ACM Trans. Math. Software*, **16**, pp. 1–17.

Dongarra, J. J., J. Du Croz, S. Hammarling, and R. Hanson (1988), "An extended set of Fortran basic linear algebra subprograms," *ACM Trans. Math. Software*, **14**, pp. 1–17.

Dongarra, J. J., and E. Grosse (1987), "Distribution of mathematical software via electronic mail," *Comm. ACM*, **30**, pp. 403–407.

Dongarra, J. J., F. G. Gustavson, and A. Karp (1984), "Implementing linear algorithms for dense matrices on a vector pipeline machine," *SIAM Review*, **26**, pp. 91–112.

Dongarra, J. J., and D. C. Sorensen (1987), "A fully parallel algorithm for the symmetric eigenvalue problem," *SIAM J. Sci. Stat. Comput.*, **8**, pp. s139–s154.

Duff, I. S., A. M. Erisman, and J. K. Reid (1986), *Direct Methods for Sparse Matrices*, Oxford Univ. Press, London/New York.

Forsythe, G. E., and P. Henrici (1960), "The cyclic Jacobi method for computing the principal values of a complex matrix," *Trans. Amer. Math. Soc.*, **94**, pp. 1–23.

Gantmacher, F. R. (1959), *Matrix Theory*, Chelsea, New York.

Hadlock, C. R. (1978), *Field Theory and its Classical Problems*, Carus Mathematical Monographs, Mathematical Association of America.

Hageman, L. A., and D. M. Young (1981), *Applied Iterative Methods*, Academic Press, New York.

Hager, W. W. (1984), "Condition estimates," *SIAM J. Sci. Stat. Comput.*, **5**, pp. 311–316.

Hansen, E. R. (1963), "On cyclic Jacobi methods," *J. SIAM*, **11**, pp. 448–459.

Heller, D. (1978), "A survey of parallel algorithms in numerical linear algebra," *SIAM Review*, **20**, pp. 740–777.

Henrici, P. (1958), "On the speed of convergence of cyclic and quasicyclic Jacobi methods for computing eigenvalues of Hermitian matrices," *J. SIAM*, **6**, pp. 144–162.

Higham, N. J. (1987), "A survey of condition number estimation for triangular matrices," *SIAM Rev.*, **29**, pp. 575–596.

Higham, N. J. (1988), "FORTRAN codes for estimating the one-norm of a real or complex matrix, with applications to condition estimation," *ACM Trans. Math. Software*, **14**, pp. 381–396.

Higham, N. J. (1990), "Experience with a matrix norm estimator," *SIAM J. Sci. Stat. Comput.*, **11**, pp. 804–809.

Higham, N. J., and D. J. Higham (1989), "Large growth factors in Gaussian elimination with pivoting," *SIAM J. Matrix Anal. Appl.*, **10**, pp. 155–164.

Jacobi, C. G. J. (1846), "Über ein leichtes Verfahren die in der Theorie der Säcularstörungen vorkommenden Gleichungen numerisch aufzulösen," *J. reine Angew. Math.*, **30**, pp. 51–94.

Jankowski, M., and M. Woźniakowski (1977), "Iterative refinement implies numerical stability," *BIT*, **17**, pp. 303–311.

Jennings, A. (1977), *Matrix Computations for Engineers and Scientists*, Wiley, New York.

Johnson, R. A., and D. W. Wichern (1988), *Applied Multivariate Statistical Analysis*, 2nd ed., Prentice–Hall, Englewood Cliffs, N.J.

Kreyszig, E. (1978), *Introductory Functional Analysis with Applications*, Wiley, New York.

Lancaster, P., and M. Tisminetsky (1985), *The Theory of Matrices*, 2nd ed., Academic Press, New York.

Lawson, C., R. Hanson, D. Kincaid, and F. Krogh (1979), "Basic linear algebra subprograms for Fortran usage," *ACM Trans. Math. Software*, **5**, pp. 308–323.

Lo, S.-S., B. Philippe, and A. Sameh (1987), "A multiprocessor algorithm for the symmetric tridiagonal eigenvalue problem," *SIAM J. Sci. Stat. Comput.*, **8**, pp. s155–s165.

Luk, F. T., and H. Park (1987), "On the equivalence and convergence of parallel Jacobi SVD algorithms," *Proc. SPIE*, vol. 826, *Advanced Algorithms and Architectures for Signal Processing*, pp. 152–159.

Luk, F. T., and H. Park (1989), "On parallel Jacobi orderings," *SIAM J. Sci. Stat. Comput.*, **10**, pp. 18–26.

McCormick, S. F. (1987), *Multigrid Methods*, SIAM, Philadelphia.

Meyer, C. D., and G. W. Stewart (1988), "Derivatives and perturbations of eigenvectors," *SIAM J. Numer. Anal.*, **25**, pp. 679–691.

Modi, J. J., and J. D. Pryce (1985), "Efficient implementation of Jacobi's diagonalization method on the DAP," *Numer. Math.*, **46**, pp. 443–454.

Ortega, J. M. (1988), *Introduction to Parallel and Vector Solution of Linear Systems*, Plenum, New York.

Ortega, J. M., and R. G. Voigt (1985), "Solution of partial differential equations on vector and parallel computers," *SIAM Review*, **27**, pp. 149–240.

Parlett, B. N., and W. G. Poole, Jr. (1973), "A geometric theory for the *QR*, *LU*, and power iterations," *SIAM J. Numer. Anal.*, **8**, pp. 389–412.

Skeel, R. D. (1979), "Scaling for numerical stability in Gaussian elimination," *J. Assoc. Comput. Mach.*, **26**, pp. 494–526.

———— (1980), "Iterative refinement implies numerical stability for Gaussian elimination," *Math. Comp.*, **35**, pp. 817–832.

Stewart, G. W. (1971), "Error bounds for approximate invariant subspaces of closed linear operators," *SIAM J. Numer. Anal.*, **8**, pp. 796–808.

———— (1972), "On the sensitivity of the eigenvalue problem $Ax = \lambda Bx$," *SIAM J. Numer. Anal.*, **9**, pp. 669–686.

———— (1973), "Error and perturbation bounds for subspaces associated with certain eigenvalue problems," *SIAM Review*, **15**, pp. 727–764.

———— (1976), "Simultaneous iteration for computing invariant subspaces of non-Hermitian matrices," *Numer. Math.*, **25**, pp. 123–136.

———— (1977), "On the perturbation of pseudo-inverses, projections and least squares problems," *SIAM Review*, **19**, pp. 634–662.

Trefethen, L. N., and R. S. Schreiber (1990), "Average-case stability of Gaussian elimination," *SIAM J. Matrix Anal. Appl.*, **11**, pp. 335–360.

Watkins, D. S. (1982), "Understanding the QR algorithm," *SIAM Review*, **24**, pp. 427–440.

Watkins, D. S., and L. Elsner (1990), "Convergence of algorithms of decomposition type for the eigenvalue problem," *Linear Algebra Appl.*, to appear.

Watkins, D. S., and L. Elsner (1991), "Chasing algorithms for the eigenvalue problem," *SIAM J. Matrix Anal. Appl.*, to appear.

Whiteside, R. A., N. S. Ostlund, and P. G. Hibbard (1984), "A parallel Jacobi diagonalization algorithm for a loop multiple processor system," *IEEE Trans. Comput.*, **C-33**, pp. 409–413.

Wilkinson, J. H. (1961), "Error analysis of direct methods of matrix inversion," *J. Assoc. Comp. Mach.*, **8**, pp. 281–330.

Index